Robust Engineering Designs of Partial Differential Systems and Their Applications

Robust Engineering Designs of Partial Differential Systems and Their Applications

Bor-Sen Chen

CRC Press
Taylor & Francis Group
Boca Raton London New York

CRC Press is an imprint of the
Taylor & Francis Group, an **informa** business

First edition published 2022
by CRC Press
6000 Broken Sound Parkway NW, Suite 300, Boca Raton, FL 33487-2742

and by CRC Press
4 Park Square, Milton Park, Abingdon, Oxon, OX14 4RN

CRC Press is an imprint of Taylor & Francis Group, LLC

Library of Congress Cataloging-in-Publication Data
Names: Chen, Bor-Sen, author.
Title: Robust engineering designs of partial differential systems and their applications / Bor-Sen Chen.
Description: First edition. | Boca Raton, FL : CRC Press, 2022. | Includes bibliographical references and index.
Identifiers: LCCN 2021033602 (print) | LCCN 2021033603 (ebook) | ISBN 9781032134437 (hbk) | ISBN 9781032134451 (pbk) | ISBN 9781003229230 (ebk)
Subjects: LCSH: Reliability (Engineering)—Mathematics. | Fuzzy systems—Design. | Differential equations, Partial.
Classification: LCC TA169 .C4685 2022 (print) | LCC TA169 (ebook) | DDC 620/.00452—dc23
LC record available at https://lccn.loc.gov/2021033602
LC ebook record available at https://lccn.loc.gov/2021033603

ISBN: 978-1-032-13443-7 (hbk)
ISBN: 978-1-032-13445-1 (pbk)
ISBN: 978-1-003-22923-0 (ebk)

DOI: 10.1201/9781003229230

Typeset in Times
by codeMantra

Contents

Preface

Many phenomena in science and engineering have been modeled by partial differential systems (PDSs), for example, some phenomena related to heat flows, fluid flows, elastic wave, flexible structure, etc., in mechanical systems, chemical engineering systems, molecular transmission systems, biological systems, population dynamical systems, neurophysiological systems, biodynamicical systems, etc. In recent years, a great deal of concern has been raised regarding the study of PDSs. In the past decades, the stabilization design problems of linear PDSs have been widely studied by control engineers. The linear parabolic PDSs can be transferred to an equivalent infinite-dimensional ordinary differential system (ODS). According to the separation of eigenvalues based on Galerkin's method, the infinite-dimensional ODS can be separated into a finite dimensional slow mode system and an infinite-dimensional fast mode system. So a robust finite-dimensional controller based on the appropriate finite-dimensional slow mode system could be designed to stabilize the linear PDS with the robust stability to tolerate the infinite-dimensional fast mode system. Recently, unlike the infinite-dimensional ODS to represent the PDS, the partial differential operator on space in PDS could be approximated by a finite-difference operator. Then, we can obtain a set of finite-difference systems with the approximation error and truncation error to represent the PDS. When all finite-difference grid points are represented by a spatial vector, the set of finite-difference systems is represented by an equivalent spatial state space system. Then a robust controller based on spatial state space system can be developed to treat the stabilization problem of linear PDSs with the ability to tolerate the truncation error. More recently, the divergence theorem can be utilized to transform the diffusion matrix inequality (DMI) to bilinear matrix inequality (BMI) and then the BMI could be transformed to linear matrix inequality (LMI) by Poincare inequality method. Therefore the control design of partial differential system could be significantly simplified by divergence theorem and Poincare inequality. After the robust control design of linear PDSs, the state estimation and filter design of linear stochastic PDS in signal processing have been developed by engineers for signal processing application. Based on fuzzy interpolation and global linearization method to interpolate several local linear PDSs to approximate a nonlinear PDS, the robust control and filter design methods in linear PDSs can be also extended to the nonlinear stochastic PDSs for more practical applications. These control and filter designs of PDSs and their applications are the main topics of the book.

The book is divided into three parts. Part I is about some backgrounds of partial differential systems. In Part I, Galerkin and finite-difference methods are introduced to approximated PDS by a finite-dimension ODS with a residual system. Robust control design is also introduced to stabilize finite-dimension ODS with the ability to tolerate the residual system. Then, for nonlinear PDSs, fuzzy system is also introduced to interpolate several local linearized systems by fuzzy bases to approximate a nonlinear PDS by adaptive fuzzy algorithm. In Part II, robust filter and estimator designs are introduced for signal processing in PDSs under measurement noise and external disturbance in the spatiotemporal domain. First, based on a set of sensor measurements, a robust filter design is introduced to linear stochastic PDSs with measurement noise and external disturbance. Then, a robust filter design is introduced to nonlinear stochastic PDSs in sensor signal processing. In Part III, we will introduce some robust control design problems of PDSs. We first introduce the robust stabilization control design problem of large flexible structure systems, and then the robust stabilization design problem of stochastic linear PDSs under spatiotemporal disturbance and sensor measurement noise. Based on fuzzy state-space modeling, robust observer-based control design is also introduced for the stabilization of nonlinear PDSs and nonlinear stochastic PDSs. Finally, robust reference tracking control design is introduced for nonlinear distributed parameter time-delay systems.

This book not only introduces recent robust filtering, robust stabilization and robust reference tracking control design techniques in linear and nonlinear stochastic partial differential systems but also applies these filter and control design techniques to mechanical systems, aerospace systems, and control engineering systems. Consequently, this book is useful for postgraduate students, engineers and scientists in mechanic engineering, electrical engineering, aerospace engineering, bioengineering, and ecology.

I first touched to partial differential equations in the course of electromagnetics when I was a sophomore. When most students were still in confusion with the operators like Curl and Div, I could always finish the homeworks of this course because I had read a book of vector analysis in the previous summer vocations. Consequently, I also took the course of electromagnetic wave in junior, which discussed Maxwell electromagnetic wave equation in the spatiotemporal domain. Then I had taken the graduate courses of fluid dynamics, elastic dynamics, numerical forecasting of meteorology, electrodynamics, etc., in my M.S. of geophysics. In my PhD in electrical engineering at USC, I had a roommate of architecture. He had taken a course of finite element, which was a required but hard subject for him and his classmates. I was always enthusiastic to help them solving homeworks of this course. At his convenience, I also learned some techniques of finite element. Therefore, I had more background on partial differential systems than conventional control engineers. When I became a professor, I had a co-supervised PhD student in the department of aerospace engineering. He was interested in the robust control design of large flexible structure and therefore I was involved in the robust control and filter designs of partial differential systems. Consequently, I had widely applied these knowledges to signaling processing and control engineering design of partial differential systems in the last three decades.

Last, but not least, I would like to thank Dr. Gagandeep Singh, a publisher in engineering at CRC Press, who has given me a lot of help during the course of this book. Finally, I would like to thank Dr. Shih-Ju Ho for his help and Ms. Ci-Jun Wang for her careful typing.

The MathWorks, Inc.
3 Apple Hill Drive
Natick, MA 01760-2098 USA
Tel: 508-647-7000
Fax: 508-647-7001
E-mail: info@mathworks.com
Web: www.mathworks.com

Author

Bor-Sen Chen earned a BS in electrical engineering at Tatung Institute of Technology, Taipei, Taiwan, in 1970, and an MS in geophysics at National Central University, Chungli, Taiwan in 1973, and a PhD at the University of Southern California, Los Angeles, USA, in 1973. From 1973 to 1987, he was a lecturer, associate professor, and professor at Tatung Institute of Technology. Since 1987 he has been a professor, chair professor, and Tsing Hua Distinguished Chair Professor with the Department of Electrical Engineering of National Tsing Hua University, Hsinchu, Taiwan. His research interests include robust control theory and engineering design, robust signal processing and communication system design, and systems biology. He has published more than 300 journal papers, including 130 papers in control, 70 papers in signal processing and communication, and 100 papers in systems biology. He has also published 10 monographs. He was the recipient of numerous awards for his academic accomplishments in robust control, fuzzy control, H_∞ control, stochastic control, signal processing and systems biology, including four Outstanding Research Awards of National Science Council, Academic Award in Engineering from the Ministry of Education, National Chair Professor of the Ministry of Education, Best Impact Award of IEEE Taiwan Section for his most SCI citations of IEEE members in Taiwan, etc. His current research focuses on the H_∞ team formation network tracking control of large-scale UAVs, large-scale biped robots and their team cooperation, deep neural network-based control design of nonlinear dynamic systems, systems medicine design based on design specifications and deep learning schemes, etc. He is a life fellow of IEEE.

Part I

Background

1 Introduction to Partial Differential Systems

1.1 PARTIAL DIFFERENTIAL EQUATIONS AND PARTIAL DIFFERENTIAL SYSTEMS

In this section, we introduce some partial differential equations (PDEs) at first, especially the second-order PDEs with boundary conditions [1–3].Some fundamental properties such as existence, uniqueness and stability of PDEs are also introduced. Then some corresponding partial differential systems (PDSs) are also given to describe some nonlinear PDSs related to the robust system design problems in this book [4–16].

1.1.1 Partial Differential Equations

A differential equation that contains, in addition to the dependent variable and the independent variables, one or more partial derivatives of the dependent variable is called a PDE. In general, it can be written as [1–3]

$$F\left(x_1, x_2, \ldots, x_N, y, \frac{\partial y}{\partial x_1}, \frac{\partial y}{\partial x_2}, \ldots, \frac{\partial^2 y}{\partial x_1^2}, \frac{\partial^2 y}{\partial x_1 \partial x_2}, \ldots\right) = 0 \tag{1.1}$$

where $x_1, x_2, \ldots, x_N$ are the independent variables with $N \in \mathbb{N}$ and $y = y(x_1, x_2, \ldots, x_N) \in \mathbb{R}$ is an unknown scalar function of these independent variables. The dependent variable y and its partial derivatives, which are considered to be in a suitable bounded domain Ω of the N-dimensional space R^N in the independent variables $x_1, x_2, \ldots, x_N$, satisfy equation (1.1) identically in Ω. The order of a PDE is the highest derivative that appears in the equation (1.1). For example, the most general second-order 2-dimensional PDE is

$$c_a \frac{\partial^2 y}{\partial x_1^2} + c_b \frac{\partial^2 y}{\partial x_1 \partial x_2} + c_c \frac{\partial^2 y}{\partial x_2^2} + c_d \frac{\partial y}{\partial x_1} + c_e \frac{\partial y}{\partial x_2} + c_f y = c \tag{1.2}$$

where the coefficients c_a, c_b, c_c, c_d, c_e, and c_f are all allowed to be functions of (x_1, x_2), and the inhomogeneity or forcing function $c = c(x_1, x_2)$.

The classification of the second-order 2-dimensional PDE equation (1.2) is based upon the possibility of reducing the equation (1.2) by a coordinate transformation to canonical or discriminant form at a point, such as [1]

$$c_b^2(x_1^p, x_2^p) - 4c_a(x_1^p, x_2^p)c_c(x_1^p, x_2^p) \tag{1.3}$$

DOI: 10.1201/9781003229230-2

at a point (x_1^p, x_2^p). An equation is said to be hyperbolic, parabolic, or elliptic at a point (x_1^p, x_2^p) accordingly as equation (1.3) is positive, zero, or negative. If this is true for all points $(x_1, x_2) \subset \Omega$, then the equation (1.2) is said to be hyperbolic, parabolic, or elliptic in the domain Ω; namely, second-order 2-dimensional hyperbolic PDE, second-order 2-dimensional parabolic PDE, or second-order 2-dimensional elliptic PDE. For examples, the wave equation

$$\frac{\partial^2 y}{\partial x_1^2} - \frac{\partial^2 y}{\partial x_2^2} = 0 \tag{1.4}$$

has discriminant $4 > 0$, and is hyperbolic; the heat equation

$$\frac{\partial^2 y}{\partial x_1^2} - \frac{\partial y}{\partial x_2} = 0 \tag{1.5}$$

has discriminant $0 = 0$, and is parabolic; the Poisson equation [3]

$$\frac{\partial^2 y}{\partial x_1^2} + \frac{\partial^2 y}{\partial x_2^2} = -c \tag{1.6}$$

has discriminant $-4 < 0$, and is elliptic.

Obviously, it is necessary to specify a boundary condition if the solution of PDE is to be determined. The two most important kinds of boundary conditions are [1–3]

i. Dirichlet boundary condition (D): $y(x_1, x_2, \cdots, x_N) = 0, \forall x_1, x_2, \cdots, x_N \in \partial\Omega$
ii. Neumann boundary condition (N):

$$\frac{\partial y(x_1, x_2, \cdots, x_N)}{\partial \vec{n}_{base}} = 0, \forall x_1, x_2, \cdots, x_N \in \partial\Omega \tag{1.7}$$

where $\frac{\partial}{\partial \vec{n}_{base}}$ is the outward normal derivative, $\vec{n}_{base}$ is the normal vector to the boundary $\partial\Omega$. In considering time t, the initial condition $y(t = 0, x_1, x_2)$ is necessary as well. The mathematical problem to solve such PDE is said well-posed if it satisfies the following fundamental properties [1–3]

i. Existence: There exists at least one solution satisfying both boundary and initial conditions.
ii. Uniqueness: There is at most one solution.
iii. Stability: The unique solution depends continuously on the data of the problem.

The first property is an obvious logical condition. The same can be said about the uniqueness property. In order to really reflect the physical problem that has a unique solution, the mathematical problem must have a unique solution. For the last property, if the solution is to have physical significance, a small change in the initial data must produce a small change in the solution. The data in a physical problem are normally obtained from experiment, and are approximated in order to solve the problem by numerical or approximate methods. It is essential to know that the process of making an approximation, such as the finite difference approximation, to the data produces only a small change in the solution. These properties for the oncoming analysis discussed in the following sections are assumed for simplicity. Surely, when simulating the PDE examples by using the numerical approximate method, i.e., the central finite difference scheme, these properties are guaranteed.

1.1.2 Partial Differential System

A PDS is a compound dynamic system constructed from PDE. Since most physical phenomena in reality can be described as PDE, the investigation on PDS has become more and more popular nowadays. Among three types of the PDSs, which are hyperbolic, parabolic, and elliptic PDSs, the most attractive one to us is the following N-dimensional nonlinear parabolic PDS [1–2]

$$\frac{\partial y(x,t)}{\partial t} = D\Delta y(x,t) + f(y(x,t)) \tag{1.8}$$

on the spatio-temporal domain $\Omega \times [0,T]$, where $x = [x_1, x_2, \cdots, x_N]^T \in \Omega \subset \mathbb{R}^N$ is the space variable with a smooth boundary $\partial\Omega$ and $T \in \mathbb{R}_+$ is the terminal time. Here, we let Ω be a bounded convex domain for simplicity. $f(\cdot)$ is a smooth function with $f(0) = 0, y(x,t) \triangleq \left[y_1(x,t), y_2(x,t), \cdots, y_{n_y}(x,t) \right]^T \in \mathbb{R}^{n_y}$ and $y \in D(A)$ is the state variable. Finally, $\Delta = \nabla^2 = \nabla \cdot \nabla$ is the Laplace operator for the N-dimensional spatial domain, and $D\Delta y(x,t)$, which represents the effect of diffusion on the system, can be written as

$$D\Delta y(x,t) = [d_{ij}]_{n_y \times n_y} \begin{bmatrix} \frac{\partial^2 y_1(x,t)}{\partial x_1^2} + \frac{\partial^2 y_1(x,t)}{\partial x_2^2} + \cdots + \frac{\partial^2 y_1(x,t)}{\partial x_N^2} \\ \frac{\partial^2 y_2(x,t)}{\partial x_1^2} + \frac{\partial^2 y_2(x,t)}{\partial x_2^2} + \cdots + \frac{\partial^2 y_2(x,t)}{\partial x_N^2} \\ \vdots \\ \frac{\partial^2 y_{n_y}(x,t)}{\partial x_1^2} + \frac{\partial^2 y_{n_y}(x,t)}{\partial x_2^2} + \cdots + \frac{\partial^2 y n_y(x,t)}{\partial x_N^2} \end{bmatrix} \tag{1.9}$$

where $\frac{\partial}{\partial \vec{n}}$ is the outward normal derivative, n is the normal vector to the boundary $\partial\Omega$, and the initial condition is $y(x,0) \triangleq y_0(x)$.

where $y \in D(A)$ is a known diffusion coefficient matrix. Similar to the boundary conditions in equation (1.7), the boundary conditions we are interested in are [1–3]

i. Dirichlet boundary condition (D):

$$y(x,t) = 0, \forall x \in \partial\Omega \tag{1.10}$$

ii. Neumann boundary condition (N): $\dfrac{\partial y(x,t)}{\partial \vec{n}} = 0, \forall x \in \partial\Omega$

Remark 1.1.1

i. Diffusion coefficient D, also called diffusivity, is an important parameter indicative of the diffusion mobility. It is not only encountered in Fick's law, but also in numerous other equations of physics and chemistry. Diffusion coefficient D is generally prescribed for a given pair of species. For a multicomponent system, it is prescribed for each pair of species in the system. The higher the diffusivity (of one substance with respect to another), the faster they diffuse into each other. Many physical processes have this diffusion behavior, such as fluid momentum diffusion (D is kinematic viscosity, Newton's law), thermal energy diffusion (D is thermal diffusivity, Fourier's law), fluid turbulent eddy diffusion (D is eddy viscosity), porous media solute dispersion caused by spatially varying velocity (D is aquifer dispersion coefficient), and aquifer flow (D is hydraulic diffusivity).
ii. For broadly well-posedness of the PDS, forward diffusion, i.e., $D \geq 0$, is necessary in general. For example, the solution at $t \to \infty$ is called the steady-state solution, after the system reaches an equilibrium state. For steady states, we have

$$\frac{\partial y}{\partial t} = 0$$

Therefore, for the linear diffusion equation $\dfrac{\partial y}{\partial t} = \Delta y$, we have

$$\Delta y = 0$$

which is the Laplace equation. The solution depends on the boundary conditions in equation (1.10). In the 1-dimensional case, $\dfrac{\partial^2 y}{\partial x^2}$, which means the solution is of the form $ax+b$, where a, b are constants. The solution can be computed analytically, depending on the boundary conditions. However, consider the linear inverse diffusion equation

$$\frac{\partial y}{\partial t} = -\Delta y$$

This is an unstable ill-posed process. By changing the sign of the time differentiation $\partial t \to -\partial t$, one can view this also as reversing in time the forward diffusion process. Any chemist knows that diffusion is a smoothing process since the concentration of a substance tends to flatten out. Going backward, the situation becomes more and more chaotic. Hence, one would not expect well-posedness of the backward-in-time problem for the diffusion equation. That is, going from a blurred signal to a sharp one (not a physical process). To avoid the possible ill-posedness of the concerned PDS like equation (1.8), the positive semidefinite diffusion coefficient matrix D is commonly used.

When applying a control force to the PDS equation (1.8), we have

$$\frac{\partial y(x,t)}{\partial t} = D\Delta y(x,t) + f(y(x,t)) + Bu(x,t) \tag{1.11}$$

where $u(x,t) \triangleq [u_1(x,t), u_2(x,t), \cdots, u_{n_u}(x,t)]^T \in \mathbb{R}^{n_u}$ is the control force to be designed with the influence matrix $B \in \mathbb{R}^{n_y \times n_u}$. The PDS in equation (1.11) is so-called the N-dimensional noise-free nonlinear parabolic PDS. Generally, systems suffer from external noise in practice. Hence, we have the following N-dimensional nonlinear parabolic PDS

$$\frac{\partial y(x,t)}{\partial t} = D\Delta y(x,t) + f(y(x,t)) + Bu(x,t) + E_n n(x,t) \tag{1.12}$$

where $n(x,t) \triangleq [n_1(x,t), n_2(x,t), \cdots, n_{n_n}(x,t)]^T \in L_2(\Omega \times \mathbb{R}_+; \mathbb{R}^{n_n})$ represents the external noise to the PDS with the influence matrix $E_n \in \mathbb{R}^{n_y \times n_n}$. However, the system state y may not be available in reality. Therefore, the N-dimensional nonlinear parabolic PDS equation (1.12) is reformulated as [4,5]

$$\frac{\partial y(x,t)}{\partial t} = D\Delta y(x,t) + f(y(x,t)) + Bu(x,t) + E_n n(x,t) \tag{1.13}$$

$$z(x,t) = h(y(x,t)) + E_m m(x,t) \tag{1.14}$$

where $h(\cdot)$ is a smooth measure function with $h(0) = 0$ and the rest of settings are the same as the PDS equation (1.12). $z(x,t) \triangleq [z_1(x,t), z_2(x,t), \cdots, z_{n_z}(x,t)]^T \in \mathbb{R}^{n_z}$ represents the measurement output affected by the measurement noise $m(x,t) \triangleq [m_1(x,t), m_2(x,t), \cdots, m_{n_m}(x,t)]^T \in L_2(\Omega \times \mathbb{R}_+; \mathbb{R}^{n_m})$ with the influence matrix $E_m \in \mathbb{R}^{n_z \times n_m}$. The control designs for the aforementioned PDSs in equations (1.11) and (1.12), and the state observation and control designs in equations (1.13) and (1.14), are quite interesting topics.

1.2 NUMERICAL AND APPROXIMATION METHODS FOR PDEs AND PDSs

In dealing with many equations arising from the modeling of physical problems, the determination of such exact solutions in a simple domain is almost a formidable task even when the boundary and/or initial data are simple. In view of the wide-spread accessibility of today's high speed electronic computers, numerical and approximation methods are becoming increasingly important and useful in real applications. This section is devoted to the analytical treatment of PDEs and PDSs. Three main analytical methods to find the exact analytical solution of these equations within simple domains will be discussed. In the following sections, the Galerkin method, also called the Galerkin finite element method for PDEs and PDSs will be presented at first. Then, the finite difference methods (FDMs), including the methods with forward, central, and backward finite difference schemes, for PDEs and PDSs will be provided. Finally, the divergence theorem and Poincaré inequality methods to deal with the PDEs and PDSs with Dirichlet and Neumann boundary condition in equation (1.10) will be discussed. These analytic methods of PDEs and PDSs will provide numerical and approximation methods for robust system designs of PDEs and PDSs in the following chapters.

1.2.1 Galerkin Method

Galerkin formulated a numerical approximation method that may be applied to a problem, in which no simple variational principle exists [3]. In the area of numerical analysis, the Galerkin method is a class of methods for converting a continuous operator problem (such as a differential equation) to a discrete problem. In principle, it is equivalent to applying the method of variation of parameters to a function space, by converting the equation to a weak formulation. Consider the boundary value problem governed by the 2-dimensional differential equation

$$Ay = f, \text{ in } \Omega \tag{1.15}$$

with the boundary condition

$$B(y) = 0, \text{on } \partial Q \tag{1.16}$$

where A is a self-adjoint differential operator in the Hilbert space $L_2(\Omega;\mathbb{R}^{n_y})$, $f \in L_2(\Omega;\mathbb{R}^{n_y})$, and $y \in D(A)$. In order to solve the boundary value problem (1.15) and (1.16), we construct an approximate solution y in the form

$$y_n = y_0(x_1,x_2) + \sum_{i=1}^{n} a_i\phi_i(x_1,x_2) \tag{1.17}$$

which is called the Galerkin solution of equations (1.15) and (1.16), where $\phi_i(x_1,x_2) \in L_2(\Omega;\mathbb{R}^{n_y})$ are known functions, y_0 is introduced to satisfy the boundary conditions, and the coefficients a_i are to be determined. Substituting equation (1.17) into (1.15) gives a nonzero residual R_n

$$R_n(a_1,a_2,\cdots a_n,x_1,x_2) = Ay_n = Ay_0 + \sum_{i=1}^{n} a_i A\phi_i \tag{1.18}$$

In this method, the unknown coefficients a_i are determined by solving the following equations [3,4]

$$\langle R_n,\phi_j \rangle_{L_2(\Omega;\mathbb{R}^{n_y})} = 0, \text{for } j = 1,2,\cdots,n \tag{1.19}$$

Since A is linear, this equation (1.19) can be written as

$$\sum_{i=1}^{n} a_i \langle A\phi_i,\phi_j \rangle_{L_2\left(\Omega;\mathbb{R}^{n_y}\right)} = -\langle Ay_0,\phi_j \rangle_{L_2\left(\Omega;\mathbb{R}^{n_y}\right)} \tag{1.20}$$

which determines a_i 's. Substitution of the a_i 's obtained from the solution of equation (1.20) into (1.17) gives the required approximate solution y_n. However, the general solution y_n in equation (1.17) with the determination of the coefficients a_i calculated from equation (1.20) may be difficult. An interesting connection between the Galerkin solution and the Fourier representation of y is found. We seek a Galerkin solution in the following form:

$$y_n = \sum_{i=1}^{n} a_i\phi_i(x_1,x_2) \tag{1.21}$$

with a special restriction

$$\langle \phi_i,\phi_j \rangle_{L_2\left(\Omega;\mathbb{R}^{n_y}\right)} = \begin{cases} 0, i \neq j \\ 1, i = j \end{cases} \tag{1.22}$$

Thus, the application of the Galerkin method to equation (1.15) gives

$$\sum_{i=1}^{n} a_i\phi_i \langle A\phi_i,\phi_j \rangle_{L_2\left(\Omega;\mathbb{R}^{n_y}\right)} = \langle f,\phi_j \rangle_{L_2\left(\Omega;\mathbb{R}^{n_y}\right)} \tag{1.23}$$

which is, by equation (1.22),

$$a_i = \langle f,\phi_i \rangle_{L_2\left(\Omega;\mathbb{R}^{n_y}\right)} \tag{1.24}$$

so the Galerkin solution equation (1.21) becomes

$$y_n = \sum_{i=1}^{n} \langle f, \phi_i \rangle_{L_2\left(\Omega;\mathbb{R}^{n_y}\right)} \phi_i(x_1, x_2) \tag{1.25}$$

Evidently, the Galerkin solution (1.21) is just the finite Fourier series solution. When n goes to infinity, we will obtain the perfect approximation

$$y = \lim_{n\to\infty} y_n = \lim_{n\to\infty} \sum_{i=1}^{n} \langle f, \phi_i \rangle_{L_2\left(\Omega;\mathbb{R}^{n_y}\right)} \phi_i(x_1, x_2) \tag{1.26}$$

Therefore, the perfect Galerkin solution like equation (1.26) results in the infinite-dimensional ($n \to \infty$) problem. That is, we have to use infinitely many equations in the linear combination form to obtain the perfectly exact solution of PDEs or PDSs. This Galerkin method makes it possible simulate solutions of PDEs and PDSs with sufficient large number n. To guarantee the well-posedness of the PDEs or PDSs with the Galerkin method, both the property of boundedness

$$\left|\left\langle \langle A_{y_n}, \phi_j \rangle_{L_2\left(\Omega;\mathbb{R}^{n_y}\right)} \right\rangle\right| \le c_1 \left\|A_{y_n}\right\|_{L_2\left(\Omega;\mathbb{R}^{n_y}\right)} \left\|\phi_j\right\|_{L_2\left(\Omega;\mathbb{R}^{n_y}\right)}, \forall j \tag{1.27}$$

and the property of ellipticity

$$\left|\left\langle \langle A_{y_n}, \phi_j \rangle_{L_2\left(\Omega;\mathbb{R}^{n_y}\right)} \right\rangle\right| \ge c_2 \left\|y_n\right\|^2_{L_2\left(\Omega;\mathbb{R}^{n_y}\right)} \tag{1.28}$$

for some constants $c_1>0$ and $c_2>0$, are necessary. These two conditions imply well-posedness by the Lax-Milgram theorem.

Here, we take a kind of Poisson equation for example to find the approximate solution by using the Galerkin method. Consider the following Poisson equation:

$$\frac{\partial^2 y}{\partial x_1^2} + \frac{\partial^2 y}{\partial x_2^2} = -1, in\ \Omega = \left\{(x_1, x_2) \middle| |x_1| < a, |x_2| < b\right\} \tag{1.29}$$

with the boundary condition

$$y = 0 \text{ on } \partial\Omega = \{x_1 = \pm a, x_2 = \pm b\} \tag{1.30}$$

Now, we seek a trial solution in the for [3,4]

$$y_M(x_1, x_2) = \sum_{m=1,3,5\cdots}^{M} \sum_{n=1,3,5\cdots}^{M} a_{mn} \phi_{mn}(x_1, x_2) \tag{1.31}$$

where

$$\phi_{mn}(x_1,x_2)=\cos\left(\frac{m\pi x_1}{2a}\right)\cos\left(\frac{n\pi x_2}{2b}\right) \tag{1.32}$$

In this case $A=\Delta=\nabla^2$ and the residual R_M is

$$R_M = Ay_M + 1 = -\left[\sum_{m=1}^{M}\sum_{n=1}^{M}\left(\frac{m^2\pi^2}{4a^2}+\frac{n^2\pi^2}{4b^2}\right)a_{mn}\phi_{mn}\right]+1 \tag{1.33}$$

According to the Galerkin method, we obtain

$$\begin{aligned}\langle R_M,\phi_{kl}\rangle_{L_2(\Omega;\mathbb{R}^{n_y})} &= \int_{-b}^{b}\int_{-a}^{a} R_M\cos\left(\frac{k\pi x_1}{2a}\right)\cos\left(\frac{l\pi x_2}{2b}\right)dx_1dx_2\\ &= \frac{ab\pi^2}{4}\left(\frac{k^2}{a^2}+\frac{l^2}{b^2}\right)a_{kl}-\frac{16ab}{\pi^2 kl}(-1)^{(k+l)/2-1}=0\end{aligned} \tag{1.34}$$

or

$$a_{kl}=\left(\frac{8ab}{\pi^2}\right)^2\frac{(-1)^{(k+l)/2-1}}{(b^2k^2+a^2l^2)} \tag{1.35}$$

Thus, the solution of the problem is

$$y_M(x_1,x_2)=\left(\frac{8ab}{\pi^2}\right)^2\sum_{m=1,3,5\cdots}^{M}\sum_{n=1,3,5\cdots}^{M}\frac{(-1)^{(m+l)/2-1}\phi_{mn}}{(b^2m^2+a^2n^2)} \tag{1.36}$$

Moreover, in the limit $M\to\infty$, the solution $y_M(x_1,x_2)$ is in perfect agreement with those obtained by the double Fourier series.

The Galerkin method can be used not only to simulate solutions of PDEs and PDSs, but also to deal with the diffusion effect on the parabolic PDSs for the control design purposes. Taking a simple 1-dimensional linear parabolic PDS for example, we consider the PDS [3]

$$\frac{\partial y}{\partial t}=\Delta y+ay+bu \tag{1.37}$$

where $y\in D(A)$ is a constant function, $u\in\mathbb{R}$ is a constant control input, $a\in\mathbb{R}$, and $b\in\mathbb{R}$. With the Hilbert-Schmidt theorem, the solution of equation (1.37) can be represented as [3]

$$y=\sum_{l=1}^{\infty}y_1(t)\phi_l(x) \tag{1.38}$$

which is also called spectral expansion, where $y_l(t) = \langle y, \phi_l(x) \rangle_{L_2(\Omega;\mathbb{R})}$ and $\Delta\phi_l(x) = \lambda_l \phi_l(x)$ with the eigenvalues λ_l as well as the corresponding orthonormal eigenfunctions $\phi_l(x)$. The orthonormal eigenfunctions $\phi_l(x)$ have the property

$$\langle \phi_k(x), \phi_l(x) \rangle_{L_2(\Omega;\mathbb{R})} = \begin{cases} 0, k \neq l \\ 1, k = l \end{cases} \tag{1.39}$$

and constitute a set of bases for $L_2(\Omega;\mathbb{R})$. λ_l is ordered so that $Re\{\lambda_{l+1}\} \leq Re\{\lambda_l\}$, where $Re\{\lambda_l\}$ denotes the real part of λ_l. Hence, we have

$$\left\langle \frac{\partial y}{\partial t}, \phi_l(x) \right\rangle_{L_2(\Omega;\mathbb{R})} = \frac{\partial y_l(t)}{\partial t}, u_l(t) \triangleq \langle u, \phi_l(x) \rangle_{L_2(\Omega;\mathbb{R})} \tag{1.40}$$

and

$$\langle \Delta y, \phi_l(x) \rangle_{L_2(\Omega;\mathbb{R})} = \lambda_l y_l(t) \tag{1.41}$$

From equations (1.40) and (1.41), we can rewrite equation (1.37) as

$$\frac{\partial y_l(t)}{\partial t} = \lambda_l y_l(t) + a y_l(t) + b u_l(t), \quad \text{for } l = 1, 2, \cdots \tag{1.42}$$

Then, we collect all $y_l(t)$ to construct the following ordinary differential system (ODS)

$$\dot{y}(t) = Ay(t) + BU(t) \tag{1.43}$$

$y(t) \triangleq [y_1(t), \cdots, y_\infty(t)]^T \in \mathbb{R}^\infty$, $A \triangleq diag(\lambda_1 + a, \cdots, \lambda_\infty + a) \in \mathbb{R}^{\infty \times \infty}$, $U(t) = [u_1(t), \cdots, u_\infty(t)]^T \in \mathbb{R}^\infty$, and $B \triangleq diag(b, \cdots, b) \in \mathbb{R}^{\infty \times \infty}$. To avoid infinite-dimensional problem of the ODS, based on the property of the eigenvalues λ_l and the orthonormal eigenfunctions $\phi_l(x)$, the system state of the PDS equation (1.37) should be partitioned into a finite-dimensional slow subsystem and an infinite-dimensional fast residual subsystem. Hence, we define

$$y(x,t) = y_c(x,t) + y_r(x,t) \tag{1.44}$$

where

$$y_c(x,t) \triangleq \sum_{l=1}^{n_c} y_l(t)\phi_l(x) \text{ and } y_r(x,t) \triangleq \sum_{l=n_c+1}^{\infty} y_l(t)\phi_l(x) \tag{1.45}$$

Similar to the analysis in equations (1.40)–(1.43), we can obtain [4]

$$\dot{y}_c(t) = A_c y_c(t) + A_{cr} y_r(t) + B_c U_c(t) \tag{1.46}$$

$$\dot{y}_r(t) = A_{rc}y_c(t) + A_r y_r(t) + B_r U_r(t) \tag{1.47}$$

where $U_c(t) \triangleq [u_1(t),\cdots,u_{n_c}(t)]^T$, $U_r(t) \triangleq [u_{n_c+1}(t),\cdots,u_\infty(t)]^T$, $y_c(t) \triangleq [y_1(t),\cdots,y_{n_c}(t)]^T$ is the system state of the finite-dimensional slow controllable subsystem, and $y_r(t) \triangleq [y_{n_c+1}(t),\cdots,y_\infty(t)]^T$ is the system state of the coupled infinite-dimensional fast residual subsystem. The matrices A_c, A_r, B_c and B_r, with A and B as follows:

$$A = \begin{bmatrix} A_c & A_{cr} \\ A_{rc} & A_r \end{bmatrix} \text{ and } B = \begin{bmatrix} B_c & 0 \\ 0 & B_r \end{bmatrix} \tag{1.48}$$

Generally, the residual subsystem equation (1.47) is asymptotically stable because one always chooses a negative definite matrix A_r, which means $\lambda_{n_c+1} + a < 0,\ldots,$ $\lambda_\infty + a < 0$ based on the property $Re\{\lambda_{l+1}\} \le Re\{\lambda_l\}$, i.e., the unstable models should be put into A_c. Moreover, in this simple 1-dimensional linear parabolic PDS equation (1.37), we have $A_{cr} = 0$ and $A_{rc} = 0$; however, if the PDS is a complex one, then the coupled matrices A_{cr} and A_{rc} would be nonzero matrices. Similarly, $B_r U_r(t) = 0$, which is so-called the control spillover, when considering this simple PDS. Consequently, by choosing a sufficient large number n_c, the control designs for the simple PDS equation (1.37) can be reformulated as ones for the ODS equation (1.46) that can be realized via the traditional algebraic matrix techniques. However, for the complex PDSs in the studies, the effects of the coupled matrices A_{cr} and A_{rc}; the control spillover $B_r U_r(t)$; the observation spillover; external and measurement noises need to be considered. That is, not only the finite-dimensional slow subsystem like equation (1.46) needs to be controlled, but also the infinite-dimensional fast residual subsystem like equation (1.47) needs to be tolerated. Therefore, the main disadvantage of using the Galerkin method to deal with the diffusion effect on the parabolic PDSs for the control designs is that the coupled infinite-dimensional fast residual subsystem needs to be considered in the design procedure, which means the effect of the subsystem needs to be tolerated to avoid unexpected control performance. Besides, the size of $y_c(t)$, i.e., n_c, is large normally, which means dealing with the diffusion effect on the parabolic PDSs by using the Galerkin method consumes a great deal of computational resources. In this book, the Galerkin method will be applied to the control and state observation design of linear and nonlinear PDSs in Chapters 6, 7 and 9.

1.2.2 Finite Difference Method

In mathematics, FDMs are numerical methods for approximating the solutions to differential equations using finite difference equations to approximate derivatives. By using the FDMs, one can approximate the derivatives appearing in a PDE or PDS by sums and differences of algebraic function values at a set of discrete points, usually uniformly spaced with respect to each independent variable. Briefly, the FDM can be characterized as follows. FDM utilizes uniformly spaced

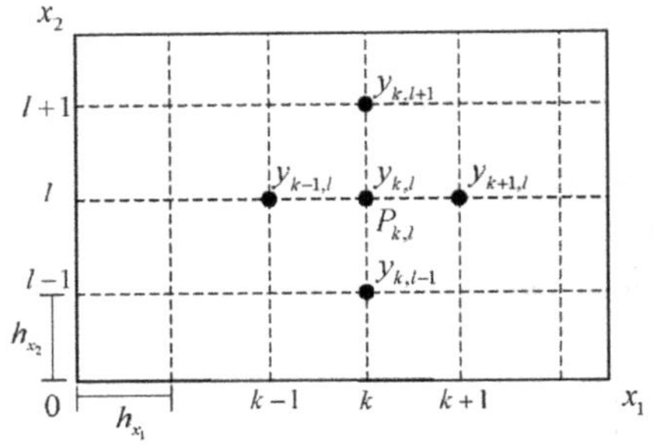

FIGURE 1.1 An example of finite difference grids of the central finite difference scheme on the 2-dimensional spatial domain Ω.

grids (Figure 1.1, e.g., the central finite difference scheme on the 2-dimensional spatial domain Ω).

At each node $y_{k,l}$, each derivative is approximated by an algebraic expression which references the adjacent nodes, i.e., $y_{k-1,l}, y_{k+1,l}, y_{k,1-l}$ and $y_{k,1+l}$. A system of algebraic equations is obtained by evaluating the previous step for each node and the system is solved for the dependent variable. For clarity, we illustrate the principle of the FDM on the 1-dimensional spatial domain Ω, for example. The system of algebraic equations with the sums and differences of algebraic function values is based on the famous Taylor Series from elementary calculus [5]

$$y(x) = y(h) + \frac{(x-h)}{1!} y'(h) + \frac{(x-h)^2}{2!} y''(h) + \cdots = \sum_{n=0}^{\infty} \frac{(x-h)^n}{n!} y^{(n)}(h) \quad (1.49)$$

where $y^{(n)}$ denotes the derivative taken n times of y with respect to the space variable x, $n!$ denotes the factorial of n, and h is denoted as an increment in x and is assumed positive. Now, we replace x and h in the previous formula (1.49) by $x+h$ and x, respectively. Then, we have

$$y(x+h) = y(x) + \frac{h}{1!} y'(x) + \frac{h^2}{2!} y''(x) + \cdots = \sum_{n=0}^{\infty} \frac{h^n}{n!} y^{(n)}(x) \quad (1.50)$$

For the treatment of many problems, it is convenient to take only the first two terms of the right hand side of the previous equation,

$$y(x+h) = y(x) + hy'(x) + O(h^2) \quad (1.51)$$

where the expression $O(h^2)$ represents the error of the approximation, which is proportional to h^2, and $O(h^2)$ is the famous big O notation that describes the limiting behavior of a function when the argument h^2 tends towards a particular value or infinity. From this relationship, we can easily define what is known as the first order, forward finite difference scheme to $\frac{dy}{dx}$, i.e.,

$$\frac{dy}{dx} = \frac{y(x+h) - y(x)}{h} + O(h) \tag{1.52}$$

Likewise, we can define the first order, backward finite difference scheme to $\frac{dy}{dx}$

$$\frac{dy}{dx} = \frac{y(x) - y(x-h)}{h} + O(h) \tag{1.53}$$

by

$$y(x-h) = y(x) + hy'(x) + O(h^2) \tag{1.54}$$

If instead, taking up to the third order, then we have

$$y(x+h) = y(x) + hy'(x) + \frac{h^2}{2} y''(x) + O(h^3) \tag{1.55}$$

$$y(x-h) = y(x) - hy'(x) + \frac{h^2}{2} y''(x) + O(h^3) \tag{1.56}$$

By subtracting equation (1.56) from (1.55), we can obtain the first order, central finite difference scheme to $\frac{dy}{dx}$, i.e.,

$$\frac{dy}{dx} = \frac{y(x+h) - y(x-h)}{2h} + O(h^2) \tag{1.57}$$

The central finite difference scheme obviously has the best approximation since the increment h is always chosen as a small positive number which is definitely far less than one ($0 < h \ll 1$). Therefore, the central finite difference scheme is normally used to simulate solutions of PDEs and PDSs; thus, the illustration of the followings focuses on the FDM with the central finite difference scheme. As to the second order, central finite difference scheme to $\frac{dy}{dx}$, we take up to the forth order as follows:

$$y(x+h) = y(x) + hy'(x) + \frac{h^2}{2} y''(x) + \frac{h^3}{6} y'''(x) + O(h^4) \tag{1.58}$$

$$y(x-h) = y(x) - hy'(x) + \frac{h^2}{2} y''(x) - \frac{h^3}{6} y'''(x) + O(h^4) \tag{1.59}$$

and

$$y(x-h) + y(x+h) = 2y(x) + h^2 y''(x) + O(h^4) \tag{1.60}$$

Now, the second-order central finite difference scheme to $\dfrac{d^2 y}{dx^2}$ can be obtained as follows:

$$\frac{d^2 y}{dx^2} = \frac{y(x+h) - 2y(x) + y(x-h)}{h^2} + O(h^2) \qquad (1.61)$$

Hence, for the numerical solutions of PDEs and PDSs on the 1-dimensional spatial domain Ω, one can utilize the following approximation:

$$\frac{\partial y(x,t)}{\partial t} = \frac{y(x, t+h_t) - y(x,t)}{h_t} + O(h_t) \qquad (1.62)$$

$$\frac{\partial y(x,t)}{\partial x} = \frac{y(x+h_x, t) - y(x-h_x, t)}{2h_x} + O(h_x^2) \qquad (1.63)$$

$$\frac{\partial^2 y(x,t)}{\partial x^2} = \frac{y(x+h_x) - 2y(x) + y(x-h_x)}{h_x^2} + O(h_x^2) \qquad (1.64)$$

where h_{x_1} and h_{x_2} are the increment of the time and space variables x_1 and x_2, respectively. Moreover, for the numerical solutions of PDEs and PDSs on the 2-dimensional spatial domain Ω, one can use equations (1.63) and (1.64) to obtain the following:

$$\frac{\partial^2 y\big((x_1, x_2), t\big)}{\partial x_1 \partial x_2} = \frac{y\big((x_1 + h_{x_1}, x_2 + h_{x_2}), t\big) + y\big((x_1 - h_{x_1}, x_2 - h_{x_2}), t\big)}{4 h_{x_1} h_{x_2}}$$

$$- \frac{y\big((x_1 + h_{x_1}, x_2 - h_{x_2}), t\big) + y\big((x_1 - h_{x_1}, x_2 + h_{x_2}), t\big)}{4 h_{x_1} h_{x_2}}$$

$$+ O\left(\frac{(h_{x_1} + h_{x_2})^4}{h_{x_1} h_{x_2}} \right) \qquad (1.65)$$

$$\Delta y = \frac{y\big((x_1 + h_{x_1}, x_2), t\big) - 2y\big((x_1 + x_2), t\big) + y\big((x_1 - h_{x_1}, x_2), t\big)}{h_{x_1}^2}$$

$$+ \frac{y\big((x_1, x_2 + h_{x_2}), t\big) - 2y\big((x_1 + x_2), t\big) + y\big(y(x_1, x_2 - h_{x_2}), t\big)t\big)}{h_{x_2}^2} + O(\cdot) \qquad (1.66)$$

where h_{x_1} and h_{x_2} are the increment of the space variables x_1 and x_2, respectively. $O(\cdot)$ is proportional to the complicated function constructed from $h_{x_1}^2$ and $h_{x_2}^2$. One can use the results in equations (1.62)–(1.66) to obtain the numerical solutions of PDEs and PDSs. Since the parabolic PDEs and PDSs are concerned in this book, the followings focus on the approximation equations (1.62) and (1.66).

For the parabolic PDEs and PDSs on the 2-dimensional spatial domain Ω (Figure 1.1), we first impose a mesh of grid points; hence, we have the interior grid points $\left(x_1^{(k)}, x_2^{(l)}\right) = \left(kh_{x1}, lh_{x2}\right)$, where $k = 1, 2, \cdots, N_{x1}$, and $l = 1, 2, \cdots, N_{x2}$. With the result of equation (1.66), but without considering $O(\cdot)$, we have

$$\Delta y = \frac{y\left(\left(x_1^{(k+1)}, x_2^{(l)}\right), t\right) - 2y\left(\left(x_1^{(k)} + x_2^{(l)}\right), t\right) + y\left(\left(x_1^{(k-1)}, x_2^{(l)}\right), t\right)}{h_{x1}^2}$$

$$+ \frac{y\left(\left(x_1^{(k)}, x_2^{(l+1)}\right), t\right) - 2y\left(\left(x_1^{(k)} + x_2^{(l)}\right), t\right) + y\left(\left(x_1^{(k)}, x_2^{(l-1)}\right), t\right)}{h_{x2}^2} \quad (1.67)$$

For simplicity, one always defines h_x to be the distance between mesh points, i.e., $h_{x1} = h_{x2} = h_x$. This yields the simplification

$$\Delta y_{k,l} = \frac{1}{h_x^2}\left[y_{k+1,l} + y_{k-1,l} + y_{k,l+1} + y_{k,l-1} - 4y_{k,l}\right] \quad (1.68)$$

where $y_{k,l} \triangleq y\left(x_1^{(k)}, x_2^{(l)}, t\right) = y\left(kh_{x1}, lh_{x2}\right)$. If the higher order term is considered, then we could find out that the local truncation error for $\Delta y_{k,l}$ is

$$\frac{1}{12} h_x^2 \left(\frac{\partial^4 y}{\partial x_1^4} + \frac{\partial^4 y}{\partial x_2^4} \right) \Bigg|_{x_1 = x_1^{(k)}, x_2 = x_2^{(l)}, y = y\left(x_1^{(k)}, x_2^{(l)}\right)} \quad (1.69)$$

which is usually written as $O\left(h_x^2\right)$ such that the error is proportional to the space constant square $O\left(h_x^2\right)$. That is, the local truncation error $O\left(h_x^2\right)$, which will be discussed later, can be neglected if one chooses a sufficient small grid size h_x. Note that five points (points to the left, right, above and below of a central point, $\left(x_1^{(k)}, x_2^{(l)}\right)$) are involved in the approximation of $\Delta y_{k,l}$; hence, the FDM with the central finite difference scheme is also called the five point formula. Obviously, the FDMs with the forward, central, and backward finite difference schemes are spatial-discretized methods.

Recall that whatever types of PDEs or PDSs, i.e., the hyperbolic, parabolic, and elliptic types of PDEs or PDSs, can be numerically simulated by employing the FDM mentioned above. In order to guarantee the well-posedness of the PDEs or PDSs with the FDM, we give the following discussion. Suppose $y(x_1, x_2)$ represents the exact solution of a PDE or PDS $L\left(y(x_1, x_2)\right) = 0$ with independent variables x_1 as well as x_2, and $y_{k,l}$ is the exact solution of the corresponding spatial-discretized algebraic equation $F\left(y_{k,l}\right) = 0$, which is constructed from equations (1.62)–(1.66). Then, the FDM is said to be convergent if $y_{k,l}$ tends to $y(x_1, x_2)$ as h_{x1} and h_{x2} tend to zero. The difference $d_{k,l} \triangleq y\left(x_1^{(k)}, x_2^{(l)}\right) - y_{k,l} \triangleq Y_{k,l} - y_{k,l}$ is **the truncation error** or the so-called **discretization error.** This error can be generally be minimized by decreasing the grid sizes h_{x1} and h_{x2}. Moreover, this error depends on the number of terms in the truncated series equations (1.62)–(1.66), which are used to approximate each partial derivative. Another kind of error is introduced when a PDE or PDS is approximated by the spatial-discretized algebraic equation. If the solution $y_{k,l}$ is replaced by the solution $y(x_1,x_2)$ at the grid points (see Figure 1.1), then the value $F(Y_{k,l})$ is called the **local truncation error** at $P_{k,l}$. The FDM and the PDE or PDS are said to be consistent if $F(Y_{k,l})$ tends to zero as h_{x_1} and h_{x_2} tend to zero. In general, the spatial-discretized algebraic equations cannot be solved exactly because the numerical computation is carried out only up to a finite number of decimal places. Consequently, another kind of error is introduced in the solution $y_{k,l}$ during the actual process of computation. This error is called the round-off error, and also depends upon the type of computer used. In practice, the actual computational solution is $y_{k,l}^*$, but not $y_{k,l}$, so that the difference $r_{k,l} \triangleq y_{k,l} - y_{k,l}^*$ is the round-off error at $P_{k,l}$. In reality, the **round-off error** depends mainly on the actual computational process and the FDM itself; therefore, this error cannot be made small by allowing h_{x_1} and h_{x_2} to tend to zero. Hence, the total error involved in the finite difference analysis at the point $P_{k,l}$ is

$$Y_{k,l} - y_{k,l}^* = (Y_{k,l} - y_{k,l}) + (y_{k,l} - y_{k,l}^*) = d_{k,l} - r_{k,l} \tag{1.70}$$

Usually, the truncation error $d_{k,l}$ is bounded when $y_{k,l}$ is bounded because the value of $Y_{k,l}$ is fixed for a given PDE or PDS with the prescribed boundary and initial condition. This fact is used or assumed in order to introduce the concept of stability. The FDM is said to be stable if the round-off errors are sufficiently small for all k as $l \to \infty$; that is, the growth of $r_{k,l}$ can be controlled. Lax (1954) proved a remarkable theorem which establishes the relationship between consistency, stability, and convergence for the FDM.

Theorem 1.2.1 (Lax's Equivalence Theorem [1,2])

Given a properly posed linear initial-value problem and a FDM approximation to it that satisfies the consistency criterion, stability is the necessary and sufficient condition for convergence.

In other words, Lax states that for a consistent FDM for a well-posed linear initial-value problem, the method is convergent if and only if it is stable. This theorem is important because the convergence, which is desired, is ordinarily difficult to establish. Nevertheless, the consistency is straightforward to verify, which means the FDM and the PDE or PDS are consistent based on equations (1.62)–(1.66) with $h_{x_1} \to 0$ and $h_{x_2} \to 0$. In addition, the stability is typically much easier to show than the convergence. Thus, the convergence is usually established via the Lax's equivalence theorem. For showing the stability, Von Neumann stability analysis (1947) is employed. Von Neumann stability analysis (also known as Fourier stability analysis) is a procedure used to check the stability of FDM as applied to a PDE or PDS. Generally, the stability of numerical schemes is closely associated with numerical error in software engineering and mathematics, which is the combination of the truncation and round-off errors. Von Neumann stability analysis states that a FDM is stable if the errors made at one time step h_t of the calculation do not cause the errors to increase as the computations are continued. On the contrary, if the errors grow with time, the FDM is unstable. The stability analysis is based on the decomposition of the errors into Fourier series. For examples, consider 1-dimensional and 2-dimensional heat equations as

$$\frac{\partial y(x,t)}{\partial t} = \alpha \frac{\partial^2 y(x,t)}{\partial x^2} \tag{1.71}$$

and

$$\frac{\partial y((x_1,x_2),t)}{\partial t} = \alpha_1 \frac{\partial^2 y((x_1,x_2),t)}{\partial x_1^2} + \alpha_2 \frac{\partial^2 y((x_1,x_2),t)}{\partial x_2^2} \tag{1.72}$$

where $\alpha \in \mathbb{R}_+, \alpha_1 \in \mathbb{R}_+$, and $\alpha_2 \in \mathbb{R}_+$. Then, the stability conditions of equations (1.71) and (1.72) are

$$\frac{\alpha h_t}{h_x^2} \le \frac{1}{2}, \text{ and } \frac{\alpha_1 h_t}{h_{x1}^2} + \frac{\alpha_2 h_t}{h_{x2}^2} \le \frac{1}{2} \tag{1.73}$$

respectively, where h_{x1}^2 and h_{x2}^2 have been defined in equations (1.65) and (1.66). As to the effect of diffusion on the concerned PDS equation (1.8), we choose a conservative stability condition as

$$\lambda_{\max}^{D}\left(\frac{h_t}{h_{x1}^2} + \frac{h_t}{h_{x2}^2} + \cdots + \frac{h_t}{h_{xN}^2}\right) \le \frac{1}{2} \tag{1.74}$$

where $\lambda_{\max}^{D}$ is the maximum eigenvalue of the diffusion coefficient matrix D. With the stability conditions (1.73) and (1.74) and the consistency based on $h_{x1} \to 0$ and $h_{x2} \to 0$, the convergence can be guaranteed via the Lax's equivalence theorem. Surely, the FDM with the central finite difference scheme under the stability condition (1.74) is utilized in our simulation results for the concerned PDSs throughout this book.

After introducing the solutions of PDEs and PDSs via using the FDMs, we briefly demonstrate the treatments of the diffusion effect on the parabolic PDSs for the control design purposes. With the central finite difference scheme, the similar methods will be used in the followings to deal with the diffusion effect. The primary idea is to collect the system states at all grid nodes on spatial domain as a spatial state vector. Then, the PDSs can be transformed into the ODSs with the local truncation errors. Accordingly, the control designs for the PDSs can be formulated as ones for the ODSs that can be realized via the traditional algebraic matrix techniques. Taking a simple 2-dimensional linear parabolic PDS for example, we consider the PDS

$$\frac{\partial y}{\partial t} = D\triangle y + ay + bu \tag{1.75}$$

where the only difference between equations (1.75) and (1.37) is that $x = [x_1, x_2]^T \in \Omega \subset \mathbb{R}^2$ with $D \in \mathbb{R}_+$. With the typical grid mesh shown in Figure 1.1, the system state y is represented by $y_{k,l}(t)$ at the grid node $x_{k,l}$ ($x_1 = kh_{x_1} = kh_x, x_2 = lh_{x_2} = lh_x$), where $k = 1,2,\cdots,N_{x_1}$, and $l = 1,2,\cdots,N_{x_2}$, i.e., $y(x,t)|_{x=x_{k,l}} \triangleq y_{k,l}(t)$. At the grid points $x_{k,l}$, the central finite difference approximation for the Laplace operator Δ can be written as follows [3]:

$$\left(\frac{\partial^2 y(x,t)}{\partial x^2} \right)_{x=x_{k,l}} = \frac{y_{k+1,l}(t) + y_{k-1,l}(t) - 2y_{k,l}(t)}{h_x^2}$$

$$+ \frac{y_{k,l+1}(t) + y_{k,l-1}(t) - 2y_{k,l}(t)}{h_x^2} + O_{k,l}(h_x^2) \qquad (1.76)$$

where $O_{k,l}(h_x^2)$ is referred to as the local truncation error, which converges to zero if the grid size h_x is chosen to be as small as possible by applying the Lax's equivalence theorem in Theorem 1.2.1, i.e., $O_{k,l}(h_x^2) \to 0$ as $h_x \to 0$. Therefore, the PDS equation (1.75) can be represented as follows:

$$y_{k,l}(t) = ay_{k,l}(t) + bu_{k,l}(t)$$

$$+ \frac{D}{h_x^2} \left[y_{k+1,l}(t) + y_{k-1,l}(t) - 2y_{k,l}(t) + y_{k,l+1}(t) + y_{k,l-1}(t) - 2y_{k,l}(t) \right] (1.77)$$

where $u_{k,l}(t)$ is the control effort at the grid node $x_{k,l}$. To simplify the control design, we define a spatial state vector $Y(t)$ to collect the states $y_{k,l}(t)$ at all grid nodes in Figure 1.1, and $Y(t) \in \mathbb{R}^{N_Y}$ is given as follows:

$$Y(t) = \left[y_{1,1}(t),\cdots,y_{N_{x_1},1}(t),\cdots,y_{k,l}(t),\cdots,y_{1,N_{x_2}}(t),\cdots,y_{N_{x_1},N_{x_2}}(t) \right]^T \qquad (1.78)$$

where $N_y \triangleq N_{x_1} \times N_{x_2}$. In order to simplify the index of the node $y_{k,l}(t)$, we denote the symbol $y_j(t)$ to replace $y_{k,l}(t)$. Note that the index j is from 1 to N_Y, i.e., $y_1(t) \triangleq y_{1,1}(t)$, $y_2(t) \triangleq y_{1,2}(t)$,..., $y_j(t) \triangleq y_{k,l}(t)$,..., $y_{N_Y}(t) \triangleq y_{N_{x_1},N_{x_2}}(t)$, where $j = (k-1)N_{x_1} + l$ in equation (1.78). Then, the finite difference model of two indices in equation (1.77) can be represented with only one index as follows [5,8–10]

$$\dot{y}_j(t) = ay_j(t) + bu_j(t) + DT_jY(t) \qquad (1.79)$$

where $T_j \in \mathbb{R}^{1 \times N_Y}$ expresses the interaction from the other grid nodes to the jth node as follows:

$$T_jY(t) \triangleq \left[T_{x_1,j,1} + T_{x_2,j,1},\cdots,T_{x_1,j,N_Y} + T_{x_2,j,N_Y} \right]Y(t)$$

$$= \sum_{j_2=1}^{N_Y} (T_{x_1,j,j_2} + T_{x_2,j,j_2})y_{j_2}(t) \qquad (1.80)$$

in which $T_{x1,j,j_2} \in \mathbb{R}$ and $T_{x2,j,j_2} \in \mathbb{R}$ are defined as

$$
T_{x1,j,j_2} \triangleq
\begin{cases}
\dfrac{-2}{h_x^2}I, & \text{for } j_2 = (l-1)N_{x_1}+k \\[2mm]
\dfrac{1}{h_x^2}I, & \text{for } j_2 = (l-1)N_{x_1}+k-1 \\[2mm]
\dfrac{1}{h_x^2}I, & \text{for } j_2 = (l-1)N_{x_1}+k+1 \\[2mm]
0, & \text{otherwise}
\end{cases}
$$

$$
T_{x2,j,j_2} \triangleq
\begin{cases}
\dfrac{1}{h_x^2}I, & \text{for } j_2 = (l-2)N_{x_1}+k \\[2mm]
\dfrac{1}{h_x^2}I, & \text{for } j_2 = lN_{x_1}+k \\[2mm]
\dfrac{-2}{h_x^2}I, & \text{for } j_2 = (l-1)N_{x_1}+k \\[2mm]
0, & \text{otherwise}
\end{cases}
$$

(1.81)

By applying some matrix multiplication techniques to (1.79), the PDS with the state $Y(t)$ can be constructed; thus, the control designs for the PDS (1.75) can be realized via the traditional algebraic matrix techniques when $h_x \to 0$. Note that if the state $y(x,t)$ of the PDS (1.75) is not a constant function, i.e., $y(x,t) \in \mathbb{R}^{n_y}$, then $Y(t) \in \mathbb{R}^{n_y N_Y}$, $O_{k,l}(h_x^2) \in \mathbb{R}^{n_y}$, $T_j \in \mathbb{R}^{n_y \times n_y N_Y}$, $T_{x1,j,j_2} \in \mathbb{R}^{n_y \times n_y}$, and $T_{x2,j,j_2} \in \mathbb{R}^{n_y \times n_y}$. Accordingly, the control designs for the PDSs can be realized based on control design of DDS with state $Y(t)$ as the previous analysis [4,5,9–11]. More details will be given in the following chapters. The main drawback of using the central finite difference scheme to deal with the diffusion effect on the parabolic PDSs for the control designs is that the number of grid nodes N_Y needs to be sufficient large. That is, the tradeoff problem between the grid size h_x for the central finite difference scheme and the control performance occurs. If the grid size h_x is small enough so that $O_{k,l}(h_x^2) \to 0$, then a great deal of computational resources is necessary because of the sufficient large N_Y. Otherwise, the effect of the local truncation error $O_{k,l}(h_x^2)$ needs to be considered to avoid unexpected control performance. That is, a great deal of computational resources is necessary as well.

1.2.3 Poincaré Inequality Method

Consider the PDS in equation (1.8) with Neumann boundary condition N in equation (1.10). The following equations based on Green theorem and divergence theorem [2,3] are useful to treat the system designs of PDSs with Neumann boundary condition N in equation (1.10).

$$\int_{\Omega} y^T(x)\Delta y(x)\,dx = \int_{\Omega} y^T(x)\nabla^2 y(x)\,dx$$

$$= -\int_{\Omega} \nabla^T y(x)\nabla y(x)\,dx + 2\int_{\partial\Omega} y^T(x)\frac{\partial y(x)}{\partial \vec{n}}\,ds \text{ (by Green Theorem)} \tag{1.82}$$

where $\nabla y(x)\cdot\vec{n} = \dfrac{\partial y(x)}{\partial \vec{n}}$

By the Dirichlet and Neumann boundary conditions in equation (1.10) [2,3]

$$\int_{\Omega} y^T(x)\Delta y(x)\,dx = -\int_{\Omega} \nabla^T y(x)\nabla y(x)\,dx \tag{1.83}$$

By Poincaré in equality [2,3]

$$c_p \int_{\Omega} y^T(x)y(x)\,dx \le \int_{\Omega} \nabla^T y(x)\nabla y(x)\,dx \tag{1.84}$$

for some Poincaré constant c_p, which can be chosen as the first nonzero eigenvalue of the Laplace operator Δ.

Then we obtain

$$\int_{\Omega} y^T(x)\Delta y(x)\,dx \le -c_p \int_{\Omega} y^T(x)y(x)\,dx \tag{1.85}$$

Based on equation (1.85), the Laplace operator $\Delta y(x)$ in the system design of PDSs could be treated like the conventional ordinary differential systems (PDSs). These Poincaré inequality techniques will be applied in Chapters 5, 12, and 13 for signal processing and control design of PDSs [12–14].

1.3 THE SCOPE OF THE BOOK

This book provides a perspective of new framework that can help bring together the system design theory of PDSs and practical application to engineering and bio-science. This book is divided into three parts. In part I, the background of PDE and PDS is introduced. In part II, robust signaling and synchronization of PDSs are given. In part III, several robust control designs of PDEs are introduced for linear and nonlinear PDSs.

Part I is divided into two chapters. Chapter 1 gives an introduction to PDEs and PDSs, with which this book considers robust system designs. Some fundamental mathematic properties like existence, uniqueness and stability are introduced. Three numerical and approximation methods i.e., Galerkin method, FDM, and Poincaré inequality method, are given to solve the partial differential Laplace operator, which will meet in system design problems of the following chapters. In Chapter 2, a new technique of adaptive fuzzy algorithm is introduced to solve PDEs and PDSs encountered in science and engineering [10].

Part II is divided into three chapters. Based on finite difference scheme and stochastic spatial state space model, Chapter 3 introduces a robust H_∞ filter design of linear stochastic PDSs with a set of sensor measurements [8]. With the help of fuzzy spatial state space method and finite difference scheme, in Chapter 4, a robust H_∞ filter design is also introduced for nonlinear stochastic PDSs in sensor signal processing [9]. In Chapter 5, based on Green theorem and Poincaré inequality, a robust H_∞ synchronization design is given to PDSs with diffusion coupling [12].

Part III focuses on robust control system design and is divided into eight chapters. Based on Galerkin method, a robust observer-based stabilization control design is proposed in Chapter 6 for large structural systems like supported beam problem under mode truncation, parameter perturbations and actuator saturations [6]. Chapter 7 introduces a robust observer-based control design of large flexible structures based on mode state space approach and frequency domain robustness measure method. The robust observer-based controller could stabilize the controlled mode state space model and tolerate the coupling spillovers from residual mode state space model [7]. In Chapter 8, based on finite difference scheme and spatial state space model, a robust H_∞ stabilization design for stochastic linear PDSs under spatio-temporal disturbance and sensor measurement noises. Based on Galerkin method and fuzzy state space modeling, a robust observer-based control design is proposed for nonlinear PDSs in Chapter 9. T-S fuzzy PDS can interpolate several local linear partial systems to approximation any nonlinear PDS by fuzzy bases interpolation [4]. Then the schemes in observer-based control design in Chapter 7 could be employed for robust H_∞ stabilization of nonlinear PDSs. In Chapter 10, based on fuzzy interpolation schemes, FDM and spatial state space model, a robust H_∞ reference tracking control design of nonlinear distributed parameter time-delayed systems is introduced for practical application [5]. The proposed optimal robust H_∞ reference tracking control design problem of nonlinear PDSs can be transformed to a linear matrix inequalities (LMIs) – constrained optimization problem. Based on FDM, spatial state space model and fuzzy interpolation method, a robust H_∞ stabilization design is proposed in Chapter 11 for nonlinear stochastic PDSs. A robust estimator-based controller is proposed for robust stochastic H_∞ stabilization design to efficiently attenuate the effect of random external disturbance and measurement noise in the spatio-temporal domain [11]. In Chapter 12, based on divergence theorem and Poincaré inequality to transform diffusion matrix inequality to bilinear matrix inequality (BMI), a robust fuzzy H_∞ estimator-based stabilization design is proposed for nonlinear parabolic PDS with different boundary conditions [13]. Based on the proposed decoupling method, the robust fuzzy H_∞ estimator-based stabilizing design problem for nonlinear parabolic PDSs can be effectively solved by a set of LMIs instead of BMI. Since design-cost of fuzzy controller for nonlinear PDSs is very expensive, some low design-cost fuzzy controllers are proposed for nonlinear parabolic PDSs [14]. Based on Green theorem and Poincaré inequality, robust fuzzy H_∞ stabilization area and point controllers are proposed as low design-cost controller of nonlinear PDSs in Chapter 13.

1.4 CONCLUSION

In this chapter, we first describe PDSs. Then some numerical and approximation techniques of Galerkin method, FDM and Poincaré inequality method for PDEs and PDSs are introduced to treat of Laplace operator, which is one of the most difficult problems in the signal processing and control engineering designs of PDSs in the following chapters. Finally, an overview of each chapter in three parts of the book about signal processing and control designs of the partial differential engineering systems is also outlined in this chapter.

2 Fuzzy Solutions to Partial Differential Equations

2.1 INTRODUCTION

It is well known that a lot of phenomena of nature or physical systems can be modeled by partial differential equations PDEs, such as heat equations, wave equations, and so on. Hence, studies of PDEs have become one of the main topics of modern mathematical analysis and have attracted much attention. However, the exact solutions to the PDEs cannot be easily obtained except for very simple or special cases. In recent years, many methods have been developed for solving some kinds of PDEs. For example, a spreadsheet program is used for the numerical solution of the hyperbolic equation in [17]. Moreover, some studies produce a solution in the form of an array that contains the value of the solution at a selected group of points [18]. Others use finite-element methods that are famous and widely adopted in the mechanical fields to solve some specific PDEs [19–21]. In general, a finite-element solution can only offer one discrete solution to approximate the exact solution of PDEs through minimizing the functional of the finite-element method, but this solution is often limited differentiable. In addition, the functional of variational finite-element method must be given first. In general, it is not easy to find the functional of the finite-element method for complex PDEs that cannot be easily derived from the variational calculus [22,23]. Furthermore, how close this computed solution is to the exact solution and whether it converges to the exact solution cannot easily be checked in practice. Recently, the fuzzy spline wavelets [24] were used to approach the solution of differential equations. On the other hand, a fuzzy transform technology [25] was applied to approach the numerical solutions of PDEs based on finite-difference methods in [26].

Hence, the investigation of an effective and more correct method for solving the partial differential systems (PDSs) or the systems of PDEs is an important task! As mentioned earlier, one cannot easily obtain a solution for complex PDEs (high order and nonlinear types). The main reason is that exact solutions are like unknown black boxes, and the profile of the solution cannot be easily obtained by human works or any mathematical analysis. Since the solution of PDEs can be regarded as an unknown system, an approximated method that can handle any unknown system is likely to find a suitable and accurate solution for the treated PDEs. The numerical solution via a fuzzy transform method, as in [26],

DOI: 10.1201/9781003229230-3

needs to solve a PDE via an algebraic equation through a fuzzy transform technique. However, for a nonlinear or high-dimensional PDE, the algebraic equation becomes very complex. In this chapter, a fuzzy solution via adaptive algorithm is proposed to approach the solution of PDE. Therefore, for the nonlinear or high-dimensional cases, the solution of PDE can be easily obtained by the proposed adaptive fuzzy algorithm.

In this chapter, we view the traditional problems of PDEs mentioned earlier from a different perspective with the help of fuzzy logic systems. Fuzzy logic systems have been widely used in system modeling to deal with the nonlinear unknown systems for control or filter designs in recent years [27–31]. In most of these fuzzy system designs [32–35], the fuzzy system was thought to be a universal approximator [36,37] for any nonlinear system. Fuzzy logic system has also been proved to be a very good representation for a class of nonlinear dynamic systems by the conventional schemes, and any nonlinear unknown system can be approximated to any desired accuracy [32]. For this reason, an advanced method that relies on the functional approximation capability of the fuzzy logic systems via fuzzy interpolation scheme to solve the problems of PDEs will be proposed in this chapter. This interpolation scheme employs a linear regression of fuzzy logic system as the basic approximation element, whose parameters in consequent parts are adjusted to minimize an appropriate error function via an adaptive algorithm. In addition, the upper bound of approximation error between the exact solution and the proposed fuzzy solution is derived, and this result can be easily extended to high-dimensional PDE cases. Moreover, the sufficient condition for the convergence of the proposed fuzzy solution in mesh points is also provided.

Three attractive features of the proposed method are obtained as follows:

1. The solution of PDEs via a fuzzy logic system (Gaussian membership case) is a smooth solution that can be easily used in any subsequent calculation. Most other techniques offer a discrete solution (for example, finite-difference methods in Chapter 1) or a solution with limited differentiability (for example, finite-element method).
2. The approximation error bound between the exact solution and the proposed fuzzy solution can be accurately estimated.
3. The employment of fuzzy logic system provides a solution of PDEs with superior interpolation properties. The comparative results with the finite-element method confirm this point clearly.

2.2 PROBLEM FORMULATION

2.2.1 Description of PDE Problems

Consider a class of a two-dimensional PDE formulated as follows [38]:

$$f(x, y, \phi, \phi_x, \phi_y, \phi_{xy}, \phi_{xx}, \phi_{yy}) = u(x, y) \tag{2.1}$$

where $\phi(x,y)$ is a twice-differentiable function defined on a region $\Omega = [a,b] \times [c,d] \subseteq \mathbb{R}^2$, and f is a function of variables $x, y, \phi, \phi_x, \phi_y, \phi_{xy}, \phi_{xx}, \phi_{yy}$, where $\phi_x = \dfrac{\partial \phi(x,y)}{\partial x}, \phi_y = \dfrac{\partial \phi(x,y)}{\partial y}$, and so on. Equation (2.1) can be generalized to more than two variables x, y in an obvious way (see Remark 2.2.1). In addition to equation (2.1), we must specify boundary conditions (BCs) that are most frequently encountered in scientific and engineering problems of the form:

$$g\left(x, y, \phi, \frac{\partial \phi}{\partial \vec{n}} \right)\Bigg|_{\partial\Omega} = 0 \tag{2.2}$$

where $\dfrac{\partial \phi}{\partial \vec{n}}$ is the normal derivative on $\partial\Omega$, i.e., some functions of ϕ and $\dfrac{\partial \phi}{\partial \vec{n}}$ evaluated on boundary $\partial\Omega$ are set to zero.

Assumption 2.2.1

Suppose there exists a unique solution to (2.1) and (2.2)

Remark 2.2.1

One can straightforwardly extend the aforesaid two-variable PDE problems to multivariable PDE problems as follows:

$$f(\vec{x}, \phi(\vec{x}), \nabla\phi(\vec{x}), \nabla^2\phi(\vec{x})) = u(\vec{x}) \tag{2.3}$$

subject to certain BCs as follows:

$$\phi(\vec{x}) = B(\vec{x}), \text{ (Dirichlet)} \tag{2.4}$$

or

$$\vec{n}\nabla\phi(\vec{x}) = \vec{C}(\vec{x}), \text{ (Neumann)} \tag{2.5}$$

where $\vec{x} = (x_1, \cdots x_n) \in \mathbb{R}^n, \phi(\vec{x})$ is the solution to be computed, and $\vec{n}$ is the outward unit vector normal to the boundary.

To solve the PDE problem in equations (2.1) and (2.2), we denote two functions β_0 and β_1 as follows:

$$\beta_0(x, y, \phi, \phi_x, \phi_y, \phi_{xy}, \phi_{xx}, \phi_{yy}) = f(x, y, \phi, \phi_x, \phi_y, \phi_{xy}, \phi_{xx}, \phi_{yy}) - u(x,y) \tag{2.6}$$

$$\beta_1\left(x, y, \phi, \frac{\partial \phi}{\partial n} \right) = g\left(x, y, \phi, \frac{\partial \phi}{\partial \vec{n}} \right)\Bigg|_{\partial\Omega} \tag{2.7}$$

If one exact solution $\phi(x,y)$ for the PDEs and the BCs can be found analytically, $\beta_0(*)$ and $\beta_1(*)$ will be zero. However, it is very difficult to find one exact solution from the PDEs with the BCs mentioned earlier by mathematical analysis or tools.

For this reason, we will develop one approximated fuzzy solution $\tilde{\phi}(x,y|\underline{\theta})$ in the sequel to satisfy the PDEs equation (2.1) with BCs in equation (2.2) as precisely as possible.

Problem of the PDE: Using Assumption 2.2.1, we consider the PDE equation (2.1) with boundary conditions (2.2). The design objective is to find one suitable and approximated fuzzy solution $\tilde{\phi}(x,y|\underline{\theta})$ such that $|\beta_0(*)|$ and $|\beta_1(*)|$ are as small as possible.

2.2.2 Description of Fuzzy Approximation

Practically, the unknown solution of the PDE in equation (2.1) with the BCs in equation (2.2) can be regarded as an unknown system. Inherently, the fuzzy logic system is very suitable for treating this problem with its excellent approximation ability. With the help of a fuzzy logic system, the following fuzzy solution model is introduced to approximate the unknown solution $\phi(x,y)$ of the PDEs with the BCs in equations (2.1) and (2.2).

The fuzzy solution of the PDE with the BCs in equations (2.1) and (2.2) can be described as follows:

$$\begin{aligned} &R^{\alpha}: \text{IF } x \text{ is } \mu_i^x \text{ and } y \text{ is } \mu_j^u \\ &\text{THEN } \tilde{\phi}(x,y) = \theta_{ij} \end{aligned} \tag{2.8}$$

where $R^{\alpha}(\alpha = 1,\ldots,m_x \times n_y)$ are fuzzy rules, and x and y are the input variables to the fuzzy system. $\tilde{\phi}(x,y) = \theta_{ij}$ is the solution proposed by the αth rule; $\mu_i^x(\mathrm{i} = 1,\ldots,m_x)$ and $\mu_j^u(j = 1,\ldots,n_y)$ are the membership functions that characterize the ith and jth fuzzy sets defined in the space of the variables of x and y coordinates, respectively. These membership functions in these cases have been chosen as follows:

$$\mu_i^x = \exp\left[-\frac{1}{2}\left(\frac{x - a_i^x}{\sigma_i^x}\right)^2\right] \tag{2.9}$$

$$\mu_j^u = \exp\left[-\frac{1}{2}\left(\frac{y - a_j^y}{\sigma_j^y}\right)^2\right] \tag{2.10}$$

where $a_i^x, a_j^y, \sigma_i^x, \sigma_j^y$ are the mean values and the standard deviations of the membership distributions, respectively. From the membership function shown in equations (2.9) and (2.10), it is easy to find out that they are nonlinear functions in $C\infty$. The overall fuzzy solution is given by

$$\tilde{\phi}(x,y|\underline{\theta}) = \sum_{i=1}^{m_x}\sum_{j=1}^{n_y} \theta_{ij}\xi_{ij} = \underline{\theta}^T \underline{\xi} \tag{2.11}$$

where

$$\underline{\theta} = \left[\theta_{11},\ldots,\theta_{m_x n_y}\right]^T, \underline{\xi} = \left[\xi_{11,\ldots,}\xi_{m_x n_y}\right]^T \tag{2.12}$$

$$\xi_{ij} = \frac{\mu_i^x \mu_j^y}{\sum_{i=1}^{m_x}\sum_{j=1}^{n_y} \mu_i^x \mu_j^y} \tag{2.13}$$

The physical meaning in equation (2.11) is that θ at $(i, j)=(1, 1), \ldots, (m_x, n_y)$ is interpolated by the fuzzy smooth functions ξ_{ij}, $(1, 1), \ldots, (m_x, n_y)$ to approximate the exact solution $\phi(x, y)$ of a PDE. The fuzzy solution in equation (2.11) has the following properties:

$$\sum_{i=1}^{m_x}\sum_{j=1}^{n_y} \xi_{ij} = 1, \forall (x,y) \in \Omega$$

$$\sum_{i=1}^{m_x} \mu_p^x \neq 0, \forall x \in [a,b]$$

$$\sum_{j=1}^{n_y} \mu_p^x \neq 0, \forall y \in [c,d]$$

Note that these properties are the same as for the fuzzy transform method [25].

In this chapter, the standard deviations σ_i^x and σ_j^y in equations (2.9) and (2.10) are assigned as follows:

$$\sigma_i^x = \frac{b-a}{m_x - 1} \tag{2.14}$$

$$\sigma_j^y = \frac{d-c}{n_y - 1} \tag{2.15}$$

where m_x and $n_y \geq 2$. The parameters a, b, c, and d are defined as the bounds of the region Ω, i.e., $\Omega=[a, b]\times[c, d]$.

Remark 2.2.2

i. By the arrangement in equations (2.14) and (2.15), the standard deviations σ_i^x and σ_j^y will be forced to be zero with infinite numbers of membership functions for variables x and y, i.e., $\lim_{m_x\to\infty}\sigma_i^x=0$ and $\lim_{n_y\to\infty}\sigma_j^y=0$. The physical meaning of this arrangement is that membership functions in equations (2.9) and (2.10) will degenerate into the singleton case from the Gaussian case.
ii. Since the proposed fuzzy solution $\tilde{\phi}\left(x,y|\underline{\theta}\right)$ must be continuously differentiable, the membership function must also be differentiable. Therefore, some other membership functions that are continuously differentiable functions, i.e., the sigmoidal membership function and the generalized bell membership function [36], could be used to replace the Gaussian membership function.

Remark 2.2.3

There are many choices for specifying the fuzzy solution in the consequent part in equation (2.8), e.g., a linear form as $\tilde{\phi}(x,y)=\theta_{ij}+\theta_{xij}x+\theta_{yij}y$ or a more complex bilinear form $\tilde{\phi}(x,y)=\theta_{ij}+\theta_{xij}x+\theta_{xyij}xy$ and so on. In our research, we did not adopt these solution forms in the consequent part. They are so complex and inconvenient in analysis, and huge computational effort will be needed. In order to make a more precise approximation, the nonlinearities of $\phi(x,y)$ and $\tilde{\phi}(x,y|\underline{\theta})$ should be of the same order so that a more precise approximation could be achieved by the proposed fuzzy approximation method.

2.3 SOLVING PDE PROBLEMS VIA FUZZY LOGIC SYSTEMS

2.3.1 Approximation Error Bound between Exact Solution and Proposed Fuzzy Solution

In this section, we will derive one new approximation theorem with more physical meaning than the so-called universal approximation theorem to prove that the aforesaid fuzzy solution (2.11) can be used to approximate the PDE solution to any desired degree of accuracy. Let us review the so-called universal approximation theorem before deriving the new approximation error bound for the PDE problem as follows:

Universal Approximation Theorem [30,31,39]: For any given real continuous function $g(x)$ on a compact set $U\subset\mathbb{R}^n$ and arbitrary $\varepsilon>0$, there exists a fuzzy logic system $f(x)$ such that

$$\sup_{\underline{x}\in U}\left|f(\underline{x})-g(\underline{x})\right|<\varepsilon$$

The aforesaid theorem is just an existence theorem; that is, it shows that any real, continuous function $g(\underline{x})$ over a compact set U can be approximated to any degree of accuracy by the fuzzy logic system $f(\underline{x})$ with a sufficient number of fuzzy rules. The universal approximation theorem demonstrates that such a fuzzy system $f(\underline{x})$ must exist, but it does not tell us how to find it. To find it, we need to obtain a set of proper input membership functions μ_i^x and μ_j^y, as well as proper output membership functions. Hence, the derived approximation error bound ε in the *universal approximation theorem* is physically meaningless for treating the PDE problem. More precisely, an elegant result from the rigorous mathematical point of view for the detailed relationship between the proposed fuzzy logic system and the exact system is not given yet. In the following, a modified theorem is provided for employing the fuzzy logic system in equation (2.11) to obtain an approximation solution of PDEs, and an upper bound is derived for the approximation error between the exact solution and the proposed fuzzy solution. This approximation error bound shows explicitly how the parameters of the proposed method influence their approximation capability and offers us a direction to find a suitable solution for the PDE problems.

Theorem 2.3.1

Assume that $\phi(x,y)$ is the continuously differentiable solution of a PDE problem defined on $\Omega=[a,\ b]\times[c,\ d]$. Let $\Delta_x=\dfrac{b-a}{m_x-1}$, $\Delta_y=\dfrac{d-c}{n_y-1}$, and $a_i^x=a+(i-1)\Delta_x$, $a_j^y=c+(j-1)\Delta_y$, $\sigma_i^x=\Delta_x$, $\sigma_j^y=\Delta_y$, for $i=1,\dots,m_x$ and $j=1,\dots,n_y$. Then, the approximation error bound between the exact solution $\phi(x,y)$ and the proposed fuzzy solution $\tilde{\phi}\left(x,y|\underline{\theta}\right)$ in equation (2.11) can be described as follows:

$$\left|\phi(x,y)-\tilde{\phi}\left(x,y|\underline{\theta}\right)\right|\le 0.83452\left(\phi_{x\infty}\Delta_x+\phi_{y\infty}\Delta_y\right)+\max_{i,j}\left|e^{ij}\right| \tag{2.16}$$

where Δ_x and Δ_y are defined as the distance factors, and $e^{ij}=\phi\left(a_i^x,a_j^y\right)-\theta_{ij}$ is called as the solution error factor. In addition, the two factors $\phi_{x\infty}=\sup_{(x,y)\in\Omega}\left|\dfrac{\partial\phi}{\partial x}\right|$ and $\phi_{y\infty}=\sup_{(x,y)\in\Omega}\left|\dfrac{\partial\phi}{\partial y}\right|$ are defined as smoothness factors. Note that the proposed fuzzy solution $\tilde{\phi}\left(x,y|\underline{\theta}\right)$ is continuously differentiable, because the membership functions are differentiable.

Proof

In this proof, we will establish the approximation error bound between the exact solution $\phi(x,y)$ and the proposed fuzzy solution $\tilde{\phi}\left(x,y|\underline{\theta}\right)$ with Gaussian membership functions. We consider the case that the exact solution $\phi(x,y)$ is a continuous and differentiable function

$$\left|\phi(x,y)-\tilde{\phi}(x,y|\underline{\theta})\right| = \left|\phi(x,y)-\frac{\sum_{i=1}^{m_x}\sum_{j=1}^{n_y}\theta_{ij}\mu_i^x\mu_j^y}{\sum_{i=1}^{m_x}\sum_{j=1}^{n_y}\mu_i^x\mu_j^y}\right|$$

$$= \left|\frac{\sum_{i=1}^{m_x}\sum_{j=1}^{n_y}(\phi(x,y)-\theta_{ij})\mu_i^x\mu_j^y}{\sum_{i=1}^{m_x}\sum_{j=1}^{n_y}\mu_i^x\mu_j^y}\right|$$

$$\leq \frac{\sum_{i=1}^{m_x}\sum_{j=1}^{n_y}\left|\phi(x,y)-\theta_{ij}\right|\mu_i^x\mu_j^y}{\sum_{i=1}^{m_x}\sum_{j=1}^{n_y}\mu_i^x\mu_j^y} \tag{2.17}$$

Let us analyze the term $\left|\phi(x,y)-\theta_{ij}\right|$ as

$$\left|\phi(x,y)-\theta_{ij}\right| = \left|\phi(x,y)-\phi(a_i^x,a_j^y)+\phi(a_i^x,a_j^y)-\theta_{ij}\right|$$

$$= \left|\phi(x,y)-\phi(a_i^x,a_j^y)+e^{ij}\right|$$

$$\leq \left|\phi(x,y)-\phi(a_i^x,a_j^y)\right|+\left|e^{ij}\right| \tag{2.18}$$

where

$$e^{ij} = \phi(a_i^x,a_j^y)-\theta_{ij}$$

By the mean value theorem [40], we have the following result:

$$\left|\phi(x,y)-\theta_{ij}\right| \leq \left[\left|\left.\frac{\partial\phi}{\partial x}\right|_{x=\tilde{x}_i,y=\tilde{y}_j}\left(x-a_i^x\right)\right|+\left|\left.\frac{\partial\phi}{\partial y}\right|_{x=\tilde{x}_i,y=\tilde{y}_j}\left(y-a_j^y\right)\right|\right]+\left|e^{ij}\right|$$

$$\leq \left[\left|\left.\frac{\partial\phi}{\partial x}\right|_{x=\tilde{x}_i,y=\tilde{y}_j}\right|\left|\left(x-a_i^x\right)\right|+\left|\left.\frac{\partial\phi}{\partial y}\right|_{x=\tilde{x}_i,y=\tilde{y}_j}\right|\left|\left(y-a_j^y\right)\right|\right]+\left|e^{ij}\right|$$

$$\leq \left[\phi_{x\infty}\left|\left(x-a_i^x\right)\right|+\phi_{y\infty}\left|\left(y-a_j^y\right)\right|\right]+\left|e^{ij}\right| \tag{2.19}$$

where $\tilde{x}_i$ are some values between x and a_i^x, and $\tilde{y}_j$ are some values between y and a_j^y, respectively. $\phi_{x\infty}$ and $\phi_{y\infty}$ are smoothness factors defined as follows:

$$\phi_{x\infty} = \sup_{(x,y)\in\Omega}\left|\frac{\partial\phi}{\partial x}\right| \text{ and } \phi_{y\infty} = \sup_{(x,y)\in\Omega}\left|\frac{\partial\phi}{\partial y}\right|$$

By substituting equation (2.19) into (2.17), the following equation can be obtained:

$$\begin{aligned}
&\left|\phi(x,y)-\tilde{\phi}(x,y|\underline{\theta})\right| \\
&\le \frac{\sum_{i=1}^{m_x}\sum_{j=1}^{n_y}\left[\phi_{x\infty}\left|x-a_i^x\right|\right]\mu_i^x\mu_j^y}{\sum_{i=1}^{m_x}\sum_{j=1}^{n_y}\mu_i^x\mu_j^y}+\frac{\sum_{i=1}^{m_x}\sum_{j=1}^{n_y}\left[\phi_{y\infty}\left|x-a_j^y\right|\right]\mu_i^x\mu_j^y}{\sum_{i=1}^{m_x}\sum_{j=1}^{n_y}\mu_i^x\mu_j^y}+\frac{\sum_{i=1}^{m_x}\sum_{j=1}^{n_y}\left|e^{ij}\right|\mu_i^x\mu_j^y}{\sum_{i=1}^{m_x}\sum_{j=1}^{n_y}\mu_i^x\mu_j^y} \\
&\le \frac{\sum_{i=1}^{m_x}\sum_{j=1}^{n_y}\left[\phi_{x\infty}\left|x-a_i^x\right|\right]\mu_i^x\mu_j^y}{\sum_{i=1}^{m_x}\sum_{j=1}^{n_y}\mu_i^x\mu_j^y}+\frac{\sum_{i=1}^{m_x}\sum_{j=1}^{n_y}\left[\phi_{y\infty}\left|x-a_j^y\right|\right]\mu_i^x\mu_j^y}{\sum_{i=1}^{m_x}\sum_{j=1}^{n_y}\mu_i^x\mu_j^y}+\frac{\sum_{i=1}^{m_x}\sum_{j=1}^{n_y}\max_{i,j}\left|e^{ij}\right|\mu_i^x\mu_j^y}{\sum_{i=1}^{m_x}\sum_{j=1}^{n_y}\mu_i^x\mu_j^y} \\
&\le \frac{\sum_{i=1}^{m_x}\sum_{j=1}^{n_y}\left[\phi_{x\infty}\left|x-a_i^x\right|\right]\mu_i^x\mu_j^y}{\sum_{i=1}^{m_x}\sum_{j=1}^{n_y}\mu_i^x\mu_j^y}+\frac{\sum_{i=1}^{m_x}\sum_{j=1}^{n_y}\left[\phi_{y\infty}\left|x-a_j^y\right|\right]\mu_i^x\mu_j^y}{\sum_{i=1}^{m_x}\sum_{j=1}^{n_y}\mu_i^x\mu_j^y}+\max_{i,j}\left|e^{ij}\right|.
\end{aligned} \tag{2.20}$$

The first term in the right side of equation (2.20) can be reduced to

$$\Phi_x = \phi_{x\infty}\frac{\sum_{i=1}^{m_x}\left|x-a_i^x\right|\mu_i^x}{\sum_{i=1}^{m_x}\mu_i^x} \tag{2.21}$$

where

$$\mu_i^x = \exp\left[-\frac{1}{2}\left(\frac{x-a_i^x}{\Delta_x}\right)^2\right], a_i^x = a+(i-1)\Delta_x$$

for $i = 1,\ldots,m_x$. Denote $x = a+(t-1)\Delta_x$, then

$$\Phi_x = \phi_{x\infty}\Delta_x \frac{\sum_{i=1}^{m_x} |t-i| \exp\left[-\frac{1}{2}(t-i)^2\right]}{\sum_{i=1}^{m_x} \exp\left[-\frac{1}{2}(t-i)^2\right]} \tag{2.22}$$

It can be verified numerically that for $t \in [1, m_x]$ and $m_x \geq 2$

$$\frac{\sum_{i=1}^{m_x} |t-i| \exp\left[-\frac{1}{2}(t-i)^2\right]}{\sum_{i=1}^{m_x} \exp\left[-\frac{1}{2}(t-i)^2\right]} \leq 0.83452 \tag{2.23}$$

Hence

$$\Phi_x \leq 0.83452\phi_{x\infty}\Delta_x \tag{2.24}$$

Similarly, we know that the second term in the right side of equation (2.20) is reduced to

$$\Phi_y \leq 0.83452\phi_{y\infty}\Delta_y$$

Then, Theorem 2.3.1 is proven.

Remark 2.3.1

Some different approximation error bounds in equation (2.16) could be estimated as follows:

i. For equation (2.23), a nonrigorous estimate but which can be probably turned into a rigorous one would be

$$O\left(\frac{\sum_{i=1}^{m_x} |t-i| \exp\left[-\frac{1}{2}(t-i)^2\right]}{\sum_{i=1}^{m_x} \exp\left[-\frac{1}{2}(t-i)^2\right]}\right) = \frac{\int_0^\infty |t-s| \exp\left[-\frac{1}{2}(t-s)^2\right] ds}{\int_0^\infty \exp\left[-\frac{1}{2}(t-s)^2\right] ds}$$

where

$$\begin{aligned}
&\int_0^\infty |t-s| \exp\left[-\frac{1}{2}(t-s)^2\right] ds \\
&= \int_0^t (t-s) \exp\left[-\frac{1}{2}(t-s)^2\right] ds + \int_t^\infty (s-t) \exp\left[-\frac{1}{2}(t-s)^2\right] ds \\
&= 2 - e^{-\frac{1}{2}t^2} < 2
\end{aligned}$$

and

$$\int_0^{\infty} \exp\left[-\frac{1}{2}(t-s)^2\right]ds = \int_0^{t} \exp\left[-\frac{1}{2}u^2\right]du + \int_0^{\infty} \exp\left[-\frac{1}{2}u^2\right]du$$
$$> \int_0^{\infty} \exp\left[-\frac{1}{2}u^2\right]du = \sqrt{\frac{\pi}{2}}.$$

Finally

$$\frac{\int_0^{\infty} |t-s| \exp\left[-\frac{1}{2}(t-s)^2\right]ds}{\int_0^{\infty} \exp\left[-\frac{1}{2}(t-s)^2\right]ds} \leq \frac{2\sqrt{2}}{\pi}$$

Therefore, the upper bound of Φ_x in equation (2.21) is

$$\Phi_x = \phi_{x\infty} \frac{\sum_{i=1}^{m_x} \left|x - a_i^x\right| \mu_i^x}{\sum_{i=1}^{m_x} \mu_i^x} \leq \frac{2\sqrt{2}}{\sqrt{\pi}} \phi_{x\infty}\Delta_x = 1.59577\phi_{x\infty}\Delta_x$$

In this case, we get the following approximation error bound

$$\left|\phi(x,y) - \tilde{\phi}(x,y|\underline{\theta})\right| \leq 1.59577(\phi_{x\infty}\Delta_x + \phi_{y\infty}\Delta_y) + \max_{i,j}\left|e^{ij}\right|$$

ii. The bound derived in (i) is a better choice, because it is an analytic estimate. However, we want to give a smaller error bound than that in (i). Therefore, a numerical estimate is given in the following. If we choose $\sigma_i^x = p\Delta_x, \sigma_j^y = p\Delta_y$, for $0 \leq p \leq 1$, we obtain

$$\Phi_x = \phi_{x\infty}\Delta_x \frac{\sum_{i=1}^{m_x} |t-i| \exp\left[-\frac{1}{2}\left(\frac{t-i}{p}\right)^2\right]}{\sum_{i=1}^{m_x} \exp\left[-\frac{1}{2}\left(\frac{t-i}{p}\right)^2\right]}$$

Therefore, with different p, we can compute the bound of Φ_x. For simplicity, assuming $p = 1$, we can obtain a bound of Φ_x as follows:

$$\Phi_x \leq 0.83452\phi_{x\infty}\Delta_x$$

Finally, the approximation error bound could be obtained as follows:

$$\left|\phi(x,y)-\tilde{\phi}(x,y|\underline{\theta})\right| \le 0.83452(\phi_{x\infty}\Delta_x+\phi_{y\infty}\Delta_y)+\max_{i,j}\left|e^{ij}\right|$$

From equation (2.16) in Theorem 2.3.1, the approximation error bound in Gaussian membership function case is determined by three factors: the distance factors of membership functions (Δ_x and Δ_y), the smoothness factors ($\phi_{x\infty}$ and $\phi_{y\infty}$), and the solution error factor e^{ij} between the exact solution and the solution in the consequent part in equation (2.8). From a mathematical point of view, the small distance factors of membership functions (Δ_x and Δ_y) mean that a large number of rules are used, and this will reduce the approximation error bound. The smoothness factors ($\phi_{x\infty}$ and $\phi_{y\infty}$) show us the smoothness of the PDE solution. It is easy to realize that the smoothness of the solution means that the degree of difficulty of a solution can be approximated. A solution with a sharper profile cannot be easily approximated basically.

Remark 2.3.2

The aforesaid description gives us a direction for dealing with the complex solution-finding problem of PDE. If we reduce the distance factors of membership functions (Δ_x and Δ_y) by increasing the number of membership functions and solution error factor e^{ij} by suitable tuning algorithms, then the approximation error bound will be reduced to as small as possible. As for the smoothness factors $\phi_{x\infty}$ and $\phi_{y\infty}$, these reveal the physical characteristics of the solution $\phi(x,y)$.

The convergent property for the case $m_x \to \infty$ and $n_y \to \infty$, i.e., all points (x, y) in the domain Ω are considered, will be discussed in the following theorem based on the result given in Theorem 2.3.1.

Assumption 2.3.1

Suppose there exists a set of parameters $\theta_{ij} = \phi\left(a_i^x, a_j^y\right)$ as $m_x \to \infty$ and $n_y \to \infty$. Based on Assumption 2.3.1 and Theorem 2.3.1, we get the following theorem.

Theorem 2.3.2

$$\lim_{n_y\to\infty}\lim_{m_x\to\infty}\tilde{\phi}\left(x,y|\underline{\theta}\right)=\phi(x,y) \tag{2.25}$$

Let $m_x \to \infty$ and $n_y \to \infty$. Then, the proposed fuzzy solution will converge to the exact solution as follows:

Proof

In this proof, we will prove that the proposed fuzzy solution will converge to the exact solution according to Theorem 2.3.1 and Assumption 2.3.1 for $m_x \to \infty$ and $n_y \to \infty$. From equation (2.16), the approximation upper bound between the exact solution and the proposed fuzzy solution can be described by the following equation:

$$\left|\phi(x,y)-\tilde{\phi}\left(x,y\middle|\underline{\theta}\right)\right| \le 0.83452\left(\phi_{x\infty}\Delta_x+\phi_{y\infty}\Delta_y\right)+\max_{i,j}\left|e^{ij}\right| \tag{2.26}$$

Take the limit for both sides of equation (2.26) with m_x and n_y approaching infinity and note the fact that $\phi_{x\infty}$ and $\phi_{y\infty}$ are bound and $\Delta_x \to 0$, $\Delta_y \to 0$. We have

$$\left|\phi(x,y)-\lim_{n_y\to\infty}\lim_{m_x\to\infty}\tilde{\phi}\left(x,y\middle|\underline{\theta}\right)\right| \le \lim_{n_y\to\infty}\lim_{m_x\to\infty}\max_{i,j}\left|e^{ij}\right| \tag{2.27}$$

From equations (2.14) and (2.15), the standard deviations σ_i^x and σ_j^y will be forced to be zero with infinite numbers of rules (i.e., m_x, $n_y \to \infty$). By this arrangement, membership functions in equations (2.9) and (2.10) will degenerate into the singleton case from the Gaussian case. Since if there exists a set of $\theta_{ij}=\phi\left(a_i^x,a_j^y\right)$ such that $e^{ij}=0$, as m_x, $n_y \to \infty$, the following result can be derived from equation (2.27)

$$\left|\phi(x,y)-\lim_{n_y\to\infty}\lim_{m_x\to\infty}\tilde{\phi}\left(x,y\middle|\underline{\theta}\right)\right| = 0 \tag{2.28}$$

i.e., $\lim_{n_y\to\infty}\lim_{m_x\to\infty}\tilde{\phi}\left(x,y\middle|\underline{\theta}\right)=\phi(x,y)$. Theorem 2.3.2 is proven.

2.3.2 Adaptive Law for Adjustable Parameters $\underline{\theta}$ in Mesh Points

In Theorem 2.3.2, the numerical solution with all points (x, y) in domain Ω is impractical. In actual practice, only a finite number of mesh points $\left(x(i)=a_i^x, y(j)=a_j^y\right)$ are needed to be approached. To deal with this kind of problem, the following two approximated error equations with the fuzzy solution are defined first

$$\tilde{\beta}_0\left(x,y\middle|\underline{\theta}\right)=f\left(x,y,\tilde{\phi}_x\left(x,y\middle|\underline{\theta}\right),\tilde{\phi}_y\left(x,y\middle|\underline{\theta}\right),\tilde{\phi}_{xy}\left(x,y\middle|\underline{\theta}\right)),\tilde{\phi}_{xx}\left(x,y\middle|\underline{\theta}\right),\tilde{\phi}_{yy}\left(x,y\middle|\underline{\theta}\right)\right)$$
$$-u(x,y) \tag{2.29}$$

$$\tilde{\beta}_1\left(x,y\middle|\underline{\theta}\right)=g\left(x,y,\tilde{\phi}\left(x,y\middle|\underline{\theta}\right),\frac{\partial\tilde{\phi}\left(x,y\middle|\underline{\theta}\right)}{\partial\vec{n}}\right)\Bigg|_{\partial\Omega} \tag{2.30}$$

To obtain an approximated solution to the aforesaid error equations (2.29) and (2.30), the collocation method [2] that assumes the discretization of the domain

Ω and its boundary $\partial\Omega$ into a set of points $\hat{\Omega}$ and $\partial\hat{\Omega}$ (these points are denoted by $(x(i), y(j))$, for $i = 1,\ldots,m_x$ and $j = 1,\ldots,n_y$), respectively, is adopted. For the convenience of design, these points $x(i), y(j)$ are located in $x(i) = a+(i-1)\Delta_x$ and $y(j) = c+(i-1)\Delta_y$, respectively. The problem is then transformed into the following error equations:

$$\tilde{\beta}_0\left(x(i),y(j)|\underline{\theta}\right) = f\left(x(i),y(j)\right), \tilde{\phi}\left(x(i),y(j)|\underline{\theta}\right), \tilde{\phi}_x\left(x(i),y(j)|\underline{\theta}\right),$$
$$\tilde{\phi}_y\left(x(i),y(j)|\underline{\theta}\right), \tilde{\phi}_{xy}\left(x(i),y(j)|\underline{\theta}\right), \tilde{\phi}_{xx}\left(x(i),y(j)|\underline{\theta}\right),$$
$$\tilde{\phi}_{yy}\left(x(i),y(j)|\underline{\theta}\right) - u\left(x(i),y(j)\right), \forall x(i),y(j) \in \hat{\Omega} \tag{2.31}$$

$$\tilde{\beta}_1\left(x(i),y(j)|\underline{\theta}\right) = g\left(x(i),y(j),\tilde{\phi}\left(x(i),y(j)|\underline{\theta}\right),\frac{\partial\tilde{\phi}\left(x(i),y(j)\|\underline{\theta}\right)}{\partial\vec{n}}\right)\Bigg|_{\partial\hat{\Omega}} \tag{2.32}$$

Let $\tilde{\phi}\left(x(i),y(j)|\underline{\theta}\right)$ denote a fuzzy approximate solution to the PDE problem in equations (2.31) and (2.32), where $\underline{\theta}$ stands for a set of fuzzy parameters to be adjusted. The PDE problem in equations (2.31) and (2.32) is then transformed into the following constrained minimization problem

$$\min_{\underline{\theta}} \sum_{i=1}^{m_x}\sum_{j=1}^{n_y} \tilde{\beta}_0\left(x(i),y(j)|\underline{\theta}\right)^2 \tag{2.33}$$

subject to the constraints imposed by the BCs

$$\sum_{(x(i),y(j))\in\partial\hat{\Omega}} \tilde{\beta}_1\left(x(i),y(j)|\underline{\theta}\right)^2 = 0 \tag{2.34}$$

The nonlinear constrained optimization problem in the aforesaid equations (2.31) and (2.32) may be tackled in a number of ways, for example, Lagrange multipliers, active set methods, a penalty function approach, and so on [41].

Assumption 2.3.2

The PDE and BCs encountered in equations (2.31) and (2.32) are smooth, twice continuously differentiable with respect to parameters θ in the consequent part, and the first and second derivatives of the PDE and BCs with respect to the adjustable parameters θ are bounded, i.e., $\left\|\frac{\partial f}{\partial\underline{\theta}}\right\|, \left\|\frac{\partial g}{\partial\underline{\theta}}\right\|, \left\|\frac{\partial^2 f}{\partial\underline{\theta}^2}\right\|$, and $\left\|\frac{\partial^2 g}{\partial\underline{\theta}^2}\right\| < \infty$.

In this chapter, we handle the solution finding of PDEs problem from an energy perspective for the convenience of analyzing the convergent property of the solution in mesh points. Note at this point that the problem has been reduced from the original constrained optimization problem to an unconstrained one (that is much easier to handle).

From equation (2.11), if m_x is the number of membership functions for variable x and n_y is the number of membership functions for the variable y, then a mesh of PDEs with $m = m_x n_y$ points can be obtained for training from the domain Ω.

In this way, the problem in equations (2.31) and (2.32) is transformed into the following unconstrained minimization problem:

$$E = \sum_{i=1}^{m_x} \sum_{j=1}^{n_y} \left\{ \frac{1}{2} \left[\tilde{\beta}_0 \left(x(i), y(j) \middle| \underline{\theta} \right) \right]^2 \right\} + \sum_{(x(i),y(j)) \in \partial\hat{\Omega}} \frac{1}{2} \left[\kappa_{ij} \tilde{\beta}_1 \left(x(i), y(j) \middle| \underline{\theta} \right) \right]^2 \quad (2.35)$$

where κ_{ij} is the weighting factor.

Then, our design aims to specify the parameter θ of fuzzy logical system in equation (2.11) to minimize the energy function E in equation (2.35). Equation (2.35) can be easily transformed into the following form

$$E = \frac{1}{2} \underline{S}^T \underline{S} \quad (2.36)$$

where $\underline{S}$ is the vector of the collection of error equations in mesh points

$$\underline{S} = \left[\tilde{\beta}_0 (x(1), y(1) | \underline{\theta}), \tilde{\beta}_0 (x(1), y(2) | \underline{\theta}), \cdots, \kappa_{m_x n_y} \tilde{\beta}_1 (x(m_x), y(n_y) | \underline{\theta}) \right]^T_{1 \times (m_x n_y + 2(m_x + n_y))}$$
(2.37)

From the energy function in equation (2.36), a learning algorithm for the parameter $\underline{\theta}$ of the fuzzy logic system is derived from a set of ordinary differential equations. For the solution problem of PDEs, we have the following main result after some mathematical manipulations.

Theorem 2.3.3

By the following adaptive laws, the fuzzy solution for the problem of the PDEs in equation (2.1) *with BCs in equation* (2.2) *can be found by*

$$\frac{d\underline{\theta}}{dt} = -\eta_\theta \frac{\partial \underline{S}^T}{\partial \underline{\theta}} \underline{S} \quad (2.38)$$

where the learning rate $\eta_\theta > 0$. *Furthermore, the errors equations in mesh points equations* (2.32) *and* (2.33) *can be proven to converge to zero when the following sufficient condition is satisfied:*

$$\frac{\partial \underline{S}}{\partial \underline{\theta}} \frac{\partial \underline{S}^T}{\partial \underline{\theta}} > 0 \tag{2.39}$$

Otherwise, the error equations in mesh points can only be proven to converge to a constant.

Proof

To prove the convergence of the error equations in mesh points, we need the following lemma from [32,42,43].

Lemma 2.3.1 [32,42,43]

If a scalar function $E(t)$ satisfies the following conditions:

1. $E(t)$ is bounded from below;
2. $\dot{E}(t)$ is negative semidefinite;
3. $\dot{E}(t)$ is uniformly continuous in time; then $\lim_{t\to\infty} \dot{E}(t) = 0$.

It is straightforward to demonstrate the convergence of the proposed fuzzy solution $\tilde{\phi}_x\left(x, y \middle| \underline{\theta}\right)$ in mesh points via the following derivation. By differentiating equation (2.36) with Assumption 2.3.2, the following equation can be obtained:

$$\frac{dE}{dt} = \left(\frac{\partial \underline{S}}{\partial \underline{\theta}} \frac{d\underline{\theta}}{dt}\right)^T \underline{S} \tag{2.40}$$

where the derivative of a vector $\underline{S}$ with respect to a vector $\underline{\theta}$ is defined by

$$\frac{\partial \underline{S}}{\partial \underline{\theta}} = \left[\frac{\partial s_q}{\partial \theta_{ij}}\right]_{(m_x n_y + 2(m_x + n_y)) \times m_x n_y}$$

for $q = 1, \ldots, (m_x n_y + 2(m_x + n_y))$, $i = 1, \ldots, m_x$ and $j = 1, \ldots, n_y$. By choosing adaptive laws as Theorem 2.3.3 and the fact $\eta_\theta > 0$, we have

$$\frac{dE}{dt} = -\eta_\theta \underline{S}^T \frac{\partial \underline{S}}{\partial \underline{\theta}} \frac{\partial \underline{S}^T}{\partial \underline{\theta}} \underline{S} \leq 0 \tag{2.41}$$

From the facts that E in equation (2.35) is positive definite and $\dot{E}$ in equation (2.41) is negative semidefinite, it follows that E is upper bounded, i.e., $0 \leq E(t) \leq E(0)$. The derivative of $\dot{E}$ in equation (2.41) is with respect to t.

$$\ddot{E} = -\eta_\theta\left(\underline{\dot{S}}^T \frac{\partial \underline{S}}{\partial \underline{\theta}}\frac{\partial \underline{S}^T}{\partial \underline{\theta}}\underline{S} + \underline{S}^T \frac{\partial \underline{\dot{S}}}{\partial \underline{\theta}}\frac{\partial \underline{S}^T}{\partial \underline{\theta}}\underline{S} + \underline{S}^T \frac{\partial \underline{S}}{\partial \underline{\theta}}\frac{\partial \underline{\dot{S}}^T}{\partial \underline{\theta}}\underline{S} + \underline{S}^T \frac{\partial \underline{S}}{\partial \underline{\theta}}\frac{\partial \underline{S}^T}{\partial \underline{\theta}}\underline{\dot{S}}\right)$$

$$= -\eta_\theta\left(-\eta_\theta \underline{S}^T \frac{\partial \underline{S}}{\partial \underline{\theta}}\frac{\partial \underline{S}}{\partial \underline{\theta}}\frac{\partial \underline{S}^T}{\partial \underline{\theta}}\underline{S} - \eta_\theta \underline{S}^T \frac{\partial^2 \underline{S}}{\partial \underline{\theta}^2}\frac{\partial \underline{S}^T}{\partial \underline{\theta}}\underline{S}\frac{\partial \underline{S}^T}{\partial \underline{\theta}}\underline{S}\right.$$

$$\left. - \eta_\theta \underline{S}^T \frac{\partial \underline{S}}{\partial \underline{\theta}}\underline{S}^T \frac{\partial \underline{S}}{\partial \underline{\theta}}\frac{\partial^2 \underline{S}^T}{\partial \underline{\theta}^2}\underline{S} - \eta_\theta \underline{S}^T \frac{\partial \underline{S}}{\partial \underline{\theta}}\frac{\partial \underline{S}^T}{\partial \underline{\theta}}\frac{\partial \underline{S}^T}{\partial \underline{\theta}}\underline{S}\right)$$

$$= \eta_\theta^2 \underline{S}^T \left(\frac{\partial \underline{S}^T}{\partial \underline{\theta}}\frac{\partial \underline{S}}{\partial \underline{\theta}}\frac{\partial \underline{S}^T}{\partial \underline{\theta}} + \frac{\partial^2 \underline{S}}{\partial \underline{\theta}^2}\frac{\partial \underline{S}^T}{\partial \underline{\theta}}\underline{S}\frac{\partial \underline{S}^T}{\partial \underline{\theta}} + \frac{\partial \underline{S}}{\partial \underline{\theta}}\underline{S}^T \frac{\partial \underline{S}}{\partial \underline{\theta}}\frac{\partial^2 \underline{S}^T}{\partial \underline{\theta}^2} + \frac{\partial \underline{S}}{\partial \underline{\theta}}\frac{\partial \underline{S}^T}{\partial \underline{\theta}}\frac{\partial \underline{S}^T}{\partial \underline{\theta}}\right)\underline{S} \tag{2.42}$$

By Assumption 2.3.2, we have $\ddot{E} < \infty$. Hence, $\dot{E}$ $^{\cdot}$ is uniformly continuous in time. Then, with the help of Lemma 2.3.1, we have

$$\lim_{t\to\infty} \dot{E}(t) = -\eta_\theta \underline{S}^T \frac{\partial \underline{S}}{\partial \underline{\theta}}\frac{\partial \underline{S}^T}{\partial \underline{\theta}}\underline{S} = 0 \tag{2.43}$$

From equation (2.43), one sufficient condition for ensuring the convergence of error equations in mesh points for a class of PDEs can be obtained. That is

$$\frac{\partial \underline{S}}{\partial \underline{\theta}}\frac{\partial \underline{S}^T}{\partial \underline{\theta}} > 0 \tag{2.44}$$

Suppose there exists a class of PDEs and BCs in equations (2.1) and (2.2) that satisfy the sufficient condition in equation (2.44). Then, the following result can be obtained from equations (2.43) and (2.44) [44,45]

$$\lim_{t\to\infty} \underline{S} = 0 \tag{2.45}$$

which implies

$$\lim_{t\to\infty} \tilde{\beta}_0(x(1), y(1)\big|\underline{\theta}) = 0, \lim_{t\to\infty} \tilde{\beta}_0(x(1), y(2)\big|\underline{\theta}) = 0, \cdots, \lim_{t\to\infty} \tilde{\beta}_1(x(m_x), y(n_y)\big|\underline{\theta}) = 0$$

which further implies

$$\lim_{t\to\infty} \tilde{\phi}(x(1), y(1)\big|\underline{\theta}) = \phi(x(1), y(1)), \lim_{t\to\infty} \tilde{\phi}(x(1), y(2)\big|\underline{\theta}) = \phi(x(1), y(2)), \cdots,$$

and $\lim_{t\to\infty} \tilde{\phi}(x(m_x), y(n_y)\big|\underline{\theta}) = \phi(x(m_x), y(n_y))$

According to the aforesaid analysis, the adaptive design for fuzzy logical system to solve PDE is summarized in the following.

Design Procedure:

1. Construct the fuzzy rule bases for the approximated fuzzy solution $\tilde{\phi}(x, y|\underline{\theta})$, which consists of $m_x \times n_y$ rules whose IF parts comprise all the possible combinations of the fuzzy sets μ_i^x and μ_i^y

$$\begin{aligned} R^{\alpha} &: \text{ IF } x \text{ is } \mu_i^x \text{ and } y \text{ is } \mu_j^y \\ &\text{THEN } \tilde{\phi}(x, y) = \theta_{ij} \end{aligned} \tag{2.46}$$

2. Construct the fuzzy basis functions

$$\xi_{ij} = \frac{\mu_i^x \mu_j^y}{\sum_{i=1}^{m_x} \sum_{j=1}^{n_y} \mu_i^x \mu_j^y} \tag{2.47}$$

and collect them into a vector $\underline{\xi}$ in a natural ordering for $\alpha = 1, \ldots, m_x \times n_y$. Collect the parameters θ_{ij} in the consequent part, in the same ordering as $\underline{\xi}$, into vector $\underline{\theta}$. Then, construct $\tilde{\phi}(x, y|\underline{\theta})$ as

$$\tilde{\phi}(x, y|\underline{\theta}) = \underline{\theta}^T \underline{\xi} \tag{2.48}$$

3. Use the following adaptive laws to adjust the parameter vector $\underline{\theta}$

$$\frac{d\theta}{dt} = -\eta_\theta \frac{\partial \underline{S}^T}{\partial \underline{\theta}} \underline{S} \tag{2.49}$$

2.4 SIMULATION RESULTS

Example 2.4.1 (Poisson Equation)

We consider the following boundary value problem with Dirichlet BCs:

$$\frac{\partial^2 \phi(x, y)}{\partial x^2} + \frac{\partial^2 \phi(x, y)}{\partial y^2} = e^{-x}(x - 2 + y^3 + 6y) \tag{2.50}$$

with $x, y \in [0,10]$ and the Dirichlet BCs: $\phi(0, y) = y^3$, and $\phi(10, y) = (10 + y^3)e^{-10}$, and $\phi(x,0) = xe^{-x}$, $\phi(x,10) = e^{-x}(x+10)$. The analytic solution is $\phi(x, y) = e^{-x}(x + y^3)$ and is displayed in Figure 2.1. The aforesaid problem was defined on the domain $\Omega = [0,10] \times [0,10]$, and in order to perform training, a mesh of 121 points is obtained by considering 11 equidistant points of domain $[0,10]$ of each variable as $x=0, 1, \ldots, 10$, and $y=0, 1, \ldots, 10$, respectively.

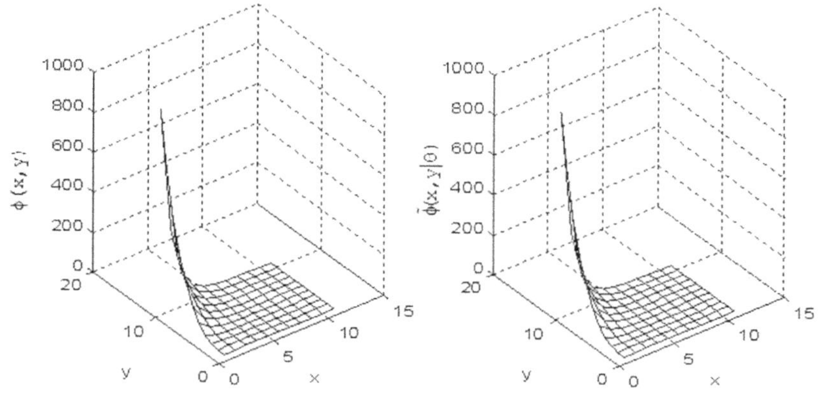

FIGURE 2.1 Profiles of (a) the exact solution and (b) the fuzzy solution $\tilde{\phi}(x,y|\underline{\theta})$ of Poisson equation.

From the design procedure in equations (2.46)–(2.49), an adaptive fuzzy solution can be obtained via the following steps.

Step 1: Construct the fuzzy rule bases for the approximated fuzzy solution $\tilde{\phi}(x,y|\underline{\theta})$. The means of memberships for variables x and y can be directly located in $a_i^x = 0,1,\ldots,10$, and $a_j^y = 0,1,\ldots,10$, as the aforesaid arrangements and the values of the standard deviations of memberships are assigned to be $\sigma_0^x = \sigma_1^x = \cdots = \sigma_{10}^x = 1$, and $\sigma_0^y = \sigma_1^y = \cdots = \sigma_{10}^y = 1$.

The αth rule is described as

$$R^{\alpha}: \text{ IF } x \text{ is } \mu_i^x \text{ and } y \text{ is } \mu_j^y$$

$$\text{THEN } \tilde{\phi}(x,y) = \theta_{ij},\ for\ i,j = 0,1,\ldots,10,$$

where $\alpha = 1,\ldots,121$.

Step 2: Construct the basis functions

$$\xi_{ij} = \frac{\mu_i^x \mu_j^y}{\sum_{i=0}^{10} \sum_{j=0}^{10} \mu_i^x \mu_j^y} \tag{2.51}$$

where μ_i^x and μ_j^y are Gaussian functions in equations (2.9) and (2.10), and collect them into a vector $\underline{\xi}$ in a natural ordering for $\alpha = 1,\ldots,121$. Collect the parameters θ_{ij} in the consequent part (the parameters are initially assigned as $\theta_{00}=\theta_{01}=\ldots=\theta_{1010}=0.5$), in the same ordering as $\underline{\xi}$, into vector θ. Then, $\tilde{\phi}(x,y|\underline{\theta})$ is constructed as

$$\tilde{\phi}(x,y|\underline{\theta}) = \underline{\theta}^T \underline{\xi} \tag{2.52}$$

By substituting equation (2.52) into (2.50) and the Dirichlet conditions, equation (2.50) becomes

$$\underline{\theta}^T F(x,y) = e^{-x}(x - 2 + y^3 + 6y) \tag{2.53}$$

and boundary conditions are reduced to

$$\underline{\theta}^T \xi(0,y) = y^3 \tag{2.54}$$

$$\underline{\theta}^T \xi(10,y) = (10 + y^3)e^{-10} \tag{2.55}$$

$$\underline{\theta}^T \xi(x,0) = xe^{-x} \tag{2.56}$$

$$\underline{\theta}^T \xi(x,10) = e^{-x}(x + 10) \tag{2.57}$$

where

$$F(x,y) = \begin{bmatrix} \xi_{xx}^{00}(x,y) + \xi_{yy}^{00}(x,y) \\ \xi_{xx}^{01}(x,y) + \xi_{yy}^{01}(x,y) \\ \vdots \\ \xi_{xx}^{1010}(x,y) + \xi_{yy}^{1010}(x,y) \end{bmatrix}$$

and

$$\xi_{xx}^{ij}(x,y) = \frac{\partial \left(\dfrac{\mu_i^x \mu_j^y}{\sum_{i=0}^{10} \sum_{j=0}^{10} \mu_i^x \mu_j^y} \right)^2}{\partial x^2} = \frac{\left(\dfrac{\mu_j^y}{\sum_{j=0}^{10} \mu_j^y} \right)^2 \partial \left(\dfrac{\mu_i^x}{\sum_{i=0}^{10} \mu_i^x} \right)^2}{\partial x^2}$$

$$\xi_{yy}^{ij}(x,y) = \frac{\partial \left(\dfrac{\mu_i^x \mu_j^y}{\sum_{i=0}^{10} \sum_{j=0}^{10} \mu_i^x \mu_j^y} \right)^2}{\partial y^2} = \frac{\left(\dfrac{\mu_i^x}{\sum_{i=0}^{10} \mu_i^x} \right)^2 \partial \left(\dfrac{\mu_j^y}{\sum_{j=0}^{10} \mu_j^y} \right)^2}{\partial x^2}$$

Then, we define the following five error functions for treating this problem:

$$\tilde{\beta}_0(x,y,\underline{\theta}) = \underline{\theta}^T F(x,y) - e^{-x}(x - 2 + y^3 + 6y) \tag{2.58}$$

$$\tilde{\beta}_1(y,\underline{\theta}) = \underline{\theta}^T \underline{\xi}(0,y) - y^3 \tag{2.59}$$

$$\tilde{\beta}_2(y,\underline{\theta}) = \underline{\theta}^T \underline{\xi}(10,y) - (10 + y^3)e^{-10} \tag{2.60}$$

$$\tilde{\beta}_3(x,\underline{\theta}) = \underline{\theta}^T \underline{\xi}(x,0) - xe^{-x} \tag{2.61}$$

$$\tilde{\beta}_4(x,\underline{\theta}) = \underline{\theta}^T \underline{\xi}(x,10) - e^{-x}(x + 10) \tag{2.62}$$

By the collocation method [18], a convex computation energy function is described as follows:

$$E=\sum_{i=0}^{10}\sum_{j=0}^{10}\frac{1}{2}\left[\tilde{\beta}_0\left(x(i),y(j),\underline{\theta}\right)\right]^2+\sum_{j=0}^{10}\frac{1}{2}\left[\kappa_1\tilde{\beta}_1(y(j),\underline{\theta})\right]^2+\sum_{j=0}^{10}\frac{1}{2}\left[\kappa_2\tilde{\beta}_2(y(j),\underline{\theta})\right]^2$$

$$+\sum_{i=0}^{10}\frac{1}{2}\left[\kappa_3\tilde{\beta}_3(x(j),\underline{\theta})\right]^2+\sum_{i=0}^{10}\frac{1}{2}\left[\kappa_4\tilde{\beta}_4(x(i),\underline{\theta})\right]^2$$

$$=\frac{1}{2}\underline{S}^T\underline{S} \tag{2.63}$$

where

$$\underline{S}=\left[\tilde{\beta}_0\left(x(0),y(0)\right),\tilde{\beta}_0\left(x(0),y(1)\right),\cdots,\kappa_1\tilde{\beta}_1(y(10),\underline{\theta}),\cdots,\kappa_4\tilde{\beta}_4(x(10),\underline{\theta})\right]^T$$

and $\kappa_1=\kappa_2=\kappa_3=\kappa_4=5$ are weighting factors.

Step 3: Using the following adaptive law to adjust the parameter vector θ of the proposed solution $\tilde{\phi}(x,y|\underline{\theta})$

$$\frac{d\underline{\theta}}{dt}=-\eta_\theta\frac{\partial\underline{S}^T}{\partial\underline{\theta}}\underline{S} \tag{2.64}$$

where the learning rate η_θ is chosen as 5.

With Step 1 to Step 3, one fuzzy solution $\tilde{\phi}(x,y|\underline{\theta})$ for the Poisson equation can be obtained and is displayed in Figure 2.1b.

As seen in Figure 2.1, it is clear that the solution is very accurate at all mesh points, and the profile of Figure 2.1a and b is almost the same. Figure 2.2 is the history of the error energy function E. The history of energy function in Figure 2.2 and the error profile between the exact solution $\phi(x,y)$ and the proposed fuzzy solution $\tilde{\phi}(x,y|\underline{\theta})$ in Figure 2.3, and the perfect results that error equations in mesh points converge to a quite small value, imply that the proposed fuzzy solution $\tilde{\phi}(x,y|\underline{\theta})$ is almost the same as the exact solution $\phi(x,y)$ in mesh points with very high accuracy.

Preliminary Comparison with Finite Elements: The aforesaid PDE problem was also solved with the variational finite-element method that has been widely acknowledged as one of the most effective approaches to the solution of PDEs [22,23]. By the derivation in a similar manner, as that in [22], it can be shown that for solving this problem, the following function should be minimized:

$$\chi=\frac{1}{2}\int\int_\Omega\left[\left(\frac{\partial\phi(x,y)}{\partial x}\right)^2+\left(\frac{\partial\phi(x,y)}{\partial y}\right)^2+2e^{-x}(x-2+y^3+6y)\phi(x,y)\right]dxdy \tag{2.65}$$

Then, we can obtain the approximation solution $\bar{\phi}(x,y)$ in mesh points by minimizing equation (2.65). The profile of the finite-element solution $\bar{\phi}(x,y)$

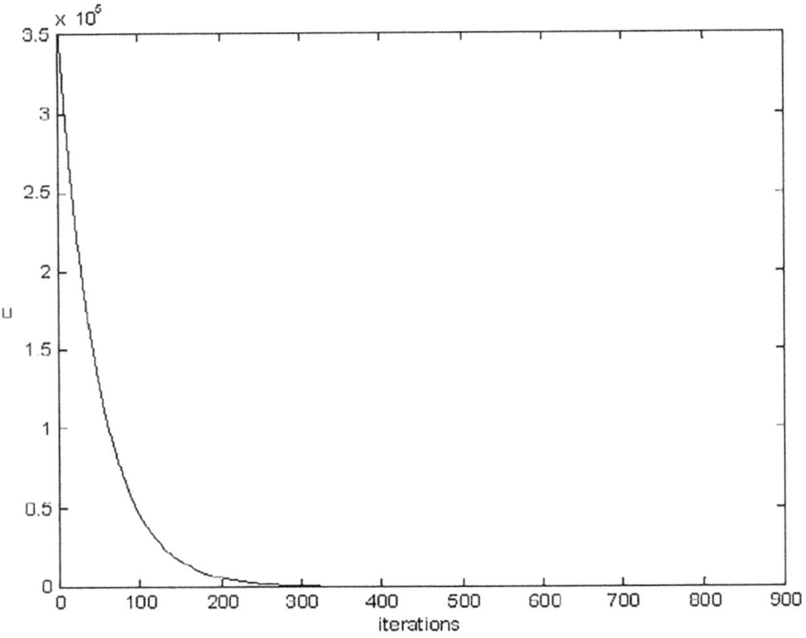

FIGURE 2.2 The history of the energy function E for Example 2.4.1.

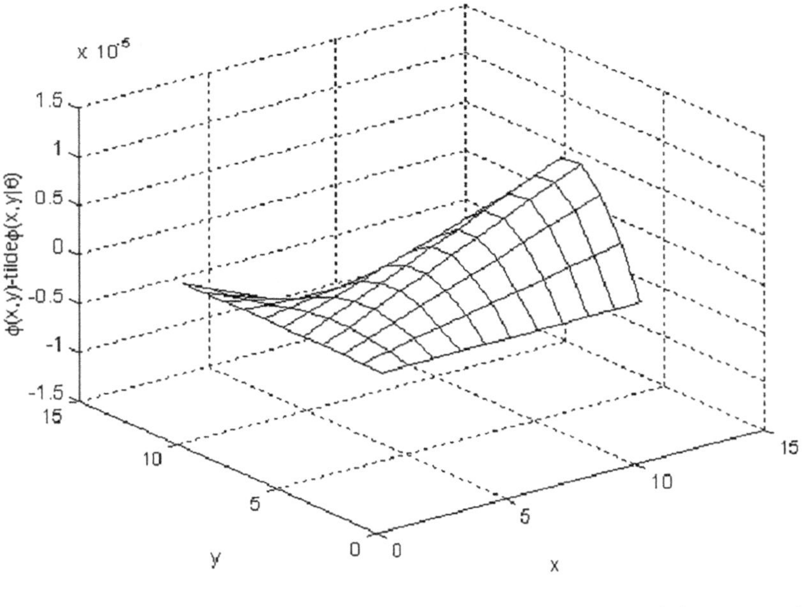

FIGURE 2.3 Error profile between the exact solution $\phi(x, y)$ and the proposed fuzzy solution $\tilde{\phi}\left(x, y \middle| \underline{\theta}\right)$.

within the range $x \in [0, 10]$ and $y \in [0, 10]$ is shown in Figure 2.4. The error profile between the exact solution $\phi(x,y)$ and the finite-element solution $\bar{\phi}(x,y)$ in the mesh points is displayed in Figure 2.5. In general, the finite-element method is a finite-difference method based on mesh points.

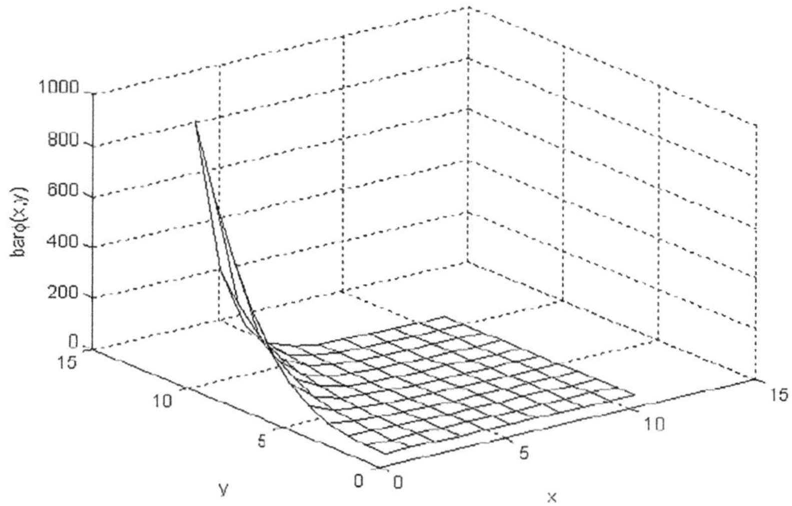

FIGURE 2.4 Profile of the finite-element solution $\bar{\phi}(x,y)$ of Poisson equation.

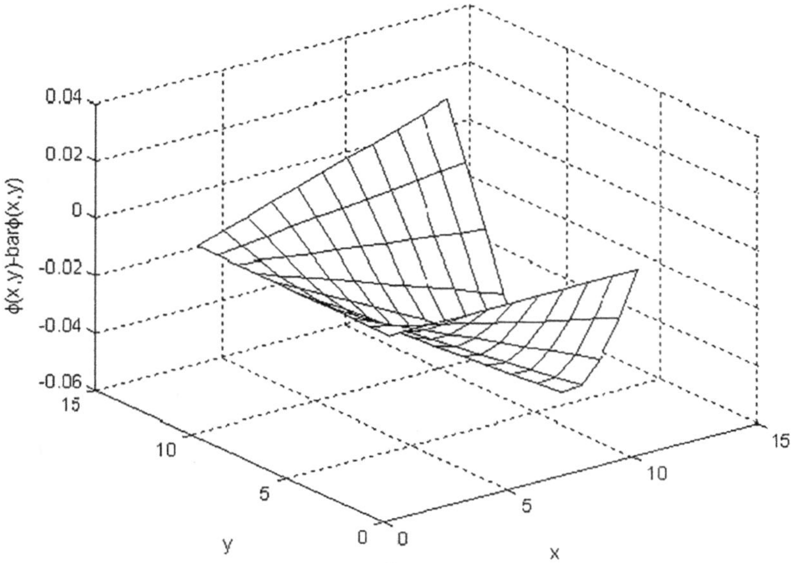

FIGURE 2.5 Error profile between the exact solution $\phi(x,y)$ and the finite-element solution $\bar{\phi}(x,y)$.

The proposed fuzzy method provides a smooth solution with a superior interpolation property. As seen in Figures 2.3 and 2.5, the proposed method yields better performance than the finite-element method because it offers a much precise solution by the adaptive fuzzy method for the PDE problem than the finite-element method. The comparison between the proposed method and the finite-element method is based on the same computational complexity (i.e., on the same grid points). If the number of grid points is increased, the errors are all decreased. However, the computation complexity will increase. There is a tradeoff between complexity and accuracy in these approximation methods.

Example 2.4.2 (Nonlinear PDE Case)

In this case, we will not compare the simulation results of the proposed method with those of the finite-element method because the functional of the finite-element method for this problem cannot be found. Consider the following nonlinear PDE problem:

$$\nabla^2\phi(x,y)+\phi(x,y)\frac{\partial}{\partial y}\phi(x,y)=\sin(\pi x)\left(2-\pi^2y^2+2y^3\sin(\pi x)\right) \tag{2.66}$$

with mixed BCs:

$$\phi(0,y)=0,\ \phi(10,y)=0,\ \phi(x,0)=0$$

and

$$\frac{\phi(x,10)}{\partial y}=20\sin(\pi x)$$

where

$$\nabla^2\phi(x,y)=\frac{\partial^2\phi}{\partial x^2}+\frac{\partial^2\phi}{\partial y^2}$$

The analytic solution is $\phi(x,y)=y^2\sin(\pi x)$ and is presented in Figure 2.6a. This problem is defined on the domain $\Omega=\left[1.5,11.5\right]\times\left[1.5,11.5\right]$. In order to perform training, a mesh of 121 points is obtained. In this problem, the initial conditions for $\underline{\theta}$, σ_i^x, and σ_j^y are given as those in Example 2.4.1. By the similar design procedure in the aforesaid example, we have the following simulation results. Figure 2.6b is the profile of the proposed fuzzy solution. Figure 2.7 is the error profile between the exact solution $\phi(x,y)$ and the proposed fuzzy solution $\tilde{\phi}(x,y|\underline{\theta})$ in the mesh points. From the error profile and the history of the energy function E as shown in Figures 2.7 and 2.8, respectively, one precise fuzzy solution $\tilde{\phi}(x,y|\underline{\theta})$ can be ensured for this nonlinear PDE.

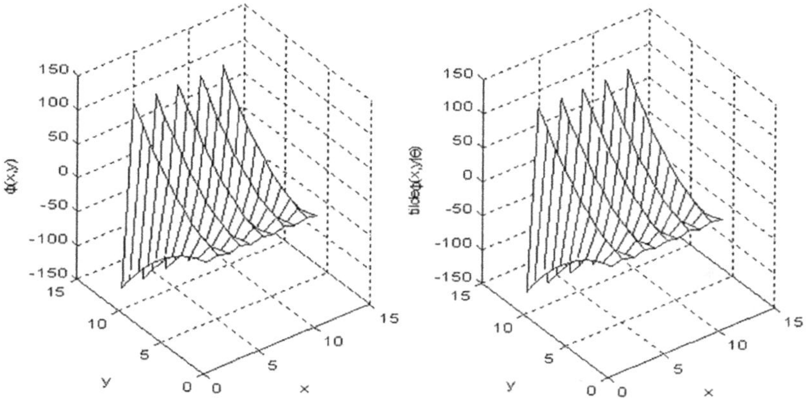

FIGURE 2.6 Profiles of (a) the exact solution $\phi(x,y)$ and (b) the proposed fuzzy solution $\tilde{\phi}\left(x,y\middle|\underline{\theta}\right)$ for Example 2.4.2.

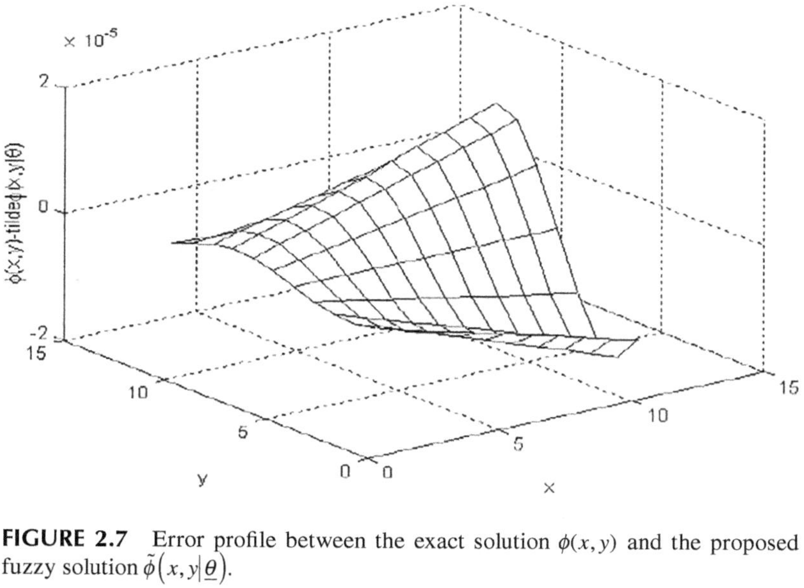

FIGURE 2.7 Error profile between the exact solution $\phi(x,y)$ and the proposed fuzzy solution $\tilde{\phi}\left(x,y\middle|\underline{\theta}\right)$.

2.5 CONCLUSION

An advanced methodology based on the fuzzy logic system and the adaptive algorithm is successfully adopted to solve the PDE problems. First of all, the approximation error bound between the exact solution and the proposed fuzzy solution is elegantly derived in this chapter. The error bound is dependent on the physical

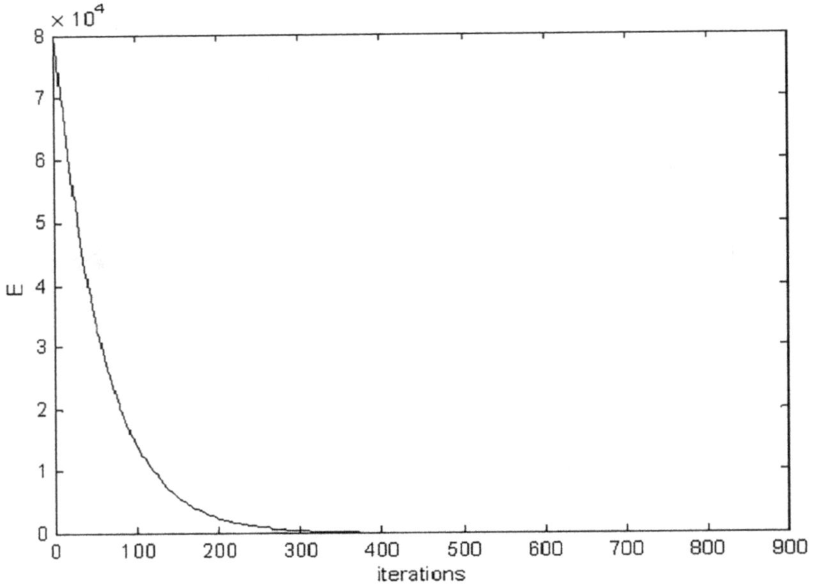

FIGURE 2.8 History of the energy function E for Example 2.4.2.

effects of the number of membership functions and the specification of parameters in the consequent parts. We found that the accuracy of the proposed method can be improved by the following ways: (i) by increasing the number of membership functions (which results in more complex fuzzy systems); and (ii) finding a set of adaptive laws for the adjustable parameters in the proposed fuzzy solution.

In this chapter, a set of adaptive laws for the adjustable parameters of the proposed fuzzy solution can be derived from minimizing a convex and bounded error energy function, and one precise fuzzy solution for the encountered PDE problems can be found. Furthermore, one sufficient condition to ensure the convergence of error equations in mesh points is obtained. The solution of PDEs between these mesh points can be obtained by fuzzy interpolation via membership functions. In addition, two simulation examples with comparison are adopted to verify the performance of the proposed adaptive fuzzy method for the solution of PDEs. The simulation results reveal that the proposed method is fairly effective in solving a wide class of quite general PDE problems. This chapter has extended the fuzzy approximation method from the conventional ordinary nonlinear systems to nonlinear partial differential systems with potential practical applications to signal processing and control engineering design in the following chapters.

Part II

Robust Signal Processing Design

3 Robust Filter Design for Linear Stochastic Partial Differential Systems via a Set of Sensor Measurements

3.1 INTRODUCTION

The H_∞ filtering problem is to design a state estimator to estimate the unavailable state under external disturbance via the noisy output measurement, which can guarantee the L_2 gain (from the external disturbance to the estimation error) i.e., the effect of external disturbance on filtering error to be less than a prescribed level [46–57]. In contrast with the well-known Kalman filter, one of the main advantages of H_∞ filtering is that it is not necessary to know exactly the statistical properties of external disturbance but only assumes the external disturbance to have bounded energy; see [58] for the practical application of H_∞ filtering in signal processing.

In recent years, H_∞ filtering and control problems have become a popular research topic and have gained extensive attentions; see [59–71] and the reference therein. We summarize below the recent development on stochastic H_∞ filtering problem. In [59], a bounded real lemma was presented for linear continuous-time stochastic systems, and accordingly the full- and reduced-order robust H_∞ estimation problems for stationary continuous-time linear stochastic uncertain systems were discussed by [68,69], respectively. Recently, there are some works on H_∞ control or state estimation in the nonlinear stochastic systems governed by Itô-type stochastic differential equations [61,62].

In recent years, a great deal of concern has been raised regarding the study of partial differential systems (PDSs) [72–74]. Many phenomena in science and engineering have been modeled by deterministic PDSs, for example, some mechanical systems related to heat flows, fluid flows, elastic wave, flexible structures [75], chemical engineering [76], biology, population dynamics, neurophysiology, and biodynamics [77–79], etc. Since the most of phenomena have some uncertainties due to the existence of different stochastic fluctuations, the more accurate representation of the behaviors should be modeled by stochastic partial differential system (SPDS) [72,73]. Some of the examples are turbulent flow in fluid dynamics [80], diffusion in random media, and molecule signal transduction in biology [81,82]. In chemical engineering, many chemical processes are characterized by the presence

DOI: 10.1201/9781003229230-5

of spatial variations and time delays [74,76]. Therefore, we shall address the study of LSPDSs. In the past decade, the stabilization design problems of LSPDSs have been widely studied [72]. The LSPDS can be transferred to an equivalent infinite-dimensional ordinary differential system based on Galerkin method in Section 1.2.1. According to the separation of eigenvalues, the infinite-dimensional ordinary differential system can be separated into a finite-dimensional slow mode and an infinite-dimensional fast mode. A finite-dimensional controller based on the finite-dimensional mode could be designed to stabilize the LSPDS. In this chapter, we can focus on the robust filtering problem of LSPDSs, which is an important topic for signal processing and state estimation in system control designs.

In this chapter, the general robust filtering theory for LSPDSs with a set of sensor measurements is studied from the spatio-temporal H_∞ disturbance attenuation point of view. Then the H_∞ linear partial differential state estimator is designed for LSPDSs based on the general robust filtering theory, which needs to solve a complex HJII and is not easy to implement. In this chapter, in order to simplify the filter design, based on the finite difference scheme in Section 1.2.2 and the Kronecker product, the states of all finite difference grids in the spatio-domain are formulated as a spatial state vector so that a spatial state space system can represent an LSPDS with the finite difference truncation error. In this situation, the spatio-temporal H_∞ filtering performance can be transformed to an equivalent temporal H_∞ filtering performance so that an implementable H_∞ spatial state estimator can be designed to robustly estimate the state variables of LSPDSs. For LSPDSs, the H_∞ filter design needs to solve an algebraic Riccati inequality (ARI). Then, by Schur complement, an equivalent LMI is used to replace the ARI to solve the robust state estimation problem of LSPDSs. Finally, a robust H_∞ state estimation example for a heat transfer system by a set of noisy sensor measurements is given to illustrate the robust filter design procedure and to confirm the H_∞ filtering performance of the proposed robust state estimation for LSPDSs. The main features of the proposed method include: (i) A spatio-temporal H_∞ filtering performance is employed to treat the robust state estimation problem for LSPDSs with disturbance $v(x,t)$ and noise $w(x,t)$ in the spatial-temporal domain. (ii) Based on HJII, a more general spatio-temporal H_∞ filter design theory is developed for LSPDSs. (iii) Finite difference method and the Kronecker product are used to transform the LSPDS to an equivalent spatial state space system so that an implementable H_∞ spatial state estimator could be easily designed for LSPDSs via solving an LMI through the help of MATLAB LMI toolbox.

3.2 A GENERAL H_∞ FILTERING SETTING FOR LINEAR STOCHASTIC PARTIAL DIFFERENTIAL SYSTEMS

Consider the following LSPDS with a set of noisy sensor measurements:

$$\frac{\partial y(x,t)}{\partial t} = \kappa\nabla^2 y(x,t) + Ay(x,t) + g_v v(x,t) + Hy(x,t)w(x,t)$$

$$z(t) = \left[h_1\left(y\left(q_1,t\right) + D_n n\left(q_1,t\right)\right),\ldots,h_{n_z}\left(y\left(q_{n_z},t\right) + D_n n\left(q_{n_z},t\right)\right)\right]^T \quad (3.1)$$

for $x = [x_1, x_2]^T \in U = [x_{0,1}, x_{f,1}] \times [x_{0,2}, x_{f,2}] \in \mathbb{R}^2$ and $t > 0$, where $y(x,t) \triangleq [y_1(x,t), \ldots, y_n(x,t)]^T \in \mathbb{R}^n$ is the state variable; x and t are the space and time variables, respectively. The space domain U is a bounded domain. The system coefficients $\kappa \in \mathbb{R}^{n \times n}$, $A \in \mathbb{R}^{n \times n}$, and $H \in \mathbb{R}^{n \times n}$. The measurement output $z(t) \triangleq [z_1(t), \ldots, z_{n_z}(t)]^T \in \mathbb{R}^{n_z}$ may be interpreted as sensor observations, where n_z is the number of sensors. $h_i \triangleq h(q_i) \in \mathbb{R}^{1 \times n}$ is the observation influence vector at the sensor location q_i of observation, where $i = 1, \ldots, n_z$. $y(q_i, t) \in \mathbb{R}^{1 \times n}$ is the state variable at the sensor location q_i of observation, where $i = 1, \ldots, n_z$ $n(q_i, t) \in L_2(U \times \mathbb{R}_+; \mathbb{R}^n)$ is the stochastic spatial-temporal measurement noise at the sensor location q_i, where $i = 1, \ldots, n_z$ and $D_n \in \mathbb{R}^{n \times n}$ is the noise influence matrix. Therefore, in the case with free measurement noise, we have $z_i(t) = h_i y(q_i, t)$, where $i = 1, \ldots, n_z$. $v(x,t) \in L_2(U \times \mathbb{R}_+; \mathbb{R}^{n_v})$ is the vector of the external disturbance in the spatio-temporal domain and $g_v \in \mathbb{R}^{n \times n_v}$ is the disturbance influence matrix. $w(x,t) \in \mathbb{R}$ is a stationary spatio-temporal white noise. The Laplace operator ∇^2 is defined as follows:

$$\nabla^2 y(x,t) \triangleq \sum_{k=1}^{2} \frac{\partial^2}{\partial x_k^2} y(x,t)$$

where $\frac{\partial^2}{\partial x_k^2} y(x,t) \triangleq \left[\frac{\partial^2}{\partial x_k^2} y_1(x,t), \ldots, \frac{\partial^2}{\partial x_k^2} y_n(x,t) \right]^T \in \mathbb{R}^n$. The initial value is given by $y(x,0) = y_0(x)$. For simplicity, the boundary condition is usually given by Dirichlet boundary condition, i.e. $y(x,t) = a$ constant on ∂U, or by Neumann boundary condition $\nabla y(x,t) \cdot \vec{n} = 0$ on ∂U, where $\vec{n}$ is the normal vector to the boundary ∂U.

Remark 3.2.1

The stochastic state equation in (3.1) can be written in the Itô differential form as follows:

$$dy(x,t) = \left(\kappa \nabla^2 y(x,t) + \mathrm{A} y(x,t) + g_v v(x,t) \right) dt + H y(x,t) dW(x,t)$$

$$z(t) = \left[h_1(y(q_1,t) + D_n n(q_1,t)), \ldots, h_{n_z}(y(q_{n_z},t) + D_n n(q_{n_z},t)) \right]^T \tag{3.2}$$

where $dW(x,t) = w(x,t)dt$ with $W(x,t)$ being the Wiener process or Brownian motion which is a zero mean Gaussian random field with variance t at each location x [73].

Next, let us consider the following LSPDS without external disturbance $v(x,t)$:

$$dy(x,t) = \left(\kappa \nabla^2 y(x,t) + \mathrm{A} y(x,t) \right) dt + H y(x,t) dW(x,t) \tag{3.3}$$

Definition 3.2.1

The equilibrium point $y(x,t)=0$ of the LSPDS in equation (3.3) is said to be stochastically asymptotically stable if $E\{y^T(x,t)y(x,t)\}\to 0$ as $t\to\infty$, $\forall x\in U$.

Suppose that $V(y(x,t))=0$ is a strong Itô functional [73]. Then we have [72,73,83]

$$\int_U V(y(x,t))dx=\int_U V(y(x,0))dx+\int_0^t LV(y(x,s))ds$$
$$+\int_0^t\int_U \frac{\partial V(y(x,s))}{\partial y}^T Hy(x,s)dW(x,s) \tag{3.4}$$

where the infinitesimal operator $LV(y(x,s))$ is denoted as

$$LV(y(x,s))\triangleq\int_U\left(\frac{\partial V(y(x,s))}{\partial y}\right)^T\left(\kappa\nabla^2 y(x,s)+Ay(x,s)\right)dx$$
$$+\int_U\frac{1}{2}(Hy(x,s))^T\frac{\partial^2 V(y(x,s))}{\partial y^2}(Hy(x,s))dx \tag{3.5}$$

Lemma 3.2.1 [72,73,83]

A strong Itô functional $V(y(x,t))$: $\mathbb{R}^n\to\mathbb{R}$ is said to be a Lyapunov functional for the system in equation (3.3), if (i) $V(0)=0$ and $V(y(x,t))>0$ for $y(x,t)\neq 0$; (ii) $ELV(y(x,t))\leq 0$.

Theorem 3.2.1 [73,83]

Suppose the system in equation (3.3) has a solution $y(x,t)$ and there exists a Lyaponov functional $V(y(x,t))$. Furthermore, $ELV(y(x,t))<0$, then the equilibrium point $y(x,t)=0$ of the system in equation (3.3) is stochastically asymptotically stable.

Since this chapter deals with the asymptotical stochastic filtering problem, it is inevitably related to the stochastic stability, i.e., the asymptotical stability in probability or the exponentially mean square stability [72,73] of the estimation error system. For convenience of filtering design, the following proposition on the stochastically asymptotical stability of the equilibrium point $y(x,t)=0$ of the system in equation (3.3) is given first.

Consider the case in equation (3.2) with $v(x,t)=0$ and $n(x,t)=0$

$$\begin{aligned} dy(x,t)&=\left(\kappa\nabla^2 y(x,t)+Ay(x,t)\right)dt+Hy(x,t)dW(x,t)\\ \eta(x,t)&=Qy(x,t)\end{aligned} \tag{3.6}$$

then we have the following result.

Theorem 3.2.2

For the system equation (3.6), if there exists a positive functional $V(y(x,t)) \in C^2(\mathbb{R}^n)$ with $V(0)=0$ solving the following HJII:

$$\int_U \begin{bmatrix} \left(\frac{\partial V(y(x,s))}{\partial y}\right)^T \left(\kappa\nabla^2 y(x,t) + Ay(x,t) + \eta^T(x,t)\eta(x,t)\right) + \\ \frac{1}{2}(Hy(x,t))^T \frac{\partial^2 V(y(x,t))}{\partial y^2}(Hy(x,t)) \end{bmatrix} dx < 0 \quad (3.7)$$

then (i) the equilibrium point $y(x,t)=0$ of the system in equation (3.6) is stochastically asymptotically stable and (ii) the output variance $E\int_U \int_0^t \|\eta(x,t)\|^2 dt dx < E\int_U V(y(x,0))dx$.

Proof

Let us choose the Lyapunov functional $V(y(x,t)) > 0$, $\forall y(x,t) \neq 0$ and $V(0)=0$ for the linear partial differential system in equation (3.6). First, from the inequality equation (3.7), we have

$$ELV(y(x,t)) < E\int_U -\|\eta(x,t)\|^2 dx \leq 0$$

i.e.,

$$ELV(y(x,t)) < 0$$

So by Theorem 3.2.1, the equilibrium point $y(x,t)=0$ of the system in equation (3.6) is stochastically asymptotically stable. Next we have

$$E\int_U \int_0^{t_f} \|\eta(x,t)\|^2 dt dx$$

$$= E\int_U \left[V(y(x,0)) - V(y(x,t_f)) + \int_0^{t_f} \left(\|\eta(x,t)\|^2 + \frac{\partial V(y(x,t))}{\partial t} \right) dt \right] dx \quad (3.8)$$

By the fact $V(y(x,t_f)) \geq 0$ and Ito formula [61,62], we get

$$E\int_U\int_0^{tf}\|\eta(x,t)\|^2 dtdx \le E\int_U V(y(x,0))dx + E\int_U\int_0^{tf}\|\eta(x,t)\|^2 dtdx$$
$$+E\int_U\int_0^{tf}\left[\left(\frac{\partial V(y(x,t))}{\partial y}\right)^T\left(\kappa\nabla^2 y(x,t)+Ay(x,t)\right)\right.$$
$$\left.+\frac{1}{2}(Hy(x,t))^T\frac{\partial^2 V(y(x,t))}{\partial y^2}(Hy(x,t))\right]dtdx \qquad (3.9)$$

By the inequality equations (3.7) and (3.9), we get

$$E\int_U\int_0^{tf}\|\eta(x,t)\|^2 dtdx \le E\int_U V(y(x,0))\,dx \qquad (3.10)$$

Remark 3.2.2

Similarly, if the following Hamilton Jacobi inequality (HJI) holds:

$$\left(\frac{\partial V(y(x,t))}{\partial y}\right)^T\left(\kappa\nabla^2 y(x,t)+Ay(x,t)+\eta^T(x,t)\eta(x,t)\right)$$
$$+\frac{1}{2}(Hy(x,t))^T\frac{\partial^2 V(y(x,t))}{\partial y^2}(Hy(x,t))<0 \qquad (3.11)$$

for all $x \in U$, then the result in Theorem 3.2.2 still holds but with stricter condition than equation (3.7).

Lemma 3.2.2 [84]

For any matrices (or vectors) M_1 and M_2 with appropriate dimensions, we have

$$M_1^T M_2 + M_2^T M_1 \le M_1^T M M_1 + M_2^T M^{-1} M_2$$

where M is any positive-definite symmetric matrix.

Corollary 3.2.1

$X^T PY + Y^T PX \le \xi X^T PX + \xi^{-1} Y^T PY$ for any constant ξ, a symmetric $P = P^T \ge 0$ and two matrices (or vectors)X and Y with appropriate dimensions.

Proof

This corollary is obviously true when M_1, M_2, and M in Lemma 3.2.2 are substituted with $P^{\frac{1}{2}}X$, $P^{\frac{1}{2}}Y$, and ξI with I being an identity matrix, respectively.

After discussing the stochastically asymptotic stability of SPDS, the H_∞ robust ability of the following linear stochastic system is discussed:

$$\begin{aligned} dy(x,t) &= \kappa\nabla^2 y(x,t) + Ay(x,t) + g_v v(x,t) + Hy(x,t)dW(x,t) \\ \eta(x,t) &= Qy(x,t) \end{aligned} \tag{3.12}$$

Theorem 3.2.3

For the system equation (3.12), if there exists a positive functional $V(y(x,t)) \in C^2(\mathbb{R}^n)$ with $V(0)=0$ solving the following HJII:

$$\begin{aligned} &\int_U \Bigg[\left(\frac{\partial V(y(x,t))}{\partial y}\right)^T \left(\kappa\nabla^2 y(x,t) + Ay(x,t)\right) + y^T(x,t)Q^T Q y(x,t) \\ &+ \frac{1}{4\gamma^2}\left(\frac{\partial V(y(x,t))}{\partial y}\right)^T g_v g_v^T \left(\frac{\partial V(y(x,t))}{\partial y}\right) \\ &+ \frac{1}{2}(Hy(x,t))^T \frac{\partial^2 V(y(x,t))}{\partial y^2}(Hy(x,t)) \Bigg] dx < 0 \end{aligned} \tag{3.13}$$

then (i) the equilibrium point $y(x,t)=0$ of the system in equation (3.12) is stochastically asymptotically stable in the case of $v(x,t)=0$, and (ii) the following spatio-temporal H_∞ robust performance holds:

$$E\int_U \int_0^{t_f} \|\eta(x,t)\|^2 dtdx < E\int_U V(y(x,0))\, dx + \gamma^2 E\int_U \int_0^{t_f} \|v(x,t)\|^2 dtdx \tag{3.14}$$

$\forall v(x,t) \in L_2(U \times \mathbb{R}_+; \mathbb{R}^{n_v})$, $v(x,t) \neq 0$ for some $\gamma > 0$ if the initial state $y(x,0)=0$ and

$$E\int_U \int_0^{t_f} \|\eta(x,t)\|^2 dtdx < \gamma^2 E\int_U \int_0^{t_f} \|v(x,t)\|^2 dtdx \tag{3.15}$$

holds if the initial state $y(x,0)=0$.

Proof

Let us choose the Lyapunov functional $V(y(x,t))>0$, $\forall y(x,t)\neq 0$ and $V(0)=0$. First, for the system in equation (3.12) in the case of $v(x,t)=0$, from the inequality equation (3.13), we have

$$ELV(y(x,t))<E\int_U\left[-y^T(x,t)Q^TQy(x,t)-\frac{1}{4\gamma^2}\left(\frac{\partial V(y(x,t))}{\partial y}\right)^T g_v g_v^T\left(\frac{\partial V(y(x,t))}{\partial y}\right)\right]$$

$$dx\le 0$$

i.e.,

$$ELV(y(x,t))<0$$

So by Theorem 3.2.1, the equilibrium point $y(x,t)=0$ of the system in equation (3.12) in the case of $v(x,t)\neq 0$ is stochastically asymptotically stable. Next, we have

$$E\int_U\int_0^{t_f}\|\eta(x,t)\|^2 dtdx$$

$$=E\int_U\left[V(y(x,0))-V(y(x,t_f))+\int_0^{t_f}\left(\|\eta(x,t)\|^2+\frac{\partial V(y(x,t))}{\partial t}\right)dt\right]dx$$

$$\le E\int_U\int_0^{t_f}\left[\left(\frac{\partial V(y(x,t))}{\partial y}\right)^T(\kappa\nabla^2 y(x,t)+Ay(x,t)+g_v v(x,t)+\|\eta(x,t)\|^2\right.$$

$$\left.+\frac{1}{2}(Hy(x,t))^T\frac{\partial^2 V(y(x,t))}{\partial y^2}(Hy(x,t))\right]dtdx+E\int_U V(y(x,0))dx$$

$$=E\int_U\int_0^{t_f}\left[\left(\frac{\partial V(y(x,t))}{\partial y}\right)^T\left(\kappa\nabla^2 y(x,t)+Ay(x,t)\right)+\left(\frac{\partial V(y(x,t))}{\partial y}\right)^T\right.$$

$$\left.+g_v v(x,t)+y^T(x,t)Q^TQy(x,t)+\frac{1}{2}(Hy(x,t))^T\frac{\partial^2 V(y(x,t))}{\partial y^2}(Hy(x,t))\right]dtdx$$

$$+E\int_U V(y(x,0))dx \tag{3.16}$$

$$\frac{1}{2}\left(A^TB+B^TA\right)\le\frac{1}{4\gamma^2}A^TA+\gamma^2B^TB \tag{3.17}$$

By Corollary 3.2.1, we have the fact that for all matrices, vectors A, B, and a positive scalar γ. Then, by equation (3.17) with $A = \left(\frac{\partial V(y(x,t))}{\partial y}\right)^T g_v$ and $B = v(x,t)$, the inequality equation (3.16) can be modified as the following inequality:

$$E\int_U\int_0^{t_f} \|\eta(x,t)\|^2 dtdx$$

$$\leq E\int_U\int_0^{t_f}\left[\left(\frac{\partial V(y(x,t))}{\partial y}\right)^T\left(\kappa\nabla^2 y(x,t) + \mathrm{A}y(x,t)\right) + \frac{1}{4\gamma^2}\left(\frac{\partial V(y(x,t))}{\partial y}\right)^T\right.$$

$$+ g_v g_v^T\left(\frac{\partial V(y(x,t))}{\partial y}\right) + \gamma^2 v^T(x,t)v(x,t) + y^T(x,t)Q^T Q y(x,t)$$

$$\left.+\frac{1}{2}(Hy(x,t))^T\frac{\partial^2 V(y(x,t))}{\partial y^2}(Hy(x,t))\right]dtdx + E\int_U V(y(x,0))dx \tag{3.18}$$

Therefore, by the inequality in equation (3.13), we get

$$E\int_U\int_0^{t_f}\|\eta(x,t)\|^2 dtdx < E\int_U V(y(x,0))\,dx + \gamma^2 E\int_U\int_0^{t_f}\|v(x,t)\|^2 dtdx$$

Remark 3.2.3

(i) If $v(x,t)$ is of deterministic external disturbance, then the expectation operator on in equations (3.14) and (3.15) should be neglected. (ii) If the following HJI holds:

$$\left(\frac{\partial V(y(x,t))}{\partial y}\right)^T\left(\kappa\nabla^2 y(x,t) + Ay(x,t)\right) + y^T(x,t)Q^T Q y(x,t)$$

$$+\frac{1}{4\gamma^2}\left(\frac{\partial V(y(x,t))}{\partial y}\right)^T g_v g_v^T\left(\frac{\partial V(y(x,t))}{\partial y}\right) + \frac{1}{2}(Hy(x,t))^T\frac{\partial^2 V(y(x,t))}{\partial y^2}(Hy(x,t)) < 0 \tag{3.19}$$

then the H_∞ robust performance in equations (3.14) or (3.15) also holds but with a stricter condition than equation (3.13); (iii) If we let

$$J(v) = \frac{1}{2}E\int_U\int_0^{t_f}\left(\|\eta(x,t)\|^2 - \gamma^2\|v(x,t)\|^2\right)dtdx \tag{3.20}$$

One can see that, for any $v(x,t)$ and $v^*(x,t) \in L_2(U \times \mathbb{R}_+;\mathbb{R}^{n_v}) \cap \tilde{\Omega}$ with $\tilde{\Omega} \triangleq \left\{ v(x,t): \lim_{t\to\infty} EV\left(y(x,t)\right) = 0 \right\}$, $J(v) \le J(v^*)$, where

$$v^*(x,t) = \frac{1}{2\gamma^2} g_v^T \left(\frac{\partial V\left(y(x,t)\right)}{\partial y} \right) \tag{3.21}$$

that is, $v^*(x,t)$, the worst-case disturbance, results in the maximal possible energy gain from the disturbance to the output $\eta(x,t)$ in equation (3.6).

In the LSPDS in equations (3.1) or (3.2), if the state $y(x,t)$ is unavailable, we can only measure the system via $z(x,t)$ by a set of sensors at some locations. Therefore, how to estimate $y(x,t)$ from the measurement $z(x,t)$ is an important topic of robust filtering problem for linear stochastic PDSs. Suppose the following state estimation is proposed for linear stochastic PDS in equation (3.1):

$$d\hat{y}(x,t) = \left[\kappa\nabla^2 \hat{y}(x,t) + A\hat{y}(x,t) + L(x)\left(z(t) - \left[h_1\hat{y}(q_1,t),\ldots,h_{n_z}\hat{y}(q_{n_z},t)\right]^T \right)\right] dt \tag{3.22}$$

$\forall x \in U$, where $L(x)$ denotes the filter gain to be designed to achieve the H_∞ state estimation performance.

Let us denote the estimation error as

$$e(x,t) = y(x,t) - \hat{y}(x,t) \tag{3.23}$$

From equations (3.1) and (3.22), we get the following estimation error:

$$\begin{aligned} de(x,t) = &\left\{ \begin{aligned} &\kappa\nabla^2 e(x,t) + Ae(x,t) - L(x)\left[h_1 e(q_1,t),\ldots,h_{n_z} e(q_{n_z},t)\right]^T \\ &\qquad - L(x)\left[h_1 D_n n(q_1,t),\ldots,h_{n_z} D_n n(q_{n_z},t)\right]^T + g_v v(x,t) \end{aligned} \right\} dt \\ &+ Hy(x,t)dW(x,t) \end{aligned} \tag{3.24}$$

$\forall x \in U$. Then, we get the following augmented system:

$$\begin{aligned} \begin{bmatrix} dy(x,t) \\ de(x,t) \end{bmatrix} = &\begin{bmatrix} \kappa\nabla^2 y(x,t) + Ay(x,t) \\ \kappa\nabla^2 e(x,t) + Ae(x,t) - L(x)[h_1 e(q_1,t),\ldots,h_{n_z} e(q_{n_z},t)]^T \end{bmatrix} dt \\ &+ \begin{bmatrix} g_v & 0_n & \cdots & 0_n \\ g_v & -L(x)\begin{bmatrix} h_1 \\ 0_{1\times n} \\ \vdots \\ 0_{1\times n} \end{bmatrix} Dn & \cdots & -L(x)\begin{bmatrix} 0_{1\times n} \\ 0_{1\times n} \\ \vdots \\ h_{n_z} \end{bmatrix} Dn \end{bmatrix} \begin{bmatrix} v(x,t) \\ n(q_1,t) \\ \vdots \\ n(q_{n_z},t) \end{bmatrix} \\ &+ \begin{bmatrix} Hy(x,t) \\ Hy(x,t) \end{bmatrix} dW(x,t) \end{aligned} \tag{3.25}$$

Let us denote

$$\bar{y}(x,t) \triangleq \begin{bmatrix} y(x,t) \\ e(x,t) \end{bmatrix}, \tilde{\kappa} \triangleq \begin{bmatrix} \kappa & 0_n \\ 0_n & \kappa \end{bmatrix}, \tilde{A} \triangleq \begin{bmatrix} A & 0_n \\ 0_n & A \end{bmatrix}, \tilde{H} \triangleq \begin{bmatrix} H & 0_n \\ H & 0_n \end{bmatrix},$$

$$\tilde{G}_v \triangleq \begin{bmatrix} g_v & 0_n & \cdots & 0_n \\ g_v & -L(x)\begin{bmatrix} h_1 \\ 0_{1\times n} \\ \vdots \\ 0_{1\times n} \end{bmatrix} Dn & \cdots & -L(x)\begin{bmatrix} 0_{1\times n} \\ 0_{1\times n} \\ \vdots \\ h_{n_z} \end{bmatrix} Dn \end{bmatrix}, \bar{v}(x,t) \triangleq \begin{bmatrix} v(x,t) \\ n(q_1,t) \\ \vdots \\ n(q_{n_z},t) \end{bmatrix}$$

Then, the augmented system can be represented as

$$d\bar{y}(x,t) = \{\kappa\nabla^2\bar{y}(x,t) + \mathrm{A}\bar{y}(x,t) + \begin{bmatrix} 0_{n\times 1} \\ -L(x)[h_1 e(q_1,t),\ldots,h_{n_z} e(q_{n_z},t)]^T \end{bmatrix}$$
$$+ \tilde{G}_v\bar{v}(x,t)\}dt + \tilde{H}\bar{y}(x,t)dW(x,t),$$

$$\mathrm{e(x,t)} = [0_n, I_n]\bar{y}(x,t) \triangleq Q\bar{y}(x,t), \quad \forall x \in U \tag{3.26}$$

Definition 3.2.2

In the case $v(x,t) = 0$, the LSPDSs are said to be asymptotically estimated in probability if $e(x,t) \to 0$ in probability for all $x \in U$ as $t \to \infty$.

Based on Theorem 3.2.3, we get the following H_∞ estimation result for LSPDS.

Theorem 3.2.4

For the augmented system in equation (3.26), if we can specify the filter gain L(x) for the partial differential state estimator in equation (3.22) such that the following HJII holds:

$$\int_U \left\{ \left(\frac{\partial V(\bar{y}(x,t))}{\partial \bar{y}} \right)^T (\tilde{\kappa}\nabla^2\bar{y}(x,t) + \tilde{\mathrm{A}}\bar{y}(x,t) \right.$$
$$+ \begin{bmatrix} 0_{n\times 1} \\ -L(x)[h_1 e(q_1,t),\ldots,h_{n_z} e(q_{n_z},t)]^T \end{bmatrix} + \bar{y}^T(x,t)Q^T Q\bar{y}(x,t)$$

$$+\frac{1}{4\gamma^2}\left(\frac{\partial V(\bar{y}(x,t))}{\partial \bar{y}}\right)^T \tilde{G}_v\tilde{G}_v^T\left(\frac{\partial V(\bar{y}(x,t))}{\partial \bar{y}}\right)$$

$$+\frac{1}{2}(\tilde{H}\bar{y}(x,t))^T\frac{\partial^2 V(\bar{y}(x,t))}{\partial \bar{y}^2}(\tilde{H}\bar{y}(x,t))\Bigg]dt<0 \tag{3.27}$$

then (i) the equilibrium point $\bar{y}(x,t)=0$ *of the augmented system in equation (3.26) is stochastically asymptotically stable in the case of* $\bar{v}(x,t)=0$, *i.e.,* $E\{\bar{y}^T(x,t)\bar{y}(x,t)\}\to 0$ *and* $E\{e^T(x,t)e(x,t)\}\to 0$ *as* $t_f\to\infty$, *and (ii) for some* $\gamma>0$, $\forall\bar{v}(x,t)\in L_2(U\times\mathbb{R}_+;\mathbb{R}^{n_v+n_z})$, $\bar{v}(x,t)\neq 0$, *the following spatio-temporal stochastic* H_∞ *state estimation performance holds:*

$$E\int_U\int_0^{t_f}\|e(x,t)\|^2dtdx<E\int_U V(\bar{y}(x,0))dx+\gamma^2E\int_U\int_0^{t_f}\|\bar{v}(x,t)\|^2dtdx \tag{3.28}$$

if the initial state $\bar{y}(x,0)\neq 0$ or

$$E\int_U\int_0^{t_f}\|e(x,t)\|^2dtdx<\gamma^2E\int_U\int_0^{t_f}\|\bar{v}(x,t)\|^2dtdx \tag{3.29}$$

if the initial state $\bar{y}(x,0)=0$.

Proof

By the similar procedure in the proof of Theorem 3.2.3 with $y(x,t)$, $v(x,t)$, and $\eta(x,t)$ in equation (3.12) being replaced by $\bar{y}(x,t)$, $\bar{v}(x,t)$, and $e(x,t)$ in equation (3.26), respectively.

From the analysis above, the stochastic H_∞ state estimator design needs to specify the filter gain $L(x)$ in the linear partial differential estimator in equation (3.22) so that there exists a positive solution $V(\bar{y}(x,t))$ with $V(0)=0$ for equation (3.27). However, even the filter gain $L(x)$ is solved from the complex HJII in equation (3.27), it is still difficult to implement the partial differential state estimator in equation (3.22). Therefore, a simpler and easier implementable H_∞ robust estimator for linear PDSs is proposed via the semi-discretization finite difference scheme and the Kronecker product in the following section.

3.3 IMPLEMENTABLE H_∞ FILTER FOR LINEAR STOCHASTIC PARTIAL DIFFERENTIAL SYSTEMS

For the implementation of robust H_∞ state estimator of LSPDSs in equation (3.2), the main problem is to obtain a more suitable state space model to represent the LSPDSs. For this purpose, the semi-discretization finite difference scheme and the Kronecker product are employed. The finite difference scheme is employed to approximate the partial differential operator $\nabla^2 y(x,t)$ in equation (3.2) to simplify the design procedure of state estimator of LSPDS. Consider a typical mesh as shown in Figure 3.1. The state $y(x,t)$ is represented by $y_{k,l}(t) \in \mathbb{R}^n$ at the grid node $x_{k,l}(x_1 = k\Delta, x_2 = l\Delta)$, where $k=1,\ldots, N_1$ and $l=1,\ldots, N_2$, i.e., $y(x,t)\big|_{x=x_{k,l}} \triangleq y_{k,l}(t)$. At the grid point $x_{k,l}$, the central finite difference approximation in Section 1.2.2 for the partial differential operator can be written as follows [85–88]:

$$\kappa\nabla^2 y(x,t) = \kappa\left(\frac{y_{k+1,l}(t) + y_{k-1,l}(t) - 2y_{k,l}(t)}{\Delta^2} + \frac{y_{k,l+1}(t) + y_{k,l-1}(t) - 2y_{k,l}(t)}{\Delta^2}\right) + O_{k,l}(\Delta^2) \tag{3.30}$$

The remainder term $O_{k,l}(\Delta^2) \in \mathbb{R}^n$ is called the local truncation error.

Remark 3.3.1

$O_{k,l}(\Delta^2) \approx -(\Delta^2/12)(\partial^4 y/\partial^4 x)\big|_{x=x_{k,l}}$ [88], i.e., $O_{k,l}(\Delta^2)$ is related to $\partial^4 y/\partial^4 x\big|_{x=x_{k,l}}$. Maybe we could estimate $O_{k,l}(\Delta^2)$ with some estimation of $\partial^4 y/\partial^4 x\big|_{x=x_{k,l}}$. In general, it has been proven [89] that if grid points in a spatial domain are dense enough then the truncation error $O_{k,l}(\Delta^2)$ will be small enough.

Based on the finite difference approximation in equation (3.30), the LSPDS in equation (3.2) can be represented by the following stochastic finite difference system:

$$dy_{k,l}(t) = \kappa\frac{1}{\Delta^2}\left[y_{k+1,l}(t) + y_{k-1,l}(t) + y_{k,l+1}(t) + y_{k,l-1}(t) - 4y_{k,l}(t)\right]dt$$

$$Ay_{k,l}(t) + g_v v_{k,l}(t)dt + Hy_{k,l}(t)dW_{k,l}(t) + O_{k,l}(\Delta^2)dt \tag{3.31}$$

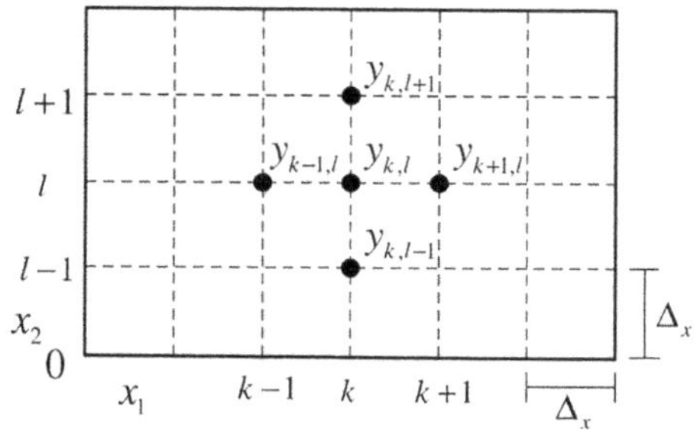

FIGURE 3.1 Finite difference grids on the spatio-domain.

where $y_{k,l}(t) = y(x,t)\big|_{x=x_{k,l}}$, $v_{k,l}(t) = v(x,t)\big|_{x=x_{k,l}}$, and $W_{k,l}(t) = W(x,t)\big|_{x=x_{k,l}}$. Let us denote

$$T_{k,l}y_{k,l}(t) = \frac{1}{\Delta^2}\left[y_{k+1,l}(t) + y_{k-1,l}(t) + y_{k,l+1}(t) + y_{k,l-1}(t) - 4y_{k,l}(t)\right] \quad (3.32)$$

Then, we get

$$dy_{k,l}(t) = \left[\kappa T_{k,l}y_{k,l}(t) + Ay_{k,l}(t) + g_v v_{k,l}(t)\right]dt + Hy_{k,l}(t)dW_{k,l}(t) + O_{k,l}(\Delta^2)dt \quad (3.33)$$

For the simplification of state estimation design for LSPDSs, we define a spatial state vector $y(t)$ to collect the state variable $y_{k,l}(t) \in \mathbb{R}^n$ at all grid nodes in Figure 3.1. For Dirichlet boundary conditions [89], the values of $y_{k,l}(t)$ at the boundary are fixed, for example, $y(x,t) = 0$ on ∂U. We have $y_{k,l}(t) = 0$ at $k=0$, N_1+1 or $l=0$, N_2+1. Therefore, the spatial state vector $y(t) \in \mathbb{R}^{nN}$ for state variables at all grid nodes is defined as follows:

$$y(t) = \left[y_{1,1}^T(t), y_{k,1}^T(t), \ldots, y_{N_1,1}^T(t), \ldots, y_{k,l}^T(t), \ldots, y_{1,N_2}^T(t), \ldots, y_{k,N_2}^T(t), \ldots, y_{N_1,N_2}^T(t)\right]^T \quad (3.34)$$

where $N \triangleq N_1 \times N_2$. Note that n is the dimension of the vector $y_{k,l}(t)$ for each grid node and $N_1 \times N_2$ is the number of grid nodes. For example, let $N_1 = 2$ and $N_2 = 2$, we have $y(t) = \left[y_{1,1}^T(t), y_{2,1}^T(t), y_{1,2}^T(t), y_{2,2}^T(t)\right]^T \in \mathbb{R}^{4n}$. In order to simplify the index of the node $y_{k,l}(t) \in \mathbb{R}^n$ in the spatial state vector $y(t) \in \mathbb{R}^{nN}$, we denote the symbol $y_j(t) \in \mathbb{R}^n$ to replace $y_{k,l}(t)$. Note that the index j is from 1 to N, i.e., $y_1(t) \triangleq y_{1,1}(t)$, $y_2(t) \triangleq y_{2,1}(t)$,..., $y_j(t) \triangleq y_{k,l}(t)$,..., $y_N(t) \triangleq y_{N_1,N_2}(t)$, where $j = (l-1)N_1 + k$ in equation (3.34). The stochastic difference model of two indices in equation (3.33) could be represented with only one index as follows:

$$dy_j(t) = \left[\kappa T_j y_{k,l}(t) + Ay_j(t) + g_v v_j(t)\right]dt + Hy_j(t)dW_j(t) + O_j(\Delta^2)dt \quad (3.35)$$

where $v_j(t) = v_{k,l}(t)$, $dW_j(t) = dW_{k,l}(t)$, $O_j\left(\Delta^2\right) = O_{k,l}\left(\Delta^2\right)$ with $j = (l-1)N_1 + k$ and the matrix T_j is defined as

$$T_j y(t) = \frac{1}{\Delta^2}\left[0_n \cdots 0_n \; I_n \quad 0_n \cdots 0_n \; I_n - 4I_n \; I_n \; 0_n \cdots 0_n \; I_n \; 0_n \cdots 0_n\right] y(t),$$

$$\text{position} \rightarrow \qquad 1 \quad j-N_1 \quad j-1 \quad j+1 \quad j+N_1 \quad N_1N_2$$

in which 0_n denotes the $n \times n$ zero matrix and I_n denotes the $n \times n$ identity matrix.

The measurement output $z(t)$ in equations (3.1) or (3.2) can be represented as follows:

$$z(t)=\sum_{j=1}^{N}[C_j y_j(t)+D_j n_j(t)] \tag{3.36}$$

where the matrix C_j is defined as $C_j=C_{\{k,l\}}=\left[C_{\{k,l\},1}^T,\ldots,C_{\{k,l\},n_z}^T\right]^T\in\mathbb{R}^{n_z\times n}$ with $j=(l-1)N_1+k$, in which the element is given as $C_{\{k,l\},i}=h_i$ for $x_{k,l}=q_i$ or $C_{\{k,l\},i}=0$ for $x_{k,l}\neq q_i$; the matrix D_j is defined as $D_j=D_{\{k,l\}}=\left[D_{\{k,l\},1}^T,\ldots,D_{\{k,l\},n_z}^T\right]^T\in\mathbb{R}^{n_z\times n}$ with $j=(l-1)N_1+k$, in which the element is given as $D_{\{k,l\},i}=h_i D_n$ for $x_{k,l}=q_i$ or $D_{\{k,l\},i}=0$ for $x_{k,l}\neq q_i$. Note that q_i is the ith sensor location as defined in equation (3.1). The jth measurement noise is defined as $n_j(t)=n(x,t)\big|_{x=x_{k,l}}$ where $j=(l-1)N_1+k$.

We collect all states $y_i(t)$ of grid nodes in equations (3.35) and (3.36) to the state $y(t)$ in equation (3.34). The Kronecker product can be used to simplify the representation. Using the Kronecker product, the system in equation (3.35) can be represented as the following stochastic spatial state space system:

$$dy(t)=[I_N\otimes\kappa]Ty(t)dt+[I_N\otimes A]y(t)dt+G_v v(t)dt+[I_N\otimes H]y(t)\circ d\tilde{W}(t)+O(\Delta^2)dt$$

$$z(t)=Cy(t)+Dn(t) \tag{3.37}$$

where $d\tilde{W}(t)\triangleq\left[J^T dW_1(t),\ldots,J^T dW_N(t)\right]^T\in\mathbb{R}^{nN}$ with $J\triangleq[1,\ldots,1]^T\in\mathbb{R}^n$, i.e., all the elements in J are equal to 1, $O(\Delta^2)\triangleq[O_1^T(\Delta^2),\ldots,O_N^T(\Delta^2)]^T$ and the spatial state vector $y(t)$ denoted in equation (3.34) represents $y(x,t)$ at all grid points on the spatial domain in Figure 3.1. The corresponding matrices are defined as $T=\left[T_1^T,\ldots,T_N^T\right]^T\in\mathbb{R}^{n_z\times nN}$, $G_v=I_N\otimes g_v\in\mathbb{R}^{nN\times n_v N}$, $C=[C_1,\ldots,C_N]\in\mathbb{R}^{n_z\times nN}$ and $D=[D_1,\ldots,D_N]\in\mathbb{R}^{n_z\times nN}$. The external disturbance $v(t)=\left[v_1^T(t),\ldots,v_N^T(t)\right]^T\in\mathbb{R}^{n_v N}$. The measurement noise $n(t)$ is defined as $n(t)=\left[n_1^T(t),\ldots,n_N^T(t)\right]^T\in\mathbb{R}^{nN}$. The Hadamard product of matrices (or vectors) $X=\left[X_{ij}\right]_{m\times n}$ and $Y=\left[Y_{ij}\right]_{m\times n}$ of the same size is the entry wise product denoted as $X\circ Y=[X_{ij}Y_{ij}]_{m\times n}$.

Remark 3.3.2

(i) The increments of the Wiener processes $dW_1(t),\ldots, dW_N(t)$ at the N grid nodes have the property: $dW(t)\triangleq[dW_1(t),\ldots,dW_N(t)]^T\in\mathbb{R}^N$ is a zero-mean vector-process with the covariance $E\{dW(t)dW^T(t)\}=I_N dt$. (ii) The physical meaning of equation (3.37) is that LSPDS in equations (3.1) or (3.2) at all grid points on the spatial domain in Figure 3.1 can be represented by the spatial state space model in equation (3.37).

Then our robust state estimation design problem of a stochastic PDS becomes how to design a state estimator for the stochastic spatial state space system in equation (3.37) to satisfy the H_∞ filtering performance. It will be discussed in the following. An implementable state estimator based on stochastic spatial state space system to estimate $y(t)$ in equation (3.37) from the measurement $z(t)$ to guarantee the robust H_∞ filtering performance is proposed as follows:

$$d\hat{y} = \left\{[I_N \otimes \kappa]T\hat{y}(t) + [I_N \otimes A]\hat{y}(t) + \bar{L}\left(z(t) - C\hat{y}(t)\right)\right\}dt \tag{3.38}$$

where $\bar{L} \in \mathbb{R}^{nN \times n_z}$ are the estimation gains to be designed. The dynamic of the estimated error $e(t) = y(t) - \hat{y}(t)$ is given by

$$\begin{aligned} de(t) &= [I_N \otimes \kappa]Ty(t)dt + [I_N \otimes A]y(t)dt + [I_N \otimes H]y(t) \circ d\tilde{W}(t) + G_v v(t)dt \\ &\quad + O(\Delta^2)dt - \{[I_N \otimes \kappa]T\hat{y}(t) + [I_N \otimes A]\hat{y}(t) + \bar{L}(z(t) - C\hat{y}(t))\}dt \\ &= \{[I_N \otimes \kappa]T + [I_N \otimes A] - \bar{L}C\}e(t)dt + [I_N \otimes H]y(t) \circ d\tilde{W}(t) + G_v v(t)dt \\ &\quad - \bar{L}Dn(t)dt + O(\Delta^2)dt \end{aligned} \tag{3.39}$$

Combining equation (3.37) with (3.39), we get the following augmented system:

$$\begin{aligned} d\bar{y}(t) &= \bar{A}\bar{y}(t)dt + \bar{H}\bar{y}(t) \circ d\bar{W}(t) + \bar{G}_v\bar{v}(t)dt \\ e(t) &= \begin{bmatrix} 0_{nN} & I_{nN} \end{bmatrix}\bar{y}(t) \end{aligned} \tag{3.40}$$

where

$$\bar{y}(t) \triangleq \begin{bmatrix} y(t) \\ e(t) \end{bmatrix}, \bar{G}_v \triangleq \begin{bmatrix} G_v & I_{nN} & 0_{nN} \\ G_v & I_{nN} & -\bar{L}D \end{bmatrix}, \bar{v}(t) \triangleq \begin{bmatrix} v(t) \\ O(\Delta^2) \\ n(t) \end{bmatrix}, \bar{A} \triangleq \begin{bmatrix} \bar{A}_{11} & 0_{nN} \\ 0_{nN} & \bar{A}_{22} \end{bmatrix}$$

$$\begin{aligned} \bar{A}_{11} &\triangleq [I_N \otimes \kappa]T + [I_N \otimes A] \\ \bar{A}_{22} &\triangleq [I_N \otimes \kappa]T + [I_N \otimes A] - \bar{L}C \end{aligned}, \bar{H} \triangleq \begin{bmatrix} I_N \otimes H & 0_{nN} \\ I_N \otimes H & 0_{nN} \end{bmatrix}, d\bar{W}(t) \triangleq \begin{bmatrix} d\tilde{W}(t) \\ d\tilde{W}(t) \end{bmatrix},$$

0_{nN} denotes the $nN \times nN$ zero matrix and I_{nN} denotes the $nN \times nN$ identity matrix.

Remark 3.3.3

When the boundary condition is also took into account, the augmented system in equation (3.40) and its system coefficients should be modified according to what boundary condition is used. Since there are too many kinds of boundary conditions to describe, we only show Dirichlet boundary condition in the robust filter design example of Section 3.4.

Since the spatial state vector $y(t)$ in equation (3.34) is used to represent the $y(x, t)$ at all grid points, in this situation, $E\int_U \int_0^{t_f} \|e(x,t)\|^2 dtdx$, $E\int_U \int_0^{t_f} \|\bar{v}(x,t)\|^2 dtdx$ and $E\int_U V(\bar{y}(x,0))dx$ in the spatio-temporal stochastic H_∞ state estimation performance in equations (3.28) or (3.29) could be modified by the temporal forms $E\int_0^{t_f} \|e(t)\|^2 \Delta^2 dt$, $E\int_0^{t_f} \|\bar{v}(t)\|^2 \Delta^2 dt$, and $E\bar{V}(\bar{y}(0))\Delta^2$, respectively, for the stochastic spatial state space system in equation (3.27), where the Lyapunov function $\bar{V}(\bar{y}(t))$ is related with the Lyapunov function $\bar{V}(\bar{y}(x,t))$ as $\bar{V}(\bar{y}(t))=\sum_{j=1}^{N} \bar{V}(\bar{y}_i(t))$ with $(\bar{y}_i(t)) \triangleq \left[y_i^T(t), e_j^T(t) \right]^T$. Therefore, the stochastic H_∞ state estimation performance in equations (3.28) and (3.29) could be modified as

$$E\int_0^{t_f} \|e(t)\|^2 dt < E\bar{V}(\bar{y}(0)) + \gamma^2 E\int_0^{t_f} \|\bar{v}(t)\|^2 dt \tag{3.41}$$

where $\bar{y}(0) \neq 0$ and

$$E\int_0^{t_f} \|e(t)\|^2 dt < \gamma^2 E\int_0^{t_f} \|\bar{v}(t)\|^2 dt \tag{3.42}$$

when $\bar{y}(0) = 0$, where Δ^2 in both sides of equations (3.41) and (3.42) can be canceled.

Therefore, the spatio-temporal H_∞ state estimation problem in equations (3.28) or (3.29) for LSPDS in equation (3.2) is transformed to the only temporal H_∞ state estimation problem in equations (3.41) or (3.42) for the stochastic spatial state space system in equation (3.37).

Suppose the external disturbance $v(t)$ and the measurement $n(t)$ are neglected and the truncation error $O(\Delta^2)$ of the finite difference scheme approaches zero in the augmented system equation (3.40) as the number of grid nodes is large enough, i.e. $\bar{v}(t) = 0$, then the state estimator could be designed in equation (3.38) to guarantee the stochastically asymptotical estimation of the augmented system in equation (3.40).

Theorem 3.3.1

For the augmented system in equation (3.40) with $\bar{v}(t) = 0$, suppose there exist some symmetric positive definite matrix $\bar{P}$ and estimator gain $\bar{L}$ such that the following ARI holds:

$$\bar{P}\bar{A} + \bar{A}^T\bar{P} + \bar{H}^T\bar{P}\bar{H} < 0 \tag{3.43}$$

where $\bar{A} \triangleq \begin{bmatrix} [I_{N\otimes\kappa}]T & 0_{nN} \\ 0_{nN} & [I_{N\otimes\kappa}]T + [I_N \otimes A] - \bar{L}C \end{bmatrix}$ and $\bar{H} := \begin{bmatrix} I_N \otimes H & 0_{nN} \\ I_N \otimes H & 0_{nN} \end{bmatrix}$

Then, the stochastically asymptotical estimation is achieved by the state estimator in equation (3.38), i.e., $Ee^T(t)e(t) \to 0$ *as* $t \to \infty$ *as for the LSPDS with* $\bar{v}(t) = 0$.

Proof

For the augmented system in equation (3.40) with $\bar{v}(t) = 0$, the infinitesimal operator $L\bar{V}(\bar{y}(t))$ in the spatial state space model is denoted as

$$L\bar{V}(\bar{y}(t)) \triangleq \left(\frac{\partial \bar{V}(\bar{y})}{\partial \bar{y}}\right)^T \left(\bar{A}\bar{y}(t)\right) + \frac{1}{2}\left[\bar{H}\bar{y}(t)\right]^T \frac{\partial^2 \bar{V}(\bar{y})}{\partial \bar{y}^2}\left[\bar{H}\bar{y}(t)\right]$$

By the fact

$$\frac{\partial \bar{V}(\bar{y})}{\partial \bar{y}} = 2\bar{P}\bar{y}, \quad \frac{\partial^2 \bar{V}(\bar{y})}{\partial \bar{y}^2} = 2\bar{P} \tag{3.44}$$

we get

$$L\bar{V}\left(\bar{y}(t)\right) = \bar{y}^T(t)\bar{P}\left(\bar{A}\bar{y}(t)\right) + \bar{y}^T(t)\bar{A}^T\bar{P}\bar{y}(t) + \left[\bar{H}\bar{y}(t)\right]^T \bar{P}\left[\bar{H}\bar{y}(t)\right] \tag{3.45}$$

So we get

$$EL\bar{V}(\bar{y}(t)) < E\bar{y}^T(t)\left[\bar{P}\bar{A} + \bar{A}^T\bar{P} + \bar{H}^T\bar{P}\bar{H}\right]\bar{y}(t) \tag{3.46}$$

If the following ARI holds:

$$\bar{P}\bar{A} + \bar{A}^T\bar{P} + \bar{H}^T\bar{P}\bar{H} < 0 \tag{3.47}$$

then

$$EL\bar{V}\left(\bar{y}(t)\right) < 0 \tag{3.48}$$

By Theorem 3.2.1, the equilibrium point $\bar{y}(t) = 0$ of the augmented system in equation (3.40) with $\bar{v}(t) = 0$ is stochastically asymptotically stable. Therefore, the stochastically asymptotical estimation is achieved.

Theorem 3.3.2

For the augmented system in equation (3.40) with a prescribed noise attenuation level γ^2 in equations (3.41) and (3.42), suppose there exist some symmetric positive definite matrix $\bar{P}$ and estimator gain $\bar{L}$ such that the following ARI holds:

$$\bar{I} + \bar{P}\bar{A} + \bar{A}^T\bar{P} + \frac{1}{\gamma^2}\bar{P}\bar{G}_v\bar{G}_v^T\bar{P} + \bar{H}^T\bar{P}\bar{H} < 0 \tag{3.49}$$

where $\bar{I} \triangleq \begin{bmatrix} 0_{nN} & 0_{nN} \\ 0_{nN} & I_{nN} \end{bmatrix}$.

Then the robust H_∞ state estimation performance in equations (3.41) and (3.42) is guaranteed by the estimator in equation (3.38). Furthermore, the spatio-temporal stochastic H_∞ state estimation performance in equations (3.28) and (3.29) for LSPDSs in equations (3.1) or (3.2) could be achieved and the stochastically asymptotical estimation is also guaranteed when $\bar{v}(t) = 0$.

Proof

$$E\int_0^\infty e^T(t)e(t)dt = E\int_0^\infty \bar{y}^T(t)\begin{bmatrix} 0_{nN} \\ I_{nN} \end{bmatrix}\begin{bmatrix} 0_{nN} & I_{nN} \end{bmatrix}\bar{y}(t)dt$$

$$= E\bar{V}(\bar{y}(0)) - E\bar{V}(\bar{y}(\infty)) + E\int_0^\infty \left[\bar{y}^T(t)\bar{I}\bar{y}(t) + \frac{d\bar{V}(\bar{y}(t))}{dt}\right]dt \tag{3.50}$$

For the augmented system in equation (3.40) with $\bar{v}(t) \neq 0$, by Itô formula [61,62], we get

$$E\int_0^\infty e^T(t)e(t)dt$$

$$\leq E\bar{y}^T(0)\bar{P}\bar{y}(0) + E\int_0^\infty \left\{\bar{y}^T(t)\bar{I}\bar{y}(t)dt + \left(\frac{\partial \bar{V}(\bar{y})}{\partial \bar{y}}\right)^T\left[\bar{A}\bar{y}(t) + \bar{G}_v\bar{v}(t)\right]\right.$$

$$\left. + \frac{1}{2}\left[\bar{H}\bar{y}(t)\right]^T\frac{\partial^2 \bar{V}(\bar{y})}{\partial \bar{y}^2}\left[\bar{H}\bar{y}(t)\right]\right\}dt \tag{3.51}$$

By Corollary 3.2.1 and the fact

$$\frac{\partial \bar{V}(\bar{y})}{\partial \bar{y}} = 2\bar{P}\bar{y}, \quad \frac{\partial^2 \bar{V}(\bar{y})}{\partial \bar{y}^2} = 2\bar{P} \tag{3.52}$$

we get

$$E\int_0^{\infty} e^T(t)e(t)\,dt$$

$$\le E\left\{Tr\left[\bar{P}\bar{y}(0)\bar{y}^T(0)\right]\right\} + E\int_0^{\infty}\left\{\bar{y}^T(t)\bar{I}\,\bar{y}(t) + \bar{y}^T(t)\bar{P}\bar{A}\bar{y}(t)\right.$$

$$\left. +\bar{y}^T(t)\bar{A}^T\bar{P}\bar{y}(t) + \bar{y}^T(t)\bar{P}\bar{G}_v\bar{v}(t) + \bar{v}^T(t)\bar{G}_v^T\bar{P}\bar{y}(t) + \left[\bar{H}\bar{y}(t)\right]^T\bar{P}\left[\bar{H}\bar{y}(t)\right]\right\}dt$$

$$\le E\int_0^{\infty}\left\{\bar{y}^T(t)\left[\bar{I} + \bar{P}\bar{A} + \bar{A}^T\bar{P} + \bar{P}\bar{G}_v\bar{G}_v^T\bar{P} + \bar{H}^T\bar{P}\bar{H}\right]\bar{y}(t)\right\}dt$$

$$+ Tr\left[\bar{P}R_{\bar{y}(0)}\right] + E\gamma^2\int_0^{\infty}\bar{v}^T(t)\bar{v}(t)\,dt$$

where $R_{\bar{y}(0)} \triangleq E\left[\bar{y}(0)\bar{y}^T(0)\right]$. Therefore, if the following ARI holds:

$$\bar{I} + \bar{P}\bar{A} + \bar{A}^T\bar{P} + \frac{1}{\gamma^2}\bar{P}\bar{G}_v\bar{G}_v^T\bar{P} + \bar{H}^T\bar{P}\bar{H} < 0 \tag{3.53}$$

then

$$E\int_0^{\infty} e^T(t)e(t)\,dt \le Tr\left[\bar{P}R_{\bar{y}(0)}\right] + E\gamma^2\int_0^{\infty}\bar{v}^T(t)\bar{v}(t)\,dt \tag{3.54}$$

i.e., if the inequality in equation (3.53) holds, then the robust stochastic H_∞ state estimation is guaranteed. Since the equation (3.53) obviously implies the ARI in equation (3.43), the equilibrium point $\bar{y}(t) = 0$ of the augmented system in equation (3.40) with $\bar{v}(t) = 0$ is stochastically asymptotically stable. So the stochastically asymptotical estimation is guaranteed when $\bar{v}(t) = 0$.

Remark 3.3.4

If the matrix in equation (3.49) is not set to be block diagonal, then with $\bar{P} = \begin{bmatrix} P_{11} & P_{12} \\ P_{12}^T & P_{22} \end{bmatrix}$, the ARI in equation (3.49) becomes the following bilinear matrix inequality (BMI)

$$\begin{bmatrix} \Pi_{11} & \Pi_{12} & P_{11}G_v + P_{12}G_v & P_{11} + P_{12} & -P_{12}\bar{L}D \\ \Pi_{21} & \Pi_{22} & P_{12}^TG_v + P_{22}G_v & P_{12}^T + P_{22} & -P_{22}\bar{L}D \\ G_v^TP_{11} + G_v^TP_{12}^T & G_v^TP_{12} + G_v^TP_{22} & -\gamma^2 I_{n_vN} & 0_{n_vN\times nN} & 0_{n_vN\times nN} \\ P_{11} + P_{12}^T & P_{12} + P_{22} & 0_{nN\times n_vN} & -\gamma^2 I_{nN} & 0_{nN} \\ -D^T\bar{L}^TP_{12}^T & -D^T\bar{L}^TP_{22}^T & 0_{nN\times n_vN} & 0_{nN} & -\gamma^2 I_{nN} \end{bmatrix} < 0 \tag{3.55}$$

where

$$\Pi_{11} \triangleq P_{11}\Omega_1 + \Omega_1^T P_{11} + \Omega_2^T P_{11}\Omega_2 + \Omega_2^T P_{22}\Omega_2 + \Omega_2^T P_{12}^T\Omega_2 + \Omega_2^T P_{12}\Omega_2,$$

$$\Pi_{12} \triangleq \Omega_1^T P_{12} + P_{12}\Omega_1 - P_{12}\bar{L}C = \Pi_{21}^T,$$

$$\Pi_{22} \triangleq I_{nN} + P_{22}\Omega_1 + \Omega_1^T P_{22} - P_{22}\bar{L}C - C^T\bar{L}^T P_{22}^T,$$

$$\Omega_1 \triangleq [I_N \otimes \kappa]T + [I_N \otimes A],\ \Omega_2 \triangleq (I_N \otimes H).$$

Then the optimal H_∞ *filtering problem becomes how to specify* P_{11}, P_{12}, P_{22}, and $\bar{L}$ to achieve the following constrained optimization:

$$\gamma_{\min}^2 = \min_{\alpha<0, P_{11}, P_{12}, P_{22},\ \text{and}\ \bar{L}} \gamma^2 \quad \text{subject to (3.55)} \tag{3.56}$$

Based on the results in [90–92], the BMI-constrained optimization problem in equation (3.56) can be solved by an iterative LMI method as follows.

First, to improve the convergence of the iterative LMI algorithm, the ARIs in equations (3.49) and (3.55) are relaxed by introducing $\alpha\bar{P}$ in the following two inequalities [90–92], respectively.

$$\bar{I} + \bar{P}\bar{A} + \bar{A}^T\bar{P} + \frac{1}{\gamma^2}\bar{P}\bar{G}_v\bar{G}_v^T\bar{P} + \bar{H}^T\bar{P}\bar{H} - \alpha\bar{P} < 0 \tag{3.57}$$

$$\begin{bmatrix} \hat{\Pi}_{11} & \hat{\Pi}_{12} & P_{11}G_v + P_{12}G_v & P_{11} + P_{12} & -P_{12}\bar{L}D \\ \hat{\Pi}_{21} & \hat{\Pi}_{22} & P_{12}^T G_v + P_{22}G_v & P_{12}^T + P_{22} & -P_{22}\bar{L}D \\ G_v^T P_{11} + G_v^T P_{12}^T & G_v^T P_{12} + G_v^T P_{22} & -\gamma^2 I_{n_v N} & 0_{n_v N\times nN} & 0_{n_v N\times nN} \\ P_{11} + P_{12}^T & P_{12} + P_{22} & 0_{nN\times n_v N} & -\gamma^2 I_{nN} & 0_{nN} \\ -D^T\bar{L}^T P_{12}^T & -D^T\bar{L}^T P_{22}^T & 0_{nN\times n_v N} & 0_{nN} & -\gamma^2 I_{nN} \end{bmatrix} < 0 \tag{3.58}$$

where $\hat{\Pi}_{11} = \Pi_{11} - \alpha P_{11}$, $\hat{\Pi}_{12} = \Pi_{12} - \alpha P_{12}$, $\hat{\Pi}_{21} = \Pi_{21} - \alpha P_{12}^T$, and $\hat{\Pi}_{22} = \Pi_{22} - \alpha P_{22}$.

So the constrained optimization in equation (3.56) is modified as

$$\gamma_{\min}^2 = \min_{\alpha<0, P_{11}, P_{12}, P_{22},\ \text{and}\ \bar{L}} \gamma^2 \quad \text{subject to (3.58).} \tag{3.59}$$

Therefore, the iterative LMI algorithm [90] is used to achieve $\gamma_{\min}^2$ as follows.

Step 0: Set the initial value of γ^2, i.e., $\gamma^2 = \gamma_{\text{initial}}^2$.
Step 1: Decrease γ^2 and set the initial value of P_{11}, P_{12}, and P_{22} to some appropriate values, e.g. $P_{11}=I_{nN}$, $P_{12}=0_{nN}$, and $P_{22}=I_{nN}$. Set $\bar{P}_0 = \bar{P}$ and $i=1$.
Step 2: Solve the following optimization problem for $\bar{L}$ with given $P_{11,i}$, $P_{12,i}$, and $P_{22,i}$.

OP1: Minimize α_i subject to the following LMI constraint

$$\begin{bmatrix} \hat{\Pi}_{11,i} & \hat{\Pi}_{12,i} & \hat{\Pi}_{13,i} & P_{11,i}+P_{12,i} & -P_{12,i}\bar{L}D \\ \hat{\Pi}_{21,i} & \hat{\Pi}_{22,i} & \hat{\Pi}_{23,i} & P_{12,i}^T+P_{22,i} & -P_{22,i}\bar{L}D \\ \hat{\Pi}_{13,i}^T & \hat{\Pi}_{23,i}^T & -\gamma^2 I_{n_vN} & 0_{n_vN\times nN} & 0_{n_vN\times nN} \\ P_{11,i}+P_{12,i}^T & P_{12,i}+P_{22,i} & 0_{nN\times n_vN} & -\gamma^2 I_{nN} & 0_{nN} \\ -D^T\bar{L}^T P_{12,i}^T & -D^T\bar{L}^T P_{22,i}^T & 0_{nN\times n_vN} & 0_{nN} & -\gamma^2 I_{nN} \end{bmatrix} < 0 \quad (3.60)$$

where

$$\hat{\Pi}_{11,i} \triangleq P_{11,i}\Omega_1 + \Omega_1^T P_{11,i} + \Omega_2^T P_{11,i}\Omega_2 + \Omega_2^T P_{22,i}\Omega_2 + \Omega_2^T P_{12,i}^T\Omega_2 + \Omega_2^T P_{12,i}\Omega_2 - \alpha_i P_{11,i},$$

$$\hat{\Pi}_{12,i} \triangleq \Omega_1^T P_{12,i} + P_{12,i}\Omega_1 - P_{12,i}\bar{L}C - \alpha_i P_{12,i} = \hat{\Pi}_{21,i}^T,$$

$$\hat{\Pi}_{13,i} \triangleq P_{11,i}G_v + P_{12,i}G_v = \hat{\Pi}_{13,i}^T$$

$$\hat{\Pi}_{23,i} \triangleq P_{12,i}^T G_v + P_{22,i}G_v = \hat{\Pi}_{23,i}^T$$

$$\hat{\Pi}_{22,i} \triangleq I_{nN} + P_{22,i}\Omega_1 + \Omega_1^T P_{22,i} - P_{22,i}\bar{L}C - C^T\bar{L}^T P_{22,i}^T - \alpha P_{22,i}.$$

Step 3: If $\alpha_i \leq 0$, γ^2 is an appropriate H_∞ attenuation level and $\bar{L}$ is an appropriate H_∞ estimator gain. Go to Step 1.
Step 4: Solve the following optimization problem for $P_{11,i}$, $P_{12,i}$, and $P_{22,i}$ with given $\bar{L}$.

OP2: Minimize α_i subject to the LMI constraints equation (3.60) and

$$\bar{P}_i = \begin{bmatrix} P_{11,i} & P_{12,i} \\ P_{12,i}^T & P_{22,i} \end{bmatrix} = \bar{P}_i^T > 0$$

Step 5: If $\alpha_i \leq 0$, γ^2 is an appropriate H_∞ attenuation level and $\bar{L}$ is an appropriate H_∞ estimator gain. Go to Step 1.
Step 6: Solve the following optimization problem for $P_{11,i}$, $P_{12,i}$, and $P_{22,i}$ with given $\bar{L}$ and α_i.

OP3: Minimize trace ($\bar{P}_i$) subject to the above LMI constraints equation (3.60) and $\bar{P}_i = \bar{P}_i^T > 0$

Step 7: If $\|\bar{P}_{i-1} - \bar{P}_i\| / \|\bar{P}_i\| < \delta$, a prescribed tolerance, go to Step 8, else set $i = i+1$, then go to Step 2.

Step 8: The appropriate values γ^2 and $\bar{L}$ in the last iteration satisfying $\alpha_i \leq 0$ in Step 3 or 5 are just $\gamma^2{}_{\min}$ and the H_∞ estimator gain that we want, respectively. Stop.

Remark 3.3.5

Since the above iterative LMI algorithm is based on the results in [90–92], the convergence analysis and effectiveness of the above iterative LMI algorithm can be referred to [90–92].

Remark 3.3.6

For the robust stochastically asymptotical estimation problem, let us choose a Lyapunov function $\bar{V}(\bar{y})$ for the augmented system in equation (3.40) as

$$\bar{V}(\bar{y}) = \sum_{j=1}^{N} (y_j^T(t)P_{11,j}y_j(t) + e_j^T(t)P_{22,j}e_j(t)) = \bar{y}^T(t)\bar{P}\bar{y}(t) \tag{3.61}$$

where $\bar{P} = diag(P_{11}, P_{22})$, $P_{11} = P_{11}^T = diag(P_{11,1},\ldots,P_{11,N}) > 0$, and $P_{22} = P_{22}^T = diag(P_{22,1},\ldots,P_{22,N}) > 0$. After replacing $\bar{P}$ with $diag(P_{11}, P_{22})$ and using the Schur complement, the ARI in equation (3.49) becomes the following LMI:

$$\begin{bmatrix} \Pi_{11} & 0_{nN} & P_{11}G_v & P_{11} & 0_{nN} \\ 0_{nN} & \Pi_{22} & P_{22}G_v & P_{22} & -ZD \\ G_v^T P_{11} & G_v^T P_{22} & -\gamma^2 I_{n_v N} & 0_{n_v N\times nN} & 0_{n_v N\times nN} \\ P_{11} & P_{22} & 0_{nN\times n_v N} & -\gamma^2 I_{n_v N} & 0_{nN} \\ 0_{nN} & -D^T Z^T & 0_{nN\times n_v N} & 0_{nN} & -\gamma^2 I_{n_v N} \end{bmatrix} < 0 \tag{3.62}$$

where

$$\Pi_{11} \triangleq P_{11}\Omega_1 + \Omega_1^T P_{11} + \Omega_2^T P_{11}\Omega_2 + \Omega_2^T P_{22}\Omega_2,$$

$$\Pi_{22} \triangleq I_{nN} + P_{22}\Omega_1 + \Omega_1^T P_{22} - ZC - C^T Z^T,$$

$$\Omega_1 \triangleq [I_N \otimes \kappa]T + [I_N \otimes A],\ \Omega_2 \triangleq (I_N \otimes H), \text{ and } Z \triangleq P_{22}\bar{L}.$$

After solving the LMI in equation (3.62), we get P_{11}, P_{22}, and Z. So it follows that the estimator gain $\bar{L} = P_{22}^{-1}Z$.

Remark 3.3.7

In the case of $\bar{P} = diag(P_{11}, P_{22})$, based on the H_∞ state estimation analysis in Theorem 3.3.2, the optimal H_∞ state estimation method for LSPDSs can be designed by minimizing the disturbance attenuation level γ^2 as the following eigenvalue problem:

$$\gamma_{\min}^2 = \min_{P_{11}, P_{22}, \text{ and } Z} \gamma^2 \qquad (3.63)$$

$$\text{subject to LMI (3.62)}$$

The advantage of setting $\bar{P}$ to be of block diagonal form is that equation (3.62) is an LMI which can be directly solved by MATLAB LMI toolbox, whereas in the case of non-block-diagonal $\bar{P}$, the BMI in equation (3.55) should be solved by the iterative LMI procedure.

Remark 3.3.8

The relationship between the HJII in equation (3.27) and the ARI in equation (3.49), which is equivalent to the LMI in equation (3.62), is given as follows. First, let

$$\bar{L} = \begin{bmatrix} L_1 \\ \vdots \\ L_N \end{bmatrix}, L_j \triangleq L_{k,l} = L(x)\Big|_{x=x_{k,l}}, J \triangleq \begin{bmatrix} 0_{nN} & I_{nN} & 0_{nN} \\ 0_{nN} & I_{nN} & 0_{nN} \end{bmatrix}$$

Then with the finite difference approximation in equation (3.30) and approximation of integration by summation, the HJII in equation (3.27) becomes

$$\int_U \left\{ \left(\frac{\partial V(\bar{y}(x,t))}{\partial \bar{y}} \right)^T \left\{ \tilde{\kappa}\nabla^2 \bar{y}(x,t) + \tilde{A}\bar{y}(x,t) + \bar{y}^T(x,t)Q^T Q\bar{y}(x,t) \right\} \right.$$

$$+ \frac{1}{2}(\tilde{H}\bar{y}(x,t))^T \frac{\partial^2 V(\bar{y}(x,t))}{\partial \bar{y}^2}(\tilde{H}\bar{y}(x,t)) + \begin{bmatrix} 0_{n\times 1} \\ -L(x)[h_1 e(q_1,t),\dots,h_{n_z} e(q_{n_z},t)]^T \end{bmatrix}$$

$$\left. + \frac{1}{4\gamma^2} \left(\frac{\partial V(\bar{y}(x,t))}{\partial \bar{y}} \right)^T \tilde{G}_v \tilde{G}_v^T \left(\frac{\partial V(\bar{y}(x,t))}{\partial \bar{y}} \right) \right\} dx$$

$$\approx \sum_{k=1}^{N_1}\sum_{l=1}^{N_2}\left\{(2\bar{P}_{k,l}\bar{y}_{k,l})^T\left\{\begin{bmatrix} O_{k,l}(\Delta^2) \\ O_{k,l}(\Delta^2)\end{bmatrix}+\begin{bmatrix} 0_{n\times 1} \\ -L_{k,l}[h_1 e(q_1,t),\ldots,h_{n_z}e(q_{n_z},t)]^T\end{bmatrix}\right.\right.$$

$$\left.+\tilde{A}\bar{y}_{k,l}+\tilde{\kappa}\left(\frac{\bar{y}_{k+1,l}(t)+\bar{y}_{k-1,l}(t)-2\bar{y}_{k,l}(t)}{\Delta^2}+\frac{\bar{y}_{k+1,l}(t)+\bar{y}_{k-1,l}(t)-2\bar{y}_{k,l}(t)}{\Delta^2}\right)\right\}$$

$$\left.+\bar{y}_{k,l}^T Q^T Q\bar{y}_{k,l}+\frac{1}{4\gamma^2}(2\bar{P}_{k,l}\bar{y}_{k,l})^T\tilde{G}_v\tilde{G}_v^T(2\bar{P}_{k,l}\bar{y}_{k,l})+\frac{1}{2}(\tilde{H}\bar{y}_{k,l})^T(2\bar{P}_{k,l})(\tilde{H}\bar{y}_{k,l})\right\}\Delta^2$$

$$=\left\{2\bar{y}^T(t)\bar{P}\left\{\bar{A}\bar{y}(t)+\begin{bmatrix} O(\Delta^2) \\ O(\Delta^2)\end{bmatrix}\right\}+\bar{y}^T(t)\bar{I}\bar{y}(t)+\bar{y}^T(t)\bar{H}^T\bar{P}\bar{H}\bar{y}(t)\right.$$

$$\left.+\frac{1}{\gamma^2}\bar{y}^T(t)\bar{P}\left[\bar{G}_v-J\right]\left[\bar{G}_v^T-J^T\right]\bar{P}\bar{y}(t)\right\}\Delta^2$$

$$=\left\{\bar{y}^T(t)\left\{\bar{P}\bar{A}+\bar{A}^T\bar{P}+\bar{I}+\frac{1}{\gamma^2}\bar{P}\left[\bar{G}_v-J\right]\left[\bar{G}_v^T-J^T\right]\bar{P}+\bar{H}^T\bar{P}\bar{H}\right\}\bar{y}(t)\right.$$

$$\left.+2\bar{y}^T(t)\bar{P}\begin{bmatrix} O(\Delta^2) \\ O(\Delta^2)\end{bmatrix}\right\}\Delta^2$$

$$<0 \tag{3.64}$$

So let $\bar{\bar{G}}_v \triangleq \left[\bar{G}_v - J\right]$, we get the approximation of HJII in equation (3.27) as follows:

$$\left\{\bar{y}^T(t)\left\{\bar{P}\bar{A}+\bar{A}^T\bar{P}+\bar{I}+\frac{1}{\gamma^2}\bar{P}\bar{\bar{G}}_v\bar{\bar{G}}_v^T\bar{P}+\bar{H}^T\bar{P}\bar{H}\right\}\bar{y}(t)+2\bar{y}^T(t)\bar{P}\begin{bmatrix} O(\Delta^2) \\ O(\Delta^2)\end{bmatrix}\right\}\Delta^2<0 \tag{3.65}$$

If we omit the last term in the left-hand side of equation (3.65), we can get

$$\left\{\bar{y}^T(t)\left\{\bar{P}\bar{A}+\bar{A}^T\bar{P}+\bar{I}+\frac{1}{\gamma^2}\bar{P}\bar{\bar{G}}_v\bar{\bar{G}}_v^T\bar{P}+\bar{H}^T\bar{P}\bar{H}\right\}\bar{y}(t)\Delta^2<0 \tag{3.66}$$

that implies

$$\bar{P}\bar{A}+\bar{A}^T\bar{P}+\bar{I}+\frac{1}{\gamma^2}\bar{P}\bar{\bar{G}}_v\bar{\bar{G}}_v^T\bar{P}+\bar{H}^T\bar{P}\bar{H}<0 \tag{3.67}$$

which is of the similar form of the ARI in equation (3.49) except using $\bar{\bar{G}}_v$ in place of $\bar{G}_v$ in equation (3.49). Both the last redundant term

$$2\bar{y}^T(t)\bar{P}\begin{bmatrix} O(\Delta^2) \\ O(\Delta^2) \end{bmatrix} \tag{3.68}$$

in the left-hand side of equation (3.65) and the difference between $\bar{\bar{G}}_v$ and $\bar{G}_v$ are due to the fact that the derivation procedure of the ARI in equation (3.49) in Section 3.3 takes the $O(\Delta^2)$ as a part of disturbance $\bar{v}(t)$, i.e., $\bar{v}(t) \triangleq \left[v^T(t), O^T(\Delta^2), n^T(t) \right]$ whereas the derivation procedure of the HJII in equation (3.27) in Section 3.2 does not take the $O\left(\Delta^2\right)$ as a part of disturbance. In summary, it is shown that the ARI in equation (3.49) is the finite difference approximation of HJII in equation (3.27) with the approximation error in equation (3.68).

Remark 3.3.9

The filter in equation (3.38) is a Luenberger form. If the filter does not take a Luenberger form to estimate $y(t)$ in equation (3.37) from the measurement $z(t)$ to guarantee the robust H∞ filtering performance, then the non-Luenberger state estimator is

$$d\hat{y}(t) = \left\{ A_K \hat{y}(t) + B_K z(t) \right\} dt \tag{3.69}$$

where $A_K \in \mathbb{R}^{n\times n}$ and $B_K \in \mathbb{R}^{nN\times n_z}$ are the two estimator coefficients to be designed. The dynamic of the estimated error $e(t) = y(t) - \hat{y}(t)$ is given by

$$\begin{aligned} de(t) &= \left\{ [I_N \otimes \kappa] T y(t) dt + [I_N \otimes A] y(t) dt + [I_N \otimes H] y(t) \circ d\tilde{W}(t) \right\} \\ &\quad + G_v v(t) dt + O(\Delta^2) dt - \left\{ A_K \hat{y}(t) + B_K z(t) \right\} dt \\ &= \left\{ [I_N \otimes \kappa] T + [I_N \otimes A] - A_K - B_K C \right\} y(t) dt + A_K e(t) dt \\ &\quad + [I_N \otimes H] y(t) \circ d\tilde{W}(t) + G_v v(t) dt - B_K D n(t) dt + O(\Delta^2) dt \end{aligned} \tag{3.70}$$

Combining equation (3.37) with (3.70), we get the following augmented system:

$$\begin{aligned} d\bar{y}(t) &= \bar{A}\bar{y}(t)dt + \bar{H}\bar{y}(t) \circ d\bar{W} + \bar{G}_v \bar{v}(t) dt \\ e(t) &= \begin{bmatrix} 0_{nN} & I_{nN} \end{bmatrix} \bar{y}(t) \end{aligned} \tag{3.71}$$

where

$$\bar{y}(t)=\begin{bmatrix} y(t) \\ e(t) \end{bmatrix}, \bar{G}_v =, \begin{bmatrix} G_v & I_{nN} & 0_{nN} \\ G_v & I_{nN} & -B_K D \end{bmatrix}, \bar{v}(t)=\begin{bmatrix} v(t) \\ O(\Delta^2) \\ n(t) \end{bmatrix},$$

$$\bar{A} \triangleq \begin{bmatrix} \bar{A}_{11} & 0_{nN} \\ \bar{A}_{21} & \bar{A}_{22} \end{bmatrix}, \quad \begin{array}{l} \bar{A}_{11} \triangleq [I_N \otimes \kappa]T + [I_N \otimes A] \\ \bar{A}_{21} \triangleq [I_N \otimes \kappa]T + [I_N \otimes A] - A_K - B_K C, \end{array}$$

$$\bar{A}_{22} \triangleq A_K, \bar{H} = \begin{bmatrix} I_N \otimes H & 0_{nN} \\ I_N \otimes H & 0_{nN} \end{bmatrix}, d\bar{W}(t) \triangleq \begin{bmatrix} d\tilde{W}(t) \\ d\tilde{W}(t) \end{bmatrix},$$

0_{nN} denotes the $nN \times nN$ zero matrix and I_{nN} denotes the $nN \times nN$ identity matrix. For the non-Luenberger state estimator in equation (3.69), by following the proof of Theorem 3.3.2, we can also get the same result of Theorem 3.3.2 with the similar form of the ARI in equation (3.49) as follows:

$$\bar{I} + \bar{P}\bar{A} + \bar{A}^T\bar{P} + \frac{1}{\gamma^2}\bar{P}\bar{G}_v\bar{G}_v^T\bar{P} + \bar{H}^T\bar{P}\bar{H} < 0 \tag{3.72}$$

But it should be noted here that the elements of matrices $\bar{A}$ and $\bar{G}_v$ are not the same for equations (3.49) and (3.72). After replacing $\bar{P}$ with $diag(P_{11}, P_{22})$ and using the Schur complement, the inequality in equation (3.72) becomes the following LMI:

$$\begin{bmatrix} \Pi_{11} & \Pi_{12} & P_{11}G_v & P_{11} & 0_{nN} \\ \Pi_{21} & \Pi_{22} & P_{22}G_v & P_{22} & -Z_B D \\ G_v^T P_{11} & G_v^T P_{22} & -\gamma^2 I_{n_v N} & 0_{n_v N \times nN} & 0_{n_v N \times nN} \\ P_{11} & P_{22} & 0_{nN \times n_v N} & -\gamma^2 I_{n_v N} & 0_{nN} \\ 0_{nN} & -D^T Z_B^T & 0_{nN \times n_v N} & 0_{nN} & -\gamma^2 I_{n_v N} \end{bmatrix} < 0 \tag{3.73}$$

where

$$\Pi_{11} \triangleq P_{11}\Omega_1 + \Omega_1^T P_{11} + \Omega_2^T P_{11}\Omega_2 + \Omega_2^T P_{22}\Omega_2,$$

$$\Pi_{21} \triangleq P_{22}\Omega_1 - Z_A - Z_B C = \Pi_{12}^T, \Pi_{21} \triangleq I_{nN} + Z_A + Z_A^T,$$

$$\Omega_1 \triangleq [I_N \otimes \kappa]T + [I_N \otimes A], \ \Omega_2 \triangleq (I_N \otimes H),$$

$$Z_A \triangleq P_{22}A_K, Z_B \triangleq P_{22}B_K.$$

After solving the LMI in equation (3.73), we get P_{11}, P_{22}, Z_A, and Z_B. Therefore, the two estimator coefficients $A_K = P_{22}^{-1}Z_A$ and $B_K = P_{22}^{-1}Z_B$ can be obtained.

3.4 DESIGN EXAMPLE OF ROBUST H_∞ FILTER WITH SIMULATION

Consider the temperature function for a 1×0.5 m thin plate in the heat transfer system with the surrounding temperature at as follows [93]:

$$\begin{aligned}
&\frac{\partial y(x,t)}{\partial t} = \kappa\nabla^2 y(x,t) + Ay(x,t) + g_v v(x,t) + Hy(x,t)w(x,t),\\
&y(x,0) = 100\times e^{-10\times|0.5-x_1|-0.6738}\times e^{-30\times|0.5-2\times x_2|},\\
&\forall x\in U=[0,1]\times[0,0.5]\\
&y(x,t)\equiv 0^\circ\text{C}, \forall t, \forall x \text{ on the boundary of U}
\end{aligned} \tag{3.74}$$

where $y(x,t)$ is the temperature function, the location x is in meters, the time t is in seconds, $\kappa=10^{-4}$ (m^2/s) is the thermal diffusivity [93], the term $Ay(x,t)$ with $A=-0.005\text{ s}^{-1}$ means thermal dissipation when the temperature on the bar is beyond the surrounding temperature, i.e., $y(x,t)>0$ °C or thermal absorption when the temperature on the bar is below the surrounding temperature, i.e., $y(x,t)<0$ °C, the term $g_v v(x,t)$ is the environmental thermal fluctuation with $g_v=0.1$ and $v(x,t)=\sin(0.2t)e^{-0.001t-0.1x_1}$, and the term $Hy(x,t)w(x,t)$ with $H=0.002$ is due to the random parameter variation of the term $Ay(x,t)$. The grid spacing of the finite difference scheme is chosen as 0.125 m so that there are $N=7\times 3=21$ interior grid points and 24 boundary points. There are $n_z=45$ sensor locations of observation, which are at the 2-D location $q_i=[q_{i,1}, q_{i,2}]^T$ with $q_{i,1}=(k-1)\times 0.125$ m, $k=1,\ldots, 9$, $q_{i,2}=(l-1)\times 0.125$ m, $l=1,\ldots, 5$, and $i=(l-1)\times 9+k$. The observation influence vector $h_i=1$ and the noise influence matrix $D_n=2.5$ so that $h_iD_n=2.5$. So the measured output by 45 sensors is as follows:

$$z(t)=\left[y(q_1,t)+2.5n(q_1,t),\ldots,y(q_{45},t)+2.5n(q_{45},t)\right]^T \tag{3.75}$$

where $n(q_i, t)$ denotes the measurement noise at the ith sensor. All the measurement noises are zero mean with unit variance in this example.

Based on the design procedure, we construct the spatial state space system in equation (3.37) from heat transfer system in equation (3.74) and measured output of 45 sensors in equation (3.75). Then from the optimal H_∞ filtering in equation (3.56) with the help of MATLAB toolbox. We can get $\gamma^2_{\min}$, P_{11}, P_{12}, P_{22}, and the estimator gain $\bar{L}$ for the robust state estimator in equation (3.38). From the computer simulation of this example by MATLAB with the initial estimated state $\hat{y}(x,0)=20$, $\forall x\in U$ and the H_∞ state estimation performance in equation (3.41), we get

$$\frac{E\int_0^{150}\|e(t)\|^2\,dt - E\bar{V}(\bar{y}(0))}{E\int_0^{150}\|\bar{v}(t)\|^2\,dt} = (0.57)^2 < \gamma^2_{\min} = (1.003)^2$$

For each γ^2, the average number of iterations required to converge to get the appropriate values γ^2 and $\bar{L}$ is 1.05. So the iterative LMI algorithm in Remark 3.3.4 is effective. Figures 3.2–3.5 show the simulation results of the temperature $y(x,t)$, the estimated temperature $\hat{y}(x,t)$, the estimation error $e(x,t)$, and the measured output $z(x,t)$ for 0, 50, 100, 150 seconds, respectively. The estimation error $e(x,t)$ is very small in these figures. To show the HJII in equation (3.27) can be efficiently approximated by the ARI in equation (3.49) via the central finite difference method of equation (3.30) in Remark 3.3.8, by replacing with $O_{k,l}(\Delta^2),\Delta^2$, Figure 3.6 illustrates that the simulation result of the finite difference approximation error $2\bar{y}^T(t)\bar{P}\left[O^T(\Delta^2),O^T(\Delta^2)\right]^T$ between the HJII in equation (3.27) and the ARI in equation (3.49) for this thermal system example is very small.

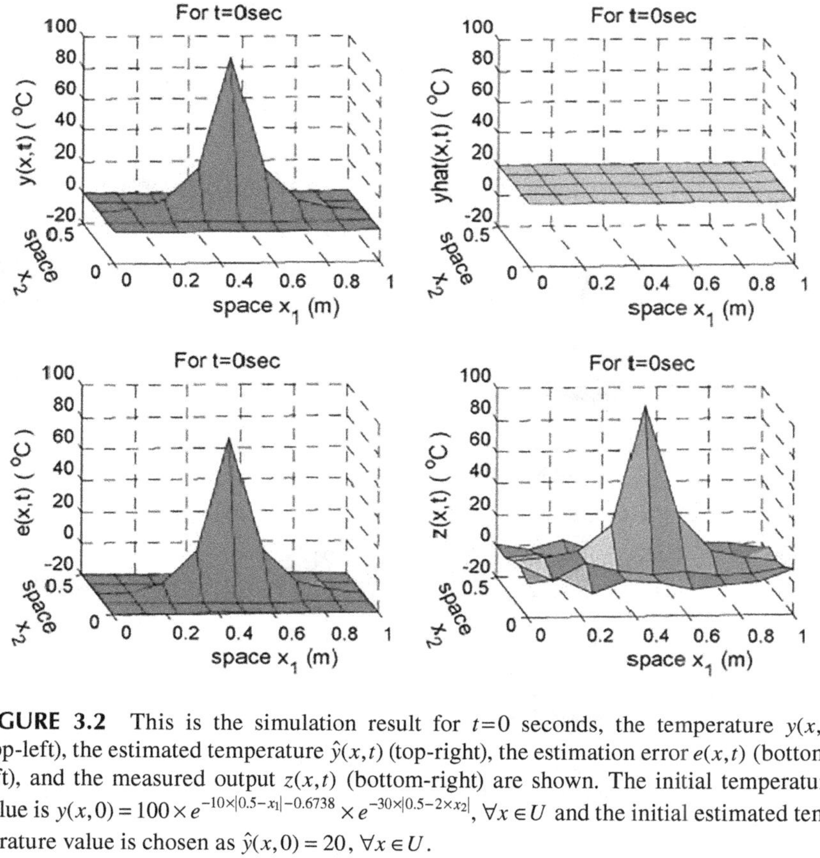

FIGURE 3.2 This is the simulation result for t=0 seconds, the temperature $y(x,t)$ (top-left), the estimated temperature $\hat{y}(x,t)$ (top-right), the estimation error $e(x,t)$ (bottom-left), and the measured output $z(x,t)$ (bottom-right) are shown. The initial temperature value is $y(x,0)=100\times e^{-10\times|0.5-x_1|-0.6738}\times e^{-30\times|0.5-2\times x_2|}$, $\forall x\in U$ and the initial estimated temperature value is chosen as $\hat{y}(x,0)=20$, $\forall x\in U$.

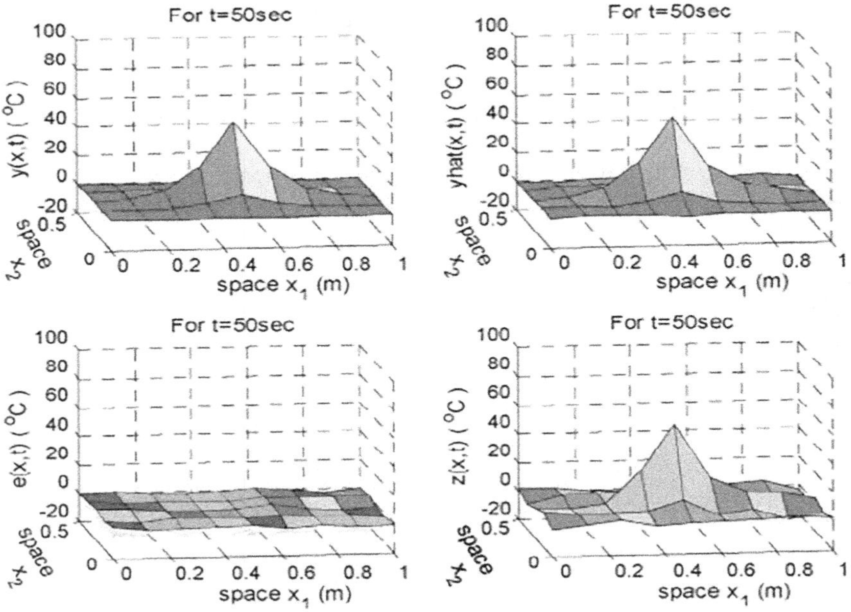

FIGURE 3.3 This is the simulation result for $t=50$ seconds.

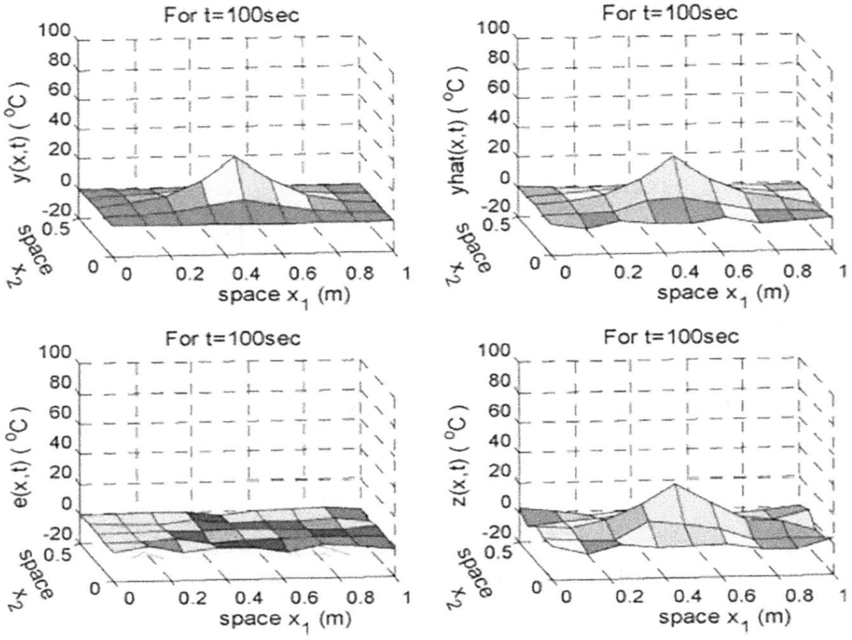

FIGURE 3.4 This is the simulation result for $t=100$ seconds.

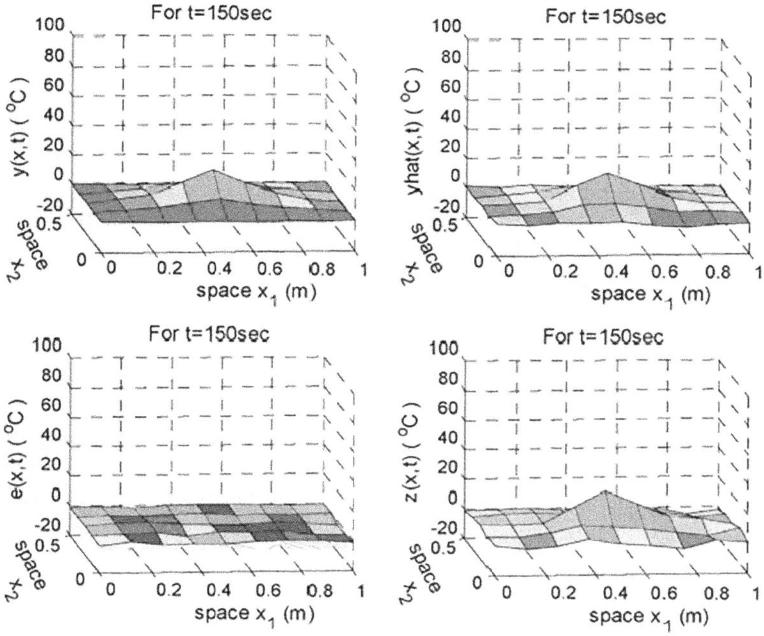

FIGURE 3.5 This is the simulation result for $t=150$ seconds.

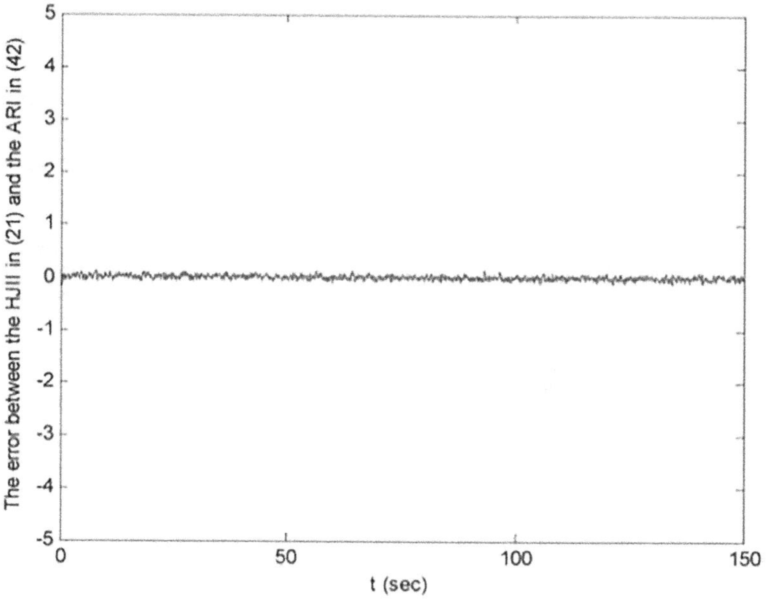

FIGURE 3.6 This is the simulation result of the redundant term $2\bar{y}^T(t)\bar{P}\left[O^T(\Delta^2),O^T(\Delta^2)\right]^T$ due to the finite difference approximation of ARI in equation (3.59) to HJII in equation (3.27).

3.5 CONCLUSION

In this chapter, the more general H_∞ filter design theory is proposed for a SPDS with spatio-temporal disturbances and state dependent noises through a set of sensor measurements. In order to make the robust H_∞ filter design easier and implementable, the semi-discretization finite difference scheme and Kronecker product are employed to represent the SPDS by a spatial state space system and then the spatio-temporal H_∞ filtering performance can be represented by the only temporal H_∞ filtering performance. Therefore, based on the more suitable spatial state space model, an implementable H_∞ filter is designed for robust state estimation of LSPDSs with spatiotemporal disturbances and state dependent noises. From the simulation example, the H_∞ filtering performance for stochastic partial differential systems can be confirmed by the proposed method.

After the robust filter design of LSPDSs has been introduced by finite difference scheme via a set of sensor measurement signals in this chapter, the more complex robust filter design of nonlinear SPDSs will be introduced in the following chapter by the fuzzy approximation method and finite difference scheme through a set of sensor measurements.

4 Robust Filter Design for Nonlinear Stochastic Partial Differential Systems in Sensor Signal Processing

4.1 INTRODUCTION

The H_∞ filtering problem is to design a state estimator to estimate unavailable state variables via noisy output measurement, and it can guarantee the L_2 gain (from the external disturbance to the estimation error) to be less than a prescribed level [94–106]. In contrast to the well-known Kalman filter, one of the main advantages of H_∞ filtering is that it is not necessary to know the exact statistical properties of external disturbances since it only assumes the external disturbances to have bounded energy, as shown in practical applications of H_∞ filtering for signal processing [101].

Recently, a great deal of concern has been raised regarding the study of partial differential systems (PDSs) [107–117]. Many phenomena in science and engineering have been modeled by deterministic PDSs, for example, some mechanical systems related to heat flows, fluid flows, elastic waves, flexible structures [113], biology, population dynamics, neurophysiology, biodynamics [114–116], etc. Since most of the phenomena have spatio-temporal uncertainties due to the existence of different stochastic fluctuations, for a more accurate representation of the behaviors, the modeling should be by SPDS [107,108]. In the past decade, the stabilization design problems of LSPDSs have been widely studied [107]. In addition, the stabilization design problem of NSPDSs has been discussed based on Galerkin's method [110] in Section 1.2.1. The robust H_∞ filter design of LSPDSs has been discussed in Chapter 3. In this chapter, we focus on the robust filter design problem of NSPDSs, which is an important topic for signal processing and state estimation in system control designs of partial differential mechanical systems, ecological systems and biodynamical systems.

For the robust state estimation of NSPDSs, the robust H_∞ filter design via a set of sensor measurements needs to solve a complex second-order HJII which has an integration of the Hamilton Jacobi inequality (HJI) over the space domain U. But it is generally very difficult to solve a corresponding second-order HJII for an H_∞

DOI: 10.1201/9781003229230-6

nonlinear state estimator. Further it is not easy to realize the H_∞ nonlinear partial differential state estimator for NSPDSs. In this situation, the fuzzy interpolation method and finite difference method in Section 1.2.2 are employed together to simplify the design procedure of a robust H_∞ state estimator for NSPDSs. Recently, the fuzzy approach [117–121] has been widely used in many fields, and it has already been applied to efficiently solving partial differential equations [117]. On the other hand, the fuzzy approach has also been applied to H_∞ filter and control [122–129] of nonlinear systems. In this chapter, we apply the fuzzy approach to investigate H_∞ filter for NSPDSs.

In this chapter, based on the fuzzy interpolation, the finite difference scheme and the Kronecker product, the states of all finite difference grids in the spatial domain are formulated as a spatial state vector so that a T-S fuzzy spatial state space system can be used to represent an NSPDS with some approximation errors. In this situation, the spatio-temporal H_∞ filtering performance can be transformed to an equivalent temporal H_∞ filtering performance so that an implementable fuzzy H_∞ spatial state estimator can be designed to robustly estimate the state variables of NSPDSs. For this, a set of LMIs are used to replace the nonlinear partial HJII to simplify the procedure of robust H_∞ filter design problem of NSPDSs through the use of fuzzy interpolation and finite difference schemes. Finally, a robust H_∞ state estimation example for a heat transfer system by a set of noisy sensor measurements is given to illustrate the filter design procedure and to confirm the H_∞ filtering performance of the proposed robust state estimation for NSPDSs. The contents of the proposed method include: (i) A spatio-temporal H_∞ filtering performance is employed to treat the robust state estimation problem for NSPDSs with disturbance $v(x, t)$ and noise $w(x, t)$ in the spatial-temporal domain. (ii) The fuzzy interpolation scheme, the finite difference method and the Kronecker product are combined to transform the NSPDS to an equivalent fuzzy spatial state space system so that an implementable H_∞ fuzzy spatial state estimator could be easily designed for practical application to NSPDSs by solving a set of LMIs using the MATLAB LMI toolbox.

Notations:

$$\begin{bmatrix} P_1 & * \\ P_2 & P_3 \end{bmatrix} \triangleq \begin{bmatrix} P_1 & P_2^T \\ P_2 & P_3 \end{bmatrix} \text{ for a symmetric matrix.}$$

$\|y\|$: the Euclidean norm of a vector y; $\|A\| \triangleq \sup_{y \neq 0} \frac{\|A_y\|}{\|y\|} = \sqrt{\lambda_{\max}\left(A^T A\right)} = \sigma_{\max}(A)$, where $\sigma_{\max}(A)$ is the maximum singular value of A. Let $L_2\left(U \times \left[0, t_f\right]; \mathbb{R}^{n_h}\right)$ denote the family of measurable functions $h(x,t): U \times \left[0, t_f\right] \rightarrow \mathbb{R}^{n_h}$) such that $E\int_0^{t_f} \int_U \|h(x,t)\|^2 dxdt \triangleq \int_\Omega \int_0^{t_f} \int_U \|h(x,t,\omega)\|^2 P(d\omega) < \infty$, where $E\{\cdot\}$ represents the mathematical expectation and $\left\{h(x,t), t \in \left[0, t_f\right]\right\}$ is a stochastic process at the space location $x \in U$ and a function of three arguments, i.e., $h(x,t,\omega), x \in U$, $t \in \left[0, t_f\right], \omega \in \Omega$, with Ω being the sample space.

4.2 PROBLEM STATEMENT AND PRELIMINARIES

Consider the following NSPDS with a set of sensor measurements:

$$\frac{\partial y(x,t)}{\partial t} = \kappa\left(y(x,t)\right)\nabla^2 y(x,t) + f\left(y(x,t)\right) + g_v v(x,t) + f_W\left(y(x,t)\right)w(x,t)$$

$$z(t) = \left[h_1\left(y(q_1,t) + D_n n(q_1,t), \ldots, h_{n_z}\right)\left(y(q_{n_z},t) + D_n n(q_{n_z},t)\right)\right]^T \tag{4.1}$$

for $x = [x_1, x_2]^T \in U = [x_{0,1}, x_{f,1}] \times [x_{0,2}, x_{f,2}] \in \mathbb{R}^2$ and $t>0$, where $y(x,t) \triangleq [y_1(x,t), \ldots, y_n(x,t)]^T \in \mathbb{R}^n$ is the system state variable; and x and t are the space and time variables, respectively. The space domain U is a bounded domain. $f(y(x,t)) \in \mathbb{R}^n$ and $f_W(y(x,t)) \in \mathbb{R}^n$ are nonlinear functions with $f(0)=0$ and $f_W(0)=0$, respectively [108]. The nonlinear diffusion function $\kappa\left(y(x,t)\right) \in \mathbb{R}^{n \times n}$ satisfies $\kappa\left(y(x,t)\right) \geq 0$ and $\kappa(0)=0$. The measured output $z(t) \triangleq \left[z_1(t), \ldots, z_{n_z}(t)\right]^T \in \mathbb{R}^{n_z}$ may be interpreted as sensor observations, where n_z is the number of sensors. $h_i \triangleq h(q_i) \in \mathbb{R}^{1 \times n}$ is the sensor observation influence vector at the sensor location q_i of observation for $i = 1, \ldots, n_z$. $y(q_i,t) \in \mathbb{R}^n$ is the state variable at the sensor location q_i of observation, where $i = 1, \ldots, n_z$. $n(q_i,t) \in L_2\left(U \times [0,t_f]; \mathbb{R}^n\right)$ is the stochastic spatial-temporal measurement noise at the location q_i of observation for $i = 1, \ldots, n_z$ and $D_n \in \mathbb{R}^{n \times n}$ is the noise influence matrix. $v(x,t) \in L_2\left(U \times [0,t_f]; \mathbb{R}^{n_v}\right)$ is the vector of the random external disturbance in the spatio-temporal domain and $g_v \in \mathbb{R}^{n \times n_v}$ is the disturbance influence matrix. $w(x,t) \in \mathbb{R}$ is a stationary spatio-temporal white noise. The Laplace operator ∇^2 is defined as follows:

$$\nabla^2 y(x,t) \triangleq \sum_{k=1}^{2} \frac{\partial^2}{\partial x_k^2} y(x,t)$$

where $\frac{\partial^2}{\partial x_k^2} y(x,t) \triangleq \left[\frac{\partial^2}{\partial x_k^2} y_1(x,t), \ldots, \frac{\partial^2}{\partial x_k^2} y_n(x,t)\right]^T \in \mathbb{R}^n$. The initial value is given by $y(x, 0)=y_0(x)$. For simplicity, the boundary condition is usually given by the Dirichlet boundary condition, i.e., $y(x,t) = a$ constant on ∂U, or by the Neumann boundary condition $\nabla y(x,t) \cdot \vec{n} = 0$ on ∂U, where $\vec{n}$ is the normal vector to the boundary ∂U.

Remark 4.2.1

(i) If the equilibrium point of interest is not at $y(x,t) = 0$, i.e., $\kappa(0) \neq 0, f(0) \neq 0, f_W(0) \neq 0$, then for the simplicity of filter design it should be shifted to $y(x,t) = 0$. (ii) If the condition $\nabla^2 y(x,t) = 0$ at $y(x,t) = 0$ can be satisfied, then the condition $\kappa(0) = 0$ can be relaxed to be $\kappa(0) \geq 0$. This is because $f(0) = 0$, $f_W(0) = 0$, and the condition $\nabla^2 y(x,t) = 0$ at $y(x,t) = 0$ together guarantee $y(x,t) = 0$ is the equilibrium point of the system equation (4.1). (iii) The stochastic state equation in (4.1) can be written in the Itô differential form as follows:

$$dy(x,t)=\kappa\left(y(x,t)\right)\nabla^2 y(x,t)+f\left(y(x,t)\right)+g_v v(x,t)dt+f_W(y(x,t))dW(x,t)$$
$$z(t)=\left[h_1\left(y(q_1,t)+D_n n(q_1,t)\right),\ldots,h_{n_z}\left(y(q_{n_z},t)+D_n n(q_{n_z},t)\right)\right]^T \tag{4.2}$$

where $dW(x,t)=w(x,t)dt$, and $W(x,t)$ is the Wiener process that is a zero mean Gaussian random field with variance at each location [108].

Next, let us consider the following NSPDS without external disturbance $v(x, t)$:

$$dy(x,t)=\kappa\left(y(x,t)\right)\nabla^2 y(x,t)+f\left(y(x,t)\right)dt+f_W\left(y(x,t)\right)dW(x,t)$$
$$\kappa(0)=0, f(0)=0, f_W(0)=0 \tag{4.3}$$

Lemma 4.2.1 (Uniqueness and Existence of Solutions [107–109])

Suppose the following conditions hold:

1. $\|\kappa(r_1)-\kappa(r_2)\|+\|f(r_1)-f(r_2)\|+\|f_W(r_1)-f_W(r_2)\|\le C\|r_1-r_2\|$ for all r_1 and r_2, where C is a positive constant.
2. $\|\kappa(r)\|+\|f(r)\|+\|f_W(r)\|\le D(1+\|r\|)$ for all r with some positive constant D.

Then there exists a unique solution of the system in equation (4.3).

Definition 4.2.1

The equilibrium point $y(x,t)=0$ of the NSPDS in equation (4.3) is said to be stochastically asymptotically stable if $E\{y^T(x,t)y(x,t)\}\to 0$ as $t\to\infty$, $\forall x\in U$.

Suppose that $V(y(x,t))$ is a positive definite function. Then we have [107,108,130]

$$\int_U V\left(y(x,t)\right)dx=\int_U V\left(y(x,0)\right)dx+\int_0^t LV(y)ds$$
$$+\int_0^t\int_U\left(\frac{\partial V\left(y(x,s)\right)}{\partial y}\right)^T f_W\left(y(x,s)\right)dW(x,s) \tag{4.4}$$

where the infinitesimal operator $LV(y)$ is denoted as

$$LV(y)\triangleq\int_U\left(\frac{\partial V\left(y(x,s)\right)}{\partial y}\right)^T(\kappa(y(x,s))\nabla^2 y(x,s)+Ay(x,s))dx$$
$$+\int_U\frac{1}{2}f_W\left(y(x,s)\right)^T\frac{\partial^2 V\left(y(x,s)\right)}{\partial y^2}f_W\left(y(x,s)\right)dx \tag{4.5}$$

Lemma 4.2.2 [107,108,130]

A positive definite function $V(y(x,t)): \mathbb{R}^n \to \mathbb{R}$ is said to be a Lyapunov function for the system in equation (4.3), if (i) $V(0) = 0$ and $V\left(y(x,t)\right) > 0$ for $y(x,t) \neq 0$, and (ii) ELV $(y) \leq 0$.

Theorem 4.2.1 [108,130]

Suppose the system in equation (4.3) has a solution *y(x, t)* and there exists a Lyaponov function *V*(*y*(*x, t*)). Furthermore, $ELV(y) < 0$, then the equilibrium point $y(x, t) = 0$ of the system in equation (4.3) is stochastically asymptotically stable.

Since this chapter deals with the asymptotical stochastic filtering problem, it is inevitably related to the stochastic stability, i.e., the asymptotical stability in probability or the exponentially mean square stability [107,108] of the estimation error system.

Lemma 4.2.3 [131]

For any matrices (or vectors) M_1 and M_2 with appropriate dimensions and any positive-definite symmetric matrix M, we have

$$M_1^T M_2 + M_2^T M_1 \leq M_1^T M M_1 + M_2^T M^{-1} M_2$$

Corollary 4.2.1

$X^T PY + Y^T PX \leq \xi X^T PX + \xi^{-1} Y^T PY$ for any constant ξ, a symmetric $P = P^T \geq 0$ and two matrices (or vectors) X and Y with appropriate dimensions.

Proof

This corollary is obviously true from Lemma 4.2.3.

In the NSPDS in equations (4.1) or (4.2), if the state *y*(*x, t*) is unavailable, we can only measure the system via *z*(*x, t*) by a set of sensors at some locations. Therefore, how to estimate *y*(*x, t*) from the measurement *z*(*x, t*) is an important topic for the robust filtering problem of NSPDSs. Suppose the following state estimation is proposed for the NSPDS in equation (4.2):

$$d\hat{y}(x,t) = \left\{ \begin{aligned} &\kappa\left(\hat{y}(x,t)\right)\nabla^2 \hat{y}(x,t) + f\left(\hat{y}(x,t)\right) + L\left(\hat{y}(x,t)\right) \\ &\qquad \times \left[z(t) - \left[h_1 \hat{y}(q_1,t), \ldots, h_{n_z} \hat{y}(q_{n_z},t) \right]^T \right] \end{aligned} \right\} dt \tag{4.6}$$

where $L(\hat{y}(x,t))$ denotes the filter gain to be designed to achieve the H_∞ state estimation performance and $\hat{y}(x,t)$ is the estimated system state and $z(t) \triangleq \left[z_1(t),\ldots,z_{n_z}(t)\right]^T$ is the measured output by n_z sensors. Let us denote

$$e(x,t) \triangleq y(x,t) - \hat{y}(x,t) \tag{4.7}$$

where $y(x, t)$ is the actual system state and $\hat{y}(x,t)$ is the estimated system state.

From equations (4.2) and (4.6), we obtain the following estimation error:

$$de(x,t) = \begin{Bmatrix} \kappa\left(y(x,t)\right)\nabla^2 y(x,t) - \kappa\left(\hat{y}(x,t)\right)\nabla^2\hat{y}(x,t) + f\left(y(x,t)\right) - f\left(\hat{y}(x,t)\right) \\ -L(\hat{y}(x,t)\left[h_1\left(y(q_1,t)-\hat{y}(q_1,t)\right),\ldots,h_{n_z}\left(y(q_{n_z},t)-\hat{y}(q_{n_z},t)\right)\right]^T \\ -L(\hat{y}(x,t)\left[h_1 D_n n(q_1,t),\ldots,h_{n_z}D_n n(q_{n_z},t)\right]^T + g_v v(x,t) \end{Bmatrix} dt$$
$$+ f_W(y(x,t))dW(x,t) \tag{4.8}$$

$\forall x \in U$. Then the augmented system can be represented as

$$d\bar{y}(x,t) = \left[F(\bar{y}(x,t),\nabla^2\bar{y}(x,t)) + F_W(\bar{y}(x,t))dW(x,t) + \tilde{G}_v\bar{v}(x,t)\right] \tag{4.9}$$

where

$$\bar{y}(x,t) \triangleq \begin{bmatrix} y(x,t) \\ e(x,t) \end{bmatrix}, \bar{v}(x,t) \triangleq \begin{bmatrix} v(x,t) \\ n(q_1,t) \\ \vdots \\ n^T(q_{n_z},t) \end{bmatrix}, F_W(\bar{y}(x,t)) = \begin{bmatrix} f_W y(x,t) \\ f_W y(x,t) \end{bmatrix},$$

$$F(\bar{y}(x,t),\nabla^2\bar{y}(x,t))$$
$$\triangleq \begin{bmatrix} \kappa(y(x,t))\nabla^2 y(x,t) + f(y(x,t)) \\ \{\kappa(y(x,t))\nabla^2 y(x,t) - \kappa(\hat{y}(x,t))\nabla^2\hat{y}(x,t) + f(y(x,t)) - f(\hat{y}(x,t)) \\ -L(\hat{y}(x,t)[h_1(y(q_1,t)-\hat{y}(q_1,t)),\ldots,h_{n_z}(y(q_{n_z},t)-\hat{y}(q_{n_z},t))]^T \end{bmatrix},$$

$$\tilde{G}_v \triangleq \begin{bmatrix} g_v & 0 & \cdots & 0 \\ g_v & -L(\hat{y}(x,t)\begin{bmatrix} h_1 \\ 0 \\ \vdots \\ 0 \end{bmatrix} D_n & \cdots & -L(\hat{y}(x,t)\begin{bmatrix} 0 \\ \vdots \\ 0 \\ h_{n_z} \end{bmatrix} D_n \end{bmatrix}.$$

Definition 4.2.2

In the case of $v(x, t)=0$, the NSPDSs are said to be stochastically asymptotically estimated if $E\{e^T(x,t)e(x,t)\} \to 0$ for all $x \in U$ as $t \to \infty$.

Now we solve the H_∞ estimation problem for NSPDS as follows.

Theorem 4.2.2

For the augmented system in equation (4.9), we may find the filter gain $L(\hat{y}(x,t)$ and a positive definite function $V(\bar{y}(x,t))$ with $V(0)=0$ for the partial differential state estimator in equation (4.6) such that the following HJII holds:

$$\int_U \Bigg\{ \left(\frac{\partial V(\bar{y}(x,t))}{\partial \bar{y}}\right)^T \left(F(\bar{y}(x,t)), \nabla^2(\bar{y}(x,t))\right) + \bar{y}^T(x,t)Q^T Q\bar{y}(x,t)$$

$$+\frac{1}{4\gamma^2}\left(\frac{\partial V(\bar{y}(x,t))}{\partial \bar{y}}\right)^T \tilde{G}_v \tilde{G}_v^T \left(\frac{\partial V(\bar{y}(x,t))}{\partial \bar{y}}\right)$$

$$+\frac{1}{2}F_W^T(\bar{y}(x,t))\frac{\partial^2 V(\bar{y}(x,t))}{\partial \bar{y}^2}F_W(\bar{y}(x,t))\Bigg\} dx < 0 \tag{4.10}$$

In this case, (i) the equilibrium point $\bar{y}(x,t)=0$ of the augmented system in equation (4.9) is stochastically asymptotically stable in the case of $\bar{v}(x,t)=0$, i.e., $E\{\bar{y}^T(x,t)\bar{y}(x,t)\} \to 0$ as $t_f \to \infty$ and the NSPDS in equation (4.2) is stochastically asymptotically estimated, i.e., $E\{e^T(x,t)e(x,t)\} \to 0$ as $t_f \to \infty$; and (ii) for some $\gamma > 0$, $\forall \bar{v}(x,t) \in L_2\left(U \times [0, t_f]; R^{nv+nn}\right)$, $\bar{v}(x,t) \neq 0$, the following spatio-temporal H_∞ state estimation performance holds:

$$E\int_U \int_0^{t_f} \|e(x,t)\|^2\, dt\, dx < E\int_U V(\bar{y}(x,0))\, dx + \gamma^2 E\int_U \int_0^{t_f} \|\bar{v}(x,t)\|^2\, dt\, dx \tag{4.11}$$

if the initial state $\bar{y}(x,0) \neq 0$ or

$$E\int_U \int_0^{t_f} \|e(x,t)\|^2\, dt\, dx < \gamma^2 E\int_U \int_0^{t_f} \|\bar{v}(x,t)\|^2\, dt\, dx \tag{4.12}$$

if the initial state $\bar{y}(x,0)=0$.

Proof

Let us choose the Lyapunov function $V(\bar{y}(x,t)) > 0, \forall \bar{y}(x,t) \neq 0$, and $V(0)=0$. First, the infinitesimal operator $LV(\bar{y})$ for the system equation (4.9) is denoted as

$$L_{\bar{v}}V(\bar{y}) = \int_U \left\{ \left(\frac{\partial V(\bar{y}(x,t))}{\partial \bar{y}} \right)^T \left(F(\bar{y}(x,t)), \nabla^2(\bar{y}(x,t)) + \tilde{G}_v \bar{v}(x,t) \right) + \frac{1}{2} F_W^T(\bar{y}(x,t)) \frac{\partial^2 V(\bar{y}(x,t))}{\partial \bar{y}^2} F_W(\bar{y}(x,t)) \right\} dx \tag{4.13}$$

From the inequality equation (4.10), we have

$$EL_{\bar{v}=0}V(\bar{y}) = E\left\{ \int_U \left\{ \left(\frac{\partial V(\bar{y}(x,t))}{\partial \bar{y}} \right)^T F(\bar{y}(x,t)), \nabla^2(\bar{y}(x,t)) + \frac{1}{2} F_W^T(\bar{y}(x,t)) \frac{\partial^2 V(\bar{y}(x,t))}{\partial \bar{y}^2} F_W(\bar{y}(x,t)) \right\} dx \right\}$$
$$< E\int_U \left\{ -\bar{y}^T(x,t)Q^T Q\bar{y}(x,t) - \frac{1}{4\gamma^2} \left(\frac{\partial V(\bar{y}(x,t))}{\partial \bar{y}} \right)^T \tilde{G}_v \tilde{G}_v^T \left(\frac{\partial V(\bar{y}(x,t))}{\partial \bar{y}} \right) \right\} dx$$
$$\leq 0 \tag{4.14}$$

i.e.,

$$EL_{\bar{v}=0}V(\bar{y}) < 0 \tag{4.15}$$

Therefore, by Theorem 4.2.1, the equilibrium point $\bar{y}(x,t) = 0$ of the system in equation (4.9) in the case of $\bar{v}(x,t) = 0$ is stochastically asymptotically stable. Hence, $E\{\bar{y}^T(x,t)\bar{y}(x,t)\} \to 0 \text{ as } t_f \to \infty$, $\forall x \in U$. Then

$$E\{e^T(x,t)e(x,t)\} = E\left\{\bar{y}^T(x,t)Q^T Q\bar{y}(x,t)\right\} \to 0$$

as $t_f \to \infty$. Next, for the augmented system in equation (4.9) with $\bar{v}(x,t) \neq 0$, by equation (4.9) and Itô formula, we have

$$E\int_U \int_0^{t_f} \|e(x,t)\|^2 \,dt\,dx = E\int_U \Big[V(\bar{y}(x,0)) - V(\bar{y}(x,t_f)) + \int_0^{t_f} \left(\|e(x,t)\|^2 + \frac{\partial V(\bar{y}(x,t))}{\partial t} \right) dt \Big] dx < E\int_U V(\bar{y}(x,0))\,dx$$
$$+E\int_U \int_0^{t_f} \bar{y}^T(x,t)Q^T Q\bar{y}(x,t) + \left(\frac{\partial V(\bar{y}(x,t)}{\partial \bar{y}} \right)^T$$
$$\left(F(\bar{y}(x,t), \nabla^2 \bar{y}(x,t)) + \tilde{G}_v \bar{v}(x,t) \right) + \frac{1}{2} F_W^T(\bar{y}(x,t)) \frac{\partial^2 V(\bar{y}(x,t))}{\partial \bar{y}^2} F_W(\bar{y}(x,t)) \Big] dt dx \tag{4.16}$$

By Corollary 4.2.1, we have the fact that

$$\frac{1}{2}\left(A^T B+B^T A\right)\leq\frac{1}{4\gamma^2}A^T A+\gamma^2 B^T B \tag{4.17}$$

for all matrices, vectors A, B, and a positive scalar γ. Then, by equation (4.17) with $A=\left(\frac{\partial V(\bar{y}(x,t))}{\partial\bar{y}}\right)^T\tilde{G}_v$ and $B=\bar{v}(x,t)$, the inequality equation (4.16) can be modified as the following inequality:

$$\begin{aligned}
&E\int_U\int_0^{t_f}\|e(x,t)\|^2\,dt\,dx\\
&\leq E\int_U V(\bar{y}(x,0))\,dx+E\int_U\int_0^{t_f}\Bigg[\left(\frac{\partial V(\bar{y}(x,t))}{\partial\bar{y}}\right)^T F(\bar{y}(x,t),\nabla^2\bar{y}(x,t))\\
&+\frac{1}{4\gamma^2}\left(\frac{\partial V(\bar{y}(x,t))}{\partial\bar{y}}\right)^T\tilde{G}_v\tilde{G}_v^T\left(\frac{\partial V(\bar{y}(x,t))}{\partial\bar{y}}\right)+\gamma^2\bar{v}^T(x,t)\bar{v}(x,t)\\
&+\bar{y}^T(x,t)Q^T Q\bar{y}(x,t)+\frac{1}{2}F_W^T(\bar{y}(x,t))\frac{\partial^2 V(\bar{y}(x,t))}{\partial\bar{y}^2}F_W(\bar{y}(x,t))\Bigg]dtdx
\end{aligned} \tag{4.18}$$

Therefore, by the inequality in equation (4.10), we get

$$E\int_U\int_0^{t_f}\|e(x,t)\|^2\,dt\,dx<E\int_U V(\bar{y}(x,0))\,dx+\gamma^2 E\int_U\int_0^{t_f}\|\bar{v}(x,t)\|^2\,dtdx \tag{4.19}$$

if the initial state $\bar{y}(x,0)\neq 0$ or

$$E\int_U\int_0^{t_f}\|e(x,t)\|^2\,dt\,dx<\gamma^2 E\int_U\int_0^{t_f}\|\bar{v}(x,t)\|^2\,dtdx \tag{4.20}$$

if the initial state $\bar{y}(x,0)=0$.

The physical meaning in equation (4.12) is that the effect of disturbances on the estimation error in spatio-temporal domain is less than the prescribed value γ. For the robust state estimation of NSPDSs, the robust stochastic H_∞ filter design needs to solve a complex second-order HJII in equation (4.10) which involves an integration of the HJI over the space domain U. In general, it is very difficult to solve the second-order nonlinear partial differential HJII. Furthermore, even though the filter gain $L(\hat{y}(x,t))$ is solved from the complex HJII in equation (4.10), it is still difficult to implement the partial differential state estimator in equation (4.6). Therefore, an H_∞ robust estimator that is simpler and more easily implemented for NSPDSs is proposed via T-S fuzzy interpolation method in the following section.

4.3 IMPLEMENTABLE H_∞ FUZZY FILTER FOR NONLINEAR STOCHASTIC PARTIAL DIFFERENTIAL SYSTEMS

To implement the robust H_∞ state estimator of NSPDSs in equation (4.2), the main problem is obtaining a suitable state space model to represent the NSPDSs. Initially, a FSPDS is proposed to approximate the NSPDS as follows:

Rule i: If $y_1(x, t)$ is F_{1i} and ⋯ and $y_n(x, t)$ is F_{ni}

then

$$dy(x,t) = (\kappa_i \nabla^2 y(x,t) + A_i y(x,t) + g_v v(x,t))dt + A_{Wi} y(x,t) dW(x,t) \tag{4.21}$$

for $i = 1,\dots, M$, where F_{ji} is the grade of the membership of $y_j(x, t)$; $\kappa_i \in \mathbb{R}^{n\times n}$, $A_i \in \mathbb{R}^{n\times n}$ and $A_{Wi} \in \mathbb{R}^{n\times n}$ are system parameters of local LSPDSs. Therefore, by equation (4.21), the overall FSPDS can be inferred as follows:

$$dy(x,t) = \sum_{i=1}^{M} \mu_i(y(x,t))[(\kappa_i \nabla^2 y(x,t) + A_i y(x,t))dt + A_{Wi} y(x,t) dW(x,t)] + g_v v(x,t)dt \tag{4.22}$$

where

$$\mu_i(y(x,t)) = \frac{\prod_{j=1}^{n} F_{ji}(y_i(x,t))}{\sum_{i=1}^{M}\prod_{j=1}^{n} F_{ji}(y_i(x,t))}$$

with $F_{ji}(y_i(x, t))$ being the grade of the membership function of $y_j(x, t)$. The denominators of $\mu_i(y(x,t))$ are only for normalization so that the total sum of the fuzzy bases $\sum_{i=1}^{M} \mu_i(y(x,t)) = 1$. Then the NSPDS in equation (4.2) can be represented as

$$\begin{aligned} dy(x,t) = &\sum_{i=1}^{M} \mu_i(y(x,t))[(\kappa_i \nabla^2 y(x,t) + A_i y(x,t))dt + A_{Wi} y(x,t) dW(x,t)] \\ &+ g_v v(x,t)dt + \varepsilon_\kappa(y(x,t))\nabla^2 y(x,t)dt + \varepsilon_f(y(x,t))dt \\ &+ \varepsilon_{fw}(y(x,t))dW(x,t). \end{aligned} \tag{4.23}$$

In equation (4.23), the fuzzy interpolation via M local LSPDSs in equation (4.22) is used to approximate the NSPDS in equation (4.2). The approximation errors $\varepsilon_\kappa(y(x,t))$, $\varepsilon_f(y(x,t))$, and $\varepsilon_{fw}(y(x,t))$ are defined as follows:

$$\varepsilon_{\kappa}(y(x,t)) \triangleq \kappa(y(x,t)) - \sum_{i=1}^{M} \mu_i(y(x,t))\kappa_i$$

$$\varepsilon_f(y(x,t)) \triangleq f(y(x,t)) - \sum_{i=1}^{M} \mu_i(y(x,t))A_i y(x,t)$$

$$\varepsilon_{fw}(y(x,t)) \triangleq f_W(y(x,t)) - \sum_{i=1}^{M} \mu_i(y(x,t))A_{Wi} y(x,t) \tag{4.24}$$

Assumption 4.3.1

Assume the fuzzy approximation errors in equation (4.24) are bounded by, $E\|\varepsilon_{\kappa}(y(x,t))\|^2 \le \sigma_{\kappa}^2$, $E\|\varepsilon_f(y(x,t))\|^2 \le \sigma_f^2 E\|y(x,t)\|^2$, *and* $E\|\varepsilon_{fw}(y(x,t))\|^2 \le \sigma_{fw}^2 E\|y(x,t)\|^2$, *where* σ_{κ}, σ_f *and* σ_{fw} *are the positive scalar values and are dependent on the number M of the fuzzy rules.*

After employing the fuzzy interpolation method to approximate the nonlinear functions of NSPDS, the finite difference scheme in Section 1.2.2 is employed to approximate the partial differential operator $\nabla^2 y(x,t)$ in equation (4.23) to simplify the design procedure for the state estimator of NSPDS. Consider a typical mesh as shown in Figure 4.1. The state $y(x,t)$ is represented by $y_{k,l}(t) \in \mathbb{R}^n$ at the grid node $x_{k,l}(x_1 = k\Delta, x_2 = l\Delta)$, where $k = 1,\ldots, N_1$ and $l = 1,\ldots, N_2$, i.e., $y(x,t)\big|_{x=x_{k,l}} \triangleq y_{k,l}(t)$. At the grid point $x_{k,l}$, the central difference approximation for the partial differential operator can be written as follows [111,132]:

$$\kappa_i \nabla^2 y(x,t) = \kappa_i \frac{y_{k+1,l}(t) + y_{k-1,l}(t) - 2y_{k,l}(t)}{\Delta^2} + \kappa_i \frac{y_{k,l+1}(t) + y_{k,l-1}(t) - 2y_{k,l}(t)}{\Delta^2} + O_{k,l}(\Delta^2) \tag{4.25}$$

where $O_{k,l}(\Delta^2) \in \mathbb{R}^n$ is called the local truncation error.

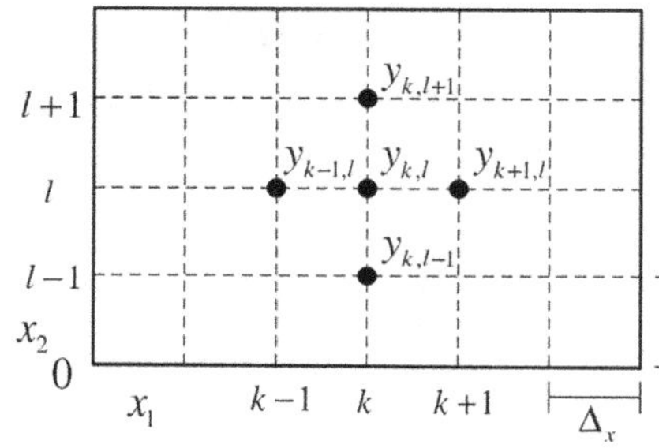

FIGURE 4.1 Finite difference grids on the spatio domain.

Remark 4.3.1

It has been proven [111] that if grid points in a spatial domain are dense enough then the truncation error $O_{k,l}(\Delta^2)$ will be small enough.

Based on the finite difference approximation in equation (4.25), the NSPDS in equation (4.23) can be represented by the following fuzzy stochastic finite difference system:

$$
\begin{aligned}
dy_{k,l}(t) = \sum_{i=1}^{M} \mu_i\left(y_{k,l}(t)\right)\Big\{ &\kappa_i \frac{1}{\Delta^2}\left[y_{k+1,l}(t) + y_{k-1,l}(t) y_{k,l+1}(t) + y_{k,l-1}(t)\right. \\
&\left. -4y_{k,l}(t)\right]dt + A_i y_{k,l}(t)dt + A_{Wi} y_{k,l}(t)dW_{k,l}(t)\Big\} + g_v v_{k,l}(t)dt \\
&+ \varepsilon_\kappa(y_{k,l}(t))\frac{1}{\Delta^2}\left[y_{k+1,l}(t) + y_{k-1,l}(t) y_{k,l+1}(t) + y_{k,l-1}(t)\right. \\
&\left. -4y_{k,l}(t)\right]dt + \varepsilon_f(y_{k,l}(t))dt + \varepsilon_{fw}(y_{k,l}(t))dW_{k,l}(t) + O_{k,l}(\Delta^2)
\end{aligned} \quad (4.26)
$$

where $y_{k,l}(t) = y(x,t)\big|_{x=x_{k,l}}$, $v_{k,l}(t) = v(x,t)\big|_{x=x_{k,l}}$, and $W_{k,l}(t) = W(x,t)\big|_{x=x_{k,l}}$. Let us denote

$$
T_{k,l} y_{k,l}(t) = \frac{1}{\Delta^2}\left[y_{k+1,l}(t) + y_{k-1,l}(t) y_{k,l+1}(t) + y_{k,l-1}(t) - 4y_{k,l}(t)\right] \quad (4.27)
$$

then

$$
\begin{aligned}
dy_{k,l}(t) = \sum_{i=1}^{M} \mu_i\left(y_{k,l}(t)\right)&\left\{\left[\kappa_i T_{k,l} y_{k,l}(t) + A_i y_{k,l}(t)\right]dt + A_{Wi} y_{k,l}(t)dW_{k,l}(t)\right\} \\
&+ g_v v_{k,l}(t)dt + \varepsilon_\kappa(y_{k,l}(t))T_{k,l} y_{k,l}(t)dt + \varepsilon_f(y_{k,l}(t))dt \\
&+ \varepsilon_{fw}(y_{k,l}(t))dW_{k,l}(t) + O_{k,l}(\Delta^2)
\end{aligned} \quad (4.28)
$$

To simplify the state estimation design for NSPDSs, we define a spatial state vector $y(t)$ to collect the state variables $y_{k,l}(t) \in \mathbb{R}^n$ at all grid nodes in Figure 4.1. For Dirichlet boundary conditions in equation (1.10) [132], the values $y_{k,l}(t)$ at the boundary are fixed, for example, $y(x,t) = 0$ on ∂U. We have $y_{k,l}(t) = 0$ at $k = 0$, N_1+1, *or* $l = 0$, N_2+1. Therefore, the spatial state vector $y(t) \in \mathbb{R}^{nN}$ for the state variables at all grid nodes is defined as follows:

$$
\begin{aligned}
y(t) = y_{1,1}^T(t), \ldots, y_{k,1}^T(t), \ldots, y_{N_1,1}^T(t), \ldots, y_{k,l}^T(t), \ldots, y_{1,N_2}^T(t), \ldots, \\
y_{k,N_2}^T(t), \ldots, y_{N_1,N_2}^T(t)]^T
\end{aligned} \quad (4.29)
$$

where $N \triangleq N_1 \times N_2$. Note that n is the dimension of the vector $y_{k,l}(t)$ for each grid node and $N_1 \times N_2$ is the number of grid nodes. For example, if $N_1 = 2$ and $N_2 = 2$, then we have $\left[y_{1,1}^T(t), y_{2,1}^T(t), y_{1,2}^T(t), y_{2,2}^T(t)\right]^T \in \mathbb{R}^{4n}$. In order to simplify the index of the node $y_{k,l}(t) \in \mathbb{R}^n$ in the spatial state vector $y(t) \in \mathbb{R}^{nN}$, we use the symbol $y_j(t) \in \mathbb{R}^n$ to replace $y_{k,l}(t)$. Note that the index j is from 1 to N, i.e., $y_1(t) \triangleq y_{1,1}(t)$, $y_2(t) \triangleq y_{2,1}(t),\dots, y_j(t) \triangleq y_{k,l}(t),\dots, y_N(t) \triangleq y_{N_1,N_2}(t)$, where $j = (l-1)N_1 + k$ in equation (4.29). The fuzzy stochastic difference model of two indices in equation (4.28) could be represented with only one index as follows:

$$
\begin{aligned}
dy_j(t) = &\sum_{i=1}^{M} \mu_i\left(y_j(t)\right)\left\{\left[\kappa_i T_j y(t) + A_i y_j(t)\right]dt + A_{Wi} y_j(t) dW_j(t)\right\} \\
&+ g_v v_j(t)dt + \varepsilon_\kappa(y_j(t))T_j y(t)dt + \varepsilon_f(y_j(t))dt \\
&+ \varepsilon_{fw}(y_j(t))dW_j(t) + O_j(\Delta^2)
\end{aligned} \tag{4.30}
$$

where $v_j(t) = v_{k,l}(t)$, $dW_j(t) = dW_{k,l}(t)$, $O_j(\Delta^2) = O_{k,l}(\Delta^2)$ with $j = (l-1)N_1 + k$ and $T_j y(t)$ is defined as follows:

$$
T_j y(t) = \frac{1}{\Delta^2}\left[0_n \cdots 0_n \; I_n \; 0_n \cdots 0_n \; I_n - 4I_n \;\; I_n \; 0_n \cdots 0_n \; I_n \; 0_n \cdots 0_n\right] y(t),
$$

$$
\text{position} \rightarrow \quad 1 \qquad j\text{-}N_1 \qquad j\text{-}1 \qquad j+1 \qquad j+N_1 \quad N_1N_2
$$

with 0_n denoting the $n \times n$ zero matrix and I_n denoting the $n \times n$ identity matrix. The measurement output in equation (4.1) can be represented as follows:

$$
z(t) = \sum_{j=1}^{N}\left[C_j y_j(t) + D_j n_j(t)\right] \tag{4.31}
$$

where the matrix C_j is defined as $C_j \triangleq C_{\{k,l\}} = \left[C_{\{k,l\},1}^T, \dots, C_{\{k,l\},n_z}^T\right]^T \in \mathbb{R}^{n_z \times n}$ with $j = (l-1)N_1 + k$, in which the element is given as $C_{\{k,l\},i} = h_i$ for $x_{k,l} = q_i$ or $C_{\{k,l\},i} = 0$ for $x_{k,l} \neq q_i$; the matrix D_j is defined as $D_j \triangleq D_{\{k,l\}} = \left[D_{\{k,l\},1}^T, \dots, D_{\{k,l\},n_z}^T\right]^T \in \mathbb{R}^{n_z \times n}$ with $j = (l-1)N_1 + k$, in which the element is given as $D_{\{k,l\},i} = h_i D_n$ for $x_{k,l} = q_i$ or $D_{\{k,l\},i} = 0$ for $x_{k,l} \neq q_i$. Note that q_i is the ith sensor location as defined in equation (4.1). The jth measurement noise is defined as $n_j \triangleq n(x,t)\big|_{x = x_{k,l}}$, where $j = (l-1)N_1 + k$.

We collect all the states $y_j(t)$ of grid nodes in equations (4.30) and (4.31) to the state $y(t)$ in (4.29). Then a diagonal fuzzy weighting matrix $\mu_i(y(t)) \in \mathbb{R}^{N \times N}$ is defined by $\mu_i(y(t)) \triangleq diag(\mu_i(y_1(t)),\dots,\mu_i(y_N(t)))$. The Kronecker product can be used to simplify the representation. Some properties of $\mu_i(y(t))$ can be obtained as follows.

Lemma 4.3.1

Using the properties of Kronecker product, we have the following properties:

Property 1:

$$\begin{aligned}\mu_i(y)\otimes A_i &= (\mu_i(y)\otimes I_n)(I_N\otimes A_i)\\ &= (I_N\otimes A_i)(\mu_i(y)\otimes I_n)\in\mathbb{R}^{nN\times nN}\end{aligned} \tag{4.32}$$

where $A_i\in\mathbb{R}^{n\times n}$.

Property 2:

$$(\mu_i(y)\otimes I_n)(\mu_j(\hat{y})\otimes I_n)=(\mu_i(y)\mu_j(\hat{y})\otimes I_n)\in\mathbb{R}^{nN\times nN} \tag{4.33}$$

Property 3:

$$\begin{aligned}&\sum_{i=1}^{M}\mu_i(y)=I_N,\sum_{i=1}^{M}(\mu_i(y)\otimes I_n)=I_{nN}\\ &and\sum_{i=1}^{M}\sum_{j=1}^{M}(\mu_i(y)\mu_j(\hat{y})\otimes I_n)=I_{nN}\end{aligned} \tag{4.34}$$

Proof

The properties 1 and 2 are the fundamental properties for Kronecker product, and property 3 can be proven via $\sum_{i=1}^{M}\mu_i(y_j(t))=1$ for $j=1,\ldots,N$.

Using the Kronecker product, the system in equation (4.30) can be represented as the following fuzzy stochastic spatial state space system:

$$\begin{aligned}dy(t)=\sum_{i=1}^{M}\{&[\mu_i(y(t))\otimes\kappa_i]Ty(t)dt+[\mu_i(y(t))\otimes A_i]y(t)dt\\ &+[\mu_i(y(t))\otimes A_{Wi}]y(t)\circ d\tilde{W}(t)\}+G_v v(t)dt+\varepsilon(y)dt\\ &+\varepsilon_W(y)\circ d\tilde{W}(t)+O(\Delta^2)\end{aligned}$$

or

$$\begin{aligned}dy(t)&=\sum_{i=1}^{M}\bar{\mu}_i(y(t))\{[I_N\otimes\kappa_i]Ty(t)dt+[I_N\otimes A_i]y(t)dt+[I_N\otimes A_{Wi}]y(t)\circ d\tilde{W}(t)\}\\ &\quad+G_v v(t)dt+\varepsilon(y)dt+\varepsilon_W(y)\circ d\tilde{W}(t)+O(\Delta^2)\\ z(t)&=Cy(t)+Dn(t)\end{aligned} \tag{4.35}$$

where $\bar{\mu}_i(y(t)) \triangleq \mu_i(y(t)) \otimes I_n$, $O(\Delta^2) \triangleq [O_1(\Delta^2)^T, ..., O_N(\Delta^2)^T]^T$, $d\tilde{W}(t) \in \mathbb{R}^{nN}$ is defined as $d\tilde{W}(t) \triangleq \left[J^T dW_1(t), ..., J^T dW_N(t)\right]^T$ with $J \triangleq [1, ..., 1]^T \in \mathbb{R}^n$, i.e., all the elements in J are equal to 1, and the spatial state vector $y(t)$ denoted in equation (4.29) represents $y(x, t)$ at all grid points on the spatial domain in Figure 4.1. The corresponding matrices are defined as $T \triangleq \left[T_1^T, ..., T_N^T\right]^T \in \mathbb{R}^{nN \times nN}$, $G_v \triangleq I_N \otimes g_v \in \mathbb{R}^{nN \times n_v N}$, $C \triangleq \left[C_1, ..., C_N\right] \in \mathbb{R}^{n_z \times nN}$, and $D \triangleq \left[D_1, ..., D_N\right] \in \mathbb{R}^{n_z \times nN}$. The random external disturbance $v(t) \triangleq \left[v_1^T(t), ..., v_N^T(t)\right]^T$. The measurement noise $n(t)$ is defined as $n(t) \triangleq \left[n_1^T(t), ..., n_N^T(t)\right]^T \in \mathbb{R}^{nN}$. The approximation error $\varepsilon(y)$ is defined as $n(t) \triangleq \left[n_1^T(t), ..., n_N^T(t)\right]^T \in \mathbb{R}^{nN}$ where $\varepsilon_f(y) \triangleq \left[\varepsilon_f(y_1)^T, ..., \varepsilon_f(y_N)^T\right]$ and $\varepsilon_\kappa(y) = diag\left(\varepsilon_\kappa(y_1), ..., \varepsilon_\kappa(y_N)\right)$. The approximation error $\varepsilon_W(y)$ is defined as $\varepsilon_W(y) \triangleq \left[\varepsilon_{fw}(y_1)^T, ..., \varepsilon_{fw}(y_N)^T\right]^T$. The Hadamard product of matrices (or vectors) $X = [X_{ij}]_{m \times n}$ and $Y = [Y_{ij}]_{m \times n}$ of the same size is the entrywise product denoted as $X \circ Y = \left[X_{ij} Y_{ij}\right]_{m \times n}$.

Remark 4.3.2

(i) The increments of the Wiener processes $dW_1(t), ..., dW_N(t)$ at the N grid nodes have the property: $dW(t) \triangleq [dW_1(t), ..., dW_N(t)]^T \in \mathbb{R}^N$ is a zero-mean vector-process with the covariance $E\{dW(t)dW^T(t)\} = I_N dt$. (ii) The physical meaning of equation (4.35) is that NSPDS in equations (4.1) or (4.2) at all grid points on the spatial domain in Figure 4.1 can be represented by the fuzzy spatial state space model in equation (4.35).

The bounds for the approximation errors $\varepsilon(y)$ and $\varepsilon_W(y)$ in equation (4.35) can be obtained by the following corollary.

Corollary 4.3.1

If the approximation errors $\varepsilon_\kappa(y(x,t)), \varepsilon_f(y(x,t))$, and $\varepsilon_{fw}(y(x,t))$ are bounded with σ_κ, σ_f, and σ_{fw} in Assumption 4.3.1, respectively, then the approximation errors $\varepsilon(y)$ and $\varepsilon_W(y)$ in equation (4.35) are bounded as

$$\begin{aligned} &E\|\varepsilon(y)\|^2 \le \sigma_\kappa^2 E\|Ty(t)\|^2 + \sigma_f^2 E\|y(t)\|^2 \\ &E\|\varepsilon_W(y)\|^2 \le \sigma_{fw}^2 E\|y(t)\|^2 \end{aligned} \tag{4.36}$$

Proof

By Assumption 4.3.1, we have $E\|\varepsilon_{\kappa_j}(y_j)\|^2 \le \sigma_\kappa^2$, $E\|\varepsilon_{f_j}(y_j)\|^2 \le \sigma_f^2 E\|y_j(t)\|^2$, and $E\|\varepsilon_{fw_j}(y_j)\|^2 \le \sigma_{fw}^2 E\|y_j(t)\|^2$. Thus

$$E\|\varepsilon(y)\|^2 = E\|\varepsilon_\kappa(y)Ty(t)+\varepsilon_f(y)\|^2 \le E\|\varepsilon_\kappa(y)Ty(t)\|^2 + E\|\varepsilon_f(y)\|^2$$

$$\le E\|\varepsilon_\kappa(y)\|^2 + E\|Ty(t)\|^2 + \sum_{j=1}^{N} E\|\varepsilon_{f_j}(y_j)\|^2$$

$$\le E\left\|diag\left(\varepsilon_{\kappa 1}(y_1),\dots,\varepsilon_{\kappa N}(y_N)\right)\right\|^2 E\|Ty(t)\|^2 + \sum_{j=1}^{N} \sigma_f^2 E\|y_j(t)\|^2$$

$$\le \sigma_\kappa^2 E\|Ty(t)\|^2 + \sigma_f^2 E\|y(t)\|^2$$

Similarly, the bound of $\varepsilon_W(y)$ can be proved.

Then our robust state estimation design problem of a SPDS becomes how to design a state estimator for the fuzzy stochastic spatial state space system in equation (4.35) to satisfy the H_∞ filtering performance. This is discussed as follows. An implementable fuzzy state estimator based on a fuzzy stochastic spatial state space system to estimate $y(t)$ in equation (4.35) from the measurement $z(t)$ to guarantee the robust H_∞ filtering performance is proposed as follows:

$$d\hat{y}(t) = \sum_{i=1}^{M} \bar{\mu}_i(\hat{y}(t))\left\{[I_N \otimes \kappa_i]T\hat{y}(t) + [I_N \otimes A_i]\hat{y}(t) + L_i(z(t) - C\hat{y}(t)\right\}dt \tag{4.37}$$

where $L_i \in \mathbb{R}^{nN\times n_z}$ are the estimation gains to be designed. The fuzzy weighting matrix $\bar{\mu}_i(\hat{y}(t))$ is defined as $\bar{\mu}_i(\hat{y}(t)) \triangleq (\mu_i(\hat{y}(t)) \otimes I_n)$. The dynamic of the estimated error $e(t) \triangleq y(t) - \hat{y}(t)$ is given by

$$\begin{aligned}
de(t) = &\sum_{i_1=1}^{M} \bar{\mu}_{i_1}(\hat{y}(t))\left\{[I_N \otimes \kappa_{i_1}]Ty(t)dt + [I_N \otimes A_{i_1}]y(t)dt\right.\\
&\left.+[I_N \otimes A_{wi_1}]y(t)\circ d\tilde{W}(t)\right\} + G_v v(t)dt + \varepsilon(y)dt + \varepsilon_W(y)\circ d\tilde{W}(t) + O(\Delta^2)\\
&-\sum_{i_2=1}^{M} \bar{\mu}_{i_2}(\hat{y}(t))\left\{[I_N \otimes \kappa_{i_2}]T\hat{y}(t) + [I_N \otimes A_{i_2}]\hat{y}(t) + L_{i_2}(z(t) - C\hat{y}(t))\right\}dt\\
= &\sum_{i_1=1}^{M}\sum_{i_2=1}^{M} \bar{\mu}_{i_1}(\hat{y}(t))\bar{\mu}_{i_2}(\hat{y}(t))\left\{\left\{[I_N \otimes \kappa_{i_2}]T + [I_N \otimes A_{i_2}] - L_{i_2}C\right\}e(t)dt\right.\\
&+[I_N \otimes (\kappa_{i_1} - \kappa_{i_2})]Ty(t)dt + [I_N \otimes (A_{i_1} - A_{i_2})]y(t)dt + [I_N \otimes A_{wi_1}]y(t)\circ d\tilde{W}(t)\\
&\left.-L_{i_2}D_n(t)dt\right\} + G_v v(t)dt + \varepsilon(y)dt + \varepsilon_W(y)\circ d\tilde{W}(t) + O(\Delta^2)
\end{aligned} \tag{4.38}$$

Combining equation (4.35) with (4.38), we get the following augmented system:

$$d\bar{y}(t)=\bar{A}(\mu,\hat{\mu})\bar{y}(t)dt+\bar{A}w(\mu)\bar{y}(t)\circ d\bar{W}+\bar{G}_v(\hat{\mu})\bar{v}(t)dt+\bar{\varepsilon}(y)dt+\bar{\varepsilon}_W(y)\circ d\bar{W}(t)$$

$$e(t)=\begin{bmatrix} 0_{nN} & I_{nN} \end{bmatrix}\bar{y}(t) \tag{4.39}$$

where

$$\bar{y}(t)\triangleq\begin{bmatrix} y(t) \\ e(t) \end{bmatrix},\bar{\varepsilon}(y)\triangleq\begin{bmatrix} \varepsilon(y) \\ \varepsilon(y) \end{bmatrix},\bar{\varepsilon}_W(y)\triangleq\begin{bmatrix} \varepsilon_W(y) \\ \varepsilon_W(y) \end{bmatrix},\bar{v}(t)\triangleq\begin{bmatrix} v(t) \\ O(\Delta^2) \\ n(t) \end{bmatrix},$$

$$\bar{G}_v(\hat{\mu})\triangleq\sum_{i_2=1}^{M}\bar{\mu}_{i_2}(\hat{y}(t))\bar{G}_{v_{i_2}},\bar{G}_{v_{i_2}}\triangleq\begin{bmatrix} G_v & I_{n\bar{N}} & 0_{n\bar{N}} \\ G_v & I_{n\bar{N}} & -L_{i_2}D \end{bmatrix},d\bar{W}(t)\triangleq\begin{bmatrix} d\tilde{W}(t) \\ d\tilde{W}(t) \end{bmatrix}.$$

$$\bar{A}(\mu,\hat{\mu})\triangleq\begin{bmatrix} \bar{A}_{11}(\mu) & 0_{nN} \\ \bar{A}_{21}(\mu,\hat{\mu}) & \bar{A}_{22}(\hat{\mu}) \end{bmatrix}=\sum_{i_1=1}^{M}\sum_{i_2=1}^{M}\bar{\mu}_{i_1}(y(t))\bar{\mu}_{i_2}(\hat{y}(t))\bar{A}_{i_1i_2} \tag{4.40}$$

$$\bar{A}_W(\mu)\triangleq\sum_{i_1=1}^{M}\bar{\mu}_{i_1}(y(t))\bar{\mu}_{i_2}(\hat{y}(t))\bar{A}_{W_{i_1}}$$

$$\bar{A}_{W_{i_1}}\triangleq\begin{bmatrix} I_N\otimes A_{W_{i_1}} & 0_{nN} \\ I_N\otimes A_{W_{i_1}} & 0_{nN} \end{bmatrix} \tag{4.41}$$

with

$$\bar{A}_{11}(\mu)\triangleq\sum_{i_1=1}^{M}\bar{\mu}_{i_1}(y(t))([I_N\otimes\kappa_{i_1}]T+[I_N\otimes A_{i_1}])$$

$$\bar{A}_{21}(\mu,\hat{\mu})\triangleq\sum_{i_1=1}^{M}\sum_{i_2=1}^{M}\bar{\mu}_{i_1}(y(t))\bar{\mu}_{i_2}(\hat{y}(t))\{[I_N\otimes(\kappa_{i_1}-\kappa_{i_2})]T+[I_N\otimes(A_{i_1}-A_{i_2})]\}$$

$$\bar{A}_{22}(\hat{\mu})\triangleq\sum_{i_2=1}^{M}\bar{\mu}_{i_2}(\hat{y}(t))\{[I_N\otimes\kappa_{i_2}]T+[I_N\otimes A_{i_2}]-L_{i_2}C\}$$

$$\bar{A}_{i_1i_2}\triangleq\begin{bmatrix} [I_N\otimes\kappa_{i_1}]T+[I_N\otimes A_{i_1}] & 0_{nN} \\ \bar{A}_{i_1i_2}^{21} & \bar{A}_{i_1i_2}^{22} \end{bmatrix}$$

where $\bar{A}_{i_1 i_2}^{21} = [I_N \otimes (\kappa_{i_1} - \kappa_{i_2})]T + [I_N \otimes (A_{i_1} - A_{i_2})]$, and $\bar{A}_{i_1 i_2}^{22} = [I_N \otimes \kappa_{i_2}]T + [I_N \otimes A_{i_2}] - L_{i_2}C$.

Since the spatial state vector $y(t)$ in equation (4.29) is used to represent $y(x,t)$ at all grid points, in this situation, $E\int_U \int_0^{t_f} \|e(x,t)\|^2 dtdx$, $E\int_U \int_0^{t_f} \|\bar{v}(x,t)\|^2 dtdx$, and $E\int_U V(\bar{y}(x,0))dx$ in the spatio-temporal stochastic H_∞ state estimation performance in equations (4.11) or (4.12) could be modified by the temporal forms $E\int_0^{t_f} \|e(t)\|^2 \Delta^2 dt$, $E\int_0^{t_f} \|\bar{v}(t)\|^2 \Delta^2 dt$, and $E\bar{V}(\bar{y}(0))\Delta^2$ respectively, for the fuzzy stochastic spatial state space system in equation (4.35), where the Lyapunov function $\bar{V}(\bar{y}(t))$ is related with the Lyapunov function $\bar{V}(\bar{y}(x,t))$ as $\bar{V}(\bar{y}(t)) = \sum_{j=1}^{N} V(\bar{y}_j(t))$ with $\bar{y}_j(t) = [y_j^T(t), e_j^T(t)]^T$. Therefore, the H_∞ state estimation performance in equations (4.11) and (4.12) could be modified as

$$E\int_0^{t_f} \|e(t)\|^2 dt < E\bar{V}(\bar{y}(0)) + \gamma^2 \int_0^{t_f} \|\bar{v}(t)\|^2 dt \tag{4.42}$$

where $\bar{y}(0) \neq 0$, and

$$E\int_0^{t_f} \|e(t)\|^2 dt < \gamma^2 E\int_0^{t_f} \|\bar{v}(t)\|^2 dt \tag{4.43}$$

where $\bar{y}(0) = 0$, where Δ^2 in both sides of equations (4.42) and (4.43) can be canceled.

Therefore, the spatio-temporal H_∞ state estimation problem in equations (4.11) or (4.12) for NSPDS in equation (4.2) is transformed to the only temporal stochastic H_∞ state estimation problem in equations (4.42) or (4.43) for the fuzzy stochastic spatial state space system in equation (4.35). From Corollary 4.3.1, we obtain the following:

$$\begin{aligned} E\|\bar{\varepsilon}_W(y)\|^2 &\le 2E\|\varepsilon_W(y)\|^2 \le 2\sigma_{fW}^2 E\|y(t)\|^2 = E\bar{y}^T R_1 \bar{y} \\ E\|\bar{\varepsilon}(y)\|^2 &\le 2E\|\varepsilon(y)\|^2 \le 2\sigma_\kappa^2 E\|Ty(t)\|^2 + 2\sigma_f^2 E\|y(t)\|^2 \\ &= E\bar{y}^T R_2 \bar{y} \end{aligned} \tag{4.44}$$

where

$$R_1 \triangleq \begin{bmatrix} 2\sigma_{fW}^2 I_{nN} & 0_{nN} \\ 0_{nN} & 0_{nN} \end{bmatrix}, R_2 \triangleq \begin{bmatrix} 2\sigma_\kappa^2 T^T T + 2\sigma_f^2 I_{nN} & 0_{nN} \\ 0_{nN} & 0_{nN} \end{bmatrix}.$$

Suppose the external disturbance $v(t)$ and the measurement $n(t)$ are neglected and the truncation error $O(\Delta^2)$ of the finite difference scheme approaches zero in the augmented system equation (4.39) as the number of grid nodes becomes large enough, i.e., $\bar{v}(t) = 0$. Then the fuzzy state estimator could be designed to guarantee the stochastically asymptotical estimation of the augmented system in equation (4.39) to tolerate the fuzzy approximation errors and the state-dependent noise. For the robust stochastically asymptotical estimation problem, let us choose a Lyapunov function $\bar{V}(\bar{y})$ for the augmented system in equation (4.39) as

$$\bar{V}(\bar{y}) = \sum_{j=1}^{N} (y_j^T(t)P_{11,j}y_j(t) + e_j^T(t)P_{22,j}e_j(t) = \bar{y}^T(t)\bar{P}\bar{y}(t) \tag{4.45}$$

where $\bar{P} = diag(P_{11}, P_{22})$, in which $P_{11} = P_{11}^T = diag(P_{11,1},...,P_{11,N}) > 0$, and $P_{22} = P_{22}^T = diag(P_{22,1},...,P_{22,N}) > 0$. Then the robust H_∞ state estimation in equations (4.42) and (4.43) can be achieved as follows.

Theorem 4.3.1

For the augmented system with a prescribed disturbance attenuation level γ^2 in equations (4.42) and (4.43), suppose there exists a symmetric positive definite matrix $\bar{P}$, a constant $\rho > 0$, and estimator gains L_i for $i = 1,...,M$, such that the following matrix inequalities hold:

$$\begin{cases} \bar{P} < \rho I_{2nN} \\ \Theta_{i_1 i_2} < 0, \end{cases} , for\ i_1, i_2 = 1,...,M \tag{4.46}$$

where

$$\Theta_{i_1 i_2} \triangleq \begin{bmatrix} \Lambda_{i_1 i_2} & \bar{P} & \bar{P}\bar{G}_{vi2} \\ \bar{P} & -I_{2nN} & 0_{2nN\times(n_vN+2nN)} \\ \bar{G}_{vi2}^T & 0_{(n_vN+2nN)\times 2nN} & -\gamma^2 I_{(n_vN+2nN)} \end{bmatrix}$$

$$\Lambda_{i_1 i_2} \triangleq \bar{I} + \bar{P}\bar{A}_{i_1 i_2} + \bar{A}_{i_1 i_2}^T \bar{P} + R_2 + 2\rho R_1 + 2\bar{A}_{Wi1}^T \bar{P}\bar{A}_{Wi1}$$

$$\bar{I} \triangleq \begin{bmatrix} 0_{nN} & 0_{nN} \\ 0_{nN} & I_{nN} \end{bmatrix}, I_{2nN} \triangleq \begin{bmatrix} I_{nN} & 0_{nN} \\ 0_{nN} & I_{nN} \end{bmatrix}, \bar{G}_{vi2} \triangleq \begin{bmatrix} G_v & I_{nN} & 0_{nN} \\ G_v & I_{nN} & -L_{i_2}D \end{bmatrix}.$$

Then the robust H_∞ state estimation performance in equations (4.42) and (4.43) is guaranteed by the fuzzy estimator in equation (4.37). Furthermore, the spatio-temporal H_∞ state estimation performance in equations (4.11) and (4.12) for NSPDSs in equations (4.1) or (4.2) could be achieved and the stochastically asymptotical estimation is also guaranteed when $\bar{v}(t) = 0$.

Proof

$$E\int_0^\infty e^T(t)e(t)\,dt = E\int_0^\infty \bar{y}^T(t)\begin{bmatrix} 0_{nN} \\ I_{nN} \end{bmatrix}\begin{bmatrix} 0_{nN} & I_{nN} \end{bmatrix}\bar{y}(t)\,dt$$

$$= E\bar{V}(\bar{y}(0)) - E\bar{V}(\bar{y}(\infty)) + E\int_0^\infty \left[\bar{y}^T(t)\bar{I}\bar{y}(t) + \frac{d\bar{V}(\bar{y}(t))}{dt}\right]dt \quad (4.47)$$

where $\bar{V}(\bar{y}(t)) = \sum_{j=1}^N V(\bar{y}_j(t))$. For the augmented system in equation (4.39) with $\bar{v}(t) \neq 0$, by the *Itô* formula [99], we get

$$E\int_0^\infty e^T(t)e(t)\,dt$$

$$\leq E\bar{y}^T(0)\bar{P}\bar{y}(0) + E\int_0^\infty \left\{\bar{y}^T(t)\bar{I}\bar{y}(t) + \left(\frac{\partial \bar{V}(\bar{y})}{\partial \bar{y}}\right)^T \left[A(\mu,\hat{\mu})\bar{y}(t) + \bar{G}_v(\hat{\mu})\bar{v}(t) + \bar{\varepsilon}(y)\right]\right.$$

$$\left. + \frac{1}{2}\left[\bar{A}_W(\mu)\bar{y}(t) + \bar{\varepsilon}_W(y)\right]^T \frac{\partial^2 \bar{V}(\bar{y})}{\partial \bar{y}^2}\left[\bar{A}_W(\mu)\bar{y}(t) + \bar{\varepsilon}_W(y)\right]\right\}dt \quad (4.48)$$

By Corollary 4.2.1 and the fact that

$$\frac{\partial \bar{V}(\bar{y})}{\partial \bar{y}} = 2\bar{P}\bar{y}, \frac{\partial^2 \bar{V}(\bar{y})}{\partial \bar{y}^2} = 2\bar{P} \quad (4.49)$$

we get

$$E\int_0^\infty e^T(t)e(t)\,dt$$

$$\leq E\{Tr[\bar{P}\bar{y}(0)\bar{y}^T(0)]\} + E\int_0^\infty \{\bar{y}^T(t)\bar{I}\bar{y}(t) + \bar{y}^T(t)\bar{P}\bar{A}(\mu,\hat{\mu})\bar{y}(t)$$

$$+ \bar{y}^T(t)\bar{A}^T(\mu,\hat{\mu})\bar{P}\bar{y}(t) + \bar{y}^T(t)\bar{P}\bar{G}_v(\hat{\mu})\bar{v}(t) + \bar{v}^T(t)\bar{G}_v^T(\hat{\mu})\bar{P}\bar{y}(t) + \bar{y}^T(t)\bar{P}\bar{\varepsilon}(y)$$

$$+ \bar{\varepsilon}^T(y)\bar{P}\bar{y}(t) + \left[\bar{A}_W(\mu)\bar{y}(t) + \bar{\varepsilon}_W(y)\right]^T \bar{P}\left[\bar{A}_W(\mu)\bar{y}(t) + \bar{\varepsilon}_W(y)\right]\}dt$$

$$\leq E\{Tr[\bar{P}\bar{y}(0)\bar{y}^T(0)]\} + E\int_0^\infty \{\bar{y}^T(t)\bar{I}\bar{y}(t) + \bar{y}^T(t)\bar{P}\bar{A}(\mu,\hat{\mu})\bar{y}(t)$$

$$+ \bar{y}^T(t)\bar{A}^T(\mu,\hat{\mu})\bar{P}\bar{y}(t) + \bar{y}^T(t)\bar{P}\bar{G}_v(\hat{\mu})\bar{v}(t) + \bar{v}^T(t)\bar{G}_v^T(\hat{\mu})\bar{P}\bar{y}(t) + \bar{y}^T(t)\bar{P}\bar{\varepsilon}(y)$$

$$+ \bar{\varepsilon}^T(y)\bar{P}\bar{y}(t) + 2\left[\bar{A}_W(\mu)\bar{y}(t)\right]^T \bar{P}\left[\bar{A}_W(\mu)\bar{y}(t)\right] + 2\left[\bar{\varepsilon}_W(y)\right]^T \bar{P}\left[\bar{\varepsilon}_W(y)\right]\}dt \quad (4.50)$$

By $\bar{P} < \rho I_{2nN}$ and the fact of equation (4.44), we have

$$\bar{y}^T(t)\bar{P}\bar{G}_v(\hat{\mu})\bar{v}(t) + \bar{v}^T(t)\bar{G}_v^T(\hat{\mu})\bar{P}\bar{y}(t) \le \gamma^2\bar{v}^T(t)\bar{v}(t) + \frac{1}{\gamma^2}\bar{y}^T(t)\bar{P}\bar{G}_v(\hat{\mu})\bar{G}_v^T(\hat{\mu})\bar{P}\bar{y}(t),$$

$$E\left\{\bar{y}^T(t)\bar{P}\bar{\varepsilon}(y) + \bar{\varepsilon}^T(y)\bar{P}\bar{y}(t)\right\} \le E\left\{\bar{y}^T(t)\bar{P}\bar{P}\bar{y}(t)\right\} + E\left\{\bar{\varepsilon}^T(y)\bar{\varepsilon}(y)\right\}$$

$$\le E\left\{\bar{y}^T(t)\bar{P}\bar{P}\bar{y}(t)\right\} + E\left\{\bar{y}^T t)R_2\bar{y}(t)\right\}$$

$$E\bar{\varepsilon}_W^T(y)\bar{P}\bar{\varepsilon}_W(y) < \rho E\left\{\bar{\varepsilon}_W^T(y)\bar{\varepsilon}_W(y)\right\} \le \rho E\left\{\bar{y}^T t)R_1\bar{y}(t)\right\} \quad (4.51)$$

and we get

$$E\int_0^\infty e^T(t)e(t)\,dt$$

$$< Tr[\bar{P}R_{\bar{y}(0)}] + E\int_0^\infty \left\{\bar{y}^T(t)\left[\bar{I} + \bar{P}\bar{A}(\mu,\hat{\mu}) + \bar{A}^T(\mu,\hat{\mu})\bar{P} + \frac{1}{\gamma^2}\bar{P}\bar{G}_v(\hat{\mu})\bar{G}_v^T(\hat{\mu})\bar{P}\right.\right.$$

$$\left.\left. + \bar{P}\bar{P} + R_2 + 2\bar{A}_W^T(\mu)\bar{P}\bar{A}_W(\mu) + 2\rho R_1\right]\bar{y}(t)\right\}dt + E\gamma^2\int_0^\infty \bar{v}^T(t)\bar{v}(t)\,dt \quad (4.52)$$

where $R_{\bar{y}(0)} \triangleq E\left\{\bar{y}(0)\bar{y}^T(0)\right\}$. Therefore, if the following inequality holds:

$$\bar{I} + \bar{P}\bar{A}(\mu,\hat{\mu}) + \bar{A}^T(\mu,\hat{\mu})\bar{P} + \bar{P}\left[\frac{1}{\gamma^2}\bar{P}\bar{G}_v(\hat{\mu})\bar{G}_v^T(\hat{\mu}) + I\right]\bar{P} + R_2$$

$$+2\bar{A}_W^T(\mu)\bar{P}\bar{A}_W(\mu) + 2\rho R_1 < 0 \quad (4.53)$$

then

$$E\int_0^\infty e^T(t)e(t)\,dt < Tr[\bar{P}R_{\bar{y}(0)}] + E\gamma^2\int_0^\infty \bar{v}^T(t)\bar{v}(t)\,dt \quad (4.54)$$

i.e., if the inequality in equation (4.53) holds, then the robust H_∞ estimation is guaranteed.

By Schur complement, equation (4.53) is equivalent to

$$\begin{bmatrix} \bar{\Lambda}_{11}(\mu,\hat{\mu}) & \bar{P} & \bar{P}\bar{G}_v(\hat{\mu}) \\ \bar{P} & -I_{2nN} & 0_{2nN\times(n_vN+2nN)} \\ \bar{G}_v^T(\hat{\mu})\bar{P} & 0_{(n_vN+2nN)\times 2nN} & -\gamma^2 I_{(n_vN+2nN)} \end{bmatrix} < 0 \quad (4.55)$$

With $\bar{\Lambda}_{11}(\mu,\hat{\mu}) \triangleq I+\bar{P}\bar{A}(\mu,\hat{\mu})+\bar{A}^T(\mu,\hat{\mu})\bar{P}+R_2+2\rho R_1+2\bar{A}_W^T(\mu)\bar{P}\bar{A}_W(\mu)$, i.e., if equation (4.55) holds, then the robust H_∞ estimation in equation (4.54) is achieved. Substituting $\bar{A}(\mu,\hat{\mu})$ in equation (4.40) and $\bar{A}_W(\mu)$ in equation (4.41) into (4.55) yields

$$\sum_{i_1=1}^{M}\sum_{i_2=1}^{M}\bar{\mu}_{i_1}(y(t))\bar{\mu}_{i_2}(\hat{y}(t))\Theta_{i_1 i_2}<0 \tag{4.56}$$

Therefore, if the matrix inequalities in equation (4.46) hold, then the inequalities in equations (4.56), (4.55) and (4.53) also hold and the robust stochastic H_∞ state estimation in equation (4.54) is guaranteed.

Now we further prove the stochastically asymptotical estimation as follows. For the augmented system in equation (4.39) with $\bar{v}(t)=0$, the infinitesimal operator $L\bar{V}(\bar{y}(t))$ in the spatial state space model is denoted as

$$L\bar{V}(\bar{y}(t)) \triangleq \left(\frac{\partial \bar{V}(\bar{y})}{\partial \bar{y}}\right)^T(\bar{A}(\mu,\hat{\mu})\bar{y}(t)+\varepsilon(\bar{y}))$$

$$+\frac{1}{2}[\bar{A}_W(\mu)\bar{y}(t)+\bar{\varepsilon}_W(y)]^T\frac{\partial^2\bar{V}(\bar{y})}{\partial\bar{y}^2}[\bar{A}_W(\mu)\bar{y}(t)+\bar{\varepsilon}_W(y)]$$

By equation (4.49), we have

$$\begin{aligned}EL\bar{V}(\bar{y}(t)) &= E\{\bar{y}^T(t)\bar{P}\bar{A}(\mu,\hat{\mu})\bar{y}(t)+\bar{y}^T(t)\bar{A}^T(\mu,\hat{\mu})\bar{P}\bar{y}(t)+\bar{y}^T(t)\bar{P}\bar{\varepsilon}(y)\\ &\quad+\bar{\varepsilon}^T(y)\bar{P}\bar{y}(t)+\left[\bar{A}_W(\mu)\bar{y}(t)+\bar{\varepsilon}_W(y)\right]^T\bar{P}\left[\bar{A}_W(\mu)\bar{y}(t)+\bar{\varepsilon}_W(y)\right]\}\\ &\le E\bar{y}^T(t)\bar{P}\bar{A}(\mu,\hat{\mu})\bar{y}(t)+\bar{y}^T(t)\bar{A}^T(\mu,\hat{\mu})\bar{P}\bar{y}(t)+\bar{y}^T(t)\bar{P}\bar{\varepsilon}(y)\\ &\quad+\bar{\varepsilon}^T(y)\bar{P}\bar{y}(t)+2\left[\bar{A}_W(\mu)\bar{y}(t)\right]^T\bar{P}\left[\bar{A}_W(\mu)\bar{y}(t)\right]\\ &\quad+2[\bar{\varepsilon}_W(y)]^T\bar{P}[\bar{\varepsilon}_W(y)]\}\end{aligned} \tag{4.57}$$

By equation (4.51), we get

$$\begin{aligned}EL\bar{V}(\bar{y}(t)) &< E\{\bar{y}^T(t)[\bar{P}\bar{A}(\mu,\hat{\mu})+\bar{A}^T(\mu,\hat{\mu})\bar{P}+\bar{P}\bar{P}+R_2+2\bar{A}_W^T(\mu)\bar{P}\bar{A}_W(\mu)\\ &\quad+2\rho R_1]\bar{y}(t)\}\end{aligned} \tag{4.58}$$

If

$$\bar{P}\bar{A}(\mu,\hat{\mu})+\bar{A}^T(\mu,\hat{\mu})\bar{P}+\bar{P}\bar{P}+R_2+2\bar{A}_W^T(\mu)\bar{P}\bar{A}_W(\mu)+2\rho R_1<0 \tag{4.59}$$

then

$$EL\bar{V}(\bar{y}(t))<0 \tag{4.60}$$

The matrix inequalities in equation (4.46) obviously imply the matrix inequalities in equation (4.53), which also imply the inequalities in equation (4.59) and $EL\bar{V}(\bar{y}(t))<0$ in equation (4.60). So by Theorem 4.2.1, the equilibrium point $\bar{y}(t)=0$ of the augmented system in equation (4.39) with $\bar{v}(t)=0$ is stochastically asymptotically stable. This shows that the stochastically asymptotical estimation is guaranteed when $\bar{v}(t)=0$.

Remark 4.3.3

If the matrix $\bar{P}$ in equation (4.46) is not set to be block diagonal, then after replacing $\bar{P}$ with

$$\bar{P}=\begin{bmatrix} P_{11} & P_{21}^T \\ P_{21} & P_{22} \end{bmatrix}$$

the matrix inequalities in equation (4.46) become

$$\begin{cases} \bar{P}=\begin{bmatrix} P_{11} & P_{21}^T \\ P_{21} & P_{22} \end{bmatrix}\le \rho I_{2nN} \\ \begin{bmatrix} \Pi_{11}^{i_1i_2} & * & * & * & * & * & * \\ \Pi_{21}^{i_1i_2} & \Pi_{22}^{i_1i_2} & * & * & * & * & * \\ P_{11} & P_{21}^T & -I_{nN} & * & * & * & * \\ P_{21} & P_{22} & 0_{nN} & -I_{nN} & * & * & * \\ \Pi_{51} & \Pi_{52} & 0_{n_vN\times nN} & 0_{n_vN\times nN} & -\gamma^2 I_{n_vN} & * & * \\ \Pi_{61} & \Pi_{62} & 0_{nN} & 0_{nN} & 0_{nN\times n_vN} & -\gamma^2 I_{nN} & * \\ \Pi_{71}^{i_2} & \Pi_{72}^{i_2} & 0_{nN} & 0_{nN} & 0_{nN\times n_vN} & 0_{nN} & -\gamma^2 I_{nN} \end{bmatrix}<0 \end{cases} \tag{4.61}$$

for $i_1,i_2=1,\ldots,M$, where

$$\Pi_{11}^{i_1 i_2} \triangleq P_{11}\Omega_{1,i_1} + \Omega_{1,i_1}^T P_{11} + P_{21}^T \Omega_{3,i_1 i_2} + \Omega_{3,i_1 i_2}^T P_{21} + 2\Omega_{2,i_1}^T P_{11}\Omega_{2,i_1}$$
$$+ 2\Omega_{2,i_1}^T P_{22}\Omega_{2,i_1} + 2\sigma_\kappa^2 T^T T + 2\sigma_f^2 I_{nN} + 4\rho\sigma_{fw}^2 I_{nN}$$
$$+ 2\Omega_{2,i_1}^T P_{21}^T \Omega_{2,i_1} + 2\Omega_{2,i_1}^T P_{21}\Omega_{2,i_1},$$
$$\Pi_{12}^{i_1 i_2} \triangleq \Omega_{1,i_1}^T P_{21}^T + \Omega_{3,i_1 i_2}^T P_{22} + P_{21}^T \Omega_{1,i_2}^T - P_{21}^T L_{i_2} C,$$
$$\Pi_{21}^{i_1 i_2} \triangleq P_{21}\Omega_{1,i_1} + P_{22}\Omega_{3,i_1 i_2} + \Omega_{1,i_2}^T P_{21} - C^T L_{i_2}^T P_{21},$$
$$\Pi_{22}^{i_2} \triangleq I_{nN} + P_{22}\Omega_{1,i_2} + \Omega_{1,i_2}^T P_{22} - P_{22} L_{i_2} C - C^T L_{i_2}^T P_{22}^T,$$
$$\Pi_{51} \triangleq G_v^T P_{11} + G_v^T P_{21}, \Pi_{52} \triangleq G_v^T P_{22} + G_v^T P_{21}^T, \Pi_{61} \triangleq P_{11} + P_{21},$$
$$\Pi_{62} \triangleq P_{22} + P_{21}^T, \Pi_{71}^{i_2} = -D^T L_{i_2}^T P_{21}, \Pi_{72}^{i_2} \triangleq -D^T L_{i_2}^T P_{22}^T,$$

with

$$\Omega_{1,i_1} \triangleq [I_N \otimes \kappa_{i_1}]T, \Omega_{1,i_2} \triangleq ([I_N \otimes \kappa_{i_2}]T + [I_N \otimes A_{i_2}]),$$
$$\Omega_{2,i_1} \triangleq (I_N \otimes A_{W_{i_1}}), \Omega_{3,i_1 i_2} \triangleq ([I_N \otimes (\kappa_{i_1} - \kappa_{i_2})]T + [I_N \otimes (A_{i_1} - A_{i_2})])$$

Then the optimal stochastic H_∞ filter design problem becomes how to specify P_{11}, P_{21}, P_{22}, and L_{i_2} to achieve the following constrained optimization:

$$\gamma_{\min}^2 = \min_{P_{11}, P_{21}, P_{22}, L_{i_2}} \gamma^2 \quad \text{subject to (4.61)} \tag{4.62}$$

The BMI-constrained optimization problem in equation (4.62) can be solved by an iteration LMI method as follows.

Step 1: Set the initial value of γ^2, i.e., $\gamma^2 = \gamma_{\text{initial}}^2$ and set the initial value of P_{11}, P_{21}, and P_{22} to some appropriate values, e.g., $P_{11} = I_{nN}, P_{21} = 0_{nN}$, and $P_{22} = I_{nN}$.

Step 2: Substitute the values of P_{11}, P_{21}, P_{22}, and γ^2 into equation (4.61), then the bilinear matrix inequalities (BMIs) in equation (4.61) become LMIs which can be solved to obtain the estimator gain L_{i_2}.

Step 3: Decrease γ^2 and substitute L_{i_2} and γ^2 into equation (4.61), then the BMIs in equation (4.61) become LMIs that can be solved to obtain P_{11}, P_{21}, and P_{22}.

Step 4: Decrease γ^2 as much as possible and repeat Steps 2 and 3 for some iterations until the LMIs in Step 2 or 3 are infeasible in the last iteration. Therefore, we get the solution $\gamma_{\min}^2, P_{11}, P_{21}, P_{22}$, and L_{i_2} when the LMIs in Steps 2 and 3 are feasible in the last second iteration.

Remark 4.3.4

After replacing $\bar{P}$ with $diag(P_{11}, P_{22})$ and using the Schur complement, the inequalities in equation (4.46) become the following LMIs:

$$\begin{cases} \begin{bmatrix} P_{11} & 0_{nN} \\ 0_{nN} & P_{22} \end{bmatrix} < \begin{bmatrix} \rho I_{nN} & 0_{nN} \\ 0_{nN} & \rho I_{nN} \end{bmatrix} \\ \begin{bmatrix} \hat{\Pi}_{11}^{i_1 i_2} & * & * & * & * & * & * \\ \hat{\Pi}_{21}^{i_1 i_2} & \Pi_{22}^{i_2} & * & * & * & * & * \\ P_{11} & 0_{nN} & -I_{nN} & * & * & * & * \\ 0_{nN} & P_{22} & 0_{nN} & -I_{nN} & * & * & * \\ \hat{\Pi}_{51} & \hat{\Pi}_{52} & 0_{n_v N \times nN} & 0_{n_v N \times nN} & -\gamma^2 I_{n_v N} & * & * \\ P_{11} & P_{22} & 0_{nN} & 0_{nN} & 0_{nN \times n_v N} & -\gamma^2 I_{nN} & * \\ 0_{nN} & \hat{\Pi}_{72}^{i_2} & 0_{nN} & 0_{nN} & 0_{nN \times n_v N} & 0_{nN} & -\gamma^2 I_{nN} \end{bmatrix} < 0 \end{cases} \quad (4.63)$$

for $i_1, i_2 = 1, \ldots, M$, where

$$\hat{\Pi}_{11}^{i_1} \triangleq P_{11}\Omega_{1,i_1} + \Omega_{1,i_1}^T P_{11} + 2\Omega_{2,i_1}^T P_{11}\Omega_{2,i_1} + 2\Omega_{2,i_1}^T P_{22}\Omega_{2,i_1}$$
$$+ 2\sigma_\kappa^2 T^T T + 2\sigma_f^2 I_{nN} + 4\rho\sigma_{fw}^2 I_{nN},$$
$$\hat{\Pi}_{22}^{i_2} \triangleq I_{nN} + P_{22}\Omega_{1,i_2} + \Omega_{1,i_2}^T P_{22} - Z_{i_2}C - C^T Z_{i_2}^T, \Pi_{21}^{i_1 i_2} \triangleq P_{22}\Omega_{3,i_1 i_2},$$
$$\hat{\Pi}_{51} \triangleq G_v^T P_{11}, \hat{\Pi}_{52} \triangleq G_v^T P_{22}, and\ \hat{\Pi}_{72}^{i_2} \triangleq -D^T Z_{i_2}^T,$$

with

$$\Omega_{1,i_1} \triangleq [I_N \otimes \kappa_{i_1}]T + [I_N \otimes A_{i_1}], \Omega_{1,i_2} \triangleq ([I_N \otimes \kappa_{i_2}]T + [I_N \otimes A_{i_2}]),$$
$$\Omega_{2,i_1} \triangleq (I_N \otimes A_{W_{i_1}}), \Omega_{3,i_1 i_2} \triangleq ([I_N \otimes (\kappa_{i_1} - \kappa_{i_2})]T + [I_N \otimes (A_{i_1} - A_{i_2})]),$$

and $Z_{i_2} \triangleq P_{22}L_{i_2}$. Thus it follows that the fuzzy estimator gain $L_{i_2} = P_{22}^{-1}Z_{i_2}$ for $i_2 = 1, \ldots, M$.

Remark 4.3.5

(i) The LMIs in equation (4.63) can be efficiently solved using the LMI toolbox in MATLAB. (ii) Based on the H_∞ state estimation analysis in Theorem 4.3.1, the optimal stochastic H_∞ state estimation method for NSPDSs can be designed by minimizing the disturbance attenuation level γ^2 as the following eigenvalue problem (EVP)

$$\gamma_{\min}^2 = \min_{\bar{P}} \gamma^2 \qquad (4.64)$$

$$\text{subject to LMIs in (4.63)}$$

4.4 SIMULATION EXAMPLE

Suppose the temperature function $y(x, t)$ for a 1×0.5 m thin plate in the heat transfer system with the surrounding temperature at 0 °C can be described by the following NSPDS [133]:

$$\frac{\partial y(x,t)}{\partial t} = \kappa\nabla^2 y(x,t) + \{F_1 y(x,t) + F_2[[273 + y(x,t)]^4 - 273^4]\} + g_v v(x,t)$$
$$+ \{H_1 y(x,t) + H_2[[273 + y(x,t)]^4 - 273^4]\} w(x,t) \qquad (4.65)$$

$$y(x,0) = 100 \times e^{-10\times|0.5-x_1|-0.6738} \times e^{-30\times|0.5-2\times x_2|}, \forall x \in U = [0,1]\times[0,0.5]$$

$y(x,t) = 0\ ^{\circ}\mathrm{C}, \forall t, \forall x$ on the boundary of U

where $y(x,t)$ is the temperature function in °C, the location x is in meters, the time t is in seconds, the term $\kappa\nabla^2 y(x,t)$ with the thermal diffusivity $\kappa = 10^{-4}\,\mathrm{m}^2/\mathrm{s}$ is due to conduction heat transfer [133], the term $F_1 y(x,t)$ with $F_1 = -0.004/\mathrm{s}$ is due to convection heat transfer [133], the term $F_2\left[[273+y(x,t)]^4 - 273^4\right]$ with $F_2 = -2\times 10^{-12}\,{}^{\circ}\mathrm{C}^{\circ}\mathrm{K}^{-4}/\mathrm{s}$ is due to radiation heat transfer [40], the term $g_v v(x,t)$ is the environmental thermal fluctuation with $g_v = 0.1$ and $v(x,t) = 10\sin(0.2t)e^{-0.001t-0.1x_1}$, the term $H_1 y(x,t)w(x,t)$ with $H_1 = 0.002/\mathrm{s}$ is due to the random parameter variation of the term $F_1 y(x,t)$, and the term $H_2\left[[273+y(x,t)]^4 - 273^4\right]$ with $H_2 = 10^{-12}\,{}^{\circ}\mathrm{C}^{\circ}\mathrm{K}^{-4}/\mathrm{s}$ is due to the random parameter variation of the term $F_2\left[[273+y(x,t)]^4 - 273^4\right]$. The grid spacing Δ of the finite difference scheme is chosen as 0.125 m so that there are $N=7\times 3=21$ interior grid points and 24 boundary points. There are n_z=45 sensor locations of observation, which are at the 2-D location $q_i = [q_{i,1}, q_{i,2}]^T$ with $q_{i,1} = (k-1)\times 0.125$ m, k=1,..., 9, $q_{i,2} = (l-1)\times 0.125$ m, and $i = (l-1)\times 9 + k$. The observation influence vector $h_i = 1$ and the noise influence matrix $D_n = 1$ so that $h_i D_n = 1$. Therefore, the measured output by 45 sensors is as follows:

$$z(t) = \left[y(q_1,t) + n(q_1,t), \ldots, y(q_{45},t) + n(q_{45},t)\right]^T$$

where $n(q_i,t)$ denotes the measurement noise at the ith sensor. All the measurement noises are zero mean with unit variance in this example.

The membership functions for the estimated temperature $\hat{y}$ are shown in Figure 4.2. The corresponding fuzzy system parameters of the nonlinear function $f(y(x,t) = F_1 y(x,t) + F_2\left[\left[273+y(x,t)\right]^4 - 273^4\right]$ are $A_1 = -0.00412256, A_2 = -0.0042616$ and $A_3 = -0.0045441$. The corresponding fuzzy system parameters of the nonlinear function $f_W\left(y(x,t)\right) = H_1 y(x,t) + H_2\left[\left[273+y(x,t)\right]^4 - 273^4\right]$ are $A_{W_1} = 0.00206$, $A_{W_2} = 0.00213$, and $A_{W_3} = 0.00227$. Based on the design procedure, we construct the spatial state space system in equation (4.35) and then solve P_{11}, P_{22}, and Z_{i_2} from the

optimal stochastic H_∞ filter design in equation (4.62) using the MATLAB toolbox to get $\gamma_{\min} = 1.01$ and the estimator gain $L_{i_2} = P_{22}^{-1} Z_{i_2}$ for the robust state estimator in equation (4.37). Then, we simulate this example by MATLAB with the initial temperature value $y(x,0) = 100 \times e^{-10\times|0.5-x_1|-0.6738} \times e^{-30\times|0.5-2\times x_2|}, \forall x \in U$ and the initial estimated temperature value is chosen as $\hat{y}(x,0) = 0, \forall x \in U$. Figures 4.3–4.6 show the simulation results for t=0, 50, 100, 150 seconds, respectively. The temperature $y(x,t)$ (top-left), the estimated temperature $\hat{y}(x,t)$ (top-right), the estimation error $e(x,t)$ (bottom-left) and the measured output $z(x,t)$ (bottom-right) are shown in these figures. At the initial instant, i.e., t=0 seconds, the initial estimation error $e(x,0)$ is large because the initial estimated temperature $\hat{y}(x,0) = 0, \forall x \in U$ is not estimated by the estimator but is chosen in advance. The large difference between the initial actual system state $y(x,0)$ and the initial chosen estimated system state $\hat{y}(x,0)$ is used to evaluate the filtering performance of the proposed method. In simulation results, the very small estimation errors $e(x,t)$ except its initial value show that our filter is effective for the NSPDS. Since

$$\frac{E\int_0^{150} \|e(t)\|^2 dt - E\bar{V}(\bar{y}(0))}{E\int_0^{150} \|\bar{v}(t)\|^2 dt} = (0.29)^2 < \gamma_{\min}^2 = (1.01)^2$$

the H_∞ state estimation performance in equation (4.42) is also satisfied.

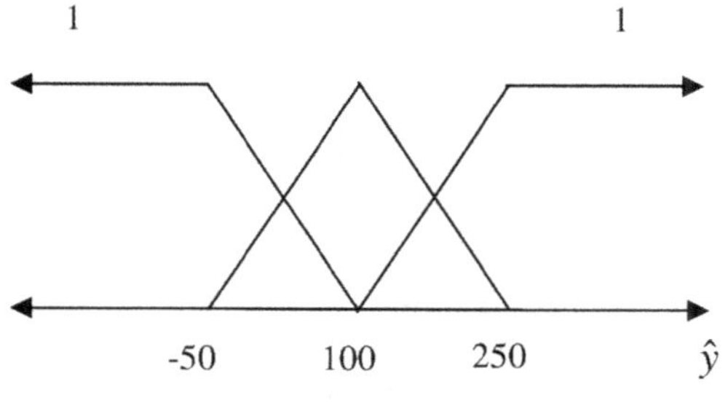

FIGURE 4.2 Membership functions for the estimated temperature $\hat{y}$ of the system equation (4.65).

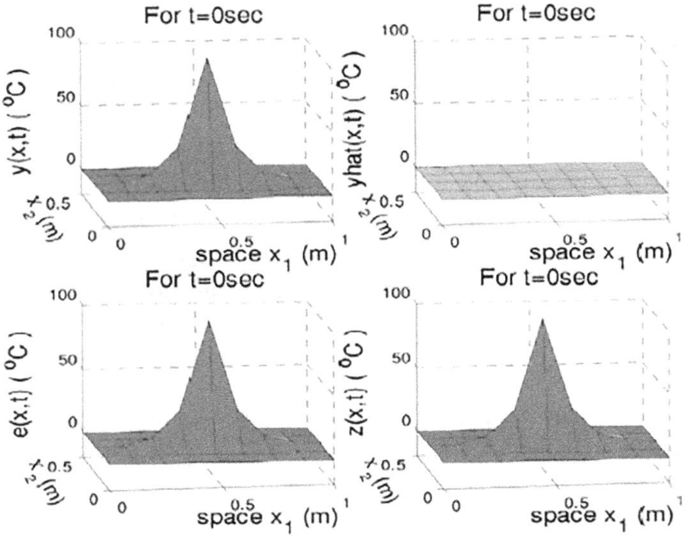

FIGURE 4.3 Simulation results for t=0 seconds: temperature $y(x,t)$ (top-left), estimated temperature $\hat{y}(x,t)$ (top-right), estimation error $e(x,t)$ (bottom-left), and measured output $z(x,t)$ (bottom-right). The initial temperature value is $y(x,0) = 100 \times e^{-10\times|0.5-x_1|-0.6738} \times e^{-30\times|0.5-2\times x_2|}$, $\forall x \in U$ and the initial estimated temperature value is chosen as $\hat{y}(x,0) = 0, \forall x \in U$.

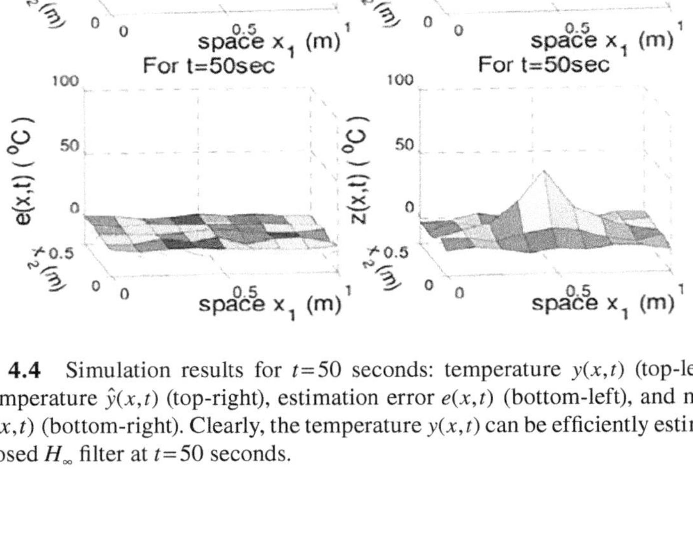

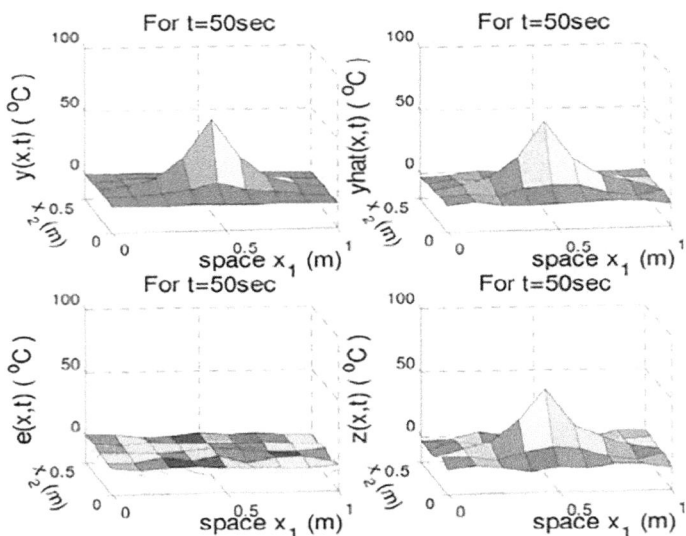

FIGURE 4.4 Simulation results for t=50 seconds: temperature $y(x,t)$ (top-left), estimated temperature $\hat{y}(x,t)$ (top-right), estimation error $e(x,t)$ (bottom-left), and measured output $z(x,t)$ (bottom-right). Clearly, the temperature $y(x,t)$ can be efficiently estimated by the proposed H_∞ filter at t=50 seconds.

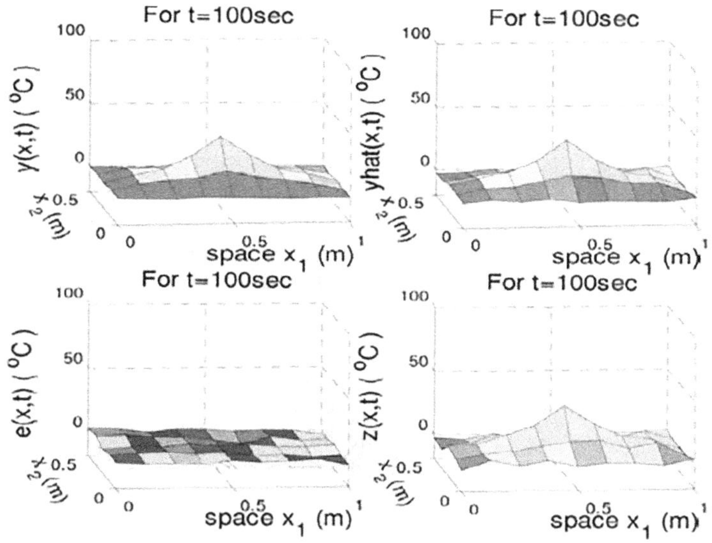

FIGURE 4.5 Simulation results for $t=100$ seconds: temperature $y(x,t)$ (top-left), estimated temperature $\hat{y}(x,t)$ (top-right), estimation error $e(x,t)$ (bottom-left), and measured output $z(x,t)$ (bottom-right). It can been seen that the estimation error $e(x,t)$ is small at $t=100$ seconds.

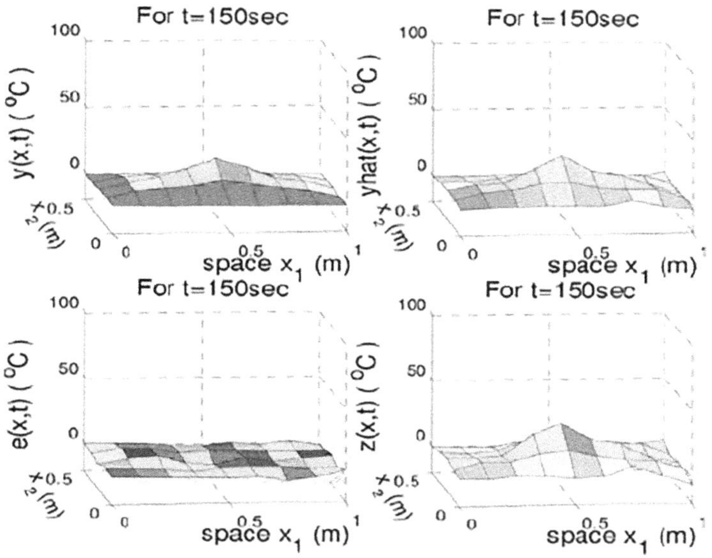

FIGURE 4.6 Simulation results for $t=150$ seconds: temperature $y(x,t)$ (top-left), estimated temperature $\hat{y}(x,t)$ (top-right), estimation error $e(x,t)$ (bottom-left), and measured output $z(x,t)$ (bottom-right). Clearly, the robust H_∞ filtering performance is still maintained at $t=150$ seconds.

4.5 CONCLUSION

This chapter proposes a robust stochastic H_∞ filter design theory for a NSPDS with spatio-temporal disturbances and state dependent noises through a set of sensor measurements. In order to make the robust stochastic H_∞ filter design easier and implementable, the fuzzy interpolation scheme, the semi-discretization finite difference method and Kronecker product are employed to represent the NSPDS by an equivalent fuzzy spatial state space system. Then the 2-D spatio-temporal H_∞ filtering performance can be represented by 1-D stochastic H_∞ filtering performance. Therefore, based on the more suitable spatial state space model, an implementable fuzzy H_∞ filter is designed for robust state estimation of NSPDSs with spatio-temporal disturbances and state-dependent noises. The proposed robust H_∞ filter design method is applicable to the robust state estimation of NSPDSs such as heat flows, fluid flows, elastic waves, flexible structures in mechanical systems, population dynamics in ecological systems, and biodynamics in biological systems. From the simulation example of heat transfer system, the robust H_∞ filtering performance for SPDSs can be confirmed by the proposed method.

5 Robust Synchronization Design of Partial Differential Systems via Diffusion Coupling

5.1 INTRODUCTION

Synchronization appears within a widespread field, ranging from natural network to artificial network, such as flushing fireflies [134], brainweb [135], yeast cell [136], semi-conductor lasers [137], sensor networks [138], and so on. Due to the ubiquitousness of synchronization, scientists attempt to understand the mechanism behind the phenomena and how it works to get its advantages. For instances, the semiconductor laser array can be synchronized to generate larger power laser, the clock of sensor network needs to be synchronized so that the sensor network can process data more correctly or, to achieve secure communication [139,140], the transmitter masks the signals with fuzzy chaotic oscillator and the receiver recovers the signals using fuzzy chaotic synchronization [141].

Recently, several studies have investigated the synchronization of coupled networks, using the master stability function (MSF) method [142] and the Lyapunov direct method [143]. Both methods provide the criteria of synchronization for nonlinear coupled networks. The criterion of MSF method can only guarantee the linearized stability of the synchronous state because the variational equation is just the linearized dynamic of coupled network along the synchronous state. The Lyapunov direct method guarantees the synchronization for coupled networks in a larger region and the range depends on the Lyapunov function that has been found. However, in general it is hard to find an appropriate Lyapunov function.

Based on the above two methods, many research works have been done on the synchronization from different points of view. In [144], the network is considered as a fixed topology and the coupled network with maximal synchronization has non-diagonal coupling matrix. In [145], the interaction of oscillators is considered as the information of diffusion process. Some research works have investigated the synchronization for networks with statistics feature, such as the small-world network [146] or the scale-free network [147]. In [148–150], the authors have shown that randomly adding shortcuts can translate a regular network to a small-world network and improve the network's synchronization [151]. In [152], Wang and Chen discussed the synchronization of scale-free network. In [153,154], the synchronization of the scale-free network with degree-degree

DOI: 10.1201/9781003229230-7

correlation is considered. The time-delayed coupling effects have also been considered for different classes of systems such as complex network with time-varying delayed coupling in [154–156]. The master-slave synchronization problems subject to communication channel noise and bandwidth limitations were studied in [157,158]. In [159,160], the authors dealt with the master-slave Lurie systems by using time-delayed feedback control and sample-feedback control, respectively.

In the real world, external disturbances are ubiquitous and may lead a given system to an unanticipated state and even destroy the synchronization. Hence, it is deserving to investigate the ability of synchronization to resist the external disturbance, i.e., the synchronization problem. The synchronization problems have been studied by many authors [161–168]. In [161,162], the authors considered the synchronization between the master and slave of chaotic oscillators. Furthermore the synchronization has been applied to time-delayed chaotic system [163,164,168], and neural systems [165]. The method is also utilized to design an antisynchronization controller between drive and response system [169].

In recent years, a great deal of concern has been raised regarding the study of partial differential systems (PDSs) [170–181]. Many phenomena in science, engineering, and biology have been modeled in the spatio-temporal domain by PDSs such as in chemical engineering, biology, population dynamics, neurophysiology and biodynamics, etc.; therefore, the synchronization problem of PDSs is addressed in this chapter.

In this chapter, N-coupled linear PDSs with constant coefficients via spatial diffusion coupling are discussed at first, then a synchronization error dynamic of coupled PDSs is defined in the spatial domain. Based on the synchronization error dynamic of the coupled PDSs, the asymptotical synchronization problem becomes an asymptotical stabilization problem. Using the Green theorem, Poincaré inequality in Section 1.2.3, and Lyapunov-Krasovskii techniques, the asymptotical synchronization criteria for these coupled linear PDSs are derived.

Aside from synchronization, the disturbance-resisting ability of synchronous partial differential systems (PDSs) has also been investigated in this chapter. Based on the disturbance attenuation theory in temporal domain [182–187], the spatio-temporal disturbance attenuation level of a coupled PDSs network is considered as the upper bound of the gain from any finite energy disturbance to the synchronization error in the spatio-temporal domain. In this chapter, based on the synchronization error dynamic with disturbances, sufficient conditions are obtained for the synchronization of N-coupled PDSs with disturbances in the spatio-temporal domain.

Except the coupled PDSs with constant coefficients, the coupled PDSs with space-dependent coefficients appear in electrostatics, stationary heat transfer and other diffusion problems for inhomogeneous media and the properties are studied by many researchers [188,189]. This chapter also considers the synchronization for N-coupled PDSs with space-dependent coefficients. In term of Lyapunov-Krasovskii method, the sufficient condition to guarantee the synchronization is obtained. Similarly, the synchronization for N-coupled PDSs with space-dependent coefficients and disturbances is studied and the criterion of synchronization is also derived by virtue of Lyapunov-Krasovskii method. These criteria need to verify the corresponding integral inequalities, which cannot be solved analytically at present, even if an appropriate Lyapunov-Krasovskii functional is given.

In order to overcome this difficulty and make these criteria to be verified easier, the semi-discrete finite difference scheme is employed to represent the synchronization error dynamic. Under this situation, the synchronization criterion of coupled PDSs needed to be investigated becomes an LMI, which can be efficiently solved with the help of LMI toolbox in MATLAB [190]. The relationship between the sufficient conditions for the synchronization, obtained by the Lyapunov-Krasovskii method and semi-discrete difference method, respectively, is also investigated.

Finally, two examples are given to illustrate the correctness of the results we got in this chapter. The criteria of asymptotical synchronization with constant coefficients are verified by the numerical simulations. The effect of the spatial domain on the synchronization is also discussed. The synchronization of two coupled PDSs with space-dependent coefficients is investigated when the spatio-temporal disturbances are considered.

For convenience, we adopt the following standard notations: $\mathbb{R}^n$ and $\mathbb{R}^{n\times m}$ denote, respectively, the n dimensional Euclidean space and the set of all $n\times m$ real matrices. $\mathbb{R}^+ = \{x \in \mathbb{R} | x \geq 0\}$. The superscript "$T$" denotes the transpose and the notation $X \geq Y$ ($X > Y$), where X and Y are symmetric matrices, means that X–Y is positive semidefinite (positive definite). I is the identity matrix with compatible dimension. C^s (U) stands for the space of functions with continuous partial derivatives of order less than or equals to s in U. $L^2(0,t_f)$ $\left(\text{or } L^2(0,\infty)\right)$ denotes the set of square integral functions on $\left[0,t_f\right]$ $\left(\text{or } [0,\infty]\right)$.

5.2 MATHEMATICAL MODEL AND SYNCHRONIZATION ERROR DYNAMIC

We consider the following N-coupled linear PDSs with constant coefficients:

$$\frac{\partial}{\partial t} y^k(x,t) = Ay^k(x,t) + B\nabla^2 y^k(x,t) + \sum_{j\neq k} D_{kj}\nabla^2\left(y^j(x,t) - y^k(x,t)\right) \quad (5.1)$$

$k = 1,\ldots, N$, for $x = (x_1,x_2)^T \in U \subset \mathbb{R}^2$ and $t > 0$, where the state variables $y^k(x,t) = \left(y_1^k(x,t),\ldots,y_n^k(x,t)\right)^T \in \mathbb{R}^n$. n and t are space and time variables, respectively, $A, B, D_{kj} \in \mathbb{R}^{n\times n}$ are constant system matrices and we assume B, D_{kj} are symmetric matrices and $D_{kj} = D_{jk}$. The Laplace operator ∇^2 for two-dimensional spatio-space is defined as follows:

$$\nabla^2 y^i(x,t) = \frac{\partial^2 y^i(x,t)}{\partial x_1^2} + \frac{\partial^2 y^i(x,t)}{\partial x_2^2} = \begin{pmatrix} \dfrac{\partial^2 y_1^i(x,t)}{\partial x_1^2} \\ \dfrac{\partial^2 y_2^i(x,t)}{\partial x_1^2} \\ \vdots \\ \dfrac{\partial^2 y_n^i(x,t)}{\partial x_1^2} \end{pmatrix} + \begin{pmatrix} \dfrac{\partial^2 y_1^i(x,t)}{\partial x_2^2} \\ \dfrac{\partial^2 y_2^i(x,t)}{\partial x_2^2} \\ \vdots \\ \dfrac{\partial^2 y_n^i(x,t)}{\partial x_2^2} \end{pmatrix} \quad (5.2)$$

for $i = 1,2,\ldots, N$. The boundary conditions and initial values are given as follows:

$$y^k(x,t) = 0, x \in \partial U, y^k(x,0) = \varphi^k(x), k = 1,\ldots,N \tag{5.3}$$

The linear PDSs in equation (5.1) are coupled through diffusion $\nabla^2\left(y^j(x,t) - y^k(x,t)\right)$ in spatio-space with constant diffusion coefficients D_{kj}. Let $y(x, t)$ be a function to which all $y^k(x, t)$, $k = 1,\ldots, N$ are expected to synchronize and $y(x, t)$ satisfies the following equation:

$$\frac{\partial y(x,t)}{\partial t} = Ay(x,t) + B\nabla^2 y(x,t) \tag{5.4}$$

subject to the boundary conditions and initial values as follows:

$$y(x,t) = 0, x \in \partial U, y(x,0) = \varphi(x) \tag{5.5}$$

Let us define the synchronization error as follows:

$$e^k(x,t) = y^k(x,t) - y(x,t) \tag{5.6}$$

Definition 5.2.1

The coupled linear PDSs in equation (5.1) are asymptotically synchronous if the synchronization error satisfies $\lim_{t\to\infty} e^k(x,t) = 0$ for all x and $k = 1,2,\ldots, N$.

By transforming the coupled PDSs in equation (5.1) with state vector $y^k(x, t)$ to synchronization errors $e^k(x, t)$ in equation (5.6), the synchronization problem of PDSs in equation (5.1) turns to the stabilization problem of the corresponding synchronization error dynamical system in the following $k = 1,2,\ldots, N$ subject to the boundary conditions and initial values as follows:

$$e^k(x,t) = 0, x \in \partial U, e^k(x,0) = \varphi^k(x) - \varphi(x),\ k = 1,\ldots,N$$

Take

$$\bar{A} = \begin{pmatrix} A & 0 & \cdots & 0 \\ 0 & A & \ddots & 0 \\ \vdots & \ddots & \ddots & \vdots \\ 0 & 0 & \cdots & A \end{pmatrix}, \bar{B} = \begin{pmatrix} B & 0 & \cdots & 0 \\ 0 & B & \ddots & 0 \\ \vdots & \ddots & \ddots & \vdots \\ 0 & 0 & \cdots & B \end{pmatrix},$$

$$D = \begin{pmatrix} -\sum_{j\neq 1} D_{1j} & D_{12} & \cdots & D_{1N} \\ D_{21} & -\sum_{j\neq 2} D_{2j} & \cdots & D_{2N} \\ \vdots & \vdots & \ddots & \vdots \\ D_{N1} & D_{N2} & \cdots & -\sum_{j\neq N} D_{Nj} \end{pmatrix}.$$

and $\bar{D} = \bar{B} + D$. Obviously, $\bar{D}$ is symmetric. Denote

$$e(x,t) = \begin{pmatrix} e^1(x,t) \\ \vdots \\ e^N(x,t) \end{pmatrix} \tag{5.7}$$

then the synchronization error dynamic of N-coupled PDSs with constant coefficients in equation (5.7) can be represented by

$$\frac{\partial e(x,t)}{\partial t} = \bar{A}e(x,t) + \bar{D}\nabla^2 e(x,t) \tag{5.8}$$

If the coefficients A, B, D_{kj} depend on the space variable, i.e., are space-dependent, then the system in equation (5.1) becomes the N-coupled PDSs in equation (5.8) with space-dependent coefficients

$$\frac{\partial}{\partial t} y^k(x,t) = A(x)y^k(x,t) + B(x)\nabla^2 y^k(x,t) + \sum_{j \neq k} D_{kj}(x)\nabla^2\left(y^j(x,t) - y^k(x,t)\right) \tag{5.9}$$

The synchronization error dynamic equation (5.7) becomes

$$\frac{\partial}{\partial t} e^k(x,t) = A(x)e^k(x,t) + B(x)\nabla^2 e^k(x,t) + \sum_{j \neq k} D_{kj}(x)\nabla^2\left(e^j(x,t) - e^k(x,t)\right) \tag{5.10}$$

Similarly, the synchronization error dynamic of N-coupled PDSs with space-dependent coefficients in equation (5.10) can be represented by

$$\frac{\partial e(x,t)}{\partial t} = \bar{A}(x)e(x,t) + \bar{D}(x)\nabla^2 e(x,t) \tag{5.11}$$

where $\bar{A}(x)$ and $\bar{D}(x)$ are defined similarly as $\bar{A}$ and $\bar{D}$, respectively.

Remark 5.2.1

Here we study the case: $x \in \mathbb{R}^2$, but the results in the sequel can be extended to the case: $x \in \mathbb{R}^n, n \in \{1,2,3,\ldots\}$ without any difficulty.

To end this section, we present a definition, which will be used in the sequel.

Definition 5.2.2

The function $V(x)$ is positive-definite and radially unbounded if $V(x)$ is positive except at $x = 0$ where $V(0) = 0$ and $V(x) \to \infty$ as $|x| \to \infty$.

5.3 SYNCHRONIZATION AND ROBUST H_∞ SYNCHRONIZATION OF COUPLED PDSs WITH CONSTANT COEFFICIENTS

Based on the synchronization error dynamic of PDSs constructed in the last section, the asymptotical synchronization problem of coupled PDSs becomes the stabilization problem of synchronization error dynamic of PDSs. For the N-coupled linear PDSs in equation (5.1) and their synchronization error dynamic in equation (5.8), the asymptotical synchronization of coupled PDSs in equation (5.1) implies the asymptotical stability of synchronization error dynamic in equation (5.8). Therefore, we will discuss the asymptotical stability of the synchronization error dynamic at first.

Definition 5.3.1

The equilibrium point $e(x, t) = 0$ of the synchronization error dynamic in equation (5.8) is asymptotically stable for all $x \in U$, if $e(x,t) \to 0$ as $t \to \infty$.

Remark 5.3.1

The Definition 5.3.1 of the asymptotical stability is essentially the strong stability in view of the distributed parameter system. For the distributed parameter systems, there are other definitions of stability, for the details, we refer to the reference [191] and the references therein.

Now we present an important lemma which is stated as Poincaré inequality in [192]. Here, we assume $U = \{x : |x_1| < l_1 \text{ and } |x_2| < l_2\}$.

Lemma 5.3.1

Let U be a cube: $|x_i| < l_i\ (i = 1,2)$ and let $h(x)$ be a real-valued function belonging to $C^1(U)$ which vanishes on the boundary ∂U of U, i.e., $h(x)|_{\partial U} = 0$. Then

$$\int_U h^2(x)dx \le l_i^2 \int_U \left|\frac{\partial h}{\partial x_i}\right|^2 dx \tag{5.12}$$

Using Lemma 5.3.1, we can get the following lemma, which will be used in the sequel.

Lemma 5.3.2

Let U be a cube: $|x_i| < l_i$ $(i=1,2)$ and let $z(x) = (z_1(x), z_2(x), \ldots, z_n(x))^T \in \mathbb{R}^n$ be a function belonging to $C^2(U)$ which vanishes on the boundary ∂U of U. Then

$$\int_U z^T(x)\nabla^2 z(x)dx \le -\left(\frac{1}{l_1^2}+\frac{1}{l_2^2}\right)\int_U z^T(x)z(x)dx$$

Proof

From Green's identity [60], using Lemma 5.3.1, we have

$$\int_U z^T(x)\nabla^2 z(x)dx$$

$$=\int_U \sum_{i=1}^{n} z_i(x)\nabla^2 z_i(x)dx$$

$$=-\int_U \sum_{i=1}^{n}\left[\left(\frac{\partial z_i}{\partial x_1}\right)^2+\left(\frac{\partial z_i}{\partial x_2}\right)^2\right]dx+2\sum_{i=1}^{n}\int_{\partial U} z_i \frac{dz_i}{d\vec{n}}dS \text{ (by Green's identity)}$$

$$=-\int_U \sum_{i=1}^{n}\left[\left(\frac{\partial z_i}{\partial x_1}\right)^2+\left(\frac{\partial z_i}{\partial x_2}\right)^2\right]dx(z_i(x)=0, x\in\partial U)$$

$$\le -\left(\frac{1}{l_1^2}+\frac{1}{l_2^2}\right)\int_U z^T(x)z(x)dx, \text{ (by Poincaré inequality in Lemma 5.3.1)}$$

where $\frac{d}{d\vec{n}}$ indicates differentiation in the direction of the exterior normal to ∂U.

Now we are in the position to state our first theorem on the asymptotical synchronization of N-coupled PDSs with constant coefficients.

Theorem 5.3.1

Suppose there exists a positive symmetric matrix P such that

$$P\bar{D}+\bar{D}P \ge 0 \tag{5.13}$$

and

$$\bar{A}^T P + P\bar{A} - \left(\frac{1}{l_1^2}+\frac{1}{l_2^2}\right)(P\bar{D}+\bar{D}P) < 0 \tag{5.14}$$

then the synchronization error system in equation (5.8) is asymptotically stable, that is, the N-coupled PDSs equation (5.1) are of asymptotical synchronization.

Proof

Define a Lyapunov-Krasovskii functional $V(e(\cdot,t)) = \int_U e^T(x,t)Pe(x,t)\,dx$. Keeping in mind that $\bar{D}$ is symmetric, we have

$$V(e(\cdot,t)) = \int_U \left[\dot{e}^T(x,t)Pe(x,t) + e^T(x,t)P\dot{e}(x,t)\right]dx$$

$$= \int_U \left[(\bar{A}e(x,t) + \bar{D}\nabla^2 e(x,t))^T Pe(x,t) + e^T(x,t)(\bar{A}e(x,t) + \bar{D}\nabla^2 e(x,t))\right]dx$$

$$= \int_U \left[e^T(x,t)(\bar{A}^T P + P\bar{A})e(x,t) + e^T(x,t)(P\bar{D} + \bar{D}P)\nabla^2 e(x,t)\right]dx \tag{5.15}$$

Since $P\bar{D} + \bar{D}P \geq 0$, there exists a matrix Q such that $Q^T Q = P\bar{D} + \bar{D}P$, then $e^T(x,t)(P\bar{D} + \bar{D}P)\nabla^2 e(x,t) = e^T(x,t)Q^T Q\nabla^2 e(x,t) = (Qe(x,t))^T \nabla^2(Qe(x,t)).$ Let $z(x,t) = Qe(x,t)$, noting $z(x,t) = Qe(x,t) = 0$ for $x \in \partial U$, using Lemma 5.3.2, we have

$$\int_U \left[e^T(x,t)(P\bar{D} + \bar{D}P)\nabla^2 e(x,t)\right]dx \leq -\left(\frac{1}{l_1^2} + \frac{1}{l_2^2}\right)\int_U \left[e^T(x,t)(P\bar{D} + \bar{D}P)e(x,t)\right]dx \tag{5.16}$$

Substituting equation (5.16) into (5.15) and using equation (5.14), we can get

$$\dot{V}(e(\cdot,t)) \leq \int_U e^T(x,t)\left[\bar{A}^T P + P\bar{A} - \left(\frac{1}{l_1^2} + \frac{1}{l_2^2}\right)(P\bar{D} + \bar{D}P)\right]e(x,t) < 0 \tag{5.17}$$

Since $V(0) = 0$, we get that, along the solution of system in equation (5.8), $V(e(\cdot,t)) \to 0$ as $t \to \infty$. Take $\lambda_{\min}(P) > 0$ to be the smallest eigenvalue of P, we have $0 < \lambda_{\min}(P)\int_U e^T(x,t)e(x,t)\,dx \leq \dot{V}(e(\cdot,t))$, then we get $\int_U e^T(x,t)e(x,t)\,dx \to 0$ as $t \to \infty$, which implies $e(x,t) \to 0$ as $t \to \infty$. We get the asymptotical stability of the system in equation (5.8), that is, the system equation (5.1) is asymptotically synchronous.

Remark 5.3.2

From equation (5.14), the asymptotical synchronization can be verified by the systems' coefficients, i.e., $\bar{A}$, $\bar{D}$ and the range of space domain, i.e., l_1 and l_2.

Since the external disturbance is inevitable, and disturbance may destroy the synchronization of the coupled PDSs, we must consider the synchronization robustness to resist external disturbance of the coupled PDSs. We consider the following coupled PDSs with disturbance

$$\frac{\partial y^k(x,t)}{\partial t} = Ay^k(x,t) + B\nabla^2 y^k(x,t) + \sum_{j\neq k} D_{kj}\nabla^2(y^j(x,t) - y^k(x,t)) + v^k(x,t) \quad (5.18)$$

where $v^k(x, t)$ stands for the external disturbance in the spatio-temporal domain and $v^k(x, t)$ is square integral, i.e., $\int_0^{t_f} \left(v^k(x,t)\right)^T v^k(x,t)dxdt < \infty$ for any $k \in \{1,2,\ldots,N\}$ and some positive constant t_f.

Let $e^k(x,t) = y^k(x,t) - y(x,t)$, then we get the following coupled synchronization error dynamic

$$\frac{\partial e^k(x,t)}{\partial t} = Ae^k(x,t) + B\nabla^2 e^k(x,t) + \sum_{j\neq k} D_{kj}\nabla^2\left(e^j(x,t) - e^k(x,t)\right) + v^k(x,t). \quad (5.19)$$

Let

$$e(x,t) = \begin{pmatrix} e^1(x,t) \\ \vdots \\ e^N(x,t) \end{pmatrix} \text{ and } v(x,t) = \begin{pmatrix} v^1(x,t) \\ \vdots \\ v^N(x,t) \end{pmatrix}$$

Take $\bar{A}$, $\bar{B}$, and $\bar{D}$ as before, then we get the disturbed synchronization error dynamic as follows:

$$\frac{\partial e(x,t)}{\partial t} = \bar{A}e(x,t) + \bar{D}\nabla^2 e(x,t) + v(x,t) \quad (5.20)$$

Assuming that the coupled linear PDSs are in the synchronization state at $t = 0$ (i.e., $e(x,0) = 0$), we consider the robust H_∞ disturbance attenuation of the synchronization error dynamic in equation (5.20) as follows:

$$\frac{\int_0^{t_f}\int_U e^T(x,t)e(x,t)dxdt}{\int_0^{t_f}\int_U v^T(x,t)v(x,t)dxdt} < \gamma^2$$

or

$$\int_0^{t_f}\int_U e^T(x,t)e(x,t)dxdt < \gamma^2 \int_0^{t_f}\int_U v^T(x,t)v(x,t)dxdt \quad (5.21)$$

If the initial condition is considered, then

$$\int_0^{t_f}\int_U e^T(x,t)e(x,t)dxdt < V(e(\cdot,0)) + \gamma^2 \int_0^{t_f}\int_U v^T(x,t)v(x,t)dxdt \quad (5.22)$$

for some positive function $V(\cdot)$. If the synchronization error dynamic in equation (5.20) satisfies the disturbance attenuation in equations (5.21) or (5.22), we say the coupled PDSs are of the robust H_∞ synchronization with a prescribed attenuation level γ. Now, we state a theorem on the robust H_∞ synchronization, which presents the sufficient conditions to guarantee the robust H_∞ synchronization of systems in equation (5.18).

Theorem 5.3.2

Given a disturbance attenuation level $\gamma > 0$, suppose that there exists a positive matrix P such that

$$P\bar{D} + \bar{D}P \geq 0 \tag{5.23}$$

and

$$I + \bar{A}^T P + P\bar{A} + \frac{1}{\gamma^2} P^T P - \left(\frac{1}{l_1^2} + \frac{1}{l_2^2}\right)\left(P\bar{D} + \bar{D}P\right) < 0 \tag{5.24}$$

then the coupled PDSs in equation (5.18) are of the H_∞ synchronization in equations (5.21) or (5.22).

Proof

Let $V(e(\cdot,t)) = \int_U e^T(x,t)Pe(x,t)dx$. Using Lemma 5.3.2 and the techniques of completing the squares, we have

$$\begin{aligned}
&\int_0^{t_f}\int_U [e^T(x,t)e(x,t) - \gamma^2 v^T(x,t)v(x,t)]dxdt \\
&= \int_0^{t_f}\left(\int_U e^T(x,t)e(x,t) - \gamma^2 v^T(x,t)v(x,t))dx + \frac{\partial V(e(\cdot,t))}{\partial t}\right)dt \\
&\quad + V(e(\cdot,0)) - V(e(\cdot,t_f)) \\
&= \int_0^{t_f}\int_U\left[e^T(x,t)e(x,t) - \gamma^2 v^T(x,t)v(x,t) + \left(\frac{\partial e(x,t)}{\partial t}\right)^T Pe(x,t)\right. \\
&\quad \left. + e^T(x,t)P\frac{\partial e(x,t)}{\partial t}\right]dxdt + V(e(\cdot,0)) - V(e(\cdot,t_f))
\end{aligned}$$

$$= \int_0^{t_f} \int_U \left[e^T(x,t)(I + \bar{A}^T P + P\bar{A})e(x,t) + e^T(x,t)(P\bar{D} + \bar{D}P)\nabla^2 e(x,t) \right.$$

$$\left. + v^T(x,t)Pe(x,t) + e^T(x,t)Pv(x,t) - \gamma^2 v^T(x,t)v(x,t) \right] dxdt$$

$$+ V(e(\cdot,0)) - V(e(\cdot,t_f))$$

$$= \int_0^{t_f} \int_U \left[-\left(\gamma v(x,t) - \frac{1}{\gamma} Pe(x,t) \right)^T \left(\gamma v(x,t) - \frac{1}{\gamma} Pe(x,t) \right) \right.$$

$$+ e^T(x,t)(I + \bar{A}^T P + P\bar{A})e(x,t) + e^T(x,t)(P\bar{D} + \bar{D}P)\nabla^2 e(x,t)$$

$$\left. + \frac{1}{\gamma^2} e^T(x,t)P^T Pe(x,t) \right] dxdt + V(e(\cdot,0)) - V(e(\cdot,t_f))$$

$$\leq V(e(\cdot,0)) + \int_0^{t_f} \int_U e^T(x,t) \left[I + \bar{A}^T P + P\bar{A} + \frac{1}{\gamma^2} P^T P \right.$$

$$\left. - \left(\frac{1}{l_1^2} + \frac{1}{l_2^2} \right)(P\bar{D} + \bar{D}P) \right] e(x,t)dxdt, \text{ (by Lemma 5.3.2)}$$

$$< V(e(\cdot,0)) \tag{5.25}$$

that is the inequality in equation (5.22). If $e(x,0) = 0$, i.e., the coupled PDSs in equation (5.18) are in the synchronization state at $t = 0$, then $V(e(\cdot,0)) = 0$, and the inequality equation (5.21) holds, therefore, we conclude that the coupled PDSs in equation (5.18) are of the robust H_∞ synchronization.

Remark 5.3.3

i. The conditions in equations (5.23) and (5.24) to guarantee the H_∞ synchronization is also sufficient for the asymptotical synchronization since

$$I + \bar{A}^T P + P\bar{A} - \left(\frac{1}{l_1^2} + \frac{1}{l_2^2} \right)(P\bar{D} + \bar{D}P) + \frac{1}{\gamma^2} P^T P < 0$$

means

$$\bar{A}^T P + P\bar{A} - \left(\frac{1}{l_1^2} + \frac{1}{l_2^2} \right)(P\bar{D} + \bar{D}P) < -I - \frac{1}{\gamma^2} P^T P < 0$$

which implies that the robust H_∞ synchronization of equation (5.18) indicates the asymptotical synchronization of equation (5.1) when $v(x,t) = 0$.

ii. The H_∞ synchronization in equations (5.23) and (5.24) can be verified by the systems coefficients $\bar{A}$ and $\bar{D}$, the range of space domain l_1 and l_2 and the prescribed disturbance attenuation level γ. In addition, the optimal disturbance attenuation level of synchronous PDSs can be measured by

$$\gamma_0 = \min_{P>0} \gamma$$

subject to (5.23) and (5.24)

iii. From the conditions in equations (5.14) and (5.24), we see that for the same system coefficients, if the constant l_i becomes smaller, i.e., the space domain U is smaller, then inequalities of equations (5.14) and (5.24) are easier to hold, i.e., it is easier to achieve the synchronization through spatial diffusion coupling in a small area. Therefore, we can get the following proposition:

Proposition 5.3.1

The asymptotical synchronization and H_∞ synchronization of coupled PDSs depend on the scale of the spatial domain: the smaller the spatial domain, the easier to achieve the asymptotical synchronization or the H_∞ synchronization.

Remark 5.3.4

In many studies, the structure of the network was considered for the synchronization problem, see, for example, [152,165]. The synchronization of a coupled network system is not only determined by the structure of the coupled network, represented by the matrix D in this chapter, but also the dynamical properties of the subsystem without coupling, represented by the matrices A and B. In this chapter, the sufficient conditions for the synchronization for the PDSs are dependent on the matrices $\bar{A}$ and $\bar{D}$, which contain the matrices A, B, D. It means the structure and the properties of the subsystem are considered both for the synchronization.

If the matrices A, B and the disturbance attenuation level are given, then we can discuss what conditions should be imposed on the matrix D to make the inequalities equations (5.23) and (5.24) have a positive solution. That is the design problem of the coupled network to achieve the synchronization. To see it clearly, we assume the matrix $D_{kj} = cd_{kj}\Gamma$, where c is a given constant, Γ is a fixed diagonal matrix, if there is a connection between node k and node $j(k \neq j), d_{kj} = d_{jk} = 1$, otherwise, $d_{kj} = d_{jk} = 0$. This form of matrix D_{kj} was adopted in [152]. Take $\tilde{D} = (d_{kj})_{N\times N}$, then $D = c(\tilde{D} \otimes \Gamma)$. Now, the topology of the coupled network can be completely presented by the matrix $\tilde{D}$. For a strong connected graph, we know that the eigenvalues of $\tilde{D}$ are negative, then if $c<0$ (positive diffusion will harm the synchronization), from inequalities equations (5.23) and (5.24), we can see the larger $|c|$ to be, which represents the strength of coupling, the easier to achieve the synchronization. Moreover, if the network is globally coupled, then the eigenvalues of the matrix $\tilde{D}$ are 0 and $-N$ (the eigenvalue 0 is simple, see [152] for the details), thus for a given $c<0$, the more nodes of the network have, the easier to achieve the synchronization.

To end this section, we figure out other results. In the literatures, many researchers have studied the synchronization of coupled systems using the error $e_{i,j}(t)=x_i(t)-x_j(t)$, where $x_i(t)$ is the state of the coupled systems, see [143]. Now, we also take the synchronization error $e(x,t)=y^1(x,t)-y^2(x,t)$, and for the simplicity, we just consider the two-coupled PDSs. We will give the criteria for the asymptotical synchronization and H_∞ synchronization for two-coupled PDSs and present the relationship between these results and the results we got before.

We consider the following two-coupled PDSs:

$$\begin{cases} \dfrac{\partial y^1(x,t)}{\partial t}=Ay^1(x,t)+B\nabla^2 y^1(x,t)+D\nabla^2(y^2(x,t)-y^1(x,t)) \\ \dfrac{\partial y^2(x,t)}{\partial t}=Ay^2(x,t)+B\nabla^2 y^2(x,t)+D\nabla^2(y^1(x,t)-y^2(x,t)) \end{cases} \tag{5.26}$$

subject to the boundary conditions and initial values

$$\begin{aligned} &y^j(x,t)=0, x\in\partial U \\ &y^j(x,0)=\varphi^j(x), j=1,2 \end{aligned} \tag{5.27}$$

and the disturbed two-coupled PDSs

$$\begin{cases} \dfrac{\partial y^1(x,t)}{\partial t}=Ay^1(x,t)+B\nabla^2 y^1(x,t)+D\nabla^2(y^2(x,t)-y^1(x,t))+v^1(x,t) \\ \dfrac{\partial y^2(x,t)}{\partial t}=Ay^2(x,t)+B\nabla^2 y^2(x,t)+D\nabla^2(y^1(x,t)-y^2(x,t))+v^2(x,t) \end{cases} \tag{5.28}$$

where $v^1(x,t)$ and $v^2(x,t)$ are the external disturbances and satisfy $v^i(x,t)\in L^2(0,t_f), x\in U$ and we also impose the boundary conditions and initial values equation (5.27).

From equations (5.26), (5.28) and keeping in mind $e(x,t)=y^1(x,t)-y^2(x,t)$, we have the following synchronization error dynamics

$$\frac{\partial e(x,t)}{\partial t}=A(x)e(x,t)+B\nabla^2 e(x,t)-2D\nabla^2 e(x,t) \tag{5.29}$$

and

$$\frac{\partial e(x,t)}{\partial t} = A(x)e(x,t) + B\nabla^2 e(x,t) - 2D\nabla^2 e(x,t) + v(x,t) \tag{5.30}$$

where $v(x,t) = v^1(x,t) - v^2(x,t)$. Obviously, $v(x,t) \in L^2(0,t_f)$. Now we present the following theorem on synchronization of system (5.26).

Theorem 5.3.3

For the synchronization error dynamic of two coupled PDSs in equation (5.29), if the following inequalities hold

$$B - 2D \geq 0 \tag{5.31}$$

$$A^T + A - 2\left(\frac{1}{l_1^2} + \frac{1}{l_2^2}\right)(B - 2D) < 0 \tag{5.32}$$

then the equilibrium point $e(x,t) = 0$ of equation (5.29) is asymptotically stable, that is, the two-coupled linear PDSs in equation (5.26) are asymptotically synchronous.

The proof is similar to the proof of Theorem 5.3.1, just take $P = I$ and $\bar{D} = B - 2D$ in the proof of Theorem 5.3.1, so we omit it.

On the H_∞ synchronization, we have the following theorem.

Theorem 5.3.4

Given a disturbance attenuation level γ, if the following inequalities hold

$$B - 2D \geq 0 \tag{5.33}$$

and

$$I + A^T + A + \frac{1}{\gamma^2}I - 2\left(\frac{1}{l_1^2} + \frac{1}{l_2^2}\right)(B - 2D) < 0 \tag{5.34}$$

then, if $e(x,0) = 0$, we have

$$\int_0^{t_f}\int_U e^T(x,t)e(x,t)dxdt < \gamma^2 \int_0^{t_f}\int_U v^T(x,t)v(x,t)dxdt$$

that is, the coupled PDSs equation (5.28) are robust H_∞ synchronization.

Proof

Take $P = I$ and $\bar{D} = B - 2D$, we can easily verify this theorem using the techniques in the proof of Theorem 5.3.2.

Now we point out the relationship of Theorems 5.3.1, 5.3.2 and the Theorems 5.3.3, 5.3.4. First, we claim that

$$\begin{pmatrix} B-D & D \\ D & B-D \end{pmatrix} \geq 0 \text{ implies } B - 2D \geq 0.$$

Now we give a brief proof. Since

$$\begin{pmatrix} x^T & y^T \end{pmatrix} \begin{pmatrix} B-D & D \\ D & B-D \end{pmatrix} \begin{pmatrix} x \\ y \end{pmatrix} \geq 0$$

holds for any x, y, take $y = -x$, we get

$$x^T (B - 2D) x \geq 0$$

which implies $B - 2D \geq 0$. Using the same technique, we get

$$\begin{pmatrix} A^T + A & 0 \\ 0 & A^T + A \end{pmatrix} - \left(\frac{1}{l_1^2} + \frac{1}{l_2^2} \right) \begin{pmatrix} B-D & D \\ D & B-D \end{pmatrix} < 0$$

which implies

$$A^T + A - 2\left(\frac{1}{l_1^2} + \frac{1}{l_2^2} \right)(B - 2D) < 0$$

From these results, we can get, without any difficulty, that if we take $P = I$ in Theorem 5.3.1, then the conditions in equations (5.13) and (5.14) imply the conditions in equations (5.31), (5.32), i.e., the Theorem 5.3.1 implies the Theorem 5.3.3. The same relationship exists for Theorem 5.3.2 and Theorem 5.3.4.

5.4 ASYMPTOTICAL SYNCHRONIZATION AND H_∞ SYNCHRONIZATION OF *N*-COUPLED PDSs WITH SPACE-DEPENDENT COEFFICIENTS

In this section, we consider the *N*-coupled PDSs with space dependent coefficients. We must point out that the technique used in equation (5.16) is not suitable under this situation since the coefficients are space-dependent. Therefore, we cannot get analogous results as the constant coefficient case. Now, we employ the Lyapunov-Krasovskii method to present a theorem on the asymptotical synchronization of *N*-coupled PDSs with space-dependent coefficients.

Theorem 5.4.1

For the synchronization error dynamic of N-coupled PDSs with space-dependent coefficients in equation (5.11), if there exists a positive-definite and radially unbounded function V(e), with V(0) = 0, such that the following integral inequality holds for all e(x, t) = 0:

$$\int_U \left(\frac{dV(e(x,t))}{\partial e}\right)^T \left[\bar{A}(x)e(x,t)+\bar{D}(x)\nabla^2 e(x,t)\right]dx<0 \tag{5.35}$$

then the equilibrium point $e(x,t)=0$ *of equation (5.11) is asymptotically stable, that is, the coupled linear PDSs in equation (5.9) are asymptotically synchronous.*

Proof

Let $L(t)=\int_U V(e(x,t))dx$.

$$\frac{dL(t)}{dt}=\int_U \left(\frac{dV(e(x,t))}{\partial e}\right)^T \left[\bar{A}(x)e(x,t)+\bar{D}(x)\nabla^2 e(x,t)\right]dx<0 \tag{5.36}$$

By virtue of $V(0)=0$, we can get $\int_U V(e(x,t))dx\to 0$ as $t\to\infty$, then $e(x,t)\to 0$ as $t\to\infty$. We can get that the equilibrium $e(x,t)=0$ of equation (5.11) is asymptotically stable, i.e., the N-coupled linear PDSs in equation (5.9) are asymptotically synchronous.

Remark 5.4.1

If the following inequality holds for some positive $V(\cdot)$,

$$\left(\frac{dV(e(x,t))}{de}\right)^T \left[\bar{A}(x)e(x,t)+\bar{D}(x)\nabla^2 e(x,t)\right]<0$$

for all $x\in U$, then the asymptotical synchronization result in Theorem 5.4.1 still holds but with stricter condition than equation (5.35).

Now we turn our attention to the H_∞ synchronization of the following N-coupled PDSs with space-dependent coefficients

$$\frac{\partial y^k(x,t)}{\partial t}=A(x)y^k(x,t)+B(x)\nabla^2 y^k(x,t)+\sum_{j\neq k} D_{kj}(x)\nabla^2(y^j(x,t)-y^k(x,t))+v^k(x,t) \tag{5.37}$$

subject to the boundary conditions and initial values

$$\begin{aligned} &y^k(x,t)=0, x\in\partial U,\\ &y^k(x,0)=\varphi^k(x), k=1,2,\ldots,N \end{aligned} \tag{5.38}$$

where $v^k(x,t)$ is the external disturbance and $v(x,t) \in L^2(0,t_f)$, for some positive constant t_f. We modify the H_∞ synchronization performance in equation (5.22) as

$$\int_0^{t_f}\int_U e^T(x,t)e(x,t)dxdt < \int_U V(e(x,0))dx + \gamma^2\int_0^{t_f}\int_U v^T(x,t)v(x,t)dxdt. \tag{5.39}$$

We can get the synchronization error dynamic as follows:

$$\frac{\partial e^k(x,t)}{\partial t} = A(x)e^k(x,t) + B(x)\nabla^2 e^k(x,t) + \sum_{j\neq k} D_{kj}(x)\nabla^2(e^j(x,t) - e^k(x,t)) + v^k(x,t).$$

Similarly, the above system can be presented by the following system:

$$\frac{\partial e(x,t)}{\partial t} = \bar{A}(x)e(x,t) + \bar{D}(x)\nabla^2 e(x,t) + v(x,t) \tag{5.40}$$

Theorem 5.4.2

Given a disturbance attenuation level γ, if there exists a positive differentiable function $V(x)$, such that the following integral inequality holds

$$\int_U\left[e^T(x,t)e(x,t) + \left(\frac{dV(e(x,t))}{de}\right)^T\left[\bar{A}(x)e(x,t) + \bar{D}(x)\nabla^2 e(x,t)\right] + \frac{1}{4\gamma^2}\left(\frac{dV(e(x,t))}{de}\right)^T\left(\frac{dV(e(x,t))}{de}\right)\right]dx < 0 \tag{5.41}$$

then the coupled PDSs equation (5.37) are of H_∞ synchronization.

Proof

Using the technique of completing the squares and equation (5.41), we have

$$\int_0^{t_f}\int_U\left[e^T(x,t)e(x,t) - \gamma^2 v^T(x,t)v(x,t)\right]dx\,dt$$

$$= \int_0^{t_f}\left(\int_U\left(e^T(x,t)e(x,t) - \gamma^2 v^T(x,t)v(x,t)\right)dx + \frac{\partial}{\partial t}\int_U Ve(x,t)dx\right)dt + \int_U [V(e(x,0)) - V(e(x,t_f))]dx$$

$$= \int_U [V(e(x,0)) - V(e(x,t_f))]dx + \int_0^{t_f} \int_U [e^T(x,t)e(x,t) - \gamma^2 v^T(x,t)v(x,t)$$

$$+ \left(\frac{dV(e(x,t))}{de}\right)^T [\bar{A}(x)e(x,t) + \bar{D}(x)\nabla^2 e(x,t) + v(x,t)]]dxdt$$

$$\leq \int_0^{t_f} \int_U \left[-\left(\gamma v(x,t) - \frac{1}{2\gamma}\frac{dV(e)}{de}\right)^T \left(\gamma v(x,t) - \frac{1}{2\gamma}\frac{dV(e)}{de}\right)\right.$$

$$+ \left(\frac{dV(e)}{de}\right)^T [\bar{A}(x)e(x,t) + \bar{D}(x)\nabla^2 e(x,t)] + e^T(x,t)e(x,t)$$

$$\left. + \frac{1}{4\gamma^2}\left(\frac{dV(e)}{de}\right)^T \left(\frac{dV(e)}{de}\right)\right] dxdt + \int_U V(e(x,0))dx$$

$$\leq \int_0^{t_f} \int_U \left[e^T(x,t)e(x,t) + \left(\frac{dV(e)}{de}\right)^T [\bar{A}(x)e(x,t) + \bar{D}(x)\nabla^2 e(x,t)]\right.$$

$$\left. + \frac{1}{4\gamma^2}\left(\frac{dV(e)}{de}\right)^T \left(\frac{dV(e)}{de}\right)\right] dxdt + \int_U V(e(x,0))dx$$

$$< \int_U V(e(x,0))dx.$$

Then the inequality in equation (5.39) holds. If $e(x,0) = 0$ and $V(0) = 0$, then the H_∞ synchronization in equation (5.21) holds, which completes the proof.

Remark 5.4.2

If the following inequality holds for some positive differentiable function *V*(*e*),

$$e^T(x,t)e(x,t) + \left(\frac{dV(e(x,t))}{de}\right)^T [\bar{A}(x)e(x,t) + \bar{D}(x)\nabla^2 e(x,t)]$$

$$+ \frac{1}{4\gamma^2}\left(\frac{dV(e(x,t))}{de}\right)^T \left(\frac{dV(e(x,t))}{de}\right) < 0$$

for all $x \in U$ and $t \in \mathbb{R}^+$, then the H_∞ synchronization result in Theorem 5.4.2 still holds but with stricter condition than equation (5.41).

Remark 5.4.3

It is easy to see inequality equation (5.41) implies inequality equation (5.35), since

$$e^T(x,t)e(x,t)+\frac{1}{4\gamma^2}\left(\frac{dV(e(x,t))}{de}\right)^T\left(\frac{dV(e(x,t))}{de}\right)\geq 0$$

that is, the H_∞ synchronization guarantees the asymptotical synchronization when $v(x,t)=0$.

Remark 5.4.4

Both the above results, Theorem 5.4.1 and Theorem 5.4.2, present sufficient conditions to guarantee the asymptotical synchronization and the H_∞ synchronization of coupled PDSs, respectively. However, as we all know, they are not practical: It is not easy to find a suitable Lyapunov-Krasovskii functional such that the inequalities in equations (5.35) and (5.41) hold, respectively. Therefore, we must proceed to find a practical method to deal with the synchronization of coupled PDSs with space-dependent coefficients.

5.5 H_∞ SYNCHRONIZATION CRITERIA BASED ON LMI

To simplify the H_∞ synchronization criteria for coupled PDSs with space-dependent coefficients, we need to develop a more suitable spatial state space model to represent the PDSs. For this purpose, the semi-discrete finite difference scheme is employed to treat the H_∞ synchronization problem of coupled PDSs with space-dependent coefficients. The finite difference scheme is used to approximate the partial differential operator ∇^2 to simplify the synchronization criteria of coupled PDSs. The finite-difference scheme has been widely used to obtain numerical solutions of PDSs [178,179]. In this chapter, we use the finite-difference method to represent the PDSs equation (5.37) to get the simpler synchronization criteria. Consider a typical grid mesh, as shown in Figure 5.1, the state $y(x, t)$ is presented by $y_{k,l}(t)$ at the grid node $x_{k,l}(x_1=k\Delta, x_2=l\Delta)$, where $k=0,\dots,N_1+1$ and $l=0,\dots,N_2+1$, i.e., $y(x,t)\big|_{x=x_{k,l}} \triangleq y_{k,l}(t)$. Note that the grid nodes $k=0$, $k=N_1+1$, $l=0$, or $l=N_2+1$ are the grid nodes at the boundary. At the interior points of grid, the central-difference approximation for the differential operator can be written as follows:

$$\begin{aligned}\nabla^2 y(x,t)\big|_{x=x_{k,l}} &= \left(\frac{\partial^2 y(x,t)}{\partial x_1^2}\right)_{x=x_{k,l}}+\left(\frac{\partial^2 y(x,t)}{\partial x_2^2}\right)_{x=x_{k,l}}\\ &= \frac{y_{k+1,l}(t)+y_{k-1,l}(t)+y_{k,l+1}(t)+y_{k,l-1}(t)-4y_{k,l}(t)}{\Delta^2}+O_{k,l}(\Delta^2) \qquad (5.42)\end{aligned}$$

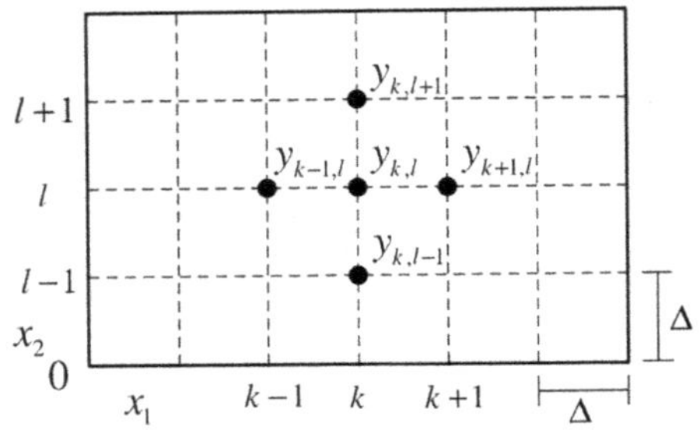

FIGURE 5.1 Finite-difference grids on the spatio-domain.

The remainder term $O_{k,l}\left(\Delta^2\right)$ is the local truncation error.

A finite-difference model can be constructed to represent $e(x, t)$ of synchronization error dynamic at $x = x_{k,l}$ in equation (5.40) as follows:

$$\left.\frac{\partial e(x,t)}{\partial t}\right|_{x=x_{k,l}}$$

$$= \frac{de_{k,l}(t)}{dt} = \bar{A}_{k,l}e_{k,l}(t) + \bar{D}_{k,l}\frac{e_{k+1,l}(t) + e_{k-1,l}(t) + e_{k,l+1}(t) + e_{k,l-1}(t) - 4e_{k,l}(t)}{\Delta^2} \quad (5.43)$$

$$+ v_{k,l}(t) + O_{k,l}(\Delta^2)$$

Let us denote

$$T_{k,l}e_{k,l}(t) = \frac{1}{\Delta^2}[e_{k+1,l}(t) + e_{k-1,l}(t) + e_{k,l+1}(t) + e_{k,l-1}(t) - 4e_{k,l}(t)] \quad (5.44)$$

then equation (5.43) can be rewritten as follows:

$$\frac{de_{k,l}(t)}{dt} = \bar{A}_{k,l}e_{k,l}(t) + \bar{D}_{k,l}T_{k,l}e_{k,l}(t) + v_{k,l}(t) + O_{k,l}(\Delta^2) \quad (5.45)$$

where $k = 1,2,\ldots,N_1$, $l = 1,2,\ldots,N_2$.

For the simplification of the asymptotical synchronization criterion for the coupled PDSs, we define a spatial vector $e(t)$ to collect the $e_{k,l}(t)$ at all gird nodes in Figure 5.1. From the boundary conditions, we know the values of $e_{k,l}(t)$ at the boundary are fixed. Therefore, the spatial state vector $e(t) \in \mathbb{R}^{nNM}$ for $e_{k,l}(t)$ at all grid nodes is defined as

$$e(t) = [e_{1,1}(t),\ldots,e_{k,1}(t),\ldots,e_{N_1,1}(t),\ldots e_{k,l}(t),\ldots e_{1,N_2}(t),\ldots e_{k,N_x}(t),\ldots e_{N_1,N_2}(t)] \quad (5.46)$$

where $M = N_1 \times N_2$. Note that nN is the dimension of the vector $e_{k,l}(t)$ for each grid node and $N_1 \times N_2$ is the number of grid nodes.

In order to simplify the index of the node $e_{k,l}(t) \in \mathbb{R}^{nN}$ in the spatial state vector $e(t) \in \mathbb{R}^{nNM}$, we denote the symbol $e_j(t) \in \mathbb{R}^{nN}$ to replace $e_{k,l}(t)$. Note that the index j from 1 to M, i.e., $e_1(t) = e_{1,1}(t)$, $e_2(t) = e_{2,1}(t)$, …, $e_j(t) = e_{k,l}(t)$, $e_M(t) = e_{N_1,N_2}(t)$, where $j = (l-1)N_1 + k$ in equation (5.46). The difference model of the two indexes in equation (5.45) could be represented with only one index as follows:

$$\frac{de_j(t)}{dt} = \bar{A}_j e_j(t) + \bar{D}_j T_j e(t) + v_j(t) + O_j(\Delta^2) \tag{5.47}$$

$j = 1,.2,...,M$, with $j = (l-1)N_1 + k$ and $T_j e(t)$ is defined as

$$T_j e(t) = \frac{1}{\Delta^2}[0,...,0,I,0,...,0,I,\ 4I,\ I,0,...,0,I,0,...,0]e(t)$$

$$\text{position} \rightarrow \qquad 1 \qquad j-N_1 \quad j-1 \quad j \quad j+1 \qquad j+N_1 \quad N_1+N_2$$

where 0 denotes the $nN \times nN$ zero matrix and I denotes the $nN \times nN$ identity matrix.

We collect all synchronization errors $e_j(t)$ of the grid nodes to the synchronization state vector in equation (5.46). Take

$$\bar{A} = \begin{pmatrix} \bar{A}_1 & 0 & \cdots & 0 \\ 0 & \bar{A}_2 & \ddots & 0 \\ \vdots & \ddots & \ddots & \vdots \\ 0 & 0 & \cdots & \bar{A}_M \end{pmatrix}, \bar{D} = \begin{pmatrix} \bar{D}_1 & 0 & \cdots & 0 \\ 0 & \bar{D}_2 & \ddots & 0 \\ \vdots & \ddots & \ddots & \vdots \\ 0 & 0 & \cdots & \bar{D}_M \end{pmatrix} \tag{5.48}$$

and

$$\bar{T} = \begin{pmatrix} T_1 \\ \vdots \\ T_M \end{pmatrix}, O(\Delta^2) = \begin{pmatrix} O_1(\Delta^2) \\ \vdots \\ O_M(\Delta^2) \end{pmatrix}$$

then equation (5.47) can be represented by the following spatial space state system

$$\begin{aligned} \frac{de(t)}{dt} &= \bar{A}e(t) + \bar{D}\bar{T}e(t) + \bar{v}(t) + O(\Delta^2) \\ &= (\bar{A} + \bar{D}T)e(t) + \bar{v}(t) + O(\Delta^2) \end{aligned} \tag{5.49}$$

where $e(t) = (e_1^T(t), e_2^T(t),..., e_M^T(t))^T$ represents $e(x,t)$ at all grid points on the spatial domain U, $\bar{v}(t) = (v_1^T(t), v_2^T(t),..., v_M^T(t))^T$. Take $v(t) = \bar{v}(t) + O(\Delta^2)$, equation (5.49) can be rewritten as follows:

$$\frac{de(t)}{dt} = (\bar{A} + \bar{D}\bar{T})e(t) + \bar{v}(t) \tag{5.50}$$

Suppose the external disturbance $\bar{v}(t)$ is absent and the truncation error $O(\Delta^2)$ of the finite difference scheme approaches zero as the number of grid nodes is large enough, i.e., $v(x,t)=0$. Then the asymptotical stability of system in equation (5.50) can be guaranteed by the following theorem.

Theorem 5.5.1

For the augmented system in equation (5.50) with $v(x,t)=0$, suppose there exists a symmetric positive matrix P such that the following inequality holds

$$\left(\bar{A}+\bar{D}\bar{T}\right)^T P + P\left(\bar{A}+\bar{D}\bar{T}\right) < 0 \tag{5.51}$$

then the equilibrium $e(x,t)=0$ of system equation (5.50) is asymptotically stable.

The proof is standard, so we omit it.

From the practical point of view, the external disturbance and the truncation error will not be zero, then we turn our attention to the H_∞ synchronization. For the spatial state space system as in equation (5.50), with some positive constant t_f, the H_∞ synchronization performance is described by

$$\int_0^{t_f} e^T(t)e(t)dt < V(e(0)) + \gamma^2 \int_0^{t_f} v^T(t)v(t)dt \tag{5.52}$$

for some positive function $V(\cdot)$.

Theorem 5.5.2

For the augmented system in equation (5.50) with a prescribed noise attenuation level γ, if there exists a positive matrix P such that the following LMI holds:

$$\begin{pmatrix} I+\left(\bar{A}+\bar{D}\bar{T}\right)^T P + P(\bar{A}+\bar{D}\bar{T}) & P^T \\ P & -\gamma^2 I \end{pmatrix} < 0 \tag{5.53}$$

then the inequality in equation (5.52) holds, and the coupled PDSs equation (5.37) are of H_∞ synchronization. Furthermore, the asymptotical synchronization could be achieved when $v(x,t)=0$.

Proof

Let $V(e(t)) = e^T(t)Pe(t)$. In term of the technique of completing the squares, we get

$$\int_0^{t_f} [e^T(t)e(t) - \gamma^2 v^T(t)v(t)]dt$$

$$= \int_0^{t_f} [e^T(t)e(t) - \gamma^2 v^T(t)v(t) + \frac{dV}{dt}]dt + V(e(0)) - V(e(t_f))$$

$$= \int_0^{t_f} [e^T(t)e(t) - \gamma^2 v^T(t)v(t) + e^T(t)((\bar{A} + \bar{D}\bar{T})^T P + P(\bar{A} + \bar{D}\bar{T}))e(t)]dt$$

$$+ V(e(0)) - V(e(t_f))$$

$$= V(e(0)) - V(e(t_f)) + \int_0^{t_f} \left[e^T(t)e(t) - \left[\gamma v(t) - \frac{1}{\gamma} Pe(t)\right]^T \left[\gamma v(t) - \frac{1}{\gamma} Pe(t)\right] \right.$$

$$\left. + e^T(t)((\bar{A} + \bar{D}\bar{T})^T P + P(\bar{A} + \bar{D}\bar{T}))e(t) + \frac{1}{\gamma^2} e^T(t)P^T Pe(t) \right] dt$$

$$= V(e(0)) - V(e(t_f)) + \int_0^{t_f} \left[-\left[\gamma v(t) - \frac{1}{\gamma} Pe(t)\right]^T \left[\gamma v(t) - \frac{1}{\gamma} Pe(t)\right] \right.$$

$$\left. + e^T(t)\left[I + (\bar{A} + \bar{D}\bar{T})^T P + P(\bar{A} + \bar{D}\bar{T}) + \frac{1}{\gamma^2} P^T P \right] e(t) \right] dt$$

$$\le V(e(0)) + \int_0^{t_f} e^T(t)\left[I + (\bar{A} + \bar{D}\bar{T})^T P + P(\bar{A} + \bar{D}\bar{T}) + \frac{1}{\gamma^2} P^T P \right] e(t)dt$$

Obviously, if $I + (\bar{A} + \bar{D}\bar{T})^T P + P(\bar{A} + \bar{D}\bar{T}) + \frac{1}{\gamma^2} P^T P < 0$, we can get the desired result in equation (5.52). By Schur complement, the LMI in equation (5.53) implies the above relationship. The last desired result is obvious, since

$$I + (\bar{A} + \bar{D}\bar{T})^T P + P(\bar{A} + \bar{D}\bar{T}) + \frac{1}{\gamma^2} P^T P < 0$$

implies

$$\left(\bar{A} + \bar{D}\bar{T}\right)^T P + P\left(\bar{A} + \bar{D}\bar{T}\right) < 0$$

From Theorem 5.5.1, the asymptotical synchronization is achieved when $v(x,t) = 0$.

The optimal noise attenuation level γ_0 can be obtained by solving the following constrained optimization

$$\gamma_0^2 = \min_{P>0} \gamma^2 \quad \text{subject to (5.53)} \tag{5.54}$$

This constrained optimization can be easily solved by decreasing γ^2 until no existence of positive solution P of the LMI in equation (5.53), which can be easily solved by the LMI toolbox in MATLAB.

Now we present the relationship between the integral inequality in equation (5.41) and the LMI in equation (5.53). First, we take $V(e(x,t)) = e^T(x,t)P(x)e(x,t)$ in equation (5.41), then with the semi-discrete difference approximation and approximation of integration by summation, the left of integral inequality equation (5.41) becomes

$$\int_U \left[e^T(x,t)e(x,t) + \left(\frac{dVe(x,t)}{de}\right)^T [\bar{A}(x)e(x,t) + \bar{D}(x)\nabla^2 e(x,t)] \right.$$

$$\left. + \frac{1}{4\gamma^2}\left(\frac{dVe(x,t)}{de}\right)^T\left(\frac{dVe(x,t)}{de}\right)\right]dx$$

$$\approx \sum_{k=1}^{N_1}\sum_{l=1}^{N_2}\left\{ e_{k,l}^T(t)e_{k,l}(t) + 2P_{k,l}e_{k,l}(t)^T \right.$$

$$\times\left[\bar{D}_{k,l}\frac{e_{k+1,l}(t)+e_{k-1,l}(t)+e_{k,l+1}(t)+e_{k,l-1}(t)-4e_{k,l}(t)}{\Delta^2}\right.$$

$$\left.\left. + \bar{A}_{k,l}e_{k,l}(t) + O_{k,l}(\Delta^2)\right] + \frac{1}{4\gamma^2}(2P_{k,l}e_{k,l}(t))^T(2P_{k,l}e_{k,l}(t))\right\}\Delta^2$$

For the simplicity, we denote the symbols $P_j, \bar{A}_j, \bar{D}_j$, and $O_j(\Delta^2)$ to replace $P_{k,l}, \bar{A}_{k,l}, \bar{D}_{k,l}$, and $O_{k,l}(\Delta^2)$, where $j = (l-1)N_1 + k$. Noting $M = N_1 \times N_2$, we have

$$\sum_{k=1}^{N_1}\sum_{l=1}^{N_2}\{e_{k,l}^T(t)e_{k,l}(t) + 2P_{k,l}e_{k,l}(t)^T\left[\bar{D}_{k,l}\frac{e_{k+1,l}(t)+e_{k-1,l}(t)+e_{k,l+1}(t)+e_{k,l-1}(t)-4e_{k,l}(t)}{\Delta^2}\right.$$

$$\left.\left. + \bar{A}_{k,l}e_{k,l}(t) + O_{k,l}(\Delta^2)\right] + \frac{1}{4\gamma^2}(2P_{k,l}e_{k,l}(t))^T(2P_{k,l}e_{k,l}(t))\right\}\Delta^2$$

$$= \sum_{j=1}^{M}\{e_j^T(t)e_j(t) + 2P_je_j(t)^T\left[\bar{D}_j\frac{e_{j+1}(t)+e_{j-1}(t)+e_{j+N_1}(t)+e_{j-N_1}(t)-4e_j(t)}{\Delta^2}\right.$$

$$\left.\left. + \bar{A}_je_j(t) + O_j(\Delta^2)\right] + \frac{1}{4\gamma^2}(2P_je_j(t))^T(2P_je_j(t))\right\}\Delta^2$$

Take $P = diag\{P_j\}$, we have

$$\sum_{j=1}^{M}\{e_j^T(t)e_j(t)+2P_je_j(t)^T\Big[\bar{D}_j\frac{e_{j+1}(t)+e_{j-1}(t)+e_{j+N_1}(t)+e_{j-N_1}(t)-4e_j(t)}{\Delta^2}$$

$$+\bar{A}_je_j(t)+O_j(\Delta^2)\Big]+\frac{1}{4\gamma^2}(2P_je_j(t))^T(2P_je_j(t))\Big\}\Delta^2$$

$$=\left[e^T(t)\left(I+\bar{A}^TP^T+P\bar{A}+(\bar{D}\bar{T})^TP+P(\bar{D}\bar{T})+\frac{1}{\gamma^2}P^TP\right)e(t)+2e^T(t)P^T\bar{D}O(\Delta^2)\right]\Delta^2$$

where $\bar{A},\bar{D},\bar{T}$, and $O(\Delta^2)$ are defined in equation (5.48) and $e(t)=(e_1^T(t),\ldots,e_M^T(t))^T$. We can read from the last equality, for small enough Δ,

$$I+\bar{A}^TP^T+P\bar{A}+(\bar{D}\bar{T})^TP+P(\bar{D}\bar{T})+\frac{1}{\gamma^2}P^TP<0$$

which results in

$$e^T(t)\left(I+\bar{A}^TP^T+P\bar{A}+(\bar{D}\bar{T})^TP+P(\bar{D}\bar{T})+\frac{1}{\gamma^2}P^TP\right)e(t)+2e^T(t)P^T\bar{D}O(\Delta^2)<0,$$

that indicates the integral inequality in equation (5.41) with $V(e)=e^TPe$. We also know that the inequality

$$I+\bar{A}^TP^T+P\bar{A}+(\bar{D}\bar{T})^TP+P(\bar{D}\bar{T})+\frac{1}{\gamma^2}P^TP<0$$

is equivalent to the LMI in equation (5.53). Then we get the following statement:

Proposition 5.5.1

For small enough Δ, if the LMI in equation (5.53) holds, then the integral inequality equation (5.41) holds and the H_∞ synchronization of the coupled disturbed PDSs equation (5.37) is guaranteed.

5.6 NUMERICAL EXAMPLES

In this section, we give two examples to illustrate the correctness of our results.

Example 5.1: First, we consider the following three-coupled PDSs with constant coefficients

$$\begin{aligned}
\frac{\partial z^1(x,y,t)}{\partial t} &= Az^1(x,y,t) + B\nabla^2 z^1(x,y,t) + D_{12}\nabla^2(z^2(x,y,t) - z^1(x,y,t)) \\
&\quad + D_{13}\nabla^2(z^3(x,y,t) - z^1(x,y,t)), \\
\frac{\partial z^2(x,y,t)}{\partial t} &= Az^2(x,y,t) + B\nabla^2 z^2(x,y,t) + D_{12}\nabla^2(z^1(x,y,t) - z^2(x,y,t)) \\
&\quad + D_{23}\nabla^2(z^3(x,y,t) - z^2(x,y,t)), \\
\frac{\partial z^3(x,y,t)}{\partial t} &= Az^3(x,y,t) + B\nabla^2 z^3(x,y,t) + D_{13}\nabla^2(z^1(x,y,t) - z^3(x,y,t)) \\
&\quad + D_{23}\nabla^2(z^2(x,y,t) - z^3(x,y,t))
\end{aligned} \tag{5.55}$$

where $z^1(x,y,t), z^2(x,y,t), z^3(x,y,t) \in \mathbb{R}^2, (x,y) \in U = [0,1] \times [0,1]$ and

$$A = \begin{pmatrix} -0.0001 & 0.0001 \\ -0.0001 & -0.0001 \end{pmatrix}, B = \begin{pmatrix} 0.0001 & 0.0001 \\ 0.0001 & 0.0001 \end{pmatrix},$$

$$D_{12} = D_{13} = D_{23} = \begin{pmatrix} -0.0001 & 0 \\ 0 & -0.0001 \end{pmatrix}.$$

We put the boundary conditions and initial conditions for three-coupled PDSs in equation (5.55) as follows:

$$z^i(0,y,t) = \begin{pmatrix} z^{i1}(0,y,t) \\ z^{i2}(0,y,t) \end{pmatrix} = z^i(x,0,t) = \begin{pmatrix} z^{i1}(x,0,t) \\ z^{i2}(x,0,t) \end{pmatrix} = \begin{pmatrix} 0 \\ 0 \end{pmatrix}$$

$$z^i(1,y,t) = \begin{pmatrix} z^{i1}(1,y,t) \\ z^{i2}(1,y,t) \end{pmatrix} = z^i(x,1,t) = \begin{pmatrix} z^{i1}(x,1,t) \\ z^{i2}(x,1,t) \end{pmatrix} = \begin{pmatrix} 0 \\ 0 \end{pmatrix}$$

$i = 1, 2.$

$$z^1(x,y,0) = \begin{pmatrix} z^{11}(x,y,0) \\ z^{12}(x,y,0) \end{pmatrix} = \begin{pmatrix} \sin(2\pi x)\sin(2\pi y)\sin(1) \\ \sin(2\pi x)\sin(2\pi y)\sin(2) \end{pmatrix}$$

$$z^2(x,y,0) = \begin{pmatrix} z^{21}(x,y,0) \\ z^{22}(x,y,0) \end{pmatrix} = \begin{pmatrix} \sin(3\pi x)\sin(3\pi y)\cos(1) \\ \sin(3\pi x)\sin(3\pi y)\cos(2) \end{pmatrix}$$

$$z^3(x,y,0) = \begin{pmatrix} z^{31}(x,y,0) \\ z^{32}(x,y,0) \end{pmatrix} = \begin{pmatrix} \sin(3\pi x)\sin(2\pi y)\cos(1) \\ \sin(2\pi x)\sin(3\pi y)\cos(2) \end{pmatrix}$$

Take $P = I$ in the Theorem 5.3.1, it is easy to verify the conditions in equations (5.13) and (5.14) hold under this situation. The states $z^{11}, z^{21}, z^{31}, z^{12}, z^{22}, z^{32}$ are shown in the Figures 5.2 and 5.3 at $t = 0$, $t = 40$, $t = 80$, respectively. We can read from these figures that the coupled PDSs are of asymptotical synchronization.

Here we also provide figures to illustrate the difference between each two states at some fixed space point. We take $y = 0.5$, from Figure 5.4, we can also see the asymptotical synchronization of the coupled PDSs in equation (5.55).

Now we will illustrate the effect of the domain U on the synchronization. We take $U = [0, 4] \times [0, 4]$, and we provide the difference between each two states at $y = 0.5$ in Figure 5.5. It can be seen that the larger domain U makes the coupled PDSs more time achieve asymptotical synchronization, and this can show the correctness of Proposition 5.3.1.

Now, we present another example of disturbed two-coupled PDSs with space-dependent coefficients.

Example 5.2: We consider the following disturbed two-coupled PDSs with space-dependent coefficients:

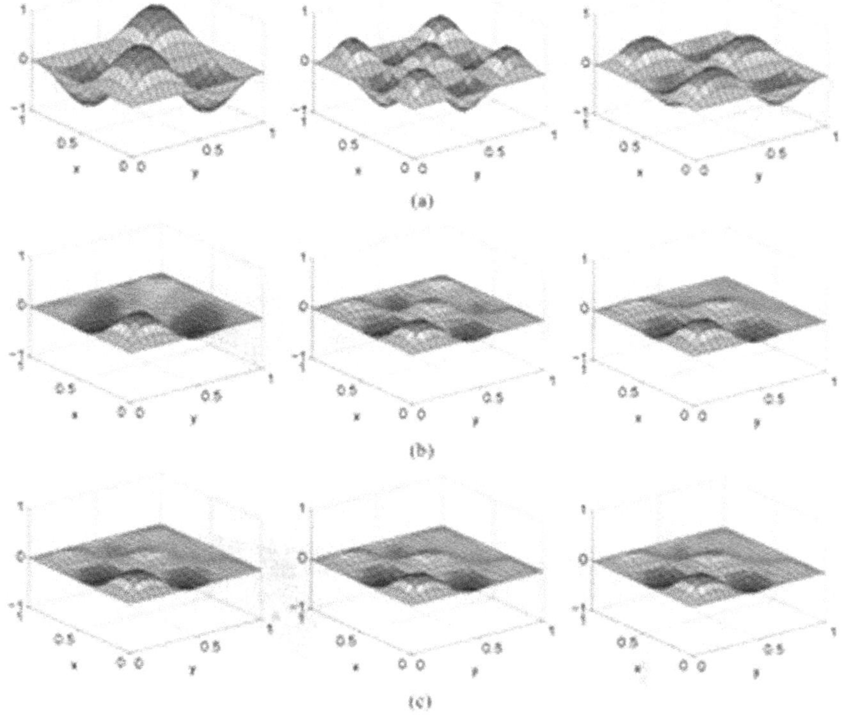

FIGURE 5.2 Simulation results of $z^{11}(x, t)$, $z^{21}(x, t)$, and $z^{31}(x, t)$ (from left to right) of coupled PDSs in Example 5.1 at different time: (a) $t = 0$, (b) $t = 40$, (c) $t = 80$.

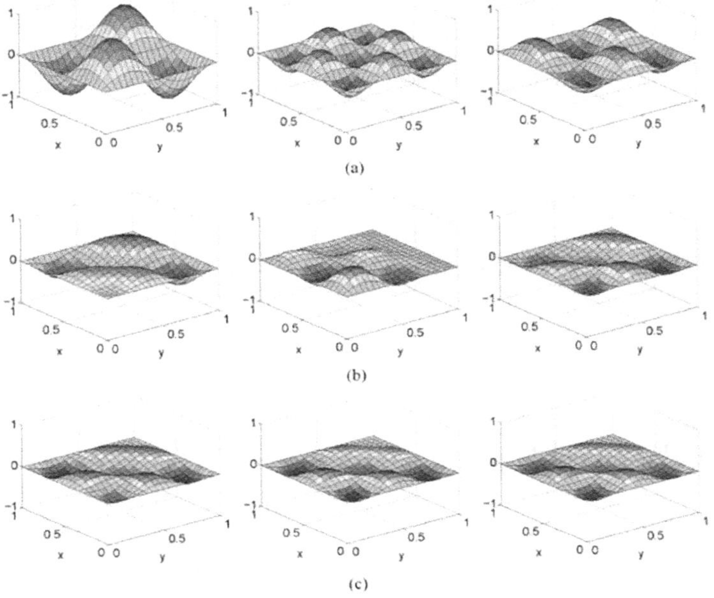

FIGURE 5.3 Simulation results of $z^{12}(x, t)$, $z^{22}(x, t)$, and $z^{32}(x, t)$ (from left to right) of coupled PDSs in Example 1 at different times: (a) $t = 0$, (b) $t = 40$, (c) $t = 80$.

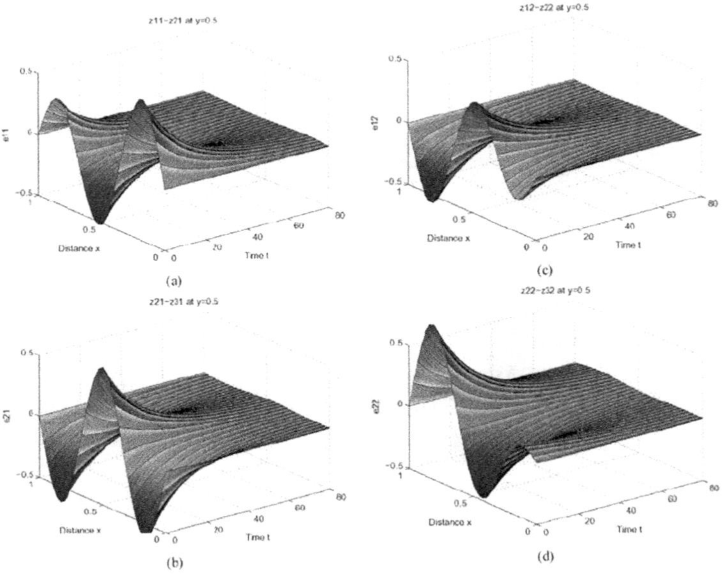

FIGURE 5.4 Difference between each two states at $y = 0.5$ in Example 5.1. (a) z^{11}–z^{21} at $y = 0.5$; (b) z^{21}–z^{31} at $y = 0.5$; (c) z^{12}–z^{22} at $y = 0.5$; (d) z^{22}–z^{32} at $y = 0.5$.

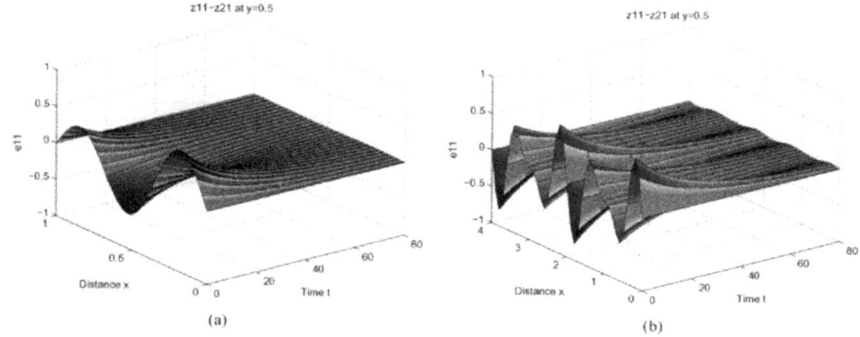

FIGURE 5.5 Effect of the space domain on the synchronization in Example 5.1. It is seen that a larger space domain makes the coupled PDSs take more time to achieve asymptotical synchronization. (a) $U = [0, 1]\times[0, 1]$; (b) $U = [0, 4]\times[0, 4]$.

$$\begin{aligned}
\frac{\partial z^1(x,y,t)}{\partial t} &= A(x,y)z^1(x,y,t) + B(x,y)\nabla^2 z^1(x,y,t) + D(x,y) \\
&\quad \times \nabla^2(z^2(x,y,t) - z^1(x,y,t)) + v^1(x,y,t) \\
\frac{\partial z^2(x,y,t)}{\partial t} &= A(x,y)z^2(x,y,t) + B(x,y)\nabla^2 z^2(x,y,t) + D(x,y) \\
&\quad \times \nabla^2(z^2(x,y,t) - z^1(x,y,t)) + v^2(x,y,t)
\end{aligned} \tag{5.56}$$

where

$$A(x,y) = \begin{pmatrix} -4+\sin(x) & \sin(x) \\ \sin(y) & -4+\sin(y) \end{pmatrix}, B(x,y) = \begin{pmatrix} 0.01\sin^2(x) & 0 \\ 0 & 0.01\sin^2(y) \end{pmatrix},$$

$$D(x,y) = \begin{pmatrix} 0.01\sin^2(x) & 0 \\ 0 & 0.01\sin^2(y) \end{pmatrix}, v^1(x,y,t) = \begin{pmatrix} 0.1\sin(xy) \\ 0.1\sin(xy) + 0.2\sin(t) \end{pmatrix},$$

$$v^2(x,y,t) = \begin{pmatrix} 0.1\sin(xy) - 0.2\sin(t) \\ 0.1\sin(xy) \end{pmatrix}.$$

We consider the H_∞ synchronization of two-coupled PDSs equation (5.56). We take $U = [0, 1]\times[0, 1]$, $\Delta = 1/6$ and $N_1 = N_2 = 5$, then $M = N_1 \times N_2 = 25$. According to equation (5.48), we get the matrix $\bar{A}, \bar{D}$, and $\bar{T}$. From the LMI in equation (5.53), we can get a positive-definite matrix P, that is, the coupled systems in equation (5.56) are of H_∞ synchronization. Moreover, it can be obtained that the optimal noise attenuation level $\gamma_0 = 0.6578$ by solving the constrained optimization problem in equation (5.54). Take the boundary conditions and initial values as follows:

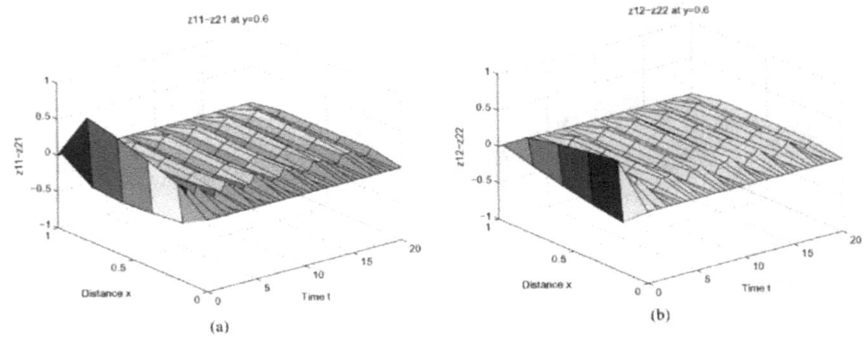

FIGURE 5.6 Difference between two coupled PDSs with space-dependent coefficients in Example 5.2. (a) Difference between z^{11} and z^{21} at $y = 0.6$; (b) Difference between z^{12} and z^{22} at $y = 0.6$.

$$z^1(x,y,0) = (x + 0.5\sin(x)\cos(y), y + 0.5\sin(y)\cos(x))^T$$

$$z^2(x,y,0) = (x - 0.5\sin(x)\cos(y), y - 0.5\sin(y)\cos(x))^T$$

$$z^1(x,0,t) = z^1(0,y,t) = z^2(x,0,t) = z^2(0,y,t) = 0$$

$$z^1(x,1,t) = z^1(1,y,t) = z^2(x,1,t) = z^2(1,y,t) = 0$$

We show the H_∞ synchronization of the coupled PDSs equation (5.56) by Figure 5.6, where the difference between state $z^1(x,y,t) = (z^{11}, z^{12})$ and $z^2(x,y,t) = (z^{21}, z^{22})$ is illustrated at $y = 0.6$.

The H_∞ synchronization performance can be computed by simulation as follows:

$$\frac{\int_0^{20} e^T(t)e(t)\,dt - V(e(0))}{\int_0^{20} v^T(t)v(t)\,dt} \approx\sim 0.4405^2 < \gamma_0^2 = 0.6578^2$$

here we take $V(e(0)) = e^T(0)Pe(0)$, P is obtained by solving the LMI in equation (5.53).

5.7 CONCLUSION

This chapter has considered the asymptotical synchronization problem and the H_∞ synchronization problem for the coupled PDSs with constant coefficients and with space-dependent coefficients, respectively. We have constructed the synchronization error dynamic and turned the synchronization problems of coupled PDSs to the stabilization problems of the synchronization error dynamic. Based on the synchronization error dynamic, we have presented sufficient conditions to guarantee the asymptotical synchronization which can be verified using the system's coefficients of the coupled PDSs and the range of the space domain. We also give the criterion of the H_∞ synchronization for N-coupled PDSs with constant coefficients. The effect of spatial domain on the synchronization can be read from these criteria. We have found that it needs to take more time for coupled PDSs to achieve synchronization in a lager rang of space domain. For the N-coupled PDSs with space-dependent coefficients, we first study the sufficient conditions for the asymptotical synchronization and the H_∞ synchronization with the Lyapunov-Krasovskii method. These sufficient conditions are given by integral inequalities. Generally, they are difficult to be solved. To overcome these difficulties, we have adopted the semi-discrete difference method to turn the coupled PDSs into an equivalent spatial state space system, then a sufficient condition is presented based on an LMI, which guarantees the H_∞ synchronization and is easy to be verified. Finally, two numerical simulations are given to illustrate the correctness of our results.

Part III

Robust Control System Design

6 Robust Stabilization Control Design of Large Structural Systems under Mode Truncation, Parameter Perturbations and Actuator Saturations

6.1 INTRODUCTION

The stabilization control problem of large structural systems (LSSs) differs from that of more conventional linear systems. The main reason is that the plant to be controlled is theoretically of infinite order and practically of very high order, while the controller and estimator must, of necessity, be of relatively lower order. In this framework the modelling error results from "mode truncation." One prominent effect due to the mode truncation in control design is that the uncontrolled high frequency modes will contaminate the observed modal amplitudes. This is called "observation spillover." When the system is excited by the feedback controller, it is known as "control spillover." If both types of the control and observation spillover are present, the controlled system may become unstable. This is referred to as the spillover phenomenon and was demonstrated by [195].

In LSSs, the system parameters, such as modal frequencies, damping coefficients, and mode shapes, are all known to be uncertain. A typical observation in structural dynamics is that the uncertainty of modes tends to increase with the mode number. But in general, the nature of this increase is not clearly known. These parameter variations can degrade the system performance and it is sometimes possible to destabilize the system. In addition, a more practical consideration in the control of an LSS is the actuator saturation. This gives rise to a nonlinear effect. Whenever a large overshoot occurs in the actuator, this nonlinearity will become significant and may induce an unstable output response. In the light of these observations it is clear that the stability is extremely desirable of any active control design proposed for LSS.

DOI: 10.1201/9781003229230-9

In the literature, the stability of large flexible structures has been investigated extensively. However, numerous reports have concentrated on the stability of reduced order design [195–197,203,211]. The problem of parameter perturbations and actuator saturations in the control design of an LSS has not received as much attention as the spillover problem. Recent results regarding both mode truncation and parameter variation problems for an LSS have been discussed by Pritchard and Townley [206] via structured stability radii in the frequency domain. Wagie and Skelton [208] adopted the time domain approach to treat the small parameter variation case. However, there still lacks stability conditions for an LSS under constrained control.

In this chapter, the robust stabilization design problem of an LSS under (i) mode truncation and linear parametric perturbations, or (ii) mode truncation and actuator saturations is treated within a state-space framework. The robust stabilization criterion in the time domain for the LSS under parameter perturbations or with saturating actuators is introduced. Robust stabilization analysis is also considered in this paper because the vector norm and its corresponding induced matrix norm are convenient in calculation. In addition, based on this, the relationships between the stability margins of controlled and residual dynamics and the spillover will be clearly observed from the derived stability conditions. A way to synthesize a stabilizing observer-based compensator such that the resulting nominal closed-loop poles can achieve certain optimal senses is provided. Finally, some numerical studies of the robust stabilizing control design on a simply supported beam are presented to illustrate the application of the proposed design methods.

6.2 MATHEMATICAL NOTATIONS AND PRELIMINARIES

Before any further analysis about the robust stabilization of an LSS under control is carried out, some relevant definitions are given as follows. The notation $*$ denotes the operation of convolution, i.e.,

$$y(t) = f(t) * g(t) = \int_{i=0}^{t} f(t-\tau)g(\tau)d\tau$$

The norm of real vector $x \in \mathbb{R}^n$ denoted by $\|x\|$ is defined as

$$\|x\| = \left(\sum_{i=1}^{n} |x_i|^2\right)^{\frac{1}{2}}$$

and for $y = Ax$, the induced matrix norm corresponding to the defined vector norm is given as [200]

$$\|A\| = \sup_{x \in \mathbb{R}^n} \frac{\|Ax\|}{\|x\|} = \sup_{\|x\|=1} \|Ax\| = \max_i \left(\lambda_i\left(A^T A\right)\right)^{\frac{1}{2}}$$

where x_i, for $i=1,2,\ldots,n$, denotes the element of the vector x, and A_{ij}, for $i,j=1,2,\ldots,n$, denotes the entry of the matrix A. $\lambda_i(A)$ is the ith eigenvalue of the matrix A. $\mathbb{C}$ –denotes the open left half complex plane, i.e.,

$$\mathbb{C}_{-} \equiv \{s \in \mathbb{C} \mid \mathrm{Re}(s<0)\}$$

where Re(·) indicates the real part.

Lemma 6.2.1 [199,202]

If an operator A generates a strongly continuous semigroup $\Phi_n(t)=\exp(At)$ such that $\|\Phi_n(t)\| \le \gamma \exp(-\alpha t)$, where $\gamma>0, \alpha>0$, and if B is a bounded operator, then (A+B) will generate a strongly continuous semigroup $\Phi_p(t)=\exp[(A+B)t]$

$$\|\Phi_p(t)\| \le \gamma \exp\left[-\left(\alpha-\gamma\|B\|t\right)\right]$$

The following lemma can be derived by extending a standard result due to Popov [205].

Lemma 6.2.2

If all elements of $g(t), g(t) \in \mathbb{R}^n$, are uniformly continuous for

$$\lim_{t\to\infty}\int_0^t \|g(\tau)\| d\tau < \infty$$

then

$$\lim_{t\to\infty}\|g(\tau)\| = 0$$

6.3 SYSTEM DESCRIPTION OF LSS

According to the Galerkin method to PDSs in Section 1.2.1, a standard modal state-space description of an LSS evaluation model with N controlled modes is considered as follows:

$$\dot{x}_c(t) = A_c x_c(t) + B_c f(t) \tag{6.1a}$$

$$\dot{x}_r(t) = A_r x_r(t) + B_r f(t) \tag{6.1b}$$

$$y(t) = C_c x_c(t) + C_r x_r(t) \tag{6.1c}$$

where the controlled mode state vector $x_c(t) \in \mathbb{R}^{2n}$ and the residual mode state vector $x_r(t) \in \mathbb{R}^{2(L-N)}$ are defined, respectively, as

$$x_c(t) = \begin{bmatrix} u_1(t) & u_2(t) & \cdots & u_N(t) & v_1(t) & v_2(t) & \cdots & v_N(t) \end{bmatrix}^T$$

$$x_r(t) = \begin{bmatrix} u_{N+1}(t) & u_{N+2}(t) & \cdots & u_L(t) & v_{N+1}(t) & v_{N+2}(t) & \cdots & v_L(t) \end{bmatrix}^T$$

where $u_k(t)$ represents the modal amplitude of the kth mode in normal modal coordinates. The reduced modal velocity $v_k(t) = u_k(t)/\omega_k$, where ω_k is the model frequency of the kth mode. $y(t) \in \mathbb{R}^z$ is a displacement sensor output vector and $f(t) \in \mathbb{R}^M$ is a control force vector. The system parameters [194,196] are

$$A_{c,r} = \begin{bmatrix} 0 & \Lambda_{c,r}^{1/2} \\ -\Lambda_{c,r}^{1/2} & \zeta_{c,r}\Lambda_{c,r}^{1/2} \end{bmatrix}, B_{c,r} = \begin{bmatrix} 0 \\ -\Lambda_{c,r}^{1/2}\bar{B}_{c,r} \end{bmatrix}, C_{c,r} = \begin{bmatrix} \bar{C}_{c,r} & 0 \end{bmatrix} \quad (6.2)$$

where subscripts "c" and "r" refer to the controlled and residual modes, respectively. Also, A_c is of dimension $2N \times 2N$, A_r is $2(L-N)\times 2(L-N)$. B_c is $2N \times M$, B_r is $2(L-N)\times M$, C_c is $Z \times 2N$, C_r is $Z \times 2(L-N)$. $\Lambda_c^{1/2} = diag\begin{bmatrix} \omega_1 & \omega_2 & \cdots & \omega_N \end{bmatrix}$ and $\Lambda_r^{1/2} = diag\begin{bmatrix} \omega_{N+1} & \omega_{N+2} & \cdots & \omega_L \end{bmatrix}$ with the modal frequency ω_k arranged in an increasing order (i.e., $\omega_k \le \omega_{k+1}$), $\varsigma_c^{1/2} = diag\begin{bmatrix} \varsigma_1 & \varsigma_2 & \cdots & \varsigma_N \end{bmatrix}$, and $\varsigma_c^{1/2} = diag\begin{bmatrix} \varsigma_1 & \varsigma_2 & \cdots & \varsigma_N \end{bmatrix}$ with ς_k being a structural damping coefficient of the kth mode. $\bar{B}_c$ and $\bar{B}_r$ are actuator influence matrices; $\bar{C}_c$ and $\bar{C}_r$ are sensor influence matrices. Note that theoretically $L \to \infty$, yet L may be large but finite in practice.

The modes of the controlled and residual parts can be separated from each other as required by standard mode-reduction techniques (e.g., aggregation, modal cost analysis, balancing, etc., see [195]), which allows for an additional freedom in the choice of the two parts.

Without loss of generality, an assumption is first made with respect to the controlled dynamics: the matrix pair (A_c, C_c) is completely observable, and the pair (A_c, B_c) is completely controllable.

6.4 STABILIZATION OF LSS WITH PARAMETER VARIATIONS

For an LSS control problem, the parameters of the plant matrix, the actuator influence matrix, and sensor influence matrix, namely, the modal frequencies and dampings, as well as the mode shapes evaluated at actuator and sensor locations, are known to be uncertain. For example, for a highly uncertain large space structure the system parameters may only be known to within ±50% accuracy [209]. Suppose the perturbed LSS of equations (6.1a, 6.1b, 6.1c) is described by

$$\dot{x}_c(t) = (A_c + \Delta A_c)x_c(t) + (B_c + \Delta B_c)f(t) \quad (6.3a)$$

$$\dot{x}_r(t) = A_r x_r(t) + B_r f(t) \tag{6.3b}$$

$$y(t) = (C_c + \Delta C_c)x_c(t) + C_r x_r(t) \tag{6.3c}$$

where ΔA_c, ΔB_c, and ΔC_c denote linear parametric perturbations of the controlled dynamics. Note that no uncertainty needs to be included for the residual dynamics that are not part of the modelling. The only information available about the uncertainty is that the uncertain elements in the LSS belong to the known compact sets: $\Delta A_c(\Delta\Sigma_c, \Delta\Omega_c) \in \Re_c, \Delta B_c(\Delta\Omega_c, \Delta\phi_c) \in \Im_c$ and $\Delta C_c(\Delta\Phi_c) \in L_c$. The most common types of the constraint sets are of the form

$$\Re_c = \begin{Bmatrix} \Delta\Sigma_c = \Delta\zeta_i \in \mathbb{R}^N, \Delta\Omega_c = [\omega_i] \in \mathbb{R}^N : \underline{\zeta}_i \le \Delta\zeta_i \le \bar{\zeta}_i, \\ \underline{\omega}_i \le \Delta\omega_i \le \bar{\omega}_i, i = 1,2,\ldots,N \end{Bmatrix}$$

$$\Im_c = \begin{Bmatrix} \Delta\Omega_c = \Delta\omega_i \in \mathbb{R}^N, \Delta\phi_c = \left[\Delta\phi_{ij}\right] \in \mathbb{R}^{N\times M} : \underline{\omega}_i \le \Delta\omega_i \le \bar{\omega}_i, \\ \underline{\phi}_{ij} \le \Delta\phi_{ij} \le \bar{\phi}_{ij}, i = 1,2,\ldots,N, j = 1,2,\ldots,M \end{Bmatrix}$$

$$L_c = \{\Delta\Phi_c = [\Delta\Phi_{ij}] \in \mathbb{R}^{Z\times N} : \underline{\Phi}_{ij} \le \Delta\phi_{ij} \le \bar{\Phi}_{ij}, i = 1,2,\ldots,Z, j = 1,2,\ldots,N\}$$

In this case, only the finite upper and lower bounds $\underline{\zeta}_i, \bar{\zeta}_i, \underline{\omega}_i, \bar{\omega}_i, \underline{\phi}_{ij}, \bar{\phi}_{ij}, \underline{\Phi}_{ij}$ and $\bar{\Phi}_{ij}$ are known. It is also known that the uncertainty of these parameters tends to increase with the increase of the mode number. In the above model, the modal damping coefficients ζ_i are perhaps the least accurately known parameters.

Before designing the active controller, it is assumed that the controllability and observability of the controlled mode subsystem are preserved under small perturbations [17]. Now consider an observer-based compensator of the form

$$\dot{\hat{x}}_c(t) = A_c\hat{x}_c(t) + B_c f(t) + K_c[y(t) - C_c\hat{x}_c(t)], \hat{x}_c(0) = 0 \tag{6.4}$$

$$f(t) = -G_c\hat{x}_c(t) \tag{6.5}$$

where $\hat{x}_c(t) \in \mathbb{R}^{2N}$ is an estimator state vector, K_c is a $(2N \times Z)$ estimator gain matrix and G_c is a feedback gain matrix of dimension $M \times 2N$. Combining equations (6.3a–c) and (6.5), the perturbed closed-loop system is given by the following composite matrix form:

$$\begin{bmatrix} \dot{x}_c(t) \\ \dot{e}_c(t) \\ \dot{x}_r(t) \end{bmatrix} = \begin{bmatrix} A_c - B_cG_c + \Delta A_c - \Delta B_cG_c & -B_cG_c - \Delta B_cG_c & 0 \\ -\Delta A_c + K_c\Delta C_c + \Delta B_cG_c & A_c - K_cC_c + \Delta B_cG_c & K_cC_r \\ -B_rG_c & -B_rG_c & A_r \end{bmatrix} \begin{bmatrix} x_c(t) \\ e_c(t) \\ x_r(t) \end{bmatrix} \tag{6.6}$$

where the estimator error $e_c(t) = \hat{x}_c(t) - x_c(t) \in \mathbb{R}^{2N}$. In the nominal closed-loop system, the terms B_rG_c and K_cC_r are the control and observation spillovers, respectively, which may cause a destabilizing effect such that the nominal eigenvalues of the system will shift away from those of $A_c - B_cG_c$, $A_c - K_cC_c$ and A_r [195]. In equation (6.6), the destabilizing effect not only is raised from the spillover terms but is caused by the linear parametric perturbations.

In dealing with the robust stabilization of the perturbed closed-loop system, we first decompose equation (6.6) and introduce equation (6.3c) as follows:

$$\dot{z}_c(t) = (H_c + \Delta H_c)z_c(t) + H_{cr}x_r(t), z_c(0) = \begin{bmatrix} x_c^T(0) & -x_c^T(0) \end{bmatrix}^T \tag{6.7a}$$

$$\dot{x}_r(t) = H_{rc}z_c(t) + H_r x_r(t) \tag{6.7b}$$

$$y(t) = (\tilde{C}_c + \Delta\tilde{C}_c)z_c(t) + C_r x_r(t) \tag{6.7c}$$

where the augmented system state vector $z_c(t) = \begin{bmatrix} x_c^T(t) & e_c^T(t) \end{bmatrix}^T \in \mathbb{R}^{4N}$, the nominal system matrices are

$$H_c = \begin{bmatrix} A_c - B_cG_c & -B_cG_c \\ 0 & A_c - K_cC_c \end{bmatrix}, H_{cr} = \begin{bmatrix} 0 \\ K_cC_r \end{bmatrix}$$

$$H_{rc} = \begin{bmatrix} -B_rG_c & -B_rG_c \end{bmatrix}, H_r = A_r, \tilde{C}_c = \begin{bmatrix} C_c & 0 \end{bmatrix}$$

and the perturbation matrices are

$$\Delta H_c = \begin{bmatrix} \Delta A_c - \Delta B_cG_c & -\Delta B_cG_c \\ -\Delta A_c + K_c\Delta C_c + \Delta B_cG_c & \Delta B_cG_c \end{bmatrix}, \Delta\tilde{C}_c = \begin{bmatrix} \Delta\ C_c & 0 \end{bmatrix}$$

Accordingly, the matrices $(H_c, \Delta H_c)$ are of dimensions $4N \times 4N$, H_{cr} is $4N \times 2(L-N)$, H_{rc} is $2(L-N) \times 4N$, H_r is $2(L-N) \times 2(L-N)$, and $(\tilde{C}_c, \Delta\tilde{C}_c)$ are $Z \times 4N$. In the absence of the spillover effect and parametric perturbations, the closed-loop controlled mode subsystem is simply described by

$$\dot{z}_c(t) = H_c z_c(t) \tag{6.8}$$

Now define the state transition matrix $\Phi_c(t)$ of equation (6.8) as

$$\Phi_c(t) = \exp(H_c t)$$

Since H_c is block triangular, we can select the regulator gains G_c and the estimator gains K_c independently, such that $\Phi_c(t)$ satisfies

$$\|\Phi_c(t)\| \le m_1 \exp(-\alpha_1 t), \text{ for } m_1 \ge 1, \alpha_i > 0, t \ge 0 \tag{6.9}$$

i.e., the eigenvalues of the nominal controlled mode subsystem $\lambda_i(A_c - B_cG_c) \in \mathbb{C}_-$, $\lambda_i(A_c - K_cG_c) \in \mathbb{C}_-, i = 1,\ldots,2N$ and $\Phi_c(t)$ is exponentially stable. Note that equation (6.9) requires all the eigenvalues of H_c to be distinct. Furthermore, we take

$$-\alpha_1 = \max_i \mathrm{Re}[\lambda_i(H_c)], i = 1,\ldots,N$$

$$= \max_i \mathrm{Re}[\lambda_i(A_c - B_cG_c)], i = 1,\ldots,2N \tag{6.10}$$

Since the estimator gains K_c must be chosen such that the eigenvalues of $A_c - K_cG_c$ (estimator poles) are slightly shifted to the left of the spectrum of $A_c - B_cG_c$ (regulator poles) for an adequate system response of estimator to converge faster than controller. Thus, the last equality holds. equation (6.10) also implies that $-\alpha_1$ is the real part of the eigenvalue of $(A_c - B_cG_c)$ which is nearest to the $j\omega-$ axis. It is noted that if there are finite unstable poles in the uncontrolled LSS, we can arrange A_c so that it contains those nominal unstable poles and the critical modes to be controlled, with A_r containing the remaining modes.

Remark 6.4.1

The bound of the type equation (6.9) can be obtained in several ways. For example, a matrix measure [200] can be used.

For the perturbed controlled mode subsystem equation (6.7a), the corresponding state transition matrix is defined as $\Phi_{cp}(t) \equiv \exp[(H_c + \Delta H_c)t]$. Suppose that the operator is bounded; by applying Lemma 1, the upper norm-bound of $\Phi_{cp}(t)$ is given by

$$\|\Phi_{cp}(t)\| \le m_1 \exp(-\beta_1 t) \tag{6.11}$$

where

$$\beta_1 = \alpha_1 - m_1 \|\Delta H_c(\Delta A_c, \Delta B_c, \Delta C_c)\|$$

It is obvious that if we define the stability margin of the nominal controlled mode subsystem equation (6.8) as α_1, requirement for obtaining an exponentially decaying norm-bound of $\Phi_{cp}(t)$ is

$$\alpha_1 > m_1 \|\Delta H_c\| \tag{6.12}$$

Since we can always select the feasible gains G_c and K_c such that $\|\Delta H_c\|$ is bounded, equation (6.12) implies that the nominal controlled mode subsystem must provide a sufficiently large stability margin α_1 so that the amount of perturbations cannot destroy the stability. For the computation of the upper norm-bound of ΔH_c, one may refer to [204,207] for the highly structured or unstructured perturbation case.

Now, we define the transition matrix of the residual mode subsystem as $\Phi_r(t) \equiv \exp[H_r t]$. Suppose the upper norm-bound for $\Phi_r(t)$ can be estimated as

$$\|\Phi_r(t)\| \leq m_2 \exp(-\beta_2 t), \text{ for } m_2 \geq 1, \beta_2 > 0, t \geq 0 \tag{6.13}$$

To estimate a conservative upper bound for $\|\Phi_r(t)\|$, we choose

$$0 < \beta_2 \leq -\max_i \mathrm{Re}[\lambda_i(H_r)], i = 1, \ldots, 2(L-N)$$

It is noted that equation (6.13) implies that all the zero-frequency modes are not included in the spectrum of A_r. That is, all the rigid body modes have been involved in the controlled mode operator A_c. This assumption is reasonable. Since the low-frequency modes in LSS are always viewed as the dominant modes and will be actively controlled. The assumption $\beta_2 > 0$ is due to the fact that most LSS have natural energy dissipating properties (with or without uncertainty).

Using the above descriptions, we can obtain the following theorem.

Theorem 6.4.1

Based on the assumption equation (6.13), if the linear feedback gains G_c and the estimator gains K_c are chosen such that equations (6.11) and (6.12) hold and the following inequality is also satisfied

$$\frac{m_1 m_2}{\beta_1 \beta_2} \|K_c C_r\| \|B_r G_c\| < 1 \tag{6.14}$$

then the perturbed closed-loop system which includes residual dynamics in equation (6.7) is also asymptotically stable.

Proof

Appendix 6.9A.

Remark 6.4.2

Some comments on various aspects of the preceding analysis are discussed as follows:

a. In the nominal case, the stability condition (6.14) can be simplified as

$$\|K_c C_r\| \|B_r G_c\| < \frac{\alpha_1 \alpha_2}{m_1 m_2} \tag{6.15}$$

which describes a permissible bound on the control and observation spillovers for the loop stability.

b. The relationship between the stability margins of the controlled and residual mode subsystems, spillovers and parameter perturbations is embodied in equation (6.14). By substituting equation (6.11) into (6.14), we can express equation (6.14) as

$$\alpha_1 - m_1\|\Delta H_c\| - \frac{m_1 m_2}{\beta_2}\|K_c C_r\|\|B_r G_c\| > 0 \tag{6.16}$$

The above equation implies that the controlled mode subsystem must provide a sufficient large stability margin α_1 so that the total amount of perturbations on the controlled dynamics (the second term of equation (6.16) and the combined spillover due to the residual dynamics (the third term) cannot destroy the stability margin. Equation (6.14) is now reformulated as

$$\beta_2 - \frac{m_1 m_2}{\alpha_1 - m_1\|\Delta H_c\|}\|K_c C_r\|\|B_r G_c\| > 0 \tag{6.17}$$

In general, for a lightly damped structure, the poles of A_r have small stability margin (i.e., a small β_2). They usually lie near the $j\omega-$ axis. Thus, equation (6.17) reveals that a large α_1 must be provided so that these residual poles cannot be brought into the right half plane due to the spillovers and perturbations.

c. For a large-size structure system, we do not wish to use explicit knowledge of $\|K_c C_r\|$ (observation spillover) and $\|B_r G_c\|$ (control spillover). Under this situation, suppose only the upper bounds on these terms can be estimated as

$$\|K_c C_r\| \le \varepsilon_1, \|B_r G_c\| \le \varepsilon_2$$

They give a measure of the effects of spillover on the closed-loop system. Thus, the robust stability inequality equation (6.14) becomes

$$\frac{m_1 m_2 \varepsilon_1 \varepsilon_2}{\beta_1 \beta_2} < 1 \tag{6.18}$$

6.4.1 Extension to Structural Models with Coupled Modes

In equation (6.1), we consider the structural models which are expressed in a decoupled state-space form. For some mechanical structures, e.g. gyroscopic systems, circulatory systems, etc. [201], these systems have asymmetric damping or stiffness coefficients. Structural modes of these systems cannot be decoupled as the previously mentioned model. Suppose these systems are expressed by the following coupled state-space representation:

$$\dot{x}_c(t) = (A_c + \Delta A_c)x_c(t) + A_{cr}x_r(t) + (B_c + \Delta B_c)f(t) \tag{6.19a}$$

$$\dot{x}_r(t) = A_{rc}x_c(t) + A_r x_r(t) + B_r f(t) \tag{6.19b}$$

$$y(t) = (C_c + \Delta C_c)x_c(t) + C_r x_r(t) \tag{6.19c}$$

where A_{cr} and A_{rc} are, respectively, $2N \times 2(L-N)$ and $2(L-N) \times 2N$ coupling matrices. Following the above analytic procedures and equation (6.1), with the matrices H_{cr} and H_{rc} in equation (6.7) replaced by

$$H_{cr} = \begin{bmatrix} A_{cr} \\ K_c C_r - A_{cr} \end{bmatrix}, H_{rc} = \begin{bmatrix} -B_r G_c + A_{rc} & -B_r G_c \end{bmatrix}$$

and other matrices in equation (6.7) remaining invariant, we have the following result.

Corollary 6.4.1

If equations (6.11)–(6.13) hold and the following inequality is satisfied

$$\frac{m_1 m_2}{\beta_1 \beta_2} \left\| \begin{bmatrix} A_{cr} \\ K_c C_r - A_{cr} \end{bmatrix} \right\| \max(\|B_r G_c - A_{rc}\|, \|B_r G_c\|) < 1 \tag{6.20}$$

then the open-loop system equation (6.19) controlled by the observer-based compensator equations (6.4) and (6.5) is also asymptotically stable.

6.5 STABILIZATION OF LSS WITH CONSTRAINED CONTROLS

For LSS control, there are always tight constraints on the actuator control efforts. Thus, the actuator nonlinearity must be considered in the practical controller design. Suppose an LSS containing saturating actuators is described as

$$\dot{x}_c(t) = A_c x_c(t) + B_c \operatorname{sat} f(t) \tag{6.21a}$$

$$\dot{x}_r(t) = A_r x_r(t) + B_r \operatorname{sat} f(t) \tag{6.21b}$$

$$y(t) = C_c x_c(t) + C_r x_r(t) \tag{6.21c}$$

where the saturation function is defined as

$$\text{sat} f(t) = \begin{bmatrix} \text{sat} f_1(t) & \text{sat} f_2(t) & \cdots & \text{sat} f_M(t) \end{bmatrix}^T \tag{6.22}$$

with

$$\text{sat} f_i(t) = \begin{cases} \underline{f_i} & f_i(t) < \underline{f_i} \\ f_i(t) & \underline{f_i} \le f_i(t) \le \overline{f_i} \\ \overline{f_i} & f_i(t) > \overline{f_i} \end{cases}$$

The observer-based compensator described by equations (6.4) and (6.5) is modified as

$$\dot{x}_c = A_c\hat{x}_c(t) + B_c \text{sat} f(t) + K_c\left[y(t) - C_c\hat{x}_c(t)\right] \tag{6.23}$$

$$f(t) = -G_c\hat{x}_c(t)$$

Combining equations (6.21)–(6.23) with the definition of $e_c(t) \equiv \hat{x}_c(t) - x_c(t) \in \mathbb{R}^{2N}$, the governing equation for this composite closed-loop system can be arranged as

$$\begin{bmatrix} \dot{x}_c(t) \\ \dot{e}_c(t) \\ \dot{x}_r(t) \end{bmatrix} = \begin{bmatrix} A_c - \frac{1}{2}B_cG_c & -\frac{1}{2}B_cG_c & 0 \\ 0 & A_c - \frac{1}{2}K_cC_c & K_cC_r \\ -\frac{1}{2}B_rG_c & -\frac{1}{2}B_rG_c & A_r \end{bmatrix} \begin{bmatrix} x_c(t) \\ e_c(t) \\ x_r(t) \end{bmatrix}$$

$$+ \begin{bmatrix} B_c \\ 0 \\ B_r \end{bmatrix} \left[\text{sat} f(t) - \frac{1}{2} f(t)\right] \tag{6.24}$$

from equation (6.24), we observe that the nominal poles of the system will be shifted from those of $A_c - \frac{1}{2}B_cG_c$, $A_c - \frac{1}{2}K_cC_c$, A_r.The magnitude of this shift depends on linear time-invariant regular perturbations $-\frac{1}{2}B_rG_c$, K_cC_r, and the additional saturated control force $\text{sat} f(t) - \frac{1}{2} f(t)$.

Now, by defining the augmented state vector $z_c(t) \equiv \begin{bmatrix} x_c^T(t) & e_c^T(t) \end{bmatrix}^T \in R^{4N}$, equation (6.24) is partitioned as follows:

$$\dot{z}_c(t) = \overline{H}_c z_c(t) + \overline{H}_{cr} x_r(t) + \tilde{B}_c\left[\text{sat} f(t) - \frac{1}{2} f(t)\right]$$

$$z_c(0) = \begin{bmatrix} x_c^T(0) & x_c^T(0) \end{bmatrix}^T \tag{6.25a}$$

$$\dot{x}_r(t) = \bar{H}_{rc} z_c(t) + \bar{H}_r x_r(t) + B_r \left[\text{sat} f(t) - \frac{1}{2} f(t) \right] \tag{6.25b}$$

where

$$\bar{H}_c = \begin{bmatrix} A_c - \frac{1}{2} B_c G_c & -\frac{1}{2} B_c G_c \\ 0 & A_c - \frac{1}{2} K_c C_c \end{bmatrix}, \bar{H}_{cr} = \begin{bmatrix} 0 \\ K_c C_r \end{bmatrix}$$

$$\bar{H}_{rc} = \begin{bmatrix} -\frac{1}{2} B_r G_c & -\frac{1}{2} B_r G_c \end{bmatrix}, \bar{H}_r = A_r, \tilde{B}_c = \begin{bmatrix} B_c \\ 0 \end{bmatrix}$$

The sensor equation (6.21c) is rewritten as

$$y(t) = \hat{C}_c z_c(t) + C_r x_r(t) \tag{6.26}$$

with

$$\hat{C} = \begin{bmatrix} C_c & 0 \end{bmatrix}$$

Since $\bar{H}_c$ is with block diagonal structure, similarly as the above, it is assumed that we can select the regulator gains G_c and estimator gains K_c such that $\lambda_i\left(A_c - \frac{1}{2} B_c G_c\right) \in C_-$, $\lambda_i\left(A_c - \frac{1}{2} K_c C_c\right) \in C_-$, $i = 1, 2, \ldots, 2N$ and hence $\bar{\Phi}_c(t) \equiv \exp(\bar{H}_c t)$ is exponentially stable. That is,

$$\bar{\Phi}_c(t) \le m_3 \exp(-\alpha_3 t), \text{ for } m_3 \ge 1, \alpha_3 > 0, t \ge 0 \tag{6.27}$$

We may take

$$\alpha_3 = -\max_i \text{Re}\left[\lambda_i(A_c - \frac{1}{2} B_c G_c) \right], i = 1, 2, \ldots, 2N$$

For the residual mode subsystem, we define $\bar{\Phi}_r(t) \equiv \exp(\bar{H}_r t)$ and make the following assumption:

$$\bar{\Phi}_r(t) \le m_4 \exp(-\beta_4 t), \text{ for } m_4 \ge 1, \beta_4 > 0, t \ge 0 \tag{6.28}$$

where β_4 is chosen as

$$\beta_4 = -\max_i \text{Re}\left[\lambda_i(\bar{H}_r) \right], i = 1, 2, \ldots, 2N$$

From these descriptions, and noting that the additional saturated control force satisfies the following sector-bound condition:

$$\left\| \text{sat}\, f(t) - \frac{1}{2} f(t) \right\| \le \frac{1}{2} \| f(t) \| = \frac{1}{2} \| -G_c \hat{x}_c(t) \| \tag{6.29}$$

then a sufficient condition for stability of the system equation (6.21a, 6.21b, 6.21c) controlled by the observer-based equations (6.22)–(6.23) will now be given as follows.

Theorem 6.5.1

If equations (6.27) and (6.28) hold and the following inequality is also satisfied:

$$\frac{m_3}{2\alpha_3} \left\{ \frac{m_4}{\beta_4} \| K_c C_r \| (\| B_r G_c \| + \| B_r \| \| G_c \|) + \| G_c \| \| B_r \| \right\} < 1 \tag{6.30}$$

then the closed-loop system equations (6.25)–(6.26) which includes saturating actuators is asymptotically stable.

Proof

Appendix 6.9B.

Remark 6.5.1

For an LSS, suppose we can only estimate the upper bounds of $\| K_c C_r \|, \| B_r G_c \|$, and $\| B_r \|$ as follows:

$$\| K_c C_r \| \le \varepsilon_3, \| B_r G_c \| \le \varepsilon_4, \| B_r \| \le \varepsilon_5$$

In this situation, equation (6.28) is reformulated as

$$\frac{m_3}{2\alpha_3} \left\{ \frac{m_4}{\beta_4} \varepsilon_3 (\varepsilon_4 + \varepsilon_5 \| G_c \|) + \| G_c \| \| B_r \| \right\} < 1 \tag{6.31}$$

In the above analysis, the non-linear saturation is considered to be inside the sector [0, 1]. This may lead to a conservative result. However, the result can be improved if we assume that the saturation is inside a smaller sector [a, 1]. Under this situation, equation (6.29) is changed to

$$\left\| \text{sat}\, f(t) - \frac{1}{2}(1+a) f(t) \right\| \le \frac{1}{2}(1-a) \| f(t) \|, 0 \le a \le 1 \tag{6.32}$$

In this case, equation (6.25) is reformulated as

$$\dot{z}_c(t) = \tilde{H}_c z_c(t) + \bar{H}_{cr} x_r(t) + \tilde{B}_c \left[\operatorname{sat} f(t) - \frac{1}{2}(1+a) f(t) \right] \tag{6.33a}$$

$$\dot{x}_r(t) = \tilde{H}_{rc} z_c(t) + \bar{H}_r x_r(t) + B_r \left[\operatorname{sat} f(t) - \frac{1}{2}(1+a) f(t) \right] \tag{6.33b}$$

where

$$\tilde{H}_c = \begin{bmatrix} A_c - \frac{1}{2}(1+a)B_c G_c & -\frac{1}{2}(1+a)B_c G_c \\ 0 & A_c - \frac{1}{2}K_c C_c \end{bmatrix}, \bar{H}_{cr} = \begin{bmatrix} 0 \\ K_c C_r \end{bmatrix}$$

$$\bar{H}_{rc} = \begin{bmatrix} -\frac{1}{2}(1+a)B_r G_c & -\frac{1}{2}(1+a)B_r G_c \end{bmatrix}, \bar{H}_r = A_r$$

Similarly as the above, we make the following assumption

$$\tilde{\Phi}_c(t) \le m_5 \exp(-\alpha_5 t), \text{ for } m_5 \ge 1, \alpha_5 > 0, t \ge 0 \tag{6.34}$$

where $\tilde{\Phi}_c(t) \equiv \exp(\tilde{H}_c t)$. We may take

$$\alpha_k = -\max_i \operatorname{Re}\left[\lambda_i \left(A_c - \frac{1}{2}(1+a)B_c G_c \right) \right], i = 1, 2, \ldots, 2N$$

The following corollary follows directly from Theorem 6.5.1.

Corollary 6.5.1

If equations (6.28) and (6.34) hold and the following inequality is also satisfied:

$$\frac{(1-a)m_5}{2\alpha_5} \left\{ \frac{m_4}{\beta_4} \|K_c C_r\| (\|B_r G_c\| + \|B_r\| \|G_c\|) + \|G_c\| \|B_r\| \right\} < 1 \tag{6.35}$$

then the closed-loop system equation (6.33) which includes saturating actuators is asymptotically stable.

6.6 CONTROLLER SYNTHESIS

The feedback gain matrix G_c is conveniently chosen to minimize the index of quadratic performance

$$J = \int_0^{\infty} \left[x_c^T(t) Q x_c(t) + \rho_c f(t)^T R f(t) \right] dt$$

where Q and R are two nonnegative definite weighting matrices. $\rho_c \in [0, \infty)$ is a weighting factor which is used to tune the regulator poles. It is natural to select $Q_c = \frac{1}{2} diag\begin{bmatrix} \omega_1^2 & \cdots & \omega_N^2 & \omega_1^2 & \cdots & \omega_N^2 \end{bmatrix}$ so that $x_c^T(t) Q x_c(t)$ represents the total energy in the controlled dynamics. The optimal control law for the controlled mode subsystem is given by

$$f(t) = -G_c \hat{x}_c(t) = -\frac{1}{\rho_c} R^{-1} B_c^T F \hat{x}_c(t) \tag{6.36}$$

where F is a $(2N \times 2N)$ positive-definite solution of the algebraic matrix Ricatti equation

$$FA_c + A_c^T F - FB_c \frac{R^{-1}}{\rho_c} B_c^T F + Q = 0 \tag{6.37}$$

The system under control is considered to be deterministic. However, by applying the technique of Kalman filtering design and adjusting the fictitious noise covariance, the estimator poles can also be determined. The estimator gains are given by

$$K_c = \frac{1}{\rho_e} PC_c^T V^{-1} \tag{6.38}$$

where P is a $(2N \times 2N)$ positive definite solution of the following estimator Ricatti equation

$$A_c P + PA_c^T - PC_c^T \frac{V^{-1}}{\rho_e} C_c P + W = 0 \tag{6.39}$$

in which W and V are, respectively, fictitious covariance matrices with assumed values. $\rho_e \in [0, \infty)$ is a weighting factor which will be employed to tune the estimator gains. In short, the designed observer-based compensator is characterized by the scalar parameters ρ_c and ρ_e. By tuning these parameters, the requirement of robust stability condition (6.14) for coupled system in equation (6.19) can be made to be satisfied.

6.7 SIMULATION EXAMPLES

The following examples describing a beam structure model are given to illustrate the robust stabilization of an LSS with parametric perturbation or actuator saturation.

Example 6.7.1

Modal control of a simply supported beam with an uncertain modal damping

The beam follows the configuration of [195]. The equations of motion of the beam are described by the following ordinary differential equations:

$$\dot{v}_k(t) + 2\zeta w_k v_k(t) + w_k u_k(t) = \sum_i 2\frac{\phi_k(p_i)}{w_k} f_i(t),\ k = 1,2,...$$

where $v_k(t) \equiv \dot{u}_k(t)/w_k$. Let the actuator be located at $p_i = 0.19$ and the displacement sensor be located at 0.71. Assume that the dominant uncertainty of this system arises from the uncertain damping coefficient ζ with $0.019 \le \zeta \le 0.021$. For the convenience of computer simulation, the first three modes are treated as controlled modes, while the fourth and fifth modes are treated as known residual modes.

Using the formulae given in equation (6.2), the modal model has the form

$$A_{c,r} = \begin{bmatrix} 0 & \Lambda_{c,r}^{1/2} \\ -\Lambda_{c,r}^{1/2} & -2\zeta\Lambda_{c,r}^{1/2} \end{bmatrix}, B_{c,r} = \begin{bmatrix} 0 \\ \hat{B}_{c,r} \end{bmatrix}, C_{c,r} = \begin{bmatrix} \bar{C}_{c,r} & 0 \end{bmatrix}$$

where for the controlled modes

$$\Lambda_c^{1/2} = diag\begin{bmatrix} \pi^2 & (2\pi)^2 & (3\pi)^2 \end{bmatrix}$$

$$\hat{B}_c = \sqrt{2}\begin{bmatrix} \frac{\sin(0.19\pi)}{\pi^2} & \frac{\sin(0.38\pi)}{(2\pi)^2} & \frac{\sin(0.57\pi)}{(3\pi)^2} \end{bmatrix}^T$$

$$\bar{C}_c = \sqrt{2}\begin{bmatrix} \sin(0.71\pi) & \sin(1.42\pi) & \sin(2.13\pi) \end{bmatrix}$$

and for the residual modes

$$\Lambda_c^{1/2} = diag\begin{bmatrix} (4\pi)^2 & (5\pi)^2 \end{bmatrix}, \hat{B}_c = \sqrt{2}\begin{bmatrix} \frac{\sin(0.76\pi)}{(4\pi)^2} & \frac{\sin(0.95\pi)}{(5\pi)^2} \end{bmatrix}^T$$

$$\bar{C}_r = \sqrt{2}\begin{bmatrix} \sin(2.84\pi) & \sin(3.55\pi) \end{bmatrix}$$

In the LQR design, the weighting matrices are selected as

$$Q = diag\begin{bmatrix} \pi^2 & (2\pi)^2 & (3\pi)^2 & \pi^2 & (2\pi)^2 & (3\pi)^2 \end{bmatrix}, R = I_6$$

While the weighting factor ρ_c is tuned to 3, the corresponding regulator gains are obtained as

$$G_c = [\ 0.0629 \quad 0.1362 \quad 0.1423 \quad 2.6139 \quad 5.1922 \quad 5.6782\ ]$$

and the regulator poles are located at $-0.3026 \pm j9.8676, -0.8760 \pm j39.4705$ and $-1.8206 \pm j88.8067$. For the estimator design, the fictitious noise covariances are selected as $W = I_6, V = I_6$. When the weighting factor ρ_e is tuned to 0.95, the resulting estimator gains are found as

$$K_c = [\ 1.1392 \quad -0.6993 \quad 0.1641 \quad 0.0064 \quad -0.0139 \quad 0.0013\ ]^T$$

and the corresponding estimator poles are located at $-0.8333 \pm j9.8676$, $-1.2689 \pm j39.4705$, $-1.8327 \pm j88.8067$. By using the selected regulator and estimator gains, and with the help of the computer calculation, we obtain $m_1 = 1.3044, a_1 = 0.3026, m_2 = 1.4250, \beta_2 = 3.0035, \|\Delta H_c\| = 0.1777$, the control spillover $\|B_r G_c\| = 0.0399$, the observation spillover $\|K_c C_r\| = 2.8271$. Using these results, the robustness condition (6.14) is checked as

$$\frac{m_1 m_2}{\beta_2} \|K_c C_r\| \|B_r G_c\| = 0.9857 < 1$$

With all nominal position states being initially set to 1.0, all nominal velocity states and all residual states are initially set to zero. The simulations of the displacement sensor signal and actuator command are shown in Figure 6.1, respectively. It is seen that a stabilizing property is achieved.

Example 6.7.2

Modal control of a simply supported beam with a constrained actuator

Consider the same beam given in Example 6.7.1, which has a light constant damping $\zeta = 0.02$ and the actuator is saturated at ±1.We want to design a stabilizing observer-based compensator such that the resulting system is asymptotically stable.

Using the same weighting matrices *Q*, *R*, *W* and *V* as given in Example 6.7.1, along with the weighting factor ρ_c tuned to 7.3, we obtain the gains G_c as

$$G_c = [\ 0.0266 \quad 0.0503 \quad 0.0523 \quad 1.2111 \quad 2.20 \quad 2.3507\ ]$$

The corresponding regulator poles are then obtained as $-0.2218 \pm j9.8677$, $-0.8079 \pm j39.4705$, $-1.7836 \pm j88.8067$. When ρ_e is tuned to 1.4, K_c is obtained as

$$K_c = [\ 1.1392 \quad -0.6993 \quad 0.1641 \quad 0.0064 \quad -0.0139 \quad 0.0013\ ]^T$$

The corresponding estimator poles are located at $-0.6960 \pm j9.8677$, $-1.1373 \pm j39.4705$, $-1.810 \pm j88.8067$. Applying a computer calculation, we have $m_3 = 1.3059, \alpha_3 = 0.2218, m_3 = 1.4284, \beta_4 = 3.1583$. The following norm-bounds are easily obtained: $\|K_c C_r\| = 2.120$, $\|B_r G_c\| = 0.0165$, $\|G_c\| = 2.3507$, $\|B_r\| = 0.0070$. We check the robust stability condition (6.30):

$$\frac{m3}{2\alpha 3}\left\{\frac{m4}{\beta 4}\|K_c C_r\|(\|B_r G_c\| + \|B_r\|\|G_c\|) + \|B_c\|\|G_c\|\right\} = 0.9855 < 1$$

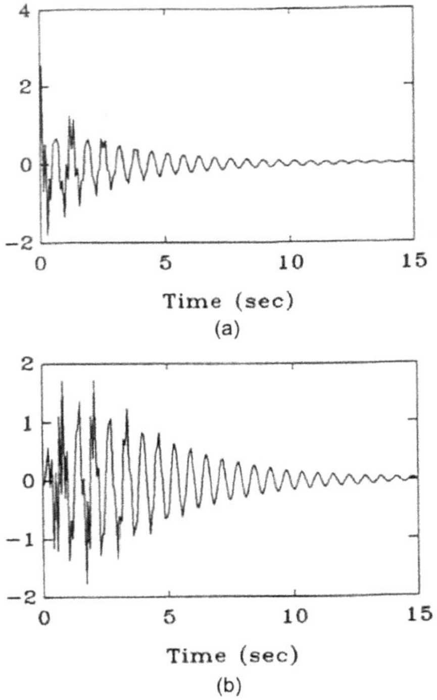

FIGURE 6.1 Simulation results of Example 6.7.1. (a) Beam deflection at sensor output; (b) actuator command.

Since the robust stability condition is satisfied, the stabilizing compensator is obtained. With the same initial conditions as used in Example 6.7.1, the simulations of the beam deflection and actuator command are shown in Figure 6.2, respectively.

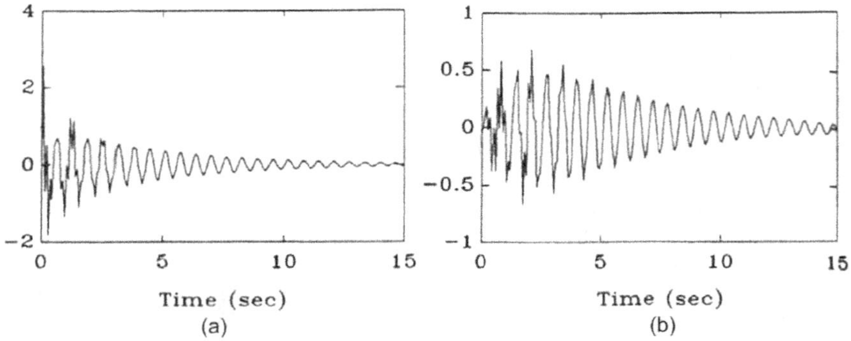

FIGURE 6.2 Simulation results of Example 6.7.2. (a) Beam deflection at sensor output; (b) actuator command.

6.8 CONCLUSIONS

This chapter presents an algorithm for robust observer-based state feedback control of an LSS by accounting for (i) mode truncation and parameter uncertainty, or (ii) mode truncation and saturating actuators in maintaining the robust stability. We have derived a sufficient condition for the robust stability of the perturbed LSS. The result has shown that it can clearly describe the relationship between the stability margins of the controlled and residual mode subsystems, spillover effect and additive parametric perturbations. This result has been also extended to the case with saturated actuators. Finally, two illustrative examples were used with simulation to confirm our results. This chapter has discussed the robust stabilization design problem of LSSs by the state space method in time domain. In the following chapter, we will continually discuss the observer-based robust control design problem of large flexible structural systems from the frequency domain perspective.

6.9 APPENDIX

6.9.A PROOF OF THEOREM 6.4.1

The solutions of equations (6.7a) and (6.7b) are given by the following coupled equations:

$$z_c(t) = \Phi_{cp}(t)z_c(0) + \Phi_{cp}(t) * H_{cr}x_r(t) \tag{6.A1}$$

$$x_r(t) = \Phi_r(t)x_r(t) + \Phi_r(t) * H_{rc}z_c(t) \tag{6.A2}$$

Taking norms on both sides of equations (6.A1) and (6.A2), we can obtain

$$\|z_c(t)\| \le \|z_c(0)\| \|\Phi_{cp}(t)\| + \|\Phi_{cp}(t) * H_{cr}x_r(t)\| \tag{6.A3}$$

$$\|z_r(t)\| \le \|z_r(0)\| \|\Phi_r(t)\| + \|\Phi_r(t) * H_{rc}z_c(t)\| \tag{6.A4}$$

Integrating both sides of equations (6.A3) and (6.A4), we obtain

$$\int_0^\infty \|z_c(t)\|dt \le \int_0^\infty \|z_c(0)\| \|\Phi_{cp}(t)\|dt + \int_0^\infty \|\Phi_{cp}(t) * H_{cr}x_r(t)\|dt \tag{6.A5}$$

$$\int_0^\infty \|x_r(t)\|dt \le \int_0^\infty \|x_r(0)\| \|\Phi_r(t)\|dt + \int_0^\infty \|\Phi_r(t) * H_{rc}z_c(t)\|dt \tag{6.A6}$$

From the fact

$$\int_0^\infty \|a(t) * b(t)\|dt \le \int_0^\infty \|a(t)\|dt + \int_0^\infty \|b(t)\|dt$$

Thus equations (6.A5) and (6.A6) can be expressed as

$$\int_0^\infty \|z_c(t)\|dt \le \|z_c(0)\|\int_0^\infty \|\Phi_{cp}(t)\|dt + \|H_{cr}\|\int_0^\infty \|\Phi_{cp}(t)\|dt\int_0^\infty \|x_r(t)\|dt \quad (6.A7)$$

$$\int_0^\infty \|x_r(t)\|dt \le \|x_r(0)\|\int_0^\infty \|\Phi_r(t)\|dt + \|H_{rc}\|\int_0^\infty \|\Phi_r(t)\|dt\int_0^\infty \|z_c(t)\|dt \quad (6.A8)$$

From equations (6.11) and (6.13), we have

$$\int_0^\infty \|\Phi_{cp}(t)\|dt \le \int_0^\infty m_1 \exp(-\beta_1 t)\,dt = \frac{m_1}{\beta_1}$$

$$\int_0^\infty \|\Phi_r(t)\|dt \le \int_0^\infty m_2 \exp(-\beta_2 t)\,dt = \frac{m_2}{\beta_2}$$

Then equations (6.A7) and (6.A8) can be expressed as

$$\int_0^\infty \|z_c(t)\|dt \le \frac{m_1}{\beta_1}\|z_c(0)\| + \frac{m_1}{\beta_1}\|H_{cr}\|\int_0^\infty \|x_r(t)\|dt \quad (6.A9)$$

$$\int_0^\infty \|x_r(t)\|dt \le \frac{m_2}{\beta_2}\|x_r(0)\| + \frac{m_2}{\beta_2}\|H_{rc}\|\int_0^\infty \|z_c(t)\|dt \quad (6.A10)$$

Substituting equation (6.A10) into (6.A9) leads to

$$\left[1 - \frac{m_1 m_2 \|H_{cr}\|\|H_{rc}\|}{\beta_1\beta_2}\right]\int_0^\infty \|z_c(t)\|dt \le \frac{m_1}{\beta_1}\|z_c(0)\| + \frac{m_1 m_2}{\beta_1\beta_2}\|H_{cr}\|\|x_r(0)\| \quad (6.A11)$$

If

$$1 - \frac{m_1 m_2 \|H_{cr}\|\|H_{rc}\|}{\beta_1\beta_2} > 0$$

or, from the equalities

$$\|H_{cr}\| = \left\|\begin{bmatrix} 0 \\ K_c C_r \end{bmatrix}\right\| = \|K_c C_r\|$$

$$\|H_{rc}\| = \left\|\begin{bmatrix} -B_r G_c & -B_r G_c \end{bmatrix}\right\| = \|B_r G_c\|$$

we may use

$$1 - \frac{m_1 m_2 \|K_c C_r\|\|B_r G_c\|}{\beta_1\beta_2} > 0 \quad (6.A12)$$

then equation (6.A11) can be expressed in the form

$$\int_0^\infty \|z_c(t)\|dt \le \frac{\dfrac{m_1}{\beta_1}\|z_c(0)\| + \dfrac{m_1 m_2 \|K_c C_r\| \|x_r(0)\|}{\beta_1 \beta_2}}{-\dfrac{m_1 m_2 \|K_c C_r\| \|B_r G_c\|}{\beta_1 \beta_2}} = M_1 \tag{6.A13}$$

where M_1 is a finite, positive constant. Combining equations (6.A10) and (6.A13), we obtain

$$\int_0^\infty \|x_r(t)\|dt \le \frac{m_2}{\beta_2}\|x_r(0)\| + \frac{m_2}{\beta_2}\|H_{rc}\| M_1 = M_2 \tag{6.A14}$$

Taking norms on both sides of equation (6.7c), then

$$\|y(t)\| \le \left(\|\tilde{C}_c\| + \|\Delta\tilde{C}_c\|\right)\|z_c(t)\| + \|C_r\|\|x_r(t)\|$$

Integrating both sides of this, we obtain

$$\begin{aligned}\int_0^\infty \|y(t)\|dt &\le \left(\|\tilde{C}_c\| + \|\Delta\tilde{C}_c\|\right)\int_0^\infty \|z_c(t)\|dt + \|C_r\|\int_0^\infty \|x_r(t)\|dt \\ &\le \left(\|\tilde{C}_c\| + \|\Delta\tilde{C}_c\|\right)M_1 + \|C_r\| M_2 \\ &= M_3 \end{aligned} \tag{6.A15}$$

Since M_1, M_2, and M_3 are finite, positive constants, and assuming these states and output signals are uniformly continuous, applying Lemma 2, equations (6.A13)–(6.A15) imply

$$\|z_c(t)\| \to 0, \|x_r(t)\| \to 0, \|y(t)\| \to 0, \text{as } t \to \infty$$

Therefore, the close-loop system equation (6.7) is asymptotically stable.

6.9.B PROOF OF THEOREM 6.5.1

We first obtain the solutions of equations (6.25a) and (6.25b), respectively, as

$$z_c(t) = \bar{\Phi}_c(t)z_c(0) + \bar{\Phi}_c(t) * H_{cr}x_r(t) + \bar{\Phi}_c(t) * B_c\left[\text{sat} f(t) - \frac{1}{2}f(t)\right] \tag{6.B1}$$

$$x_r(t) = \bar{\Phi}_r(t)x_r(0) + \bar{\Phi}_r(t) * H_{rc}x_r(t) + \bar{\Phi}_r(t) * B_r\left[\text{sat} f(t) - \frac{1}{2}f(t)\right] \tag{6.B2}$$

Taking norms on equations (6.B1) and (6.B2) yields

$$\|z_c(t)\| \le \left\|\bar{\Phi}_c(t)\right\| \|z_c(0)\| + \left\|\bar{\Phi}_c(t) * H_{cr} x_r(t)\right\| + \left\|\bar{\Phi}_c(t) * B_c\left[\operatorname{sat} f(t) - \frac{1}{2} f(t)\right]\right\| \tag{6.B3}$$

$$\|x_r(t)\| = \left\|\bar{\Phi}_r(t)\right\| \|x_r(0)\| + \left\|\bar{\Phi}_r(t) * H_{rc} x_r(t)\right\| + \left\|\bar{\Phi}_r(t) * B_r\left[\operatorname{sat} f(t) - \frac{1}{2} f(t)\right]\right\| \tag{6.B4}$$

Integrating both sides of equations (6.B3) and (6.B4), we obtain

$$\begin{aligned} \int_0^\infty \|z_c(t)\| dt \le \|z_c(0)\| \int_0^\infty \left\|\bar{\Phi}_c(t)\right\| dt + \left\|\bar{H}_{cr}\right\| \int_0^\infty \left\|\bar{\Phi}_c(t)\right\| dt \int_0^\infty \|x_r(t)\| dt \\ + \|B_c\| \int_0^\infty \left\|\bar{\Phi}_c(t)\right\| dt \int_0^\infty \left\|\left[\operatorname{sat} f(t) - \frac{1}{2} f(t)\right]\right\| dt \end{aligned} \tag{6.B5}$$

$$\begin{aligned} \int_0^\infty \|x_r(t)\| dt \le \|x_r(0)\| \int_0^\infty \left\|\bar{\Phi}_r(t)\right\| dt + \left\|\bar{H}_{rc}\right\| \int_0^\infty \left\|\bar{\Phi}_r(t)\right\| dt \int_0^\infty \|z_c(t)\| dt \\ + \|B_r\| \int_0^\infty \left\|\bar{\Phi}_r(t)\right\| dt \int_0^\infty \left\|\left[\operatorname{sat} f(t) - \frac{1}{2} f(t)\right]\right\| dt \end{aligned} \tag{6.B6}$$

Since

$$\begin{aligned} \left\|\operatorname{sat} f(t) - \frac{1}{2} f(t)\right\| &\le \frac{1}{2} \|f(t)\| \\ &= \frac{1}{2} \left\| -\begin{bmatrix} G_c & G_c \end{bmatrix} \begin{bmatrix} x_c^T(t) & e_c^T(t) \end{bmatrix}^T \right\| \\ &\le \frac{1}{2} \|G_c\| \|z_c(t)\| \end{aligned} \tag{6.B7}$$

and from equations (6.27) and (6.28)

$$\int_0^\infty \left\|\bar{\Phi}_c(t)\right\| dt \le \int_0^\infty m_3 \exp(-\alpha_3 t)\, dt = \frac{m_3}{\alpha_3} \tag{6.B8}$$

$$\int_0^\infty \left\|\bar{\Phi}_r(t)\right\| dt \le \int_0^\infty m_4 \exp(-\alpha_4 t)\, dt = \frac{m_4}{\beta_4} \tag{6.B9}$$

Substituting equations (6.B7)–(6.B9) into (6.B5) and (6.B6), we have

$$\int_0^\infty \|z_c(t)\| dt \le \frac{m_3}{\alpha_3} \|z_c(0)\| + \frac{m_3}{\alpha_3} \left\|\bar{H}_{cr}\right\| \int_0^\infty \|x_r(t)\| dt + \frac{m3}{2\alpha 3} \|B_c\| \|G_c\| \int_0^\infty \|z_c(t)\| dt$$

$$\int_0^\infty \|x_r(t)\|dt \le \frac{m4}{\beta 4}\|x_r(0)\| + \frac{m_4}{\beta_4}\left[\left\|\bar{H}_{rc} + \frac{1}{2}\|B_r\|\|G_c\|\right\|\right]\int_0^\infty \|z_c(t)\|$$

By the same procedure as given in Appendix 6.9A, we obtain the following result, if

$$\frac{m_3}{\alpha_3}\left\{\frac{m4}{\beta 4}\|\bar{H}_{cr}\|\|\bar{H}_{rc}\| + \frac{m_4}{2\beta_4}\|\bar{H}_{cr}\|\|B_r\|\|G_c\| + \frac{1}{2}\|B_c\|\|G_c\|\right\} < 1$$

or equivalently

$$\frac{m_3}{2\alpha_3}\left\{\frac{m4}{\beta 4}\|K_cC_r\|\|B_rG_c\| + \frac{m_4}{\beta_4}\|K_cC_r\|\|B_r\|\|G_c\| + \frac{1}{2}\|B_c\|\|G_c\|\right\} < 1$$

(Since $\|\bar{H}_{cr}\| = \|K_cC_r\|, \|\bar{H}_{rc}\| = \frac{1}{2}\|B_rG_c\|$), then the system (6.25) is asymptotically stable.

7 Robust Observer-Based Output Feedback Control Design of Large Flexible Structures
Mode State-Space Approach and Frequency Domain Robustness Measurement Method

7.1 INTRODUCTION

The stability problem in the control of large flexible structures differs from more conventional linear systems, because the plant to be controlled is theoretically of infinite order (and practically of very high order), while the controller/observer must, of necessity, be of relatively low order in practical applications. Higher modes, referred to as residual modes, are generally ignored in the analysis and may be excited by the controller to cause a net destabilizing effect on the large flexible structural system. This is referred to as the spillover phenomenon and is demonstrated in [212].

There are two contributions to the spillover of a large flexible structure. The first is observation spillover which entails the contamination of the sensor output signal through the presence of the residual modes. The second is control spillover which entails the excitation of the residual modes by the feedback controller. When both types of spillover are present, the closed-loop poles will be shifted and the system may become unstable. In short, this is a stability robustness problem of large flexible structures: lack of robustness may lead to a spillover effect from the neglected dynamics, which may destroy the stability of large flexible structures.

DOI: 10.1201/9781003229230-10

The issue for robust stabilization design problem of large flexible structures has been extensively investigated in the literature. It has augmented the controller with a narrow-bandpass prefilter to remove the instability mechanism [212]. By using collocated actuators and rate sensors, a robust stability test based on the positivity of operators has been proposed [213]. A time domain sufficient condition ensuring loop stability has been derived by utilizing the Bellman-Gronwall's inequality [214]. Alternative stability conditions have also been derived by using the functional inequality in the time domain [215]. The input-output stability problem and some useful stability conditions have been studied and presented by some researchers [217–219] in the frequency domain.

The main theoretical problem of robust controller designs is to find a linear time-invariant controller for the incompletely known plant and to develop robust control design strategies for a linear controller in the presence of severe model uncertainties. It is reasonable that in the large flexible structure control design, the residual modes can be considered as the modeling uncertainties and the controlled mode dynamics can be retained as the nominal plant. In this chapter, control and observation spillover will be treated as parametric uncertainties; a combination of the two spillover, termed as the total spillover, will be viewed as the unmodeled uncertainty. New robust stability conditions in the frequency domain are developed to tolerate two types of spillover described above. A design algorithm of observer-based output feedback control is also provided for large flexible structural systems to synthesize a stabilizing observer-based compensator such that the resulting nominal closed-loop system can tolerate spillover contamination. Some numerical studies on the robust stabilization control design of a simply supported beam by the proposed observer-based output feedback control are presented to illustrate the design procedure and to validate the stability robustness tests.

7.2 THE MATHEMATICAL MODEL

According to Galerkin method in Section 1.2.1, the equations of motion for most large flexible structures can be transformed into the following infinite set of second-order ordinary differential equations of modes with individual initial conditions $\dot{u}_k(0)$ and $u_k(0)$

$$\ddot{u}_k(t) + 2\zeta_k\omega_k\dot{u}_k(t) + \omega_k^2 u_k(t) = b_k f(t), k = 1,2,\ldots,\infty \tag{7.1}$$

where $u_k(t)$, ω_k, and ζ_k are the modal amplitude, mode frequency and damping coefficient of the kth mode, respectively, $f(t) \in \mathbb{R}^{1\times m}$ is the control force vector of the form $f(t) = [f_1(t), f_2(t), \ldots, f_m(t)]^T$ and $b_k \in \mathbb{R}^{1\times m}$ is the actuator influence function vector defined by $b_k = [\phi_k(p_1), \phi_k(p_2), \ldots, \phi_k(p_m)]$ with $\phi_k(\cdot)$ being the mode shape of the kth mode and $p_1, \ldots, p_m$ are the locations of m actuators. If we define the reduced modal velocity $v_k(t)$ as

$$v_k(t) = \frac{\dot{u}_k(t)}{\omega_k} \tag{7.2}$$

substituting equation (7.2) into (7.1) then yields

$$\dot{v}_k(t)+2\zeta_k\omega_k v_k(t)+\omega_k u_k(t)=\frac{b_k}{\omega_k}f(t), k=1,2,\ldots,\infty \tag{7.3}$$

In system control theory, the infinite modes have to be controlled simultaneously. However, it may not be feasible to control all of the structural modes. Ultimately only the lowest n modes will be actively controlled and the remainder be classified as residual modes. In order to give a state-space representation of mode system equation (7.3), the controlled mode state vector $x_c(t)\in\mathbb{R}^{2n}$ and the residual mode state vector $x_r(t)\in\mathbb{R}^{2(l-n)}$ are defined by

$$x_c(t)=[u_1(t),u_2(t),\ldots,u_n(t),v_1(t),v_2(t),\ldots,v_n(t)]^T \tag{7.4}$$

$$x_r(t)=[u_{n+1}(t),u_{n+2}(t),\ldots,u_l(t),v_{n+1}(t),v_{n+2}(t),\ldots,v_l(t)]^T, l\to\infty \tag{7.5}$$

respectively. In system control theory $l\to\infty$, however, in practice, l may be large but is finite. This assumption is due to the fact that the actuators and sensors cannot excite or respond to the highest frequency modes. From equations (7.2)–(7.5), it can easily be shown that the mode state-space model of the system is given by

$$\dot{x}_c(t)=A_c x_c(t)+B_c f(t) \tag{7.6}$$

$$\dot{x}_r(t)=A_r x_r(t)+B_r f(t) \tag{7.7}$$

where the system matrices are

$$A_{c,r}=\begin{bmatrix} 0 & \Lambda_{c,r}^{1/2} \\ -\Lambda_{c,r}^{1/2} & -\xi_{c,r}^{1/2} \end{bmatrix}, B_{c,r}=\begin{bmatrix} 0 \\ -\Lambda_{c,r}^{-1/2}\bar{B}_{c,r} \end{bmatrix}$$

in which subscripts c and r refer to the controlled and residual modes, respectively. Also A_c is of dimension $2n\times 2n$, A_r is $2(l-n)\times 2(l-n)$ with $l\to\infty$; B_c is $2n\times m$; B_r is $2(l-n)\times m$; $\Lambda_c^{1/2}=diag[\omega_1,\ldots,\omega_n]$ and $\Lambda_r^{1/2}=diag[\omega_{n+1},\ldots,\omega_l]$ with ω_k are ordered in increasing order of frequency $(0<\omega_k\le\omega_{k+1})$; the structural damping operator $\xi_c=2\zeta_c\Lambda_c^{1/2}$ is positive semidefinite and $\xi_r=2\zeta_r\Lambda_r^{1/2}$ is positive definite with $\zeta_c=diag[\zeta_1,\ldots,\zeta_n]$, $\zeta_r=diag[\zeta_{n+1},\ldots,\zeta_l]$, and the modal damping coefficients, $\zeta_i<1, i=1,\ldots,l$. B_c and B_r are matrices whose rows are, respectively, the controlled and residual mode shapes $\phi_k(x)$ evaluated at the m actuator positions p_i, i.e.,

$$\bar{B}_c=\begin{bmatrix} \phi_1(p_1) & \cdots & \phi_1(p_m) \\ \vdots & & \vdots \\ \phi_n(p_1) & \cdots & \phi_n(p_m) \end{bmatrix}, \bar{B}_r=\begin{bmatrix} \phi_{n+1}(p_1) & \cdots & \phi_{n+1}(p_m) \\ \vdots & & \vdots \\ \phi_l(p_1) & \cdots & \phi_l(p_m) \end{bmatrix}$$

As z point displacement sensors are located at positions q_j, the displacement sensor output vector $y(t) \in \mathbb{R}^z$ can be expressed in the form

$$y(t) = C_c x_c(t) + C_r x_r(t) \tag{7.8}$$

where $C_c = [\bar{C}_c \ \ 0]$ is of dimension $z \times 2n$ and $C_r = [\bar{C}_r \ \ 0]$ is $z \times 2(l-n)$ with

$$\bar{C}_c = \begin{bmatrix} \phi_1(q_1) & \cdots & \phi_n(q_1) \\ \vdots & & \vdots \\ \phi_1(q_z) & \cdots & \phi_n(q_z) \end{bmatrix}, \bar{C}_r = \begin{bmatrix} \phi_{n+1}(q_1) & \cdots & \phi_{n+1}(q_1) \\ \vdots & & \vdots \\ \phi_l(q_z) & \cdots & \phi_l(q_z) \end{bmatrix}$$

It is noted that the system described by equations (7.6)–(7.8) is somewhat different with the formulation given in [212]. However, other kinds of derivations are also possible. For example, one may adopt the form of partial differential equations as in [219], or by a finite element model as in [220].

7.3 PROBLEM FORMULATION

For the mode state-space system of large flexible structures described by equations (7.6)–(7.8), we assume that there is perfect knowledge of the dynamics of the n controlled modes. The residual modes are, of course, present in the mode state space system but are neglected in the design of the finite-dimensional controller. The controller, we consider, is an observer-based output feedback dynamic control law of the form

$$\dot{\hat{x}}_c(t) = A_c \hat{x}_c(t) + B_c f(t) + K_c[y(t) - \hat{y}(t)], \hat{x}_c(0) = 0 \tag{7.9a}$$

$$\hat{y}(t) = C_c \hat{x}_c(t) \tag{7.9b}$$

$$f(t) = -G_c \hat{x}_c(t) \tag{7.10}$$

where $\hat{y}(t) \in \mathbb{R}^z$ is the observer output vector, $\hat{x}_c(t) \in \mathbb{R}^{2n}$ is the observation vector, K_c is the $(2n \times z)$ observer gain matrix, and G_c is the feedback gain matrix of dimension $m \times 2n$. In state observer design we require the matrix pair (A_c, C_c) be observable. For the regulator design, it is required that the matrix pair (A_c, B_c) be controllable.

Using equations (7.6) and (7.8)–(7.10), the observation error $e_c(t) \equiv \hat{x}_c(t) - x_c(t)$ is given by

$$\dot{e}_c(t) = (A_c - K_c C_c) e_c(t) + K_c C_r x_r(t) \tag{7.11}$$

Note that additive noise terms will present in the Kalman filter version of equation (7.11), but are absent in the Luenberger observer version.

The stability robustness condition will be analyzed by considering the composite closed-loop system of the augmented matrix form:

$$\begin{bmatrix} \dot{x}_c(t) \\ \dot{e}_c(t) \\ \dot{x}_r(t) \end{bmatrix} = \begin{bmatrix} A_c - B_c G_c & -B_c G_c & 0 \\ 0 & A_c - K_c C_c & K_c C_r \\ -B_r G_c & -B_r G_c & A_r \end{bmatrix} \begin{bmatrix} x_c(t) \\ e_c(t) \\ x_r(t) \end{bmatrix} \quad (7.12)$$

It is clear that if either $C_r = 0$ and/or $B_r = 0$, the eigenvalues of the above augmented system are arised from $A_c - B_c G_c$, $A_c - K_c C_c$ and A_r. As $\lambda(A_c - B_c G_c) \in C_-$ and $A_c - K_c C_c \in C_-$, the system still remains stable. However, in most cases, both $C_r \neq 0$ and $B_r \neq 0$, the eigenvalues of the system will shift away from those of $A_c - B_c G_c$, $A_c - K_c C_c$, and A_r. From the matrix structure of equation (7.12) it can be seen that the magnitude of this shift depends entirely on $-B_r G_c$ (control spillover) and $K_c C_r$ (observation spillover). This small amount of perturbation will generally bring these poles into right complex half-plane and, thus, makes the system unstable.

Based on the above analysis, the specific control design problems to be addressed are the following:

a. For a large structural system controlled by an observer-based compensator described above, under what condition is the controlled system asymptotically stable? i.e., the spillover effect can be tolerated in our design and the controller is a robust one.
b. How to choose the control (regulator) gain G_c and the observer gain K_c such that the overall system is robustly stabilized?

7.4 ROBUST STABILIZATION WITH RESPECT TO CONTROL/OBSERVATION SPILLOVER

Now, the composite closed-loop system equation (7.12) and the sensor equation (7.8) are reformulated as follows:

$$\dot{z}_c(t) = H_c z_c(t) + H_{cr} x_r(t) \quad (7.13)$$

$$\dot{x}_r(t) = H_{rc} z_c(t) + H_r x_r(t) \quad (7.14)$$

$$y(t) = \tilde{C}_c z_c(t) + C_r x_r(t) \quad (7.15)$$

where $z_c(t) = \left[x_c^T(t) e_c^T(t) \right]^T \in \mathbb{R}^{4n}$ and the constant matrices

$$H_c = \begin{bmatrix} A_c - B_c G_c & -B_c G_c \\ 0 & A_c - K_c C_c \end{bmatrix}, H_{cr} = \begin{bmatrix} 0 \\ K_c C_r \end{bmatrix},$$

$$H_{rc} = \begin{bmatrix} -B_r G_c & -B_r G_c \end{bmatrix}, H_r = A_r, \tilde{C} = \begin{bmatrix} C_c & 0 \end{bmatrix}$$

In the absence of the spillover effect, the closed-loop system of the controlled modes is described by

$$\dot{z}_c(t) = H_c z_c(t) \tag{7.16}$$

$$y(t) = \tilde{C}_c z_c(t) \tag{7.17}$$

This can be viewed as the nominal closed-loop system. Since for any choice of modal reduction scheme it makes no practical sense, if the residuals are unstable. The unstable modes should be selected as controlled modes in the design procedure. Therefore, we make the following natural but important assumption about the residual subsystem matrix $H_r (= A_r)$:

Assumption 7.4.1 [198]

The residual subsystem matrix H_r is asymptotically stable, i.e.,

$$\hat{\Phi}_r(s) \equiv (sI - H_r)^{-1} \in C^{2(l-n)\times 2(l-n)}$$

where $\hat{\Phi}_r(s)$ denotes the resolvent matrix of H_r.

For most large flexible structures this assumption is reasonable. Although the poles of A_r may have small stability margin (due to very small damping coefficient ζ), however, the natural energy dissipating property of the flexible structure systems makes this assumption true.

It is also assumed that we can select the gain matrices G_c and K_c such that the following elementary assumption is true.

Assumption 7.4.2

The nominal closed-loop subsystem is asymptotically stable [210], i.e.,

$$\hat{\Phi}_c(s) \equiv (sI - H_c)^{-1} \in C^{4n\times 4n}$$

where $\hat{\Phi}_c(s)$ is the resolvent matrix of H_c.

This assumption implies $\lambda(A_c - B_c G_c) \in C_-$ and $\lambda(A_c - K_c C_c) \in C_-$. In practice, the observer gain K_c is chosen so that the eigenvalues of $A_c - K_c C_c$ are slightly to the left of the spectrum of $A_c - B_c G_c$ for an adequate system response.

Remark 7.4.1

If there are finitely unstable poles, we can arrange so that A_c contains those unstable poles and critical modes to be controlled, with A_r containing the remaining modes.

Before further analysis, we give the following lemmas which will be used throughout the paper.

Lemma 7.4.1 [221]

Let the matrix $A(s) \in C^{m\times n}$ then the norm function $\|\cdot\|_\infty : C^{m\times n} \to \mathbb{R}$ can be expressed by $\sup_{s\in C_+}\|A(s)\|_2 = \sup_{\omega\in[0,\infty)} \|A(j\omega)\|_2 = \|A(j\omega)\|_\infty$. That is, the norm of $A(s)$ can be computed based on the behavior of $A(s)$ on the $j\omega-$ axis alone.

Lemma 7.4.2 [222]

If $R(s) \in C^{n\times n}$ and $\|R(j\omega)\|_2 < 1, \forall\omega \in [0,\infty)$ then $[I - R(s)]^{-1} \in C^{n\times n}$, i.e., $I - R(s)$ has no zeros in C_+.

We give in the following a new result which provides a simple quantitative measure of the spillover bound under which overall loop stability can be guaranteed.

Theorem 7.4.1

Under Assumptions 7.4.1 and 7.4.2, the actual closed-loop system equations (7.13)–(7.15) involving control and observation spillover will be asymptotically stable provided

$$\delta < \inf_{\omega\in[0,\infty)} [\sigma_{\min}(j\omega I - H_c)\sigma_{\min}(j\omega I - A_r)] \equiv \mu_1 \tag{7.18}$$

where $\delta = \sqrt{2}\sigma_{\max}(B_r G_c)\sigma_{\max}(K_c C_r)$; μ_1 standards for the permissible upper bound of control/observation spillover or the robust stability measure to the spillover.

Proof

The proof of this theorem is provided in Appendix 7.8A.

From equation (7.18), it can be found, as $\omega \to \infty$, $\sigma_{\min}(j\omega I - H_c)\sigma_{\min}(j\omega I - A_r) \to \infty$ which implies that inequality equation (7.18) only needs to be tested within a finite frequency interval. To estimate the finite frequency band, we first note that the following inequalities hold:

$$\sigma_{\min}(j\omega I - H_c) \geq \omega - \sigma_{\max}(H_c), \quad \forall\omega \geq 0$$

$$\sigma_{\min}(j\omega I - A_r) \geq \omega - \sigma_{\max}(A_r), \quad \forall\omega \geq 0$$

Then

$$\begin{aligned}\sigma_{\min}(j\omega I - H_c)\sigma_{\min}(j\omega I - A_r) &\geq \omega^2 - [\sigma_{\max}(H_c) + \sigma_{\max}(A_r)]\omega \\ &+ \sigma_{\max}(H_c)\sigma_{\max}(A_r), \quad \forall\omega \geq 0\end{aligned} \tag{7.19}$$

By solving a quadratic equation of the form

$$\omega^2 - [\sigma_{\max}(H_c) + \sigma_{\max}(A_r)]\omega + \sigma_{\max}(H_c)\sigma_{\max}(A_r) - \delta = 0 \tag{7.20}$$

and noting that $\omega \geq 0$, we have

$$\omega_c = \begin{cases} \omega_1, \text{ for } \delta > \sigma_{\max}(H_c)\sigma_{\max}(A_r) \\ \omega_2, \text{ for } \delta \leq \sigma_{\max}(H_c)\sigma_{\max}(A_r) \end{cases}$$

where

$$\begin{aligned} \omega_1 &= 0.5\{\sigma_{\max}(H_c) + \sigma_{\max}(A_r) + \sqrt{[\sigma_{\max}(H_c) - \sigma_{\max}(A_r)]^2 + 4\delta}\} \\ \omega_2 &= 0.5\{\sigma_{\max}(H_c) + \sigma_{\max}(A_r) - \sqrt{[\sigma_{\max}(H_c) - \sigma_{\max}(A_r)]^2 + 4\delta}\} \end{aligned} \tag{7.21}$$

While $\omega > \omega_c$, from equations (7.19)–(7.20), it can be seen

$$\sigma_{\min}(j\omega I - H_c)\sigma_{\min}(j\omega I - A_r) > \delta, \quad \forall \omega \in [\omega_c, \infty)$$

Namely, when $\omega > \omega_c$, this inequality always holds. Therefore, the robust stability problem is reduced to checking the following inequality in the finite frequency interval, i.e.,

$$\delta < \sigma_{\min}(j\omega I - H_c)\sigma_{\min}(j\omega I - A_r), \quad \forall \omega \in [0, \omega_c)$$

or equivalently,

$$\delta < \inf_{\omega \in [0, \omega_c)} [\sigma_{\min}(j\omega I - H_c)\sigma_{\min}(j\omega I - A_r)] \tag{7.22}$$

Remark 7.4.2

Some comments on various aspects of the preceding analysis may be in order.

a. The stability robustness inequality equation (7.18) can also be expressed as

$$\mu_1 - \delta > 0$$

If μ_1 is viewed as the stability margin of a controlled large flexible structure without spillover, the above inequality reveals that the designed controller must provide a sufficiently large μ_1 so that the total amount of "control/observation spillover" cannot destroy the stability margin. To synthesize the compensator, we can examine the values of μ_1, such that the resulting one is the most robust design.

b. Since norms were used to obtain the sufficient condition (7.18), this condition may be conservative in some cases; however, it does give a simple relationship with control/observation spillover to pole shifting.

The above state-space description of a large flexible structure is considered, and a sufficient condition for stability from the time domain viewpoint is derived [214]. We give a brief description about this stability criterion. First note that $\Phi_c(t)$ and $\Phi_r(t)$ can be defined as $\Phi_c(t) \equiv L^{-1}[\hat{\Phi}_c(s)]$ (inverse Laplace transform of $\hat{\Phi}_c(s)$) and $\Phi_r(t) \equiv L^{-1}[\hat{\Phi}_r(s)]$. If H_c and A_r generate the exponentially stable semigroups $\Phi_c(t) \equiv \exp(H_c t)$ and $\Phi_r(t) \equiv \exp(A_r t)$, respectively, such that

$$\begin{aligned} \|\Phi_1(t)\| &\le m_1 \exp(-\sigma_1 t), m_1 \ge 1, \sigma_1 > 0, t \ge 0 \\ \|\Phi_2(t)\| &\le m_2 \exp(-\sigma_2 t), m_2 \ge 1, \sigma_2 > 0, t \ge 0 \end{aligned} \tag{7.23}$$

By applying the Bellman-Gronwall's inequality, the system of equations (7.13)–(7.15) is shown to be asymptotically stable if

$$m_1 m_2 \sqrt{1+\gamma+\gamma^2}\,\|H_{rc}\| < \min(\sigma_1, \sigma_2), \gamma = \frac{\|H_{cr}\|}{|\sigma_1 - \sigma_2|}, \sigma_1 \ne \sigma_2 \tag{7.24}$$

For the consistency of comparison, we take 2-norm in the above stability inequality and consider the result in finite-dimensional space. Under this setting, the above condition can be reformulated as follows:

$$\mu_B = \frac{\sqrt{2} m_1 m_2 \sigma_{\max}(B_r G_c)\sqrt{1+\gamma+\gamma^2}}{\min(\sigma_1, \sigma_2)} < 1, \gamma = \frac{\sigma_{\max}(K_c C_r)}{|\sigma_1 - \sigma_2|} \tag{7.25}$$

The proposed robust stability condition (7.18) can also be written as

$$\mu_L = \frac{\sqrt{2}\sigma_{\max}(K_c C_r)\sigma_{\max}(B_r G_c)}{\inf\limits_{\omega \ge 0}\left[\sigma_{\min}(j\omega I - H_c)\sigma_{\min}(j\omega I - A_r)\right]} < 1 \tag{7.26}$$

The following theorem will show that the proposed robustness measure μ_L is less conservative than μ_B.

Theorem 7.4.2

For the system described by equations (7.13)–(7.15), if H_c and A_r satisfy Assumptions 7.4.1 and 7.4.2 then $\mu_L < \mu_B$.

Proof

Appendix 7.8B.

Since our proposed approach is based on the state-space model, comparing the proposed robust stability criterion in [216,218,223], it is neither to compute the nominal (controlled modes), nor to compute the perturbed (including both of controlled and residual modes) transfer matrices which are required for the robustness measures in the above literature. In addition, instead of testing the robust stability condition over the whole frequency interval, we only need to check the condition within a finite frequency band.

7.5 ROBUST STABILIZATION WITH RESPECT TO THE TOTAL SPILLOVER

Although the derivation presented in Section 7.4 is able to estimate an explicit upper bound of the control/observation spillover which do not destabilize the closed-loop system; however, it exhibits some difficulties which may be inconvenient for practical control designs: (i) controller gains couple with the residual matrices B_r and C_r in the spillover terms, (ii) although the coupling matrices H_{cr}, H_{rc} may have finite ranks, the spillover bound may not be finite for certain infinite-dimensional flexible systems, (iii) in general, one may only know a roughly estimated response bound of the residual dynamics rather than the detailed knowledge about the individual matrices A_r, B_r, C_r. To avoid these drawbacks, an expanded stability criterion will be provided in this section using the minimal knowledge about the residual subsystem.

We give the following result which forms a complement to Theorem 7.4.1.

Theorem 7.5.1

Under the standing assumptions of Theorem 7.4.1, the closed-loop system equations (7.13)–(7.15) will be robustly stable provided

$$\|\Delta H(j\omega)\|_2 < \mu_2, \forall \omega \in [0,\infty) \tag{7.27}$$

where

$$\Delta H(j\omega) = C_r(j\omega I - A_r)^{-1} B_r$$

and

$$\mu_2 \equiv \left\|C_0(j\omega I - A_0)^{-1} B_0\right\|_2^{-1}$$

with $C_0 = [-G_c \;\; -G_c], A_0 = H_c, B_0 = \begin{bmatrix} 0 \\ K_c \end{bmatrix}$ or equivalently,

$$\mu_2 \equiv \left\|G_c(j\omega I - A_g)^{-1}(j\omega I - A_c)(j\omega I - A_e)^{-1} K_c\right\|_2^{-1}$$

in which $A_g = A_c - B_c G_c, A_e = A_c - K_c C_c$.

Proof

The proof is given in Appendix 7.8C.

Remark 7.5.1

If the frequency response of the residual subsystem lying within a boundary function is given by

$$\|\Delta H(j\omega)\|_2 < |\ell_a(j\omega)|, \forall\omega \in [0,\infty)$$

where $\ell_a(s)$ is a stable rational function free of poles and zeros in $\text{Re}(s) \geq 0$, then (7.27) can be expressed as

$$|\ell_a(j\omega)| < \mu_2, \forall\omega \in [0,\infty)$$

In equation (7.27), $\|\Delta H(j\omega)\|_2$ can be denoted as the total spillover bound which corresponds, in an abstract sense, to control and observation spillover bounds, i.e., $\sigma_{\max} B_r G_c$ and $\sigma_{\max} K_c C_r$, in equation (7.18). In the above analysis, the term $\Delta H(j\omega)$ in equation (7.27) is viewed as the unmodeled dynamics to the nominal plant, whereas, $B_r G_c$ and $K_c C_r$ in equation (7.18) are treated as the parametric perturbations. It is noted that $\|\Delta H(j\omega)\|_2$ contains only the matrices (A_r, B_r, C_r) which is independent on the control and observation gains (G_c, K_c). Thus an increase in (G_c, K_c) will not affect this bound. This superior feature permits one to search for the permissible gains (G_c, K_c) such that the robust stability inequality equation (7.27) holds. In the following, we propose a robust observer-based controller design by applying the modern H_∞ design technique.

To proceed, we first introduce a useful lemma which plays a central role in the subsequent controller design.

Lemma 7.5.1 [224]

Let the transfer matrix $E(s) = C(sI - A)^{-1}B + D$ where C, A, B, D be matrices of compatible dimensions. Then, for A being stable, the following H_∞-norm inequality holds

$$\|E(j\omega)\|_\infty < \gamma \tag{7.28}$$

if, and only if, $\gamma^2 I - D^T D > 0$ and there exists a symmetric positive definite matrix Z such that the algebraic Riccati equation is satisfied

$$A^T Z + ZA + (ZB + C^T D)(\gamma^2 I - D^T D)^{-1}(B^T Z + D^T C) + C^T C < 0 \tag{7.29}$$

where the inequality is in the sign-definite sense.

We now describe the procedure to find an observer-based stabilizing controller for a flexible system subject to the spillover contamination. Suppose that the upper bound of the total spillover is estimated as follows, i.e.,

$$\|\Delta H(j\omega)\|_{\infty} < \alpha \tag{7.30}$$

Then the robust stability inequality equation (7.27) is guaranteed by

$$\|C_0(j\omega I - A_0)^{-1} B_0\|_{\infty} < \beta \tag{7.31}$$

where $\beta = 1/\alpha$. It is easily seen that the above inequality corresponds to the standard state-space H_∞ optimization problem. From Lemma 7.5.1 we know that inequality equation (7.31) is met, if, and only if, there exists a real matrix $X = X^T$ with positive definite such that

$$A_0^T X + XA_0 + \frac{1}{\beta^2} XB_0B_0^T X + C_0^T C_0 < 0 \tag{7.32}$$

To create a separation between the resulting state and algebraic Riccati equations, we let $X = \begin{bmatrix} X_g & 0 \\ 0 & X_h \end{bmatrix}$ where $X_g = X_g^T > 0$ and $X_h = X_h^T > 0$. Also, let the left-hand side of equation (7.32) equal to Λ. After some simple manipulations, it can be expressed as

$$\Lambda = \begin{bmatrix} \Lambda_{11} & \Lambda_{12} \\ \Lambda_{21} & \Lambda_{22} \end{bmatrix} < 0 \tag{7.33}$$

where

$$\Lambda_{11} = (A_c - B_cG_c)^{\mathrm{T}} X_g + X_g(A_c - B_cG_c) + G_c^T G_c$$

$$\Lambda_{12} = -X_g B_c G_c + G_c^T G_c$$

$$\Lambda_{21} = \Lambda_{12}^T$$

$$\Lambda_{22} = (A_c - K_cC_c)^{\mathrm{T}} X_h + X_h(A_c - K_cC_c) + \frac{1}{\beta^2} X_hK_cK_c^T X_h + G_c^T G_c.$$

Add and subtract $X_gB_cB_c^TX_g$ to Λ_{11} and group appropriate terms to complete the square with respect to G_c then

$$\Lambda_{11} = A_c^T X_g + X_gA_c - X_gB_cB_c^T X_g + (B_c^T X_g - G_c)^T (B_c^T X_g - G_c)$$

Choosing the control gain G_c in the form

$$G_c = B_c^T X_g \tag{7.34}$$

it follows that

$$\Lambda_{11} = A_c^T X_g + X_g A_c - X_g B_c B_c^T X_g$$

and

$$\Lambda_{21} = \Lambda_{12} = 0$$

To satisfy the requirement of A being a symmetric negative definite matrix, we first require $\Lambda_{11} < 0$, i.e.,

$$A_c^T X_g + X_g A_c - X_g B_c B_c^T X_g < 0 \tag{7.35}$$

With the designed control gain G_c given by equation (7.34), Λ_{22} can be expressed as

$$\Lambda_{22} = (A_c - K_c C_c)^{\mathrm{T}} X_h + X_h (A_c - K_c C_c) + \frac{1}{\beta^2} X_h K_c K_c^T X_h + X_g B_c B_c^T X_g$$

For Λ to be negative definite, our remaining requirement is $\Lambda_{22} < 0$, i.e.,

$$(A_c - K_c C_c)^{\mathrm{T}} X_h + X_h (A_c - K_c C_c) + \frac{1}{\beta^2} X_h K_c K_c^T X_h + X_g B_c B_c^T X_g < 0 \tag{7.36}$$

Adding equations (7.35) and (7.36), and collecting appropriate terms, it yields

$$A_c^T (X_g + X_h) + (X_g + X_h) A_c - \beta^2 C_c^T C_c + \left(\frac{1}{\beta} K_c^T X_h \text{-} \beta C_c \right)^T \left(\frac{1}{\beta} K_c^T X_h \text{-} \beta C_c \right) < 0 \tag{7.37}$$

In view of equation (7.37) an obvious choice for the relationship between K_c, X_g and X_h is

$$K_c^T X_h = \beta^2 C_c$$

so that equation (7.37) takes the form of the following algebraic Riccati equation

$$A_c^T X_e + X_e A_c - \beta^2 C_c^T C_c < 0 \tag{7.38}$$

where $X_e = X_g + X_h$. Since X_h is restricted to be positive definite, thus, an admissible choice of the observer gain K_c is

$$K_c = \beta^2 X_h^{-1} C_c^T = \frac{1}{\alpha^2} \left(X_e - X_g \right)^{-1} C_c^T \tag{7.39}$$

The following theorem summarizes the result.

Theorem 7.5.2

Assume that the given total spillover $\Delta H(s)$ is bounded in the H_∞-norm sense as described by equation (7.30). Let $X_g > 0$ satisfy equation (7.35) and $X_e > X_g$ satisfy equation (7.38). If the control and observer gain matrices are, respectively, chosen by equations (7.34) and (7.39), then the observer-based controller equations (7.9)–(7.10) robustly stabilizes system equations (7.6)–(7.8).

From equations (7.34), (7.35) and (7.38), (7.39) we see that the control gain G_c is independent of the extent of perturbations; whereas the value of the observer gain K_c is closely related to the perturbation upper bound.

We now consider the case where $\alpha \to 0$. The algebraic Riccati equation $\Lambda_{11} < 0$ can be expressed as

$$(A_c - B_c G_c)^{\mathrm{T}} X_g + X_g (A_c - B_c G_c) + G_c^T G_c + Q = 0$$

for some real matrix $Q > 0$. Since $A_c - B_c G_c$ is stable and $G_c^T G_c + Q > 0$, the standard Lyapunov stability theory shows that $X_g > 0$. Therefore, there must exist a positive definite matrix X_g such that equation (7.35) holds. Since $\beta \to \infty$, thus inequality equation (7.38) holds for arbitrary $X_e > 0$. These facts together imply that there must exist an observer-based controller such that the overall system could be robustly stabilized.

The design procedure for robust observer-based controllers can now be summarized briefly as follows:

i. Estimate the upper bound of the total spillover $\Delta H(s)$ in the H_∞-norm sense.
ii. Solve equation (7.35) for positive definite solution X_g. If such solution exists, compute the control gain matrix G_c via equation (7.34).
iii. Solve equation (7.38) for positive definite solution X_e. If such solution exists, involve the matrix X_g found in step (ii) to compute the observer gain matrix K_c from equation (7.39).
iv. Synthesize the robust observer-based control law given by equations (7.9) and (7.10).

Remark 7.5.2

There is no explicit criterion which can be used to compare the conservatism between the stability robustness tests in equations (7.18) and (7.27). In principle, anyone of these may be used. However, to obtain a satisfactory system performance, one should choose suitable control gains which can effectively assign the dominant system eigenvalues to desired locations and simultaneously satisfy one of these stability inequalities.

7.6 EXAMPLE: MODAL CONTROL OF A SIMPLY SUPPORTED BEAM

In this section, we present the robust stabilization results of some numerical studies on the robust observer-based output feedback control of a standard flexible beam.

The beam's model (Figure 7.1) follows the configuration given in [212]. Its dynamics is governed by the Euler-Bernoulli partial differential equation

$$mu_{tt}(x,t) + EIu_{xxxx}(x,t) = F(x,t)$$

for $0 \le x \le L$ and $t \ge 0$, where $u(x, t)$ is the transverse displacement and $F(x, t)$ is the applied force distribution. For convenience, the beam parameters m (mass per unit length), I (moment of inertia), E (modulus of elasticity), and L (length of the beam) are set to be unity. The boundary conditions for simple support are

$$u(0,t) = u(L,t) = 0, u_{xx}(0,t) = u_{xx}(L,t) = 0$$

Let the actuator and the displacement sensor be, respectively, located at the 1/6 and 5/6 positions. All nominal position states are initially set to 1.0, and other nominal and residual states are set to zero. Following [225], the damping coefficient ζ is assumed to be 0.01. Now, the first three modes are being controlled, while the fourth to the eighth mode are acted as known residual modes. In theory, the system requires an infinite number of elastic modes to completely described its behavior, only five residual modes are considered here for the convenience of computer simulation.

Using the standard modal expansion method [226], the parameters of the controlled modal model equation (7.6) will have the form of

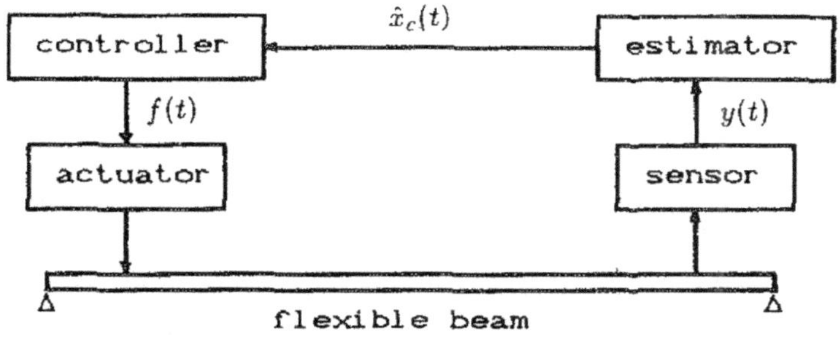

FIGURE 7.1 Observer-based output feedback control design of a simply supported (pinned-pinned) beam.

$$\Lambda_c^{1/2} = diag[\pi^2,(2\pi)^2,(3\pi)^2]$$

$$\hat{B}_c = \Lambda_c^{-1/2}\bar{B}_c = \sqrt{2}\left[\frac{\sin(\pi/6)}{\pi^2},\frac{\sin(2\pi/6)}{(2\pi)^2},\frac{\sin(3\pi/6)}{(3\pi)^2}\right]^T$$

$$\bar{C}_c = \sqrt{2}\left[\sin\frac{5\pi}{6},\sin\frac{10\pi}{6},\sin\frac{15\pi}{6}\right]$$

and for the residual model

$$\Lambda_r^{1/2} = diag[(4\pi)^2,(5\pi)^2,(6\pi)^2]$$

$$\hat{B}_r = \Lambda_r^{-1/2}\bar{B}_r = \sqrt{2}\left[\frac{\sin(4\pi/6)}{(4\pi)^2},\frac{\sin(5\pi/6)}{(5\pi)^2},\dots,\frac{\sin(8\pi/6)}{(8\pi)^2}\right]^T$$

$$\bar{C}_r = \sqrt{2}\left[\sin\frac{20\pi}{6},\sin\frac{25\pi}{6},\dots,\sin\frac{40\pi}{6}\right]$$

As the control gain is selected as

$$G_c = [0.0745\ 0.2461\ 0.3287\ 4.7466\ 12.3208\ 16.4585]\left(\|G_c\|_2 = 21.103\right).$$

the corresponding regulator poles are located at $-1.0193 \pm j88.8220$, $-0.5859 \pm j39.4764, -0.2687 \pm j9.8691$.While the observer gain is chosen as

$$K_c = [6.7724\ -6.3847\ 5.8061\ 0.3060\ -0.7294\ 1.0293]^T \left(\|K_c\|_2 = 11.046\right).$$

the corresponding observer poles are located at $-5.0511 \pm j88.8175$, $-4.2976 \pm j39.4742, -2.4411 \pm j9.8680$. With these gains, bounds for the control spillover $\|B_rG_c\|_2$ and observation spillover $\|K_cC_r\|_2$ can be obtained as follows:

$$\sigma_{\max}(B_rG_c) = 0.1818, \sigma_{\max}(K_cC_r) = 22.0895$$

Applying equation (7.22), we only need to test the robust stability condition (7.18) within a finite frequency band, i.e., $\omega \in [0,94.1015]$ rad/sec. The robust stability margin η_1 for the system without spillover is obtained as

$$\eta_1 = \inf_{\omega\in[0,94.1015)}\left[\sigma_{\min}\left(j\omega I - H_c\right)\sigma_{\min}\left(j\omega I - A_r\right)\right] = 39.595, \text{ at } \omega = 9.87 \text{ rad/sec}$$

The upper bound of control/observation spillover is given by

$$\delta = \sqrt{2}\sigma_{\max}(B_rG_c)\sigma_{\max}(K_cC_r) = 5.6802$$

The robustness test of equation (7.18) is illustrated in Figure 7.2. Clearly the requirement for robust stability is satisfied. The result of the alternative stability robustness test equation (7.27) gives

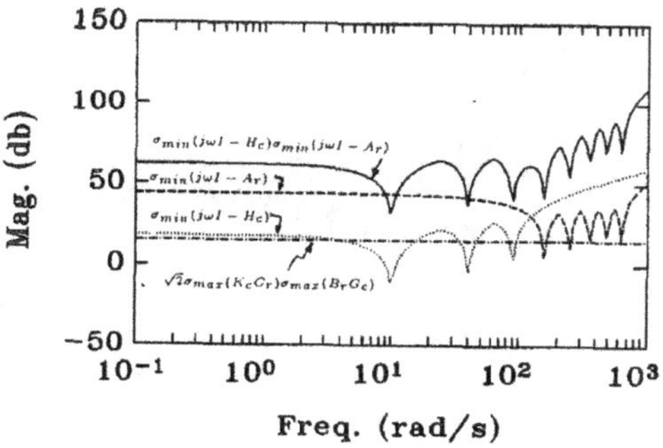

FIGURE 7.2 Robustness test of control/observation spillover in equation (7.18).

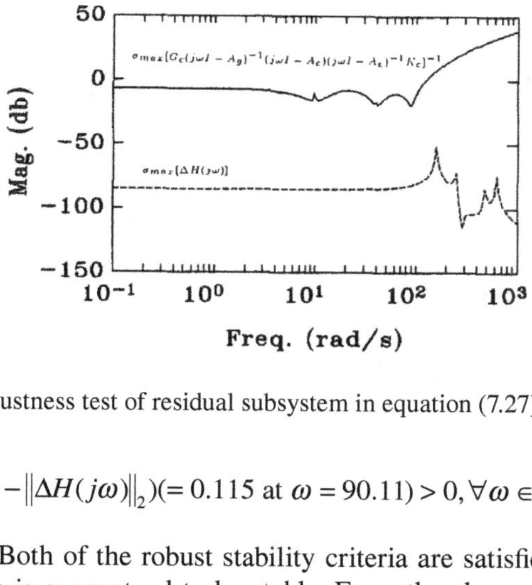

FIGURE 7.3 Robustness test of residual subsystem in equation (7.27).

$$\inf(\mu_2 - \|\Delta H(j\omega)\|_2)(= 0.115 \text{ at } \omega = 90.11) > 0, \forall \omega \in [0, \infty)$$

(see Figure 7.3). Both of the robust stability criteria are satisfied; therefore, the controlled system is guaranteed to be stable. From the above tests, we see that the chosen gains provide an adequately excessive robustness such that the controlled system could tolerate more residual modes without leading to instability. Numerical simulations for the displacement sensor signal and actuator command are, respectively, shown in Figures 7.4 and 7.5. The observer output and observation error ($\equiv \hat{y}(t) - y(t)$) are shown in Figure 7.6. From the observation about these results, it is seen that a good robust stabilization property is achieved.

To compare the conservatism between the robust stability criteria equations (7.18) and (7.27), we increase the control and observer gains, respectively, as follows:

$$G_c = [1.266\ 8.448\ 21.300\ 32.1977\ 116.000\ 220.523]\left(\|C_c\|_2 = 252.289\right)$$

$$K_c = [21.771\ -20.430\ 17.964\ 2.611\ -6.553\ 10.594]]^T\left(\|K_c\|_2 = 37.095\right)$$

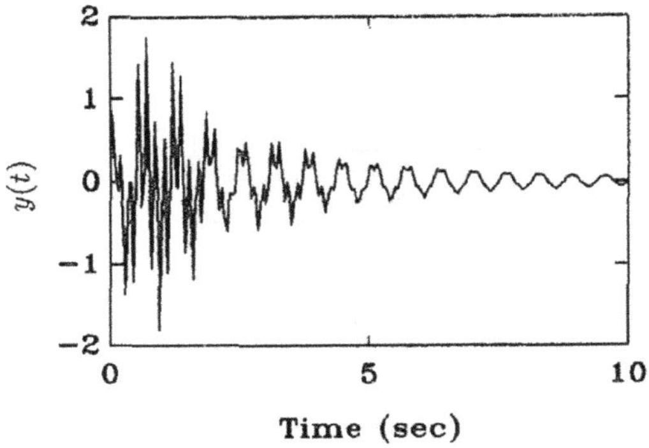

FIGURE 7.4 Beam deflection.

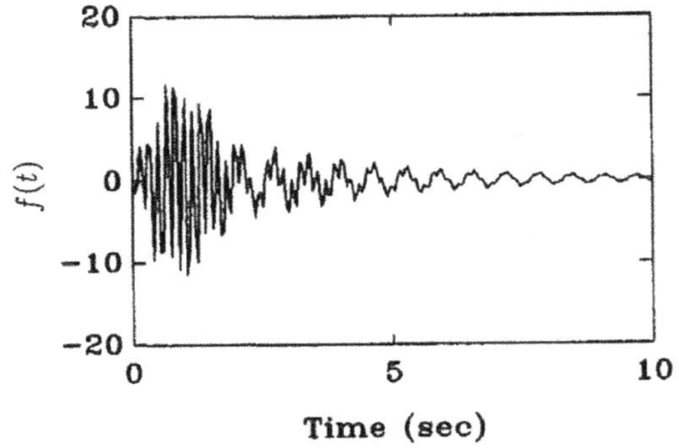

FIGURE 7.5 Actuator control force.

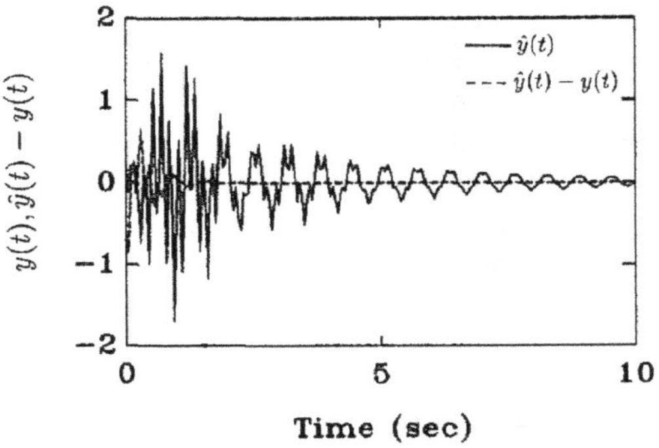

FIGURE 7.6 Observer output and observation error.

With these gains, we find that

$$\sqrt{2}\sigma_{\max}(B_rG_c)\sigma_{\max}(K_cC_r)(=228.042)>\mu_1(=116.604 \text{ at } \omega=157.03)$$

Inequality equation (7.18) is obviously not satisfied. However, the robust stability criterion equation (7.27) still holds for this situation

$$\inf(\mu_2-\|\Delta H(j\omega)\|_2)\big(=0.0074 \text{ at } \omega=82.143\big)>0, \forall\omega\in[0,\infty)$$

Hence, the flexible beam system is still stable for the larger gains chosen; inequality equation (7.27) appears to be less conservative than equation (7.18) in this example.

7.7 CONCLUSION

This chapter considers robust stabilization of the spillover phenomenon of large flexible structures in the frequency domain according to Glerkin method. Two types of frequency domain robust stability conditions for a large flexible structure system controlled by an observer-based compensator are derived based on the mode state-space model of the flexible system. In the derivation of the robust stability conditions, observation/control spillover and the total spillover are, respectively, treated as parametric and unmodeled perturbations. It is shown that an observer-based robust output feedback control law can be obtained by solving two algebraic Riccati equations in the robust stabilization design of large flexible structure. After the introduction of robust stabilization design based on Galerkin method for linear partial differential systems, a robust stabilization design based on finite difference method will be introduced in the following chapter.

7.8 APPENDIX

7.8.A Proof of Theorem 7.4.1

Taking Laplace transform of equations (7.13)–(7.15), after some trivial manipulations, yields

$$Z_c(s)=(sI-H_c)^{-1}z_c(0)+(sI-H_c)^{-1}H_{cr}X_r(s) \tag{7.A1}$$

$$X_r(s)=(sI-H_c)^{-1}x_r(0)+(sI-H_c)^{-1}H_{rc}Z_c(s) \tag{7.A2}$$

$$Y(s)=\tilde{C}_cZ_c(s)C_rX_r(s) \tag{7.A3}$$

Substituting equation (7.A2) into (7.A3), $Z_c(s)$ can be further expressed as

$$Z_c(s)=(sI-H_c)^{-1}z_c(0)+(sI-H_c)^{-1}H_{cr}[(sI-H_c)^{-1}x_r(0)+(sI-H_c)^{-1}H_{rc}Z_c(s)]$$

or equivalently,

$$[I-(sI-H_c)^{-1}H_{cr}(sI-H_c)^{-1}H_{rc}]Z_c(s)$$
$$=(sI-H_c)^{-1}z_c(0)+(sI-H_c)^{-1}H_{cr}[(sI-H_r)^{-1}x_r(0)$$

Therefore

$$Z_c(s)=\Sigma^{-1}(s)(sI-H_c)^{-1}z_c(0)$$
$$+\Sigma^{-1}(s)(sI-H_c)^{-1}H_{cr}(sI-H_r)^{-1}x_r(0) \tag{7.A4}$$

where $\Sigma(s)=I-(sI-H_c)^{-1}H_{cr}(sI-H_c)^{-1}H_{rc}$. Viewing the term $(sI-H_c)^{-1}H_{cr}$ $(sI-H_c)^{-1}H_{rc}\in C^{4n\times 4n}$ as a generalized transfer matrix function, and then applying Lemmas 7.4.1 and 7.4.2, we know that if

$$\left\|(j\omega I-H_c)^{-1}H_{cr}(j\omega I-H_r)^{-1}H_{rc}\right\|_2<1,\forall\omega\in[0,\infty) \tag{7.A5}$$

then $\Sigma^{-1}(s)\in\delta^{4n\times 4n}$. In view of equation (7.A4) and since $(sI-H_c)^{-1}\in\delta^{4n\times 4n}$ and $(sI-H_r)^{-1}\in\delta^{2(l-n)\times 2(l-n)}$, this implies that $z_c(t)$ will be asymptotically stable. Note that equation (7.A5) will be true if

$$\|H_{cr}\|_2\|H_{rc}\|_2<\left\{\sup_{\omega\in[0,\infty)}\left[\left\|(j\omega I-H_c)^{-1}\right\|_2\left\|(j\omega I-A_r)^{-1}\right\|_2\right]\right\}^{-1} \tag{7.A6}$$

Or equivalently, by invoking the notations of singular values given by [227],

$$\left\|A^{-1}\right\|_2^{-1}=[\sigma_{\max}(A^{-1})]^{-1}=\sigma_{\min}(A)$$

and

$$\|H_{cr}\|_2=\left\|\begin{bmatrix}0\\K_cC_r\end{bmatrix}\right\|_2=\sigma_{\max}(K_cC_r)$$

$$\|H_{rc}\|_2=\left\|\begin{bmatrix}-B_rG_c & -B_rG_c\end{bmatrix}\right\|_2=\sqrt{2}\sigma_{\max}(B_rG_c)$$

then equation (7.A6) can be restated in terms of singular values as follows:

$$\sqrt{2}\sigma_{\max}(B_rG_c)\sigma_{\max}(K_cC_r)<\inf_{\omega\in[0,\infty)}[\sigma_{\min}(j\omega I-H_c)\sigma_{\min}(j\omega I-A_r)]$$

Since $(sI-H_r)^{-1}\in C^{2(l-n)\times 2(l-n)}$, once $z_c(t)$ has been stabilized, from equation (7.A2) we easily see that $x_r(t)$ would be stable. As both of $z_c(t)$ and $x_r(t)$ are asymptotically stable, the robust stability of $y(t)$ can be concluded from equation (7. 3).

7.8.B Proof of Theorem 7.4.2

Let H_c be diagonalizable, then there exists a nonsingular matrix P of dimension $4n \times 4n$ such that $H_c = P^{-1}diag(\lambda_1, \lambda_2, ..., \lambda_{4n})P$, where $\lambda_i \in C\_$ are the eigenvalues of H_c. It follows that $\exp(H_c t) = P^{-1}diag(\lambda_1 t, \lambda_2 t, ..., \lambda_{4n} t)P$ and

$$\begin{aligned}
&\|\exp(H_c t)\|_2 \\
&\leq \|P^{-1}\|_2 \|P\|_2 \|diag\{\exp[\mathrm{Re}(\lambda_1)t], \exp[\mathrm{Re}(\lambda_2)t], ..., \exp[\mathrm{Re}(\lambda_{4n})t]\} \\
&\qquad \cdot diag\{\exp[j\,\mathrm{Im}(\lambda_1)t], \exp[j\,\mathrm{Im}(\lambda_2)t], ..., \exp j\,\mathrm{Im}(\lambda_{4n})t]\}\|_2 \qquad (7.B1) \\
&= m_1 \|diag\{\exp[\mathrm{Re}(\lambda_1)t], \exp[\mathrm{Re}(\lambda_2)t], ..., \exp[\mathrm{Re}(\lambda_{4n})t]\}\|_2 \\
&\leq m_1 \exp(-\sigma_1 t), t \geq 0
\end{aligned}$$

In the above $\mathrm{Im}(\cdot)$ denotes the imaginary part, the condition number $m_1 = \sigma_{\max}(P)/\sigma_{\min}(P) \geq 1$, $\sigma_1 = -\max \mathrm{Re}[\lambda_i(H_c)] > 0, i = 1, ..., 4n$. Similarly, for the residual dynamics, one may take

$$\|\exp(A_r t)\|_2 \leq m_2 \exp(-\sigma_2 t), t \geq 0 \qquad (7.B2)$$

where $m_2 \geq 1, \sigma_2 = -\max_i \mathrm{Re}[\lambda_i(A_r)] > 0, i = 1, ..., 2(l-n)$. We also have

$$\begin{aligned}
&\inf_{\omega \geq 0} \sigma_{\min}(j\omega I - H_c) \\
&= \inf_{\omega \geq 0} \sigma_{\min}[P^{-1}diag(j\omega - \lambda_1, j\omega - \lambda_2, ..., j\omega - \lambda_{4n})P] \\
&\geq [\sigma_{\min}(P)/\sigma_{\max}(P)] \inf_{\omega \geq 0} \sigma_{\min}[diag(j\omega - \lambda_1, j\omega - \lambda_2, ..., j\omega - \lambda_{4n})] \qquad (7.B3) \\
&= [\sigma_{\min}(P)/\sigma_{\max}(P)] \min_i [-\mathrm{Re}(\lambda_i(H_c))] \\
&= \sigma_1 / m_1, i = 1, 2, ..., 4n
\end{aligned}$$

Similarly one can show that

$$\inf_{\omega \geq 0} \sigma_{\min}(j\omega I - A_r) \geq \sigma_2 / m_2 \qquad (7.B4)$$

Now consider two cases (for $\sigma_1 \neq \sigma_2$):

Case i: $0 < \sigma_1 < \sigma_2$

Introducing equations (7.25) and (7.26), we have

$$\mu_B - \mu_L \geq \frac{\sqrt{2} m_1 m_2 \sigma_{\max}(B_r G_c) \sigma_{\max}(K_c C_r)}{\sigma_1 |\sigma_1 - \sigma_2|} - \frac{\sqrt{2} \sigma_{\max}(K_c C_r) \sigma_{\max}(B_r G_c)}{\inf_{\omega \geq 0} [\sigma_{\min}(j\omega I - H_c) \sigma_{\min}(j\omega I - A_r)]}$$

$$\geq \frac{\sqrt{2} m_1 m_2 \sigma_{\max}(B_r G_c) \sigma_{\max}(K_c C_r)}{\sigma_1 |\sigma_1 - \sigma_2|}$$

$$- \frac{\sqrt{2} \sigma_{\max}(K_c C_r) \sigma_{\max}(B_r G_c)}{\inf_{\omega \geq 0} [\sigma_{\min}(j\omega I - H_c)] \inf_{\omega \geq 0} [\sigma_{\min}(j\omega I - A_r)]}$$

$$\geq \frac{\sqrt{2} m_1 m_2 \sigma_{\max}(B_r G_c) \sigma_{\max}(K_c C_r)}{\sigma_1} \left(\frac{1}{|\sigma_1 - \sigma_2|} - \frac{1}{\sigma_2} \right)$$

> 0 (invoking (7.B3) and (7.B4))

Case ii: $0 < \sigma_2 < \sigma_1$
Similar as the above, it can be proved that

$$\mu_B - \mu_L \geq \frac{\sqrt{2} m_1 m_2 \sigma_{\max}(B_r G_c) \sigma_{\max}(K_c C_r)}{\sigma_2} \left(\frac{1}{|\sigma_1 - \sigma_2|} - \frac{1}{\sigma_1} \right) > 0$$

Therefore, we have $\mu_B > \mu_L$.

7.8.C Proof of Theorem 7.4.3

The proof of this theorem follows that of Theorem 7.4.1. As we can see from the proof of Theorem 7.4.1 that the matrix function $\Sigma(s)$ plays a key role in determining the stability of the closed-loop system in equations (7.13)–(7.15). If $\Sigma(s)$ has no poles lying within the closed right half-plane, then the stability is guaranteed. To investigate further, we extract the term $\Sigma(s)$ from equation (7.A4) and reformulate it as

$$\Sigma^{-1}(s) = \left\{ I - (sI - H_c)^{-1} H_{cr} (sI - H_r)^{-1} H_{rc} \right\}^{-1}$$

$$= \left\{ I - (sI - H_c)^{-1} \begin{bmatrix} 0 \\ K_c C_r \end{bmatrix} (sI - A_r)^{-1} \begin{bmatrix} -B_r G_c & -B_r G_c \end{bmatrix} \right\}^{-1}$$

$$= \left\{ I - (sI - H_c)^{-1} \begin{bmatrix} 0 \\ K_c \end{bmatrix} C_r (sI - A_r)^{-1} B_r \begin{bmatrix} -G_c & -G_c \end{bmatrix} \right\}^{-1}$$

By applying the following matrix permutation formula

$$\Sigma^{-1}(s) = [I - U(s)V]^{-1} = I + U(s)[I - VU(s)]^{-1} V \tag{7.C1}$$

where $U(s) = (sI - H_c)^{-1}\begin{bmatrix} 0 \\ K_c \end{bmatrix} C_r(sI - A_r)^{-1}B_r \in \delta^{4n\times m}$ and $V = \begin{bmatrix} -G_c & -G_c \end{bmatrix}$, $\Sigma^{-1}(s)$ can then be expanded in the form:

$$\Sigma^{-1}(s) = I + (sI - H_c)^{-1}\begin{bmatrix} 0 \\ K_c \end{bmatrix} C_r(sI - A_r)^{-1}B_r$$

$$\left\{ I - \begin{bmatrix} -G_c & -G_c \end{bmatrix} (sI - H_c)^{-1}\begin{bmatrix} 0 \\ K_c \end{bmatrix} C_r(sI - A_r)^{-1}B_r \right\}^{-1} \begin{bmatrix} -G_c & -G_c \end{bmatrix}$$

Referring to equation (7.A4), we have

$$Z_c(s) = \Sigma^{-1}(s)(sI - H_c)^{-1}\left\{ z_c(0) + \begin{bmatrix} 0 \\ K_cC_r(sI - A_r)^{-1} \end{bmatrix} x_r(0) \right\} \tag{7.C2}$$

We see that except for $\Sigma^{-1}(s)$ all resolvent matrices in equation (7.C2) are stable. From equation (7.C1), it is also found that closed-loop stability ensured by $\Sigma^{-1}(s) \in C^{4n\times 4n}$ is now replaced with $[I - VU(s)]^{-1} \in C^{m\times m}$. Thus by invoking Lemmas 7.4.1 and 7.4.2, $\Sigma^{-1}(s)$ will be analytic for all $s \in C_+$ provided

$$\|VU(j\omega)\|_2 < 1, \forall \omega \in [0, \infty)$$

or equivalently

$$\left\| \begin{bmatrix} -G_c & -G_c \end{bmatrix} (j\omega I - H_c)^{-1}\begin{bmatrix} 0 \\ K_c \end{bmatrix} C_r(j\omega I - A_r)^{-1}B_r \right\|_2 < 1, \forall \omega \in [0, \infty) \tag{7.C3}$$

and hence $z_c(t)$ is stabilized. Since we do not wish to use explicit knowledge about the residual model, a sufficient condition which guarantees equation (7.C3) is

$$\|\Delta H(j\omega)\|_2 \left\| \begin{bmatrix} -G_c & -G_c \end{bmatrix} (j\omega I - H_c)^{-1}\begin{bmatrix} 0 \\ K_c \end{bmatrix} \right\|_2 < 1, \forall \omega \in [0, \infty)$$

where $\Delta H(j\omega) = C_r(j\omega I - A_r)^{-1}B_r$. Or equivalently, by a straightforward expansion of the matrix inverse $(j\omega I - H_c)^{-1}$, this can be further expressed as $\|\Delta H(j\omega)\|_2 \|G_c(j\omega I - A_c + B_cG_c)^{-1}(j\omega I - A_c)(j\omega I - A_c + K_cC_c)^{-1}K_c\|_2 < 1$, $\forall \omega \in [0, \infty)$.

8 Robust Stabilization Design for Stochastic Linear Partial Differential Systems under Spatiotemporal Disturbances and Sensor Measurement Noises

8.1 INTRODUCTION

Partial differential systems (PDSs) [228–244] have been widely discussed recently. Many fields have been concerned with deterministic partial differential systems, for example, chemical reactor systems [229,230] electromagnetic transmission systems [231], image processing systems [232], manufacturing flow systems [236], neurophysiological systems, biodynamic systems [234,235] heat transfer systems, fluid dynamic systems, population dynamic systems, elastic wave transmission systems, flexible dynamic systems [236], etc. Since the most of phenomena have some uncertainties due to the existence of different stochastic fluctuations, the more accurate representation of the behaviors should be modeled by stochastic partial differential system (SPDS) [237–244]. Some examples of SPDSs are mobility stochastic systems in wireless networks [239], reaction stochastic systems in chemical engineering [230,240], stochastic molecular signal transduction systems in molecular biology [242,243], and distributed parameter stochastic systems in population biology [244]. Therefore, we shall address the study of LSPDSs. In the past decade, the stabilization design problems of LSPDSs have been widely studied [237]. The linear parabolic SPDS can be transferred to an equivalent infinite-dimensional ordinary differential system (ODS) based on Galerkin method in Section 1.2.1. According to the separation of eigenvalues, the infinite-dimensional ODS can be separated into a finite-dimensional slow mode system and an infinite-dimensional fast mode system. So a finite-dimensional controller based on the approximate finite-dimensional slow mode system could be designed to stabilize the LSPDS.

DOI: 10.1201/9781003229230-11

Recently, H_∞ stabilization [245–254] is an important design to efficiently attenuate the effect of the external disturbance upon the controlled output to be less than a prescribed level. The H_∞ filtering [255–258] is also needed for state estimation when state variables are unavailable. The systems concerning H_∞ stabilization in [245–254] are only ODSs, not PDSs. Even the robust control of PDSs in [230] is not of H_∞ stabilization, in its robust control design, there are some adjustable parameters needed to be chosen by try and error that is a drawback. At present, there is still no paper concerning the H_∞ robust observer-based stabilization problem of LSPDSs under system fluctuation, external disturbance, and measurement noise in the spatiotemporal domain. This motivates us to investigate the H_∞ robust observer-based stabilization problem of LSPDSs, which is an important topic in system stabilization designs. The results we propose have a wide range of applications in many fields [229–236] concerning PDSs.

In this chapter, the general robust stabilization theory for LSPDSs with a set of sensor measurements is studied from the spatiotemporal H_∞ disturbance attenuation point of view. Then the general H_∞ controller is designed for LSPDSs based on the general robust stabilization theory, which needs to solve a complex HJII and is not easy to implement, especially for the case where system matrices and noises are of spatiotemporal function. In this chapter, in order to simplify the stabilization design, based on the finite difference scheme in Section 1.2.2, the states of all finite difference grids in the spatiodomain are formulated as a spatial state vector so that a spatial state-space system can represent an LSPDS with the finite difference truncation error. In this situation, the spatiotemporal H_∞ stabilization performance can be transformed to an equivalent temporal H_∞ stabilization performance so that an implementable H_∞ spatial controller can be designed to robustly control LSPDSs. For LSPDSs, the H_∞ stabilization design needs to solve an algebraic Riccati inequality (ARI). Finally, a robust H_∞ stabilization example for a heat transfer system by a set of noisy sensor measurements is given to illustrate the robust stabilization design procedure and to confirm the H_∞ stabilization performance of the proposed robust stabilization design method for LSPDSs. The main features of the proposed method include the following: (i) A spatiotemporal H_∞ stabilization performance is employed to treat the robust stabilization problem for LSPDSs with disturbance $v(x, t)$ and noise $w(x, t)$ in the spatial-temporal domain. (ii) Based on HJII, a more general spatiotemporal H_∞ stabilization design theory is developed for LSPDSs. (iii) Finite difference method is used to transform the LSPDS to an equivalent spatial state space system so that an implementable H_∞ spatial state observer-based controller could be easily designed for LSPDSs by solving iterative linear matrix inequalities (LMIs) using the MATLAB LMI toolbox. (iv) The effect of finite difference truncation error on the stabilization can be attenuated to as small a value as possible. (v) As the grid points are dense enough, the proposed method will approach to the general theoretical H_∞ robust stabilization design. For the convenience of problem description, we define some notations as follows:

$$\|y\|^2 \triangleq \sum_{i=1}^{n} |y_i|^2, \text{where } y = [y_1, \ldots, y_n]^T$$

$L_2(U;\mathbb{R}^n)$ is the space of the n-dimensional measurable functions $y(x,t) \in \mathbb{R}^n$ defined on $x \in U$ for any t such that $\|y(x,t)\|^2_{L_2(U;\mathbb{R}^n)} \triangleq \int_U \|y(x,t)\|^2 dx < \infty$. $L_2(U \times [0,t_f];\mathbb{R}^n)$ is the space of the n-dimensional measurable functions $y(x,t) \in \mathbb{R}^n$ defined on $(x,t) \in U \times \mathbb{R}_+$ such that $\|y(x,t)\|^2_{L_2(U\times[0,t_f];\mathbb{R}^n)} \triangleq \int_0^{t_f} \int_U \|y(x,t)\|^2 dxdt < \infty$.

8.2 A GENERAL H_∞ STABILIZATION SETTING FOR LINEAR STOCHASTIC PARTIAL DIFFERENTIAL SYSTEMS

Since spatiotemporal disturbances and sensor measurement noises will affect the behavior of the control system, the perturbative LSPDS with a set of sensor measurements is described as follows:

$$\begin{aligned}
\frac{\partial y(x,t)}{\partial t} &= \kappa(x)\nabla^2 y(x,t) + A(x)y(x,t) + B(x)u(x,t) + g_v(x)v(x,t) \\
&\quad + H(x)y(x,t)w(x,t) \\
\phi(x,t) &= \begin{bmatrix} M(x)y(x,t) \\ u(x,t) \end{bmatrix} \\
z(t) &= [h_1(y(q_1,t) + D_n n(q_1,t)), \ldots, h_{n_z}(y(q_{n_z},t) + D_n n(q_{n_z},t))]^T
\end{aligned} \tag{8.1}$$

for $x = [x_1, x_2]^T \in U = [x_{0,1}, x_{f,1}] \times [x_{0,2}, x_{f,2}] \in \mathbb{R}^2$ and $t > 0$, where $y(x,t) \triangleq [y_1(x,t), \ldots, y_n(x,t)]^T \in \mathbb{R}^n$ is the state variable; x and t are the space and time variables, respectively. The space domain U is a bounded domain. The system matrices $\kappa(x) \in \mathbb{R}^{n\times n}$, $A(x) \in \mathbb{R}^{n\times n}$, $B(x) \in \mathbb{R}^{n\times n_u}$, and $H(x) \in \mathbb{R}^{n\times n}$ are compartment-dependent and therefore more suitable for practically physical or biological systems. The measured output $z(t) \triangleq [z_1(t), \ldots, z_{n_z}(t)]^T \in \mathbb{R}^{n_z}$ may be interpreted as sensor observations, where n_z is the number of sensors. $h_i \triangleq h(q_i) \in \mathbb{R}^{1\times n}$ is the observation influence vector at the sensor location q_i of observation, where $i = 1, \ldots, n_z$. $y(q_i,t) \in \mathbb{R}^n$ is the state variable at the sensor location q_i of observation, where $i = 1, \ldots, n_z$. $n(q_i,t) \in L_2[U \times [0,t_f];\mathbb{R}^n]$ is the stochastic spatial-temporal measurement noise at the sensor location q_i of observation, where $i = 1, \ldots, n_z$ and $D_n \in \mathbb{R}^{n\times n}$ is the noise influence matrix. Therefore, in the case with free measurement noise, we have $z_i(t) = h_i y(q_i,t)$, where $i = 1, \ldots, n_z$. $u(x,t) \in L_2(U \times [0,t_f];\mathbb{R}^{n_u})$ is the control input. $v(x,t) \in L_2(U \times [0,t_f];\mathbb{R}^{n_v})$ is the vector of the external disturbance in the spatiotemporal domain and $g_v(x) \in \mathbb{R}^{n\times n_v}$ is the disturbance influence matrix. $\phi(x,t)$ is the controlled output and $M(x) \in \mathbb{R}^{n_m\times n}$. $w(x,t) \in \mathbb{R}$ is a weakly stationary spatiotemporal white noise with zero mean and unit covariance as $\text{cov}(w(x,t), w(x+\Delta,t+\tau)) = \delta(\Delta)\delta(\tau)$, where $\delta(\Delta)$ and $\delta(\tau)$ are delta functions, i.e., the covariance function is independent on x and t. The Laplace operator ∇^2 is defined as follows:

$$\nabla^2 y(x,t) \triangleq \sum_{k=1}^{2} \frac{\partial^2}{\partial x_k{}^2} y(x,t)$$

where

$$\frac{\partial^2}{\partial x_k{}^2} y(x,t) \triangleq \left[\frac{\partial^2}{\partial x_k{}^2} y_1(x,t),\dots,\frac{\partial^2}{\partial x_k{}^2} y_n(x,t) \right]^T \in \mathbb{R}^n$$

The initial value is given by $y(x,0) \triangleq y_0(x)$. For simplicity, the boundary condition is usually given by the Dirichlet boundary condition, i.e., $y(x,t) = a$ constant on ∂U, or by the Neumann boundary condition $\nabla y(x,t) \cdot \vec{n} = 0$ on ∂U, where $\vec{n}$ is the normal vector to the boundary ∂U.

Remark 8.2.1

The stochastic state equation in (8.1) can be written in the Itô differential form as follows:

$$\begin{aligned} dy(x,t) &= (\kappa(x)\nabla^2 y(x,t) + A(x)y(x,t) + B(x)u(x,t) + g_v(x)v(x,t))dt \\ &\quad + H(x)y(x,t)dW(x,t) \\ \phi(x,t) &= \begin{bmatrix} M(x)y(x,t) \\ u(x,t) \end{bmatrix} \\ z(t) &= [h_1(y(q_1,t) + D_n n(q_1,t)),\dots,h_{n_z}(y(q_{n_z},t) + D_n n(q_{n_z},t))]^T \end{aligned} \tag{8.2}$$

where $dW(x,t) = w(x,t)dt$ and $W(x,t)$ is the Wiener process being a zero mean Gaussian random field with variance t at each location x [238].

Next, let us consider the following LSPDS without input control $u(x,t)$ and external disturbance $v(x,t)$

$$dy(x,t) = (\kappa(x)\nabla^2 y(x,t) + A(x)y(x,t))dt + H(x)y(x,t)dW(x,t) \tag{8.3}$$

Definition 8.2.1 [237,238]

(i) The equilibrium point $y(x, t)=0$ of the LSPDS in equation (8.3) is said to be stable in probability if for any ε_1, $\varepsilon_2>0$, there exists a $\delta = \delta(\varepsilon_1,\varepsilon_2)>0$ such that if

$\|y(x,0)\|_{L_2(U;\mathbb{R}^n)} < \delta$, then $P\left\{\|y(x,t)\|_{L_2(U;\mathbb{R}^n)} > \varepsilon_1\right\} < \varepsilon_2$ for all $t > 0$. (ii) The equilibrium point $y(x, t) = 0$ of the LSPDS in equation (8.3) is said to be asymptotically stable in probability if it is stable in probability and for each $\varepsilon > 0$, there exists a $\delta = \delta(\varepsilon) > 0$ such that $\|y(x,0)\|_{L_2(U;\mathbb{R}^n)} < \delta$ guarantees $\lim_{t\to\infty} P\left\{\|y(x,t)\|_{L_2(U;\mathbb{R}^n)} > \varepsilon\right\} = 0$.

Suppose that $V(y(x,t))$ is a positive definite Itô function. Then, we have [238,270]

$$\int_U V(y(x,t))dx = \int_U V(y(x,0))dx + \int_0^t LV(y(x,s))ds + \int_0^t \int_U \left(\frac{\partial V y(x,s))}{\partial y}\right)^T H(x)y(x,s)dW(x,s) \tag{8.4}$$

where the infinitesimal operator L is denoted as

$$LV(y(x,s)) \triangleq \int_U \left(\frac{\partial V y(x,s))}{\partial y}\right)^T (\kappa(x)\nabla^2 y(x,s) + A(x)y(x,s))dx + \int_U \frac{1}{2}(H(x)y(x,s))^T \frac{\partial^2 V(y(x,s))}{\partial y^2}(H(x)y(x,s))dx \tag{8.5}$$

Lemma 8.2.1 [238,259]

A positive definite Itô function $V(y(x,t)) : \mathbb{R}^n \to \mathbb{R}$ is said to be a Lyapunov function for the system in equation (8.3), if (i) $V(0) = 0$ and $V(y(x,t)) > 0$ for $y(x,t) \neq 0$; (ii) $LV(y(x,t)) \leq 0$.

Lemma 8.2.2 [238,259]

Suppose the system in equation (8.3) has a solution $y(x,t)$ and there exists a Lyapunov function $V(y(x,t)) > 0$. Furthermore, $LV(y(x,t)) < 0$, then the equilibrium point $y(x,t) = 0$ of the system in equation (8.3) is asymptotically stable in probability.

Since this chapter deals with the asymptotical stochastic stabilization problem, it is inevitably related to the stochastic stability, i.e., the asymptotical stability in probability or the exponentially mean square stability [237,238] of the LSPDS. For convenience of control design, the asymptotical stability in probability of the equilibrium point $y(x,t) = 0$ of the system in equation (8.3) is discussed first in the following.

Consider the case in equation (8.2) with $v(x,t) = 0$ and $n(x,t) = 0$

$$dy(x,t) = (\kappa(x)\nabla^2 y(x,t) + A(x)y(x,t))dt + H(x)y(x,t)dW(x,t)$$

$$\eta(x,t) = Qy(x,t) \tag{8.6}$$

where $\eta(x,t)$ is the controlled output, then we have the following result.

Theorem 8.2.1

For the system equation (8.6), if there exists a positive definite function $V(y(x,t)) \in C^2(\mathbb{R}^n)$ with $V(0) = 0$ solving the following HJII

$$\int_U \left[\left(\frac{\partial Vy(x,t))}{\partial y} \right)^T (\kappa(x)\nabla^2 y(x,t) + A(x)y(x,t)) + y^T(x,t)Q^T Qy(x,t) \right.$$

$$\left. + \frac{1}{2}(H(x)y(x,t))^T \frac{\partial^2 Vy(x,t))}{\partial y^2}(H(x)y(x,t)) \right] dx$$

$$< 0 \tag{8.7}$$

then (i) the equilibrium point $y(x,t) = 0$ of the system in equation (8.6) is asymptotically stable in probability and (ii) the output variance

$$E\int_U \int_0^{t_f} \|\eta(x,t)\|^2 dt dx < E \int_U V(y(x,0)) dx$$

Proof

First we choose the Lyapunov function $V(y(x,t)) > 0, \forall y(x,t) \neq 0$ and $V(0) = 0$ for the linear partial differential system in equation (8.6). Then, from the inequality equation (8.7), we have $LV(y(x,t)) < \int_U -y^T(x,t)Q^T Qy(x,t) dx \leq 0$, i.e., $LV(y(x,t)) < 0$. So by Lemma 8.2.2, the equilibrium point $y(x,t) = 0$ of the system in equation (8.6) is asymptotically stable in probability. Next, we have

$$E\int_U \int_0^{t_f} \|\eta(x,t)\|^2 dt dx$$

$$= E \int_U \left[V\big(y(x,0)\big) - V\big(y(x,t_f)\big) + \int_0^{t_f} \left(\|\eta(x,t)\|^2 + \frac{\partial V\big(y(x,t)\big)}{\partial t} \right) dt \right] dx \tag{8.8}$$

By the fact $V\big(y(x,t_f)\big) \geq 0$ and Itô formula in equation (8.5), we get

$$
\begin{aligned}
&E\int_U\int_0^{t_f}\|\eta(x,t)\|^2 dtdx \\
&= E\int_U V\big(y(x,0)\big)dx \\
&+E\int_U\int_0^{t_f}\begin{bmatrix} y^T(x,t)Q^TQy(x,t)+\left(\dfrac{\partial Vy(x,t))}{\partial y}\right)^T\big(\kappa(x)\nabla^2 y(x,t)+A(x)y(x,t)\big) \\ +\dfrac{1}{2}\big(H(x)y(x,t)\big)^T\dfrac{\partial^2 Vy(x,t))}{\partial y^2}\big(H(x)y(x,t)\big)\end{bmatrix}dtdx
\end{aligned}
\tag{8.9}
$$

By the inequality equations (8.7) and (8.9), we get

$$
E\int_U\int_0^{t_f}\|\eta(x,t)\|^2 dtdx < E\int_U V(y(x,0))dx \tag{8.10}
$$

Remark 8.2.2

Similarly, if the following Hamilton Jacobi inequality (HJI) holds.

$$
\begin{aligned}
&\left(\frac{\partial Vy(x,t))}{\partial y}\right)^T(\kappa(x)\nabla^2 y(x,t)+A(x)y(x,t))+y^T(x,t)Q^TQy(x,t) \\
&+\frac{1}{2}(H(x)y(x,t))^T\frac{\partial^2 Vy(x,t))}{\partial y^2}(H(x)y(x,t)) \\
&<0, \forall x\in U
\end{aligned}
\tag{8.11}
$$

then the result in Theorem 8.2.1 still holds but with stricter condition than equation (8.7).

Lemma 8.2.3 [260]

For any matrices (or vectors) M_1 and M_2 with appropriate dimensions, we have

$$
M_1^T M_2 + M_2^T M_1 \le M_1^T M M_1 + M_2^T M^{-1} M_2
$$

where M is any positive-definite symmetric matrix.

Corollary 8.2.1

$X^T PY + Y^T PX \leq \xi X^T PX + \xi^{-1} Y^T PY$ for any constant ξ, a symmetric $P = P^T \geq 0$ and two matrices (or vectors) X and Y with appropriate dimensions.

After discussing the asymptotic stability in probability of stochastic partial differential system, the H_∞ robust ability of the following linear stochastic system is discussed:

$$
\begin{aligned}
dy(x,t) &= (\kappa(x)\nabla^2 y(x,t) + A(x)y(x,t) - g_v(x)v(x,t))dt \\
&\quad + H(x)y(x,t)dW(x,t) \\
\eta(x,t) &= Qy(x,t)
\end{aligned}
\tag{8.12}
$$

Theorem 8.2.2

For the system equation (8.12), if there exists a positive definite function $V(y(x,t)) \in C^2(\mathbb{R}^n)$ with $V(0) = 0$ solving the following HJII:

$$
\begin{aligned}
&\int_U \Bigg[\left(\frac{\partial Vy(x,t))}{\partial y}\right)^T (\kappa(x)\nabla^2 y(x,t) + A(x)y(x,t)) + y^T(x,t)Q^T Qy(x,t) \\
&+ \frac{1}{4\gamma^2}\left(\frac{\partial Vy(x,t))}{\partial y}\right)^T g_v(x)g_v^T(x)\left(\frac{\partial Vy(x,t))}{\partial y}\right) \\
&+ \frac{1}{2}(H(x)y(x,t))^T \frac{\partial^2 Vy(x,t))}{\partial y^2}(H(x)y(x,t)) \Bigg] dx < 0
\end{aligned}
\tag{8.13}
$$

then (i) the equilibrium point $y(x,t) = 0$ of the system in equation (8.12) is asymptotically stable in probability in the case of $v(x,t) = 0$, and (ii) the following spatiotemporal H_∞ robust stabilization performance holds

$$
E\int_U \int_0^{t_f} \|\eta(x,t)\|^2 dt dx < E\int_U V(y(x,0))dx + \gamma^2 E\int_U \int_0^{t_f} \|v(x,t)\|^2 dt dx \tag{8.14}
$$

$\forall v(x,t) \in L_2(U \times [0,t_f]; \mathbb{R}^{n_v}), v(x,t) \neq 0$ *for some prescribed* $\gamma > 0$ *if the initial* $y(x,0) \neq 0$ *and*

$$
E\int_U \int_0^{t_f} \|\eta(x,t)\|^2 dt dx < \gamma^2 E\int_U \int_0^{t_f} \|v(x,t)\|^2 dt dx \tag{8.15}
$$

holds if the initial state $y(x,0) = 0$.

Proof

First we choose the Lyapunov function $V(y(x,t)) > 0, \forall y(x,t) \neq 0$ and $V(0) = 0$. Then, for the system in equation (8.12) with $v(x,t) = 0$, from the inequality equation (8.13), we have

$$LV(y(x,t)) < \int_U \left[-y^T(x,t)Q^TQy(x,t) - \frac{1}{4\gamma^2}\left(\frac{\partial Vy(x,t))}{\partial y}\right)^T g_v(x)g_v^T(x)\left(\frac{\partial Vy(x,t))}{\partial y}\right)\right]$$

$dx \leq 0$ i.e., $LV\left(y(x,t)\right) < 0$. So by Lemma 8.2.2, the equilibrium point of the system in equation (8.12) in the case of $v(x,t) = 0$ is asymptotically stable in probability. Next, we have

$$\begin{aligned}
&E\int_U \int_0^{t_f} \|\eta(x,t)\|^2 dtdx \\
&= E\int_U [V(y(x,0)) - V(x,t_f)) + \int_0^{t_f} (\|\eta(x,t)\|^2 + \frac{\partial V(y(x,t))}{\partial t})dt]dx \\
&\leq E\int_U V(y(x,0))dx + E\int_U \int_0^{t_f} \Bigg[\left(\frac{\partial Vy(x,t))}{\partial y}\right)^T (\kappa(x)\nabla^2 y(x,t) + A(x)y(x,t)) \\
&+ g_v(x)v(x,t) + \|\eta(x,t)\|^2 + \frac{1}{2}(H(x)y(x,t))^T \frac{\partial^2 Vy(x,t))}{\partial y^2}(H(x)y(x,t)) \Bigg] dtdx \\
&\leq E\int_U V(y(x,0))dx + E\int_U \int_0^{t_f} \Bigg[\left(\frac{\partial Vy(x,t))}{\partial y}\right)^T (\kappa(x)\nabla^2 y(x,t) + A(x)y(x,t)) \\
&+ \frac{1}{4\gamma^2}\left(\frac{\partial Vy(x,t))}{\partial y}\right)^T g_v(x)g_v^T(x)\left(\frac{\partial Vy(x,t))}{\partial y}\right) + \gamma^2 v^T(x,t)v(x,t) \\
&+ y^T(x,t)Q^TQy(x,t) + \frac{1}{2}(H(x)y(x,t))^T \frac{\partial^2 Vy(x,t))}{\partial y^2}(H(x)y(x,t)) \Bigg] dtdx
\end{aligned} \tag{8.16}$$

Therefore, by the inequality in equation (8.13), we get

$$E\int_U \int_0^{t_f} \|\eta(x,t)\|^2 dtdx < E\int_U V(y(x,0))dx + \gamma^2 E\int_U \int_0^{t_f} \|v(x,t)\|^2 dtdx \tag{8.17}$$

Remark 8.2.3

(i) If $v(x,t)$ is a deterministic signal, then the expectation operator E on $v(x,t)$ in equations (8.14) and (8.15) should be neglected. (ii) If the following HJI holds:

$$\left(\frac{\partial Vy(x,t))}{\partial y}\right)^T (\kappa(x)\nabla^2 y(x,t) + A(x)y(x,t)) + y^T(x,t)Q^T Qy(x,t)$$

$$+\frac{1}{4\gamma^2}\left(\frac{\partial Vy(x,t))}{\partial y}\right)^T g_v(x)g_v^T(x)\left(\frac{\partial Vy(x,t))}{\partial y}\right)$$

$$+\frac{1}{2}(H(x)y(x,t))^T \frac{\partial^2 Vy(x,t))}{\partial y^2}(H(x)y(x,t)) < 0, \forall x \in U \tag{8.18}$$

then the H_∞ robust stabilization performance in equations (8.14) or (8.15) also holds but with a stricter condition than equation (8.13); (iii) If we let

$$J(v) = \frac{1}{2}E\int_U \int_{x0}^{xf} (\|\eta(x,t)\|^2 - \gamma^2 \|v(x,t)\|^2)dxdt \tag{8.19}$$

One can see that, for any $v(x,t)$ and $v^*(x,t) \in L_2(U \times [0,t_f]; \mathbb{R}^{n_v}) \cap \tilde{\Omega}$ with $\tilde{\Omega} \triangleq \{v(x,t) : \lim_{t\to\infty} EV(y(x,t)) = 0\}, J(v) \le J(v^*)$, where

$$v^*(x,t) = \frac{1}{2\gamma^2} g_v^T(x)\left(\frac{\partial V(y(x,t))}{\partial y}\right) \tag{8.20}$$

that is, $v^*(x,t)$, the worst-case disturbance, results in the maximal possible energy gain from the disturbance to the controlled output $\eta(x, t)$ in equation (8.12).

Now the H_∞ robust observer-based stabilization problem of the LSPDS in equations (8.1) or (8.2) is discussed as follows. In the LSPDS in equations (8.1) or (8.2), if the state $y(x, t)$ is unavailable, we can only measure the system via measurement output $z(t)$ by a set of sensors at some locations. Therefore, how to estimate $y(x, t)$ from the measurement $z(t)$ is an important topic of H_∞ robust observer-based stabilization problem for LSPDSs. Suppose the observer-based controller are proposed for LSPDS in equation (8.2) as follows:

$$d\hat{y}(x,t) = [\kappa(x)\nabla^2\hat{y}(x,t) + A(x)\hat{y}(x,t) + B(x)u(x,t)$$

$$+ L(x)(z(t) - [h_1\hat{y}(q_1,t),\ldots,h_{n_z}\hat{y}(q_{n_z},t)]^T)]dt, \forall x \in U \tag{8.21}$$

$$u(x,t) = K(x)\hat{y}(x,t) \tag{8.22}$$

where $L(x)$ and $K(x)$, respectively, denote the observer gain and the control gain. Let us denote the estimation error as

$$e(x,t) = y(x,t) - \hat{y}(x,t) \tag{8.23}$$

From equations (8.2), (8.21) and (8.22), we get the following estimation error:

$$de(x,t) = [\kappa(x)\nabla^2 e(x,t) + A(x)e(x,t) - L(x)[h_1 e(q_1,t),...,h_{n_z} e(q_{n_z},t)]^T$$
$$- L(x)[h_1 D_n n(q_1,t),...,h_{n_z} D_n n(q_{n_z},t)]^T + g_v(x)v(x,t)\}dt$$
$$+ H(x)y(x,t)dW(x,t), \ \forall x \in U \tag{8.24}$$

Then we obtain the following augmented system:

$$\begin{bmatrix} dy(x,t) \\ de(x,t) \end{bmatrix}$$
$$= \begin{bmatrix} \kappa(x)\nabla^2 y(x,t) + A(x)y(x,t) + B(x)K(x)y(x,t) - B(x)K(x)e(x,t) \\ \{\kappa(x)\nabla^2 e(x,t) + A(x)e(x,t) - L(x)[h_1 e(q_1,t),...,h_{n_z} e(q_{n_z},t)]^T\} \end{bmatrix} dt$$
$$+ \begin{bmatrix} H(x)y(x,t) \\ H(x)y(x,t) \end{bmatrix} dW(x,t)$$
$$+ \begin{bmatrix} g_v(x) & 0 & \cdots & 0 \\ g_v(x) & -L(x)\begin{bmatrix} h_1 \\ 0 \\ \vdots \\ 0 \end{bmatrix} D_n & \cdots & -L(x)\begin{bmatrix} 0 \\ 0 \\ \vdots \\ h_{n_z} \end{bmatrix} D_n \end{bmatrix}$$
$$\times \begin{bmatrix} v(x,t) \\ n(q_1,t) \\ \vdots \\ n(q_{n_z},t) \end{bmatrix} dt, \forall x \in U \tag{8.25}$$

Let us denote $\bar{y}(x,t) \triangleq \begin{bmatrix} y(x,t) \\ e(x,t) \end{bmatrix}$, $\tilde{\kappa}(x) \triangleq \begin{bmatrix} \kappa(x) & 0 \\ 0 & \kappa(x) \end{bmatrix}$, $\tilde{A}(x) \triangleq \begin{bmatrix} A(x) & 0 \\ 0 & A(x) \end{bmatrix}$,

$$\tilde{G}_v(x) \triangleq \begin{bmatrix} g_v(x) & 0 & \cdots & 0 \\ g_v(x) & -L(x)\begin{bmatrix} h_1 \\ 0 \\ \vdots \\ 0 \end{bmatrix} D_n & \cdots & -L(x)\begin{bmatrix} 0 \\ 0 \\ \vdots \\ h_{n_z} \end{bmatrix} D_n \end{bmatrix},$$

$\tilde{B}(x) \triangleq \begin{bmatrix} B(x)K(x) & -B(x)K(x) \\ 0 & 0 \end{bmatrix}$, $\tilde{H}(x) \triangleq \begin{bmatrix} H(x) & 0 \\ H(x) & 0 \end{bmatrix}$, and

$\bar{v}(x,t) \triangleq \left[v^T(x,t), n^T(q_1,t), \ldots, n^T(q_{n_z},t)\right]^T$ is the generic disturbance to be attenuated. Then, the augmented system equation (8.25) can be represented as

$$d\bar{y}(x,t) = \Big\{ \tilde{\kappa}(x)\nabla^2\bar{y}(x,t) + (\tilde{A}(x) + \tilde{B}(x))\bar{y}(x,t) + \tilde{G}_v(x)\bar{v}(x,t)$$

$$\left[\begin{array}{c} 0 \\ -L(x)[h_1 e(q_1,t), \ldots, h_{n_z} e(q_{n_z},t)]^T \end{array}\right]\Bigg\} dt + \tilde{H}(x)\bar{y}(x,t)dW(x,t), \forall x \in U$$

$$\phi(x,t) = \left[\begin{array}{c} M(x)y(x,t) \\ u(x,t) \end{array}\right] = \left[\begin{array}{c} M(x)y(x,t) \\ K(x)\hat{y}(x,t) \end{array}\right]$$

$$= \left[\begin{array}{cc} M(x) & 0 \\ K(x) & -K(x) \end{array}\right]\bar{y}(x,t) = Q_\phi(x)\bar{y}(x,t)$$

$$e(x,t) = \left[\begin{array}{cc} 0 & I \end{array}\right]\bar{y}(x,t) := Q_e\bar{y}(x,t) \tag{8.26}$$

Definition 8.2.2

In the case of free disturbance, i.e., $v(x,t) = 0$, the LSPDS in equations (8.25) or (8.26) is said to be asymptotically estimated in probability if the estimation error $e(x,t) \to 0$ in probability for all $x \in U$ as $t \to \infty$.

So, we get the following H_∞ robust observer-based stabilization result for LSPDS.

Theorem 8.2.3

For the augmented system in equation (8.26), if we can specify the observer gain L(x) and the control gain K(x) such that the following HJII holds:

$$\int_U \left(\frac{\partial V\bar{y}((x,t))}{\partial y}\right)^T \Big\{ \tilde{\kappa}(x)\nabla^2\bar{y}(x,t) + (\tilde{A}(x) + \tilde{B}(x))\bar{y}(x,t)$$

$$+ \left[\begin{array}{c} 0 \\ -L(x)[h_1 e(q_1,t), \ldots, h_{n_z} e(q_{n_z},t)]^T \end{array}\right]\Bigg\}$$

$$+ \frac{1}{4\gamma^2}\left(\frac{\partial V\bar{y}((x,t))}{\partial y}\right)^T \tilde{G}_v(\mathrm{x})\tilde{G}_v^T(\mathrm{x})\left(\frac{\partial V\bar{y}((x,t))}{\partial y}\right) + \bar{y}^T(x,t)Q_e^T Q_e\bar{y}(x,t)$$

$$+ \bar{y}^T(x,t)Q_\phi^T Q_\phi(x)\bar{y}(x,t) + \frac{1}{2}(\tilde{H}(x)\bar{y}(x,t))^T \frac{\partial^2 Vy((x,t))}{\partial y^2}(\tilde{H}(x)\bar{y}(x,t))\Big\} dx < 0 \tag{8.27}$$

then (i) the equilibrium point $\bar{y}(x,t)=0$ of the augmented system in equation (8.26) is asymptotically stable in probability in the case of $\bar{v}(x,t)=0$, i.e., $\bar{y}(x,t)\to 0$ in probability as $t\to\infty$ so that $e(x,t)\to 0$ and $y(x,t)\to 0$ in probability as $t\to\infty$, and (ii) for some $\gamma>0, \forall \bar{v}(x,t)\in L_2(U\times[0,t_f];\mathbb{R}^{n_v+n_z})$, $\bar{v}(x,t)\neq 0$, the following spatiotemporal H_∞ robust observer-based stabilization performance holds:

$$E\int_U\int_0^{t_f}\left\|\begin{bmatrix} e(x,t)\\ \phi(x,t)\end{bmatrix}\right\|^2 dtdx < E\int_U V(\bar{y}(x,0))dx+\gamma^2 E\int_U\int_0^{t_f}\|\bar{v}(x,t)\|^2 dtdx \quad (8.28)$$

if the initial state $\bar{y}(x,0)\neq 0$ or

$$E\int_U\int_0^{t_f}\left\|\begin{bmatrix} e(x,t)\\ \phi(x,t)\end{bmatrix}\right\|^2 dtdx < \gamma^2 E\int_U\int_0^{t_f}\|\bar{v}(x,t)\|^2 dtdx \quad (8.29)$$

if the initial state $\bar{y}(x,0)=0$.

Proof

By the similar procedure in the proof of Theorem 8.2.2 with $y(x, t)$, $v(x, t)$, and $\eta(x, t)$ in equation (8.12) being replaced by $\bar{y}(x,t)$, $\bar{v}(x,t)$, and $\begin{bmatrix} e(x,t)\\ \phi(x,t)\end{bmatrix}$ in equation (8.26), respectively.

From the analysis above, the H_∞ robust observer-based stabilization design needs to specify the observer gain $L(x)$ and the control gain $K(x)$ so that there exists a positive definite solution $V(\bar{y}(x,t))$ with $V(0)=0$ for the HJII in equation (8.27). It is the first time to achieve the spatiotemporal H_∞ observer-based stabilization performance for LSPDs. However, even though the observer gain $L(x)$ and the control gain $K(x)$ are solved from the complex HJII in equation (8.27), it is still difficult to implement the partial differential observer in equation (8.21), especially for the case where system matrices are functions of location x. Therefore, an H_∞ robust observer-based stabilization design that is simpler and easier implemented for linear partial differential systems is proposed via the semi-discretization finite difference scheme in the following section.

8.3 IMPLEMENTABLE H_∞ STABILIZATION FOR STOCHASTIC PARTIAL DIFFERENTIAL SYSTEMS

In order to implement the H_∞ robust observer-based stabilization of LSPDSs in equation (8.2), it is important to obtain a more suitable state space model to represent the LSPDSs. For this purpose, the semi-discretization finite difference scheme in Section 1.2.2 is employed. The finite difference scheme is employed to approximate the partial differential operator $\nabla^2 y(x,t)$ in equation (8.2) to simplify the design procedure of robust observer-based stabilization of LSPDS. Consider a typical mesh as shown in Figure 8.1. The state $y(x,t)$ is represented by $y_{k,l}(t) \in \mathbb{R}^n$ at the grid node $x_{k,l}(x_1 = k\Delta, x_2 = l\Delta)$, where $k = 1,\ldots,N_1$ and $l = 1,\ldots,N_2$, i.e., $y(x,t)\big|_{x=x_{k,l}} \triangleq y_{k,l}(t)$. At the grid point $x_{k,l}$, the central difference approximation for the partial differential operator can be written as follows [241,261–264]:

$$\kappa(x)\nabla^2 y(x,t) = \kappa_{k,l}\frac{y_{k+1,l}(t) + y_{k-1,l}(t) - 2y_{k,l}(t)}{\Delta^2} + \kappa_{k,l}\frac{y_{k,l+1}(t) + y_{k,l-1}(t) - 2y_{k,l}(t)}{\Delta^2} + O_{k,l}(\Delta^2) \tag{8.30}$$

where $\kappa_{k,l} = \kappa(x)\big|_{x=x_{k,l}}$ and $O_{k,l}(\Delta^2) \in \mathbb{R}^n$ is called the local truncation error.

Remark 8.3.1

It has been proven [241] that if grid points in a spatial domain are dense enough, then the truncation error $O_{k,l}(\Delta^2)$ will be small enough.

Based on the finite difference approximation in equation (8.30), the LSPDS in equation (8.2) can be represented by the following stochastic finite difference system:

$$\begin{aligned} dy_{k,l}(t) = {} & \kappa_{k,l}\frac{1}{\Delta^2}[y_{k+1,l}(t) + y_{k-1,l}(t) + y_{k,l+1}(t) + y_{k,l-1}(t) - 4y_{k,l}(t)]dt \\ & + A_{k,l}y_{k,l}(t)dt + B_{k,l}u_{k,l}(t)dt + g_{v,(k,l)}v_{k,l}(t)dt \\ & + H_{k,l}y_{k,l}(t)dW_{k,l}(t) + O_{k,l}(\Delta^2) \end{aligned} \tag{8.31}$$

where $y_{k,l}(t) = y(x,t)\big|_{x=x_{k,l}}$, $u_{k,l}(t) = u(x,t)\big|_{x=x_{k,l}}$, $v_{k,l}(t) = v(x,t)\big|_{x=x_{k,l}}$, $W_{k,l}(t) = W(x,t)\big|_{x=x_{k,l}}$, $\kappa_{k,l} = \kappa(x)\big|_{x=x_{k,l}}$, $A_{k,l} = A(x)\big|_{x=x_{k,l}}$, $B_{k,l} = B(x)\big|_{x=x_{k,l}}$, $g_{v,(k,l)} = g_v(x)\big|_{x=x_{k,l}}$, and $H_{k,l} = H(x)\big|_{x=x_{k,l}}$.

Let us denote

$$T_{k,l}y_{k,l}(t) = \frac{1}{\Delta^2}[y_{k+1,l}(t) + y_{k-1,l}(t) + y_{k,l+1}(t) + y_{k,l-1}(t) - 4y_{k,l}(t)] \tag{8.32}$$

Then, we get

$$\begin{aligned} dy_{k,l}(t) = {} & [\kappa_{k,l}T_{k,l}y_{k,l}(t) + A_{k,l}y_{k,l}(t)dt + B_{k,l}u_{k,l}(t)dt + g_{v,(k,l)}v_{k,l}(t)]dt \\ & + H_{k,l}y_{k,l}(t)dW_{k,l}(t) + O_{k,l}(\Delta^2) \end{aligned} \tag{8.33}$$

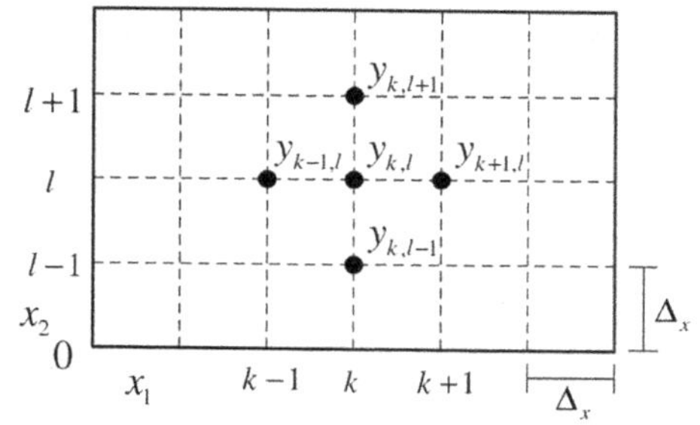

FIGURE 8.1 Finite difference grids on the spatiodomain.

In order to simplify the robust observer-based stabilization design for LSPDSs, we define a spatial state vector $y(t)$ to collect the state variable $y_{k,l}(t) \in \mathbb{R}^n$ at all grid nodes in Figure 8.1. For Dirichlet boundary conditions [264], the values of $y_{k,l}(t)$ at the boundary are fixed, for example, $y(x,t) = 0$ on ∂U. We have $y_{k,l}(t) = 0$ at $k = 0, N_1 + 1$, or $l = 0, N_2 + 1$. Therefore, the spatial state vector $y(t) \in \mathbb{R}^{nN}$ for state variables at all grid nodes is defined as follows:

$$y(t) = y_{1,1}^T(t),\ldots,y_{k,1}^T(t),\ldots,y_{N_1,1}^T(t),\ldots,y_{k,1}^T(t),\ldots,y_{1,N_2}^T(t),\ldots,y_{k,N_2}^T(t),\ldots,y_{N_1,N_2}^T(t)]^T \tag{8.34}$$

where $N \triangleq N_1 \times N_2$. Note that n is the dimension of the vector $y_{k,l}(t)$ for each grid node and $N_1 \times N_2$ is the number of grid nodes. For example, if $N_1 = 2$ and $N_2 = 2$, then we have $y(t) = \left[y_{1,1}^T(t), y_{2,1}^T(t), y_{1,2}^T(t), y_{2,2}^T(t) \right]^T \in \mathbb{R}^{4n}$. In order to simplify the index of the node $y_{k,l}(t) \in \mathbb{R}^n$ in the spatial state vector $y(t) \in \mathbb{R}^{nN}$, we use the symbol $y_j(t) \in \mathbb{R}^n$ to replace $y_{k,l}(t)$. Note that the index j is from 1 to N, i.e.,

$$y_1(t) \triangleq y_{1,1}(t), y_2(t) \triangleq y_{2,1}(t),\ldots, y_j(t) \triangleq y_{k,l}(t),\ldots, y_N(t) \triangleq y_{N_1,N_2}(t)$$

where $j = (l-1)N_1 + k$ in equation (8.34). The stochastic difference model of two indices in equation (8.33) could be represented with only one index as follows:

$$dy_j(t) = [\kappa_j T_j y(t) + A_j y_j(t) + B_j u_j(t) + g_{v,j} v_j(t)]dt + H_j y_j(t) dW_j(t) + O_j(\Delta^2) \tag{8.35}$$

where $u_j(t) = u_{k,l}(t)$, $v_j(t) = v_{k,l}(t)$, $dW_j(t) = dW_{k,l}(t)$, $O_j(\Delta^2) = O_{k,l}(\Delta^2)$, $\kappa_j = \kappa_{k,l}, T_j = T_{k,l}, A_j = A_{k,l}, B_j = B_{k,l}, g_{v,j} = g_{v,(k,l)}, H_j = H_{k,l}$ with $j = (l-1)N_1 + k$ and $T_j y(t)$ is defined as follows:

$$T_j y(t) = \frac{1}{\Delta^2}[0_n \cdots 0_n \; I_n \; 0_n \cdots 0_n \; I_n \; -4I_n \; I_n \; 0_n \cdots 0_n \; I_n \; 0_n \cdots 0_n] y(t)$$

$$\text{position} \rightarrow \quad 1 \quad j-N_1 \quad j-1 \quad j+1 \quad j+N_1 \quad N_1 N_2$$

with 0_n and I_n denoting the $n \times n$ zero matrix and identity matrix, respectively. The controlled output $\phi(x,t)$ in equations (8.1) or (8.2) can be represented as follows:

$$\phi_j(t) = \begin{bmatrix} M_j y_j \\ u_j \end{bmatrix} \tag{8.36}$$

where $\phi_j(t) = \phi_{k,j}(t) = \phi(x,t)\big|_{x=x_{k,l}}, M_j = M_{k,l} = M(x)\big|_{x=x_{k,l}}$. The measurement output $z(t)$ in equations (8.1) or (8.2) can be represented as follows:

$$z(t) = \sum_{j=1}^{N} \left[C_j y_j(t) + D_j n_j(t) \right] \tag{8.37}$$

where the matrix C_j is defined as $C_j \triangleq C_{\{k,l\}} = [C_{\{k,l\},1}^T, \ldots, C_{\{k,l\},n_z}^T]^T \in \mathbb{R}^{n_z \times n}$ with $j = (l-1)N_1 + k$, in which the element is given as $C_{\{k,l\},i} = h_i$ for $x_{k,l} = q_i$ or $C_{\{k,l\},i} = 0$ for $x_{k,l} \neq q_i$; the matrix D_j is defined as $D_j \triangleq D_{\{k,l\}} = [D_{\{k,l\},1}^T, \ldots, D_{\{k,l\},n_z}^T]^T \in \mathbb{R}^{n_z \times n}$ with $j = (l-1)N_1 + k$, in which the element is given as $D_{\{k,l\},i} = h_i D_n$ for $x_{k,l} = q_i$ or $D_{\{k,l\},i} = 0$ for $x_{k,l} \neq q_i$. Note that q_i is the ith sensor location as defined in equation (8.1). The jth measurement noise is defined as $n_j \triangleq n(x,t)\big|_{x=x_{k,l}}$ where $j = (l-1)N_1 + k$.

We collect all the states $y_j(t)$ of grid nodes in equations (8.35), (8.36), and (8.37) to the state $y(t)$ in equation (8.34). Then the system in equation (8.35) can be represented as the following stochastic spatial state space system:

$$dy(t) = \breve{\kappa} T y(t)dt + \breve{A} y(t)dt + \breve{B} u(t)dt + \breve{G}_v v(t)dt + \breve{H} y(t) \circ d\tilde{W}(t) + O(\Delta^2)$$

$$\phi(t) = \begin{bmatrix} \breve{M} y(t) \\ u(t) \end{bmatrix}$$

$$z(t) = Cy(t) + Dn(t) \tag{8.38}$$

where the operator $\circ$ is used for the Hadamard product $X \circ Y$ of two matrices $X = [X_{ij}]_{mn}$ and $Y = \left[Y_{ij} \right]_{mn}$ of the same size. The Hadamard product is also known as the entry product, i.e., $X \circ Y = \left[X_{ij} Y_{ij} \right]_{m \times n}$. The spatial state vector $y(t)$ denoted in equation (8.34) represents $y(x,t)$ at all grid points on the spatial domain in Figure 8.1. $u(t) \triangleq [u_1^T(t), \ldots, u_N^T(t)]^T \in \mathbb{R}^{n_u N}$, $O(\Delta^2) \triangleq \left[O(\Delta^2)^T, \ldots, O_N(\Delta^2)^T \right]^T$, and $d\tilde{W}(t) \in \mathbb{R}^{nN}$ is defined as $d\tilde{W}(t) \triangleq [J^T dW_1(t), \ldots, J^T dW_N(t)]^T$ with $J \triangleq [1, \ldots, 1]^T \in \mathbb{R}^n$, i.e., all the elements in J are equal to 1. The external disturbance $v(t) \triangleq [v_1^T(t), \ldots, v_N^T(t)]^T \in \mathbb{R}^{n_v N}$. The measurement noise $n(t)$ is defined as $n(t) \triangleq [n_1^T(t), \ldots, n_N^T(t)]^T \in \mathbb{R}^{nN}$. The corresponding matrices are defined as $T \triangleq T \triangleq [T_1^T, \ldots, T_N^T]^T \in \mathbb{R}^{nN \times nN}$, $\breve{\kappa} \triangleq diag(\kappa_1, \ldots, \kappa_N)$, $\breve{A} \triangleq diag(A_1, \ldots, A_N)$, $\breve{B} = diag(B_1, \ldots, B_N)$, $\breve{H} = diag(H_1, \ldots, H_N)$, $\breve{G}_v = diag(g_{v,1}, \ldots, g_{v,N})$, $\breve{M} = diag(M_1, \ldots, M_N)$, $C \triangleq [C_1, \ldots, C_N] \in \mathbb{R}^{n_z \times nN}$, and $D \triangleq [D_1, \ldots, D_N] \in \mathbb{R}^{n_z \times nN}$.

Remark 8.3.2

(i) The increments of the Wiener processes $dW_1(t),...,dW_N(t)$ at the N grid nodes have the property: $dW(t) \triangleq [dW_1(t),...,dW_N(t)]^T \in \mathbb{R}^N$ is a zero-mean vector-process with the covariance $E\{dW(t)dW^T(t)\} = I_N dt$. (ii) The physical meaning of equation (8.38) is that LSPDS in equations (8.1) or (8.2) at all grid points on the spatial domain in Figure 8.1 can be represented by the spatial state space model in equation (8.38).

Then our robust observer-based stabilization design problem of a stochastic partial differential system becomes how to design the robust observer and controller for the stochastic spatial state space system in equation (8.38) to satisfy the H_∞ robust observer-based stabilization performance. This is discussed in the following. The implementable observer and controller to guarantee the H_∞ robust observer-based stabilization performance are proposed as follows:

$$d\hat{y}(t) = \left\{\breve{\kappa}Ty(t) + \breve{A}y(t) + \breve{B}u(t) + \breve{L}(z(t) - C\hat{y}(t))\right\}dt \tag{8.39}$$

$$u(t) = \breve{K}\hat{y}(t) \tag{8.40}$$

with $\breve{L} \triangleq [L_1^T,...,L_N^T]^T \in \mathbb{R}^{nN\times n_z}$ and $\breve{K} \triangleq diag(K_1,...,K_N)$ being respectively the observer gain and the control gain to be designed, where $L_j = L_{k,l} = L(x)\big|_{x=x_{k,l}}$ and $K_j = K_{k,l} = K(x)\big|_{x=x_{k,l}}$. The dynamic of the estimated error $e(t) = y(t) - \hat{y}(t)$ is

$$\begin{aligned} de(t) &= \left\{\breve{\kappa}Ty(t)dt + \breve{A}y(t)dt + \breve{B}u(t)dt + \breve{H}y(t)\circ d\tilde{W}(t)\right\} + \breve{G}_v v(t)dt \\ &\quad - \left\{\breve{\kappa}Ty(t) + \breve{A}y(t) + \breve{B}u(t) + \breve{L}(z(t) - C\hat{y}(t))\right\}dt + O(\Delta^2) \\ &= \left\{\breve{\kappa}T + \breve{A} - \breve{L}C\right\}e(t)dt + \breve{H}y(t)\circ d\tilde{W}(t) + \breve{G}_v v(t)dt - \breve{L}D_n(t)dt + O(\Delta^2) \end{aligned} \tag{8.41}$$

Using the control input in equation (8.40), the system in equation (8.38) can be represented as the following stochastic spatial state-space system:

$$\begin{aligned} dy(t) &= \breve{\kappa}Ty(t)dt + \breve{A}y(t)dt + \breve{B}\breve{K}\hat{y}(t)dt + \breve{G}_v v(t)dt + \breve{H}y(t)\circ d\tilde{W}(t) + O(\Delta^2) \\ &= \breve{\kappa}Ty(t)dt + \breve{A}y(t)dt + \breve{B}\breve{K}y(t)dt - \breve{B}\breve{K}e(t)dt + \breve{G}_v v(t)dt + \breve{H}y(t)\circ d\tilde{W}(t) \\ &\quad + O(\Delta^2) \end{aligned}$$

$$\phi(t) = \begin{bmatrix} \breve{M}y(t) \\ u(t) \end{bmatrix} = \begin{bmatrix} \breve{M}y(t) \\ \breve{K}[y(t) - e(t)] \end{bmatrix}$$

$$z(t) = Cy(t) + Dn(t) \tag{8.42}$$

Combining equation (8.42) with (8.41), we get the following augmented system:

$$d\bar{y}(t) = \bar{A}\bar{y}(t)dt + \bar{H}\bar{y}(t) \circ d\bar{W} + \bar{G}_v\bar{v}(t)dt$$

$$\phi(t) = \bar{K}\bar{y}(t), e(t) = \begin{bmatrix} 0_{nN} & I_{nN} \end{bmatrix} \bar{y}(t) \tag{8.43}$$

where $\bar{y}(t) = \begin{bmatrix} y(t) \\ e(t) \end{bmatrix}$, $\bar{A} \triangleq \begin{bmatrix} \breve{\kappa}T + \breve{A} + \breve{B}\breve{K} & -\breve{B}\breve{K} \\ 0_{nN} & \breve{\kappa}T + \breve{A} - \breve{L}\breve{C} \end{bmatrix}$, $\bar{H} \triangleq \begin{bmatrix} \breve{H} & 0_{nN} \\ \breve{H} & 0_{nN} \end{bmatrix}$,

$d\bar{W}(t) \triangleq \begin{bmatrix} d\tilde{W}(t) \\ d\tilde{W}(t) \end{bmatrix}$, $\bar{G}_v \triangleq \begin{bmatrix} \breve{G}_v & I_{nN} & 0_{nN} \\ \breve{G}_v & I_{nN} & -\breve{L}\breve{D} \end{bmatrix}$, $\bar{K} \triangleq \begin{bmatrix} \breve{M} & 0_{nN} \\ \breve{K} & -\breve{K} \end{bmatrix}$, and $\bar{v}(t) \triangleq [v^T(t), O^T(\Delta^2), n^T(t)]^T$ is the generic disturbance to be attenuated.

Since the spatial state vector $y(t)$ in equation (8.34) is used to represent the $y(x,t)$ at all grid points, in this situation,

$$E\int_U \int_0^{tf} \left\| \begin{bmatrix} e(x,t) \\ \phi(x,t) \end{bmatrix} \right\|^2 dtdx, E\int_U \int_0^{tf} \left\|[\bar{v}(x,t)]\right\|^2 dtdx, \text{ and } E\int_U V(\bar{y}(x,0))dx$$

in the spatiotemporal stochastic H_∞ robust observer-based stabilization performance in equations (8.28) or (8.29) could be modified by the temporal forms

$$E\int_0^{tf} \left\| \begin{bmatrix} e(t) \\ \phi(t) \end{bmatrix} \right\|^2 \Delta^2 dt, E\int_0^{tf} \left\|[\bar{v}(t)]\right\|^2 \Delta^2 dt, \text{ and } E\bar{V}(\bar{y}(0))\Delta^2$$

respectively, for the stochastic spatial state-space system in equation (8.38), where the Lyapunov function $\bar{V}(\bar{y}(t))$ is related with the Lyapunov function $\bar{V}(\bar{y}(x,t))$ as $\bar{V}(\bar{y}(t)) = \Sigma_{j=1}^{N} V(\bar{y}_j(t))$ with $\bar{y}_j(t) = [y_j^T(t), e_j^T(t)]^T$. Therefore, the H_∞ robust observer-based stabilization performance in equations (8.28) and (8.29) could be modified as

$$E\int_0^{tf} \left\| \begin{bmatrix} e(t) \\ \phi(t) \end{bmatrix} \right\|^2 dt < E\bar{V}(\bar{y}(0)) + \gamma^2 E\int_0^{tf} \left\|[\bar{v}(t)]\right\|^2 dt \tag{8.44}$$

when $\bar{y}(0) \neq 0$ and

$$E\int_0^{tf} \left\| \begin{bmatrix} e(t) \\ \phi(t) \end{bmatrix} \right\|^2 dt < \gamma^2 E\int_0^{tf} \left\|[\bar{v}(t)]\right\|^2 dt \tag{8.45}$$

when $\bar{y}(0)=0$, where Δ^2 in both sides of equations (8.44) and (8.45) can be canceled.

Therefore, the spatiotemporal H_∞ robust observer-based stabilization problem in equations (8.28) or (8.29) for LSPDS in equation (8.2) is transformed to the only temporal H_∞ robust observer-based stabilization problem in equations (8.44) or (8.45) for the stochastic spatial state space system in equation (8.38). Suppose the external disturbance $v(t)$ and the measurement noise $n(t)$ are neglected and the truncation error $O(\Delta^2)$ of the finite difference scheme approaches zero in the augmented system equation (8.43) as the number of grid nodes becomes large enough, i.e., $\bar{v}(t)=0$. Then the robust observer-based stabilization could be designed to guarantee the asymptotical stability in probability of the augmented system in equation (8.43).

For the H_∞ robust observer-based stabilization problem, let us choose a Lyapunov function $\bar{V}(\bar{y})$ for the augmented system in equation (8.43) as

$$\bar{V}(\bar{y})=\sum_{j=1}^{N}(y_j^T(t)P_{11,j}y_j(t)+e_j^T(t)P_{22,j}e_j(t))=\bar{y}^T(t)\bar{P}\bar{y}(t) \tag{8.46}$$

where $\bar{P}=diag(P_{11},P_{22})$ in which $P_{11}=P_{11}^T=diag(P_{11,1},...,P_{11,N})>0$, and $P_{22}=P_{22}^T=diag(P_{22,1},...,P_{22,N})>0$.

Theorem 8.3.1

For the augmented system in equation (8.43) with $\bar{v}(t)=0$, suppose there exist a symmetric positive definite matrix $\bar{P}$, an observer gain $\breve{L}$ and a control gain $\breve{K}$ such that the following ARI holds:

$$\bar{P}\bar{A}+\bar{A}^T\bar{P}+\bar{H}^T\bar{P}\bar{H}<0 \tag{8.47}$$

where $\bar{A}=\begin{bmatrix}\breve{\kappa}T+\breve{A}+\breve{B}\breve{K} & -\breve{B}\breve{K}\\ 0_{nN} & \breve{\kappa}T+\breve{A}-\breve{L}C\end{bmatrix}$ *and* $\bar{H}\triangleq\begin{bmatrix}\breve{H} & 0_{nN}\\ \breve{H} & 0_{nN}\end{bmatrix}$.

Then the asymptotical estimation and asymptotical stabilization in probability are both achieved by the observer in equation (8.39) and the control input in equation (8.40), i.e., $e(t)\to 0$ and $y(t)\to 0$ in probability as $t\to\infty$ for the LSPDS with $\bar{v}(t)=0$.

Proof

For the augmented system in equation (8.43) with $\bar{v}(t)=0$, the infinitesimal operator $L\bar{V}(\bar{y}(t))$ in the spatial state-space model is denoted as

$$L\bar{V}(\bar{y}(t)) \triangleq \left(\frac{\partial \bar{V}(\bar{y})}{\partial \bar{y}}\right)^T (\bar{A}\bar{y}(t)) + \frac{1}{2}[\bar{H}\bar{y}(t)]^T \frac{\partial^2 \bar{V}(\bar{y})}{\partial \bar{y}^2}[\bar{H}\bar{y}(t)]$$

By the fact

$$\frac{\partial \bar{V}(\bar{y})}{\partial \bar{y}} = 2\bar{P}\bar{y}, \frac{\partial^2 \bar{V}(\bar{y})}{\partial \bar{y}^2} = 2\bar{P}$$

we get

$$L\bar{V}(\bar{y}(t)) = \bar{y}^T(t)\bar{P}(\bar{A}\bar{y}(t)) + \bar{y}^T(t)\bar{A}^T\bar{P}\bar{y}(t) + [\bar{H}\bar{y}(t)]^T\bar{P}[\bar{H}\bar{y}(t)] \tag{8.48}$$

So we get

$$L\bar{V}(\bar{y}(t)) = \bar{y}^T(t)[\bar{P}\bar{A} + \bar{A}^T\bar{P} + \bar{H}^T\bar{P}\bar{H}]\bar{y}(t) \tag{8.49}$$

If the following ARI holds

$$\bar{P}\bar{A} + \bar{A}^T\bar{P} + \bar{H}^T\bar{P}\bar{H} < 0 \tag{8.50}$$

then $L\bar{V}(\bar{y}(t)) < 0$.

When we consider the effects of external disturbance $v(t)$ and the measurement noise $n(t)$ simultaneously, i.e., $\bar{v}(t) \neq 0$, the observer gain $\breve{L}$ for the observer in equation (8.39) and the control gain $\breve{K}$ in equation (8.40) are specified so that the H_∞ robust observer-based stabilization performance in equations (8.44) or (8.45) is achieved. Then we get the following H_∞ robust observer-based stabilization result.

Theorem 8.3.2

For the augmented system in equation (8.43) with a prescribed noise attenuation level γ^2 in equations (8.44) and (8.45), suppose there exist a symmetric positive definite matrix $\bar{P}$, an observer gain $\breve{L}$: and a control gain $\breve{K}$ such that the following ARI holds

$$\Theta \triangleq \bar{I} + \bar{K}^T\bar{K} + \bar{P}\bar{A} + \bar{A}^T\bar{P} + \frac{1}{\gamma^2}\bar{P}\bar{G}_v\bar{G}_v^T\bar{P} + \bar{H}^T\bar{P}\bar{H} < 0 \tag{8.51}$$

or

$$\begin{bmatrix} \bar{I} + \bar{P}\bar{A} + \bar{A}^T\bar{P} + \bar{H}^T\bar{P}\bar{H} & \bar{K}^T & \bar{P}\bar{G}_v \\ \bar{K} & -I_{2nN} & 0_{2nN} \\ \bar{G}_v^T\bar{P} & 0_{2nN} & \gamma^2 I_{2nN} \end{bmatrix} < 0 \tag{8.52}$$

where $\bar{I} \triangleq \begin{bmatrix} 0_{nN} & 0_{nN} \\ 0_{nN} & I_{nN} \end{bmatrix}$ *and* $\bar{K} \triangleq \begin{bmatrix} \breve{M} & 0_{nN} \\ \breve{K} & -\breve{K} \end{bmatrix}$.

Then the H_∞ robust observer-based stabilization performance in equations (8.44) and (8.45) is guaranteed by the observer in equation (8.39) and the control input in equation (8.40), and the asymptotical estimation and asymptotical stabilization in probability are both guaranteed when $\bar{v}(t) = 0$.

Proof

$$E\int_0^{t_f} \left\| \begin{bmatrix} e(t) \\ \phi(t) \end{bmatrix} \right\|^2 dt$$

$$= E\int_0^{t_f} \left\{ \bar{y}^T(t) \begin{bmatrix} 0 \\ I_{nN} \end{bmatrix} \begin{bmatrix} 0 & I_{nN} \end{bmatrix} \bar{y}(t) + \bar{y}^T(t)\bar{K}^T\bar{K}\bar{y}(t) \right\} dt$$

$$= E\bar{V}(\bar{y}(0)) - E\bar{V}(\bar{y}(t_f)) + E\int_0^{t_f} \left[\bar{y}^T(t)(\bar{I} + \bar{K}^T\bar{K})\bar{y}(t) + \frac{dV(\bar{y}(t))}{dt} \right] dt \tag{8.53}$$

For the augmented system in equation (8.43) with $\bar{v}(t) \neq 0$, by Itô formula [253], we get

$$E\int_0^{t_f} \left\| \begin{bmatrix} e(t) \\ \phi(t) \end{bmatrix} \right\|^2 dt$$

$$\leq E\bar{y}^T(0)\bar{P}\bar{y}(0) + E\int_0^{t_f} \{\bar{y}^T(t)(\bar{I} + \bar{K}^T\bar{K})\bar{y}(t)dt]$$

$$+ \left(\frac{\partial V(\bar{y})}{\partial \bar{y}} \right)^T [\bar{A}\bar{y}(t) + \bar{G}_v\bar{v}(t) + \frac{1}{2}[\bar{H}\bar{y}(t)]^T \frac{\partial^2 V(\bar{y})}{\partial \bar{y}^2}[\bar{H}\bar{y}(t)]\}dt \tag{8.54}$$

By Corollary 8.2.1 and the fact

$$\frac{\partial V(\bar{y})}{\partial \bar{y}} = 2\bar{P}\bar{y}, \frac{\partial^2 V(\bar{y})}{\partial \bar{y}^2} = 2\bar{P} \tag{8.55}$$

we get

$$E\int_0^{t_f}\left\|\begin{bmatrix} e(t) \\ \phi(t) \end{bmatrix}\right\|^2 dt$$

$$\leq E\{Tr[\bar{P}\bar{y}(0)\bar{y}^T(0)]\} + E\int_0^{t_f}\{\bar{y}^T(t)(\bar{I}+\bar{K}^T\bar{K})\bar{y}(t)dt + \bar{y}^T(t)\bar{P}\bar{A}\bar{y}(t)$$

$$+\bar{y}^T(t)\bar{A}^T\bar{P}\bar{y}(t) + \bar{y}^T(t)\bar{P}\bar{G}_v\bar{v}(t) + \bar{v}^T(t)\bar{G}_v^T\bar{P}\bar{y}(t) + [\bar{H}\bar{y}(t)]^T\bar{P}[\bar{H}\bar{y}(t)]\}dt$$

$$\leq Tr[\bar{P}R_{\bar{y}(0)}] + E\gamma^2\int_0^{t_f}\bar{v}^T(t)\bar{v}(t)dt$$

$$+E\int_0^{t_f}\{\bar{y}^T(t)[\bar{I}+\bar{K}^T\bar{K}+\bar{P}\bar{A}+\bar{A}^T\bar{P}+\frac{1}{\gamma^2}\bar{P}\bar{G}_v\bar{G}_v^T\bar{P}+\bar{H}^T\bar{P}\bar{H}]\bar{y}(t)\}dt \quad (8.56)$$

where $R_{\bar{y}(0)} \triangleq \mathrm{E}\{\bar{y}(0)\bar{y}^T(0)\}$. Therefore, if the following ARI holds

$$\bar{I}+\bar{K}^T\bar{K}+\bar{P}\bar{A}+\bar{A}^T\bar{P}+\frac{1}{\gamma^2}\bar{P}\bar{G}_v\bar{G}_v^T\bar{P}+\bar{H}^T\bar{P}\bar{H}<0 \quad (8.57)$$

then

$$E\int_0^{t_f}\left\|\begin{bmatrix} e(t) \\ \phi(t) \end{bmatrix}\right\|^2 dt \leq Tr[\bar{P}R_{\bar{y}(0)}] + E\gamma^2\int_0^{t_f}\bar{v}^T(t)\bar{v}(t)dt \quad (8.58)$$

i.e., if the inequality in equations (8.57) or (8.51) holds, then the H_∞ robust observer-based stabilization is guaranteed. Furthermore, by Schur complement, (8.52) is equivalent to (8.51). Since the (8.57) obviously implies the ARI in (8.47), the equilibrium point $\bar{y}(t)=0$ of the augmented system in (8.43) with $\bar{v}(t)=0$ is asymptotically stable in probability. So the asymptotical estimation and asymptotical stabilization in probability are both guaranteed when $\bar{v}(t)=0$.

Proposition 8.3.1

With Θ defined in the ARI in equation (8.51), $\bar{y}^T(t)\Theta\bar{y}(t)<0$ is the finite difference approximation of HJII in equation (8.27) with the approximation error

$$2\bar{y}^T(t)\bar{P}\begin{bmatrix} O(\Delta^2) \\ O(\Delta^2) \end{bmatrix} \quad (8.59)$$

As $\Delta \to 0$, $\bar{y}^T(t)\Theta\bar{y}(t)<0$ with Θ defined in the ARI in equation (8.51) will approach the HJII in equation (8.27).

Proof

We show the relationship between the HJII in equation (8.27) and $\bar{y}^T(t)\Theta\bar{y}(t) < 0$ with Θ defined in the ARI in equation (8.51) as follows. First let

$$\breve{L} \triangleq \begin{bmatrix} L_1 \\ \vdots \\ L_N \end{bmatrix}, L_j \triangleq L_{k,l} = L(x)\big|_{x=x_{k,l}}, J \triangleq \begin{bmatrix} 0_{nN} & I_{nN} & 0_{nN} \\ 0_{nN} & I_{nN} & 0_{nN} \end{bmatrix}$$

Then with the finite difference approximation in equation (8.30) and the approximation of integration by summation, the HJII in equation (8.27) becomes

$$\int_U \left\{ \left(\frac{\partial V(\bar{y}(x,t))}{\partial \bar{y}} \right)^T \{\tilde{\kappa}(x)\nabla^2 \bar{y}(x,t) + \tilde{A}(x) + \tilde{B}(x)\bar{y}(x,t) \right.$$

$$+ \begin{bmatrix} 0_{n\times 1} \\ -L(x)[h_1 e(q_1,t),\ldots,h_{n_z} e(q_{n_z},t)]^T \end{bmatrix} \} + \bar{y}^T(x,t)Q_e^T Q_e \bar{y}(x,t)$$

$$+ \bar{y}^T(x,t)Q_\phi^T Q_\phi \bar{y}(x,t) + \frac{1}{4\gamma^2}\left(\frac{\partial V(\bar{y}(x,t))}{\partial \bar{y}} \right)^T \tilde{G}_v(x)\tilde{G}_v^T(x)\left(\frac{\partial V(\bar{y}(x,t))}{\partial \bar{y}} \right)$$

$$\left. + \frac{1}{2}(\tilde{H}(x)\bar{y}(x,t))^T \frac{\partial^2 V(\bar{y}(x,t))}{\partial \bar{y}^2} \tilde{H}(x)\bar{y}(x,t)) \right\} dx$$

$$\approx \left\{ 2\bar{y}(t)\bar{P} \left\{ \begin{bmatrix} (\breve{\kappa}T + \breve{A})y(t) \\ (\breve{\kappa}T + \breve{A})e(t) \end{bmatrix} + \begin{bmatrix} \breve{B}\breve{K} & -\breve{B}\breve{K} \\ 0_{nN} & 0_{nN} \end{bmatrix} \bar{y}(t) + \begin{bmatrix} O(\Delta^2) \\ O(\Delta^2) \end{bmatrix} \right. \right.$$

$$- \left. \begin{bmatrix} 0_{nN} & 0_{nN} \\ 0_{nN} & \breve{L}C \end{bmatrix} \bar{y}(t) \right\} \bar{y}^T(t)\bar{I}\bar{y}(t) + \bar{y}^T(t) \begin{bmatrix} \breve{M} & 0_{nN} \\ \breve{K} & -\breve{K} \end{bmatrix}^T \begin{bmatrix} \breve{M} & 0_{nN} \\ \breve{K} & -\breve{K} \end{bmatrix} \bar{y}(t)$$

$$\left. + \frac{1}{\gamma^2}\bar{y}^T(t)\bar{P}[\bar{G}_v - J][\bar{G}_v^T - J^T]\bar{P}\bar{y}(t) + \bar{y}^T(t)\bar{H}^T\bar{P}\bar{H}\bar{y}(t) \right\} \Delta^2$$

$$= \left\{ \bar{y}^T(t)\{\bar{P}\bar{A} + \bar{A}^T\bar{P} + \bar{I} + \bar{K}^T\bar{K} + \frac{1}{\gamma^2}\bar{P}\bar{G}_v\bar{G}_v^T\bar{P} + \bar{H}^T\bar{P}\bar{H}\}\bar{y}(t) \right.$$

$$\left. + 2\bar{y}^T(t)\bar{P} \begin{bmatrix} O(\Delta^2) \\ O(\Delta^2) \end{bmatrix} \right\} \Delta^2 < 0 \tag{8.60}$$

So let $\hat{\bar{G}}_v \triangleq [\bar{G}_v - J]$, we get the approximation of HJII in equation (8.27) as follows:

$$\left\{\bar{y}^T(t)\left\{\bar{P}\bar{A} + \bar{A}^T\bar{P} + \bar{I} + \bar{K}^T\bar{K} + \frac{1}{\gamma^2}\bar{P}\bar{G}_v\bar{G}_v^T\bar{P} + \bar{H}^T\bar{P}\bar{H}\right\}\bar{y}(t)\right.$$

$$\left. + 2\bar{y}^T(t)\bar{P}\begin{bmatrix} O(\Delta^2) \\ O(\Delta^2) \end{bmatrix}\right\}\Delta^2 < 0 \tag{8.61}$$

If we omit the last term in the left-hand side of equation (8.61), we can get

$$\bar{y}^T(t)\left\{\bar{P}\bar{A} + \bar{A}^T\bar{P} + \bar{I} + \bar{K}^T\bar{K} + \frac{1}{\gamma^2}\bar{P}\bar{G}_v\bar{G}_v^T\bar{P} + \bar{H}^T\bar{P}\bar{H}\right\}\bar{y}(t)\Delta^2 < 0 \tag{8.62}$$

that implies

$$\bar{I} + \bar{K}^T\bar{K} + \bar{P}\bar{A} + \bar{A}^T\bar{P} + \frac{1}{\gamma^2}\bar{P}\bar{G}_v\bar{G}_v^T\bar{P} + \bar{H}^T\bar{P}\bar{H} \tag{8.63}$$

which is of the similar form of the ARI in equation (8.51) except using $\hat{\bar{G}}_v$ in place of $\bar{G}_v$ in equation (8.51). Both the last redundant term

$$2\bar{y}^T(t)\bar{P}\begin{bmatrix} O(\Delta^2) \\ O(\Delta^2) \end{bmatrix}$$

in the left-hand side of equation (8.61) and the difference between $\hat{\bar{G}}_v$ and $\bar{G}_v$ are due to the fact that the derivation procedure of the ARI in equation (8.51) in Section 8.3 takes the $O(\Delta^2)$ as a part of the generic disturbance $\bar{v}(t)$, i.e., $\bar{v}(t) \triangleq \left[v^T(t), O^T(\Delta^2), n^T(t)\right]^T$, whereas the derivation procedure of the HJII in equation (8.27) in Section 8.2 does not take the $O(\Delta^2)$ as a part of the generic disturbance.

Remark 8.3.3

The approximation error equation (8.59) in Proposition 8.3.1 is due to the fact that the derivation procedure of the ARI in equation (8.51) in Section 8.3 takes the $O(\Delta^2)$) as a part of the generic disturbance $\bar{v}(t)$, i.e., $\bar{v}(t) \triangleq [v^T(t), O^T(\Delta^2), n^T(t)]^T$, hereas the derivation procedure of the HJII in equation (8.27) in Section 8.2 does not take the $O(\Delta^2)$ as a part of the generic disturbance.

Proposition 8.3.2

As $\Delta \rightarrow 0$, based on Proposition 8.3.1, if the ARI in equation (8.51) holds, then the spatiotemporal H_∞ robust observer-based stabilization performance in equations (8.28) and (8.29) for LSPDSs in equations (8.1) or (8.2) could be achieved and the asymptotical stability in probability for LSPDSs in equations (8.1) or (8.2) is also guaranteed when $\overline{v}(t)=0$.

Proof

From Proposition 8.3.1 and Theorem 8.2.3, this proposition is obviously proved.

Remark 8.3.4

Based on the H_∞ robust observer-based stabilization analysis in Theorem 8.3.2, the optimal H_∞ robust observer-based stabilization method for LSPDSs can be designed by minimizing the disturbance attenuation level γ^2 as the following eigenvalue problem:

$$\begin{gathered}\gamma_{\min}^2 = \min_{\overline{P},\breve{L},\breve{K}} \gamma^2 \\ \text{subject to (8.52)}\end{gathered} \tag{8.64}$$

Based on the analyses above, a design procedure for the H_∞ robust observer-based stabilization of LSPDSs is proposed as follows.

Design Procedure:

1. Perform the finite difference approximations in equations (8.30) and (8.31).
2. Construct the spatial state system in equation (8.38).
3. Propose the observer in equation (8.39), the control input in equation (8.40), and construct the augmented system in equation (8.43).
4. Since the ARI in equation (8.52) is a bilinear matrix inequality, we use the iterative LMI approach [265,266] to solve $\overline{P}$, $\breve{L}$, and $\breve{K}$ from the ARI in equation (8.52) for H_∞ robust observer-based stabilization with a prescribed disturbance attenuation level γ^2 or solve $\overline{P}$, $\breve{L}$, and $\breve{K}$ from the EVP in equation (8.64) for the optimal H_∞ robust observer-based stabilization.

Remark 8.3.5

The computation complexity for solving the ARI in equations (8.52) will increase as the number $N \triangleq N_1 \times N_2$ of grid nodes increases. There is a tradeoff between the finite difference truncation error and the computation complexity for solving the ARI in equation (8.52).

Remark 8.3.6

The effectiveness of the above iterative LMI algorithm can be referred to [265,266].

8.4 SIMULATION EXAMPLE

Consider the temperature function for a 1×0.5 m thin plate in the heat transfer system with the surrounding temperature at 0°C as follows [267]:

$$\begin{aligned}
&\frac{\partial y(x,t)}{\partial t}=\kappa\nabla^2 y(x,t)+Ay(x,t)+u(x,t)+g_v v(x,t)+Hy(x,t)w(x,t)\\
&y(x,0)=100\times e^{-10\times|0.5-x_1|-0.6738}\times e^{-30\times|0.5-2\times x_2|},\\
&\forall x\in U=[0,1]\times[0,0.5]
\end{aligned} \tag{8.65}$$

$$y(x,t)=0\text{ °C},\ \forall t,\ \forall x \text{ on the boundary of } U,\ \phi(x,t)=\begin{bmatrix} My(x,t)\\ u(x,t)\end{bmatrix}$$

where $y(x,t)$ is the temperature function, the location x is in meters, the time t is in seconds, $\kappa=10^{-4}\text{m}^2/\text{s}$ is the thermal diffusivity [267], $A=0.01\text{ s}^{-1}$, the term $g_v v(x,t)$ is the environmental thermal fluctuation with $g_v=0.1$, $v(x,t)=\sin(0.2t)e^{-0.001t-0.1x_1}$, the term $Hy(x,t)w(x,t)$ with $H=0.002$ is due to the random parameter variation of the term $Ay(x,t)$, and $M=0.1$. The grid spacing Δ of the finite difference scheme is chosen as 0.125 m so that there are $N=7\times3=21$ interior grid points and 24 boundary points. There are $n_z=45$ sensor locations of observation, which are at the 2-D location $q_i=[q_{i,1},\ q_{i,2}]^T$ with $q_{i,1}=(k-1)\times0.125$ m, $q_{i,2}=(l-1)\times0.125$ m, $l=1,\dots,5$ and $i=(l-1)\times9+k$. The observation influence vector $h_i=1$ and the noise influence matrix $D_n=1$ so that $h_iD_n=1$. So the measured output by 45 sensors is as follows:

$$z(t)=[y(q_1,t)+n(q_1,t),\dots,y(q_{45},t)+n(q_{45},t)]^T$$

where $n(q_i,t)$ denotes the measurement noise at the ith sensor. All the measurement noises are zero mean with unit variance in this example.

Based on the design procedure, we construct the spatial state space system in equation (8.38) and the augmented system in equation (8.43). From the constrained optimization problem in equation (8.64) with the help of MATLAB toolbox to get γ^2, P_{11}, P_{22}, we solve the control gain $\breve{K}$ for the control input in equation (8.40) and the observer gain $\breve{L}$ for the robust observer in equation (8.39). From the computer simulation of this example by MATLAB with the initial estimated state $\hat{y}(x,0)=20,\ \forall x\in U$ and the H_∞ robust observer-based stabilization performance in equation (8.44), we get

$$\frac{E\int_0^{100}\left\|\begin{bmatrix} e(t)\\ \phi(t)\end{bmatrix}\right\|^2 dt - E\bar{V}(\bar{y}(0))}{E\int_0^{100}\left\|[\bar{v}(t)]\right\|^2 dt} = (0.59)^2 < \gamma_{\min}^2 = (1.05)^2$$

Figures 8.2–8.5 show the simulation results of the temperature $y(x,t)$, the estimated temperature $\hat{y}(x,t)$, the estimation error $e(x,t)$, and the measured output $z(x,t)$ for $t=0$, 1, 20, 50 seconds, respectively. The estimation error $e(x,t)$ is very small in these Figures. Figure 8.6 shows the control input $u(x,t)$.

Remark 8.4.1

For each γ^2, the average number of iterations required to converge to get the appropriate values γ^2, $\breve{K}$, and $\breve{L}$ is 1.03. So this iterative LMI algorithm [265,266] is effective.

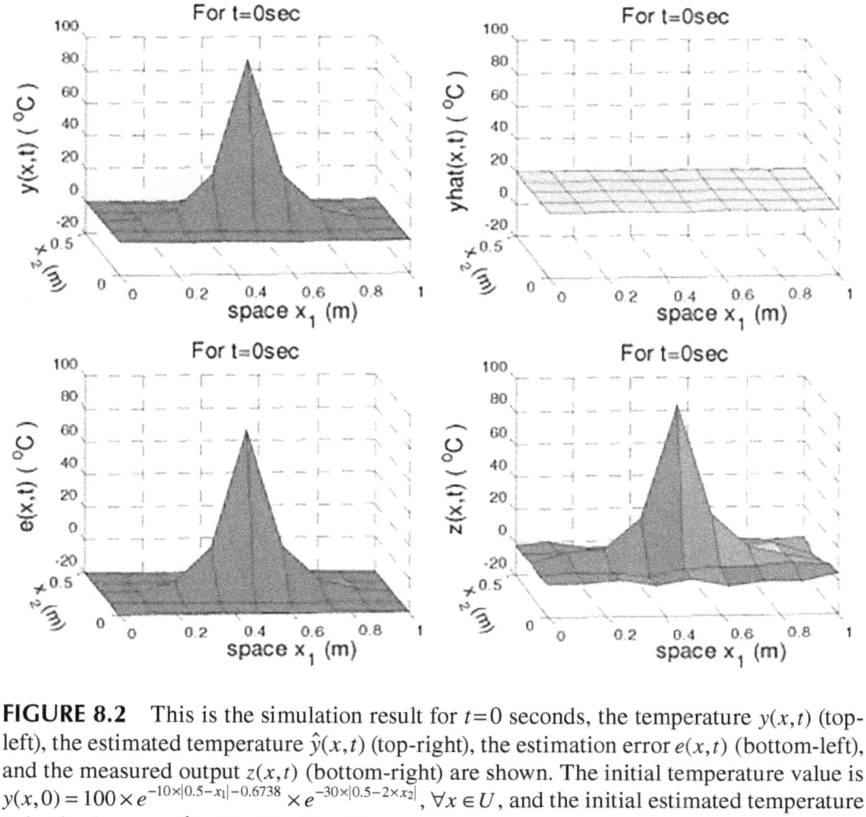

FIGURE 8.2 This is the simulation result for $t=0$ seconds, the temperature $y(x,t)$ (top-left), the estimated temperature $\hat{y}(x,t)$ (top-right), the estimation error $e(x,t)$ (bottom-left), and the measured output $z(x,t)$ (bottom-right) are shown. The initial temperature value is $y(x,0)=100\times e^{-10\times|0.5-x_1|-0.6738}\times e^{-30\times|0.5-2\times x_2|}$, $\forall x\in U$, and the initial estimated temperature value is chosen as $\hat{y}(x,0)=20$, $\forall x\in U$.

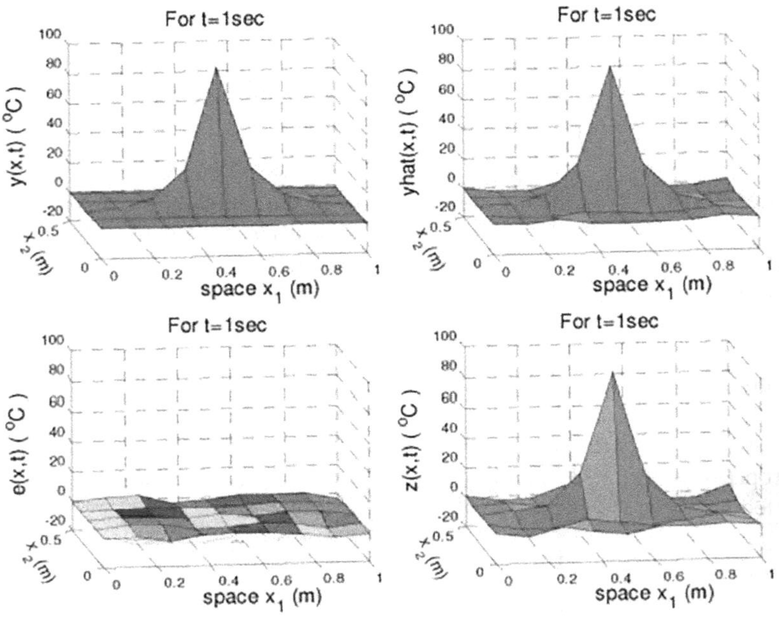

FIGURE 8.3 Simulation result for $t = 1$ seconds.

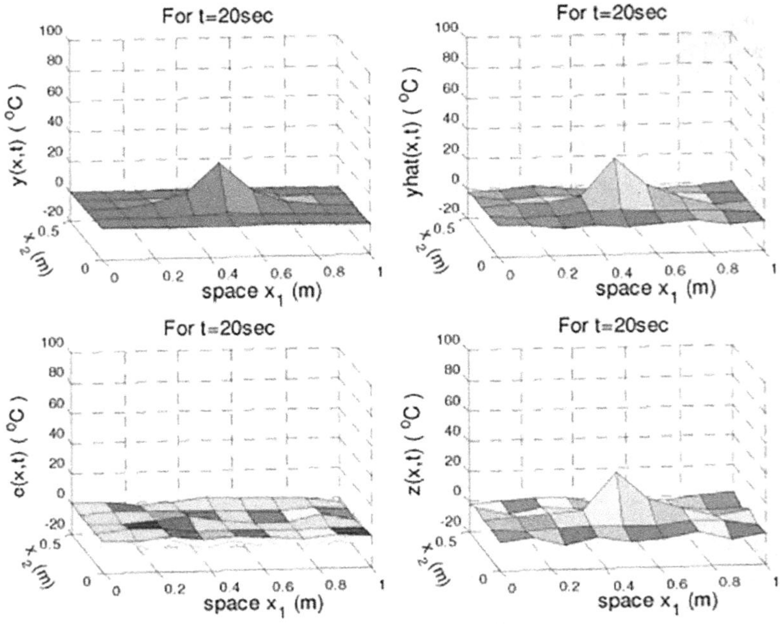

FIGURE 8.4 Simulation result for $t = 20$ seconds.

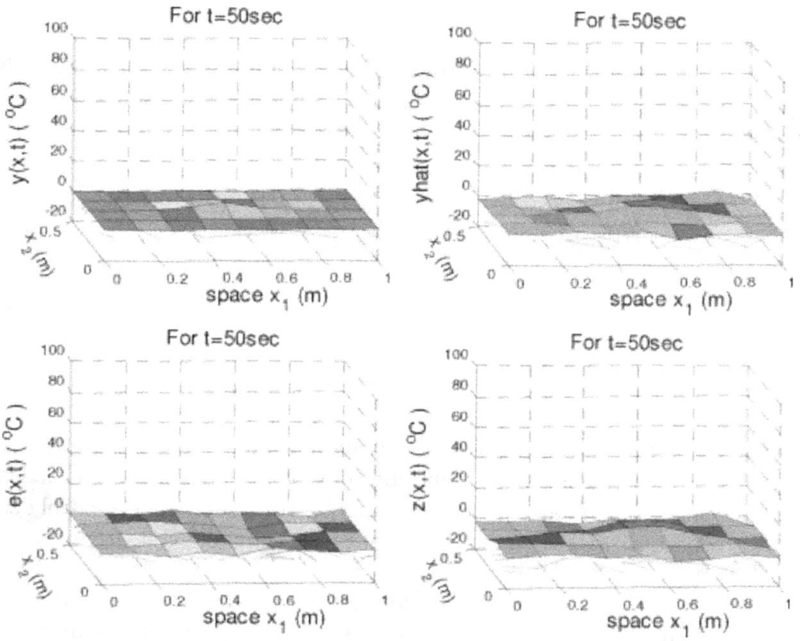

FIGURE 8.5 Simulation result for $t = 50$ seconds.

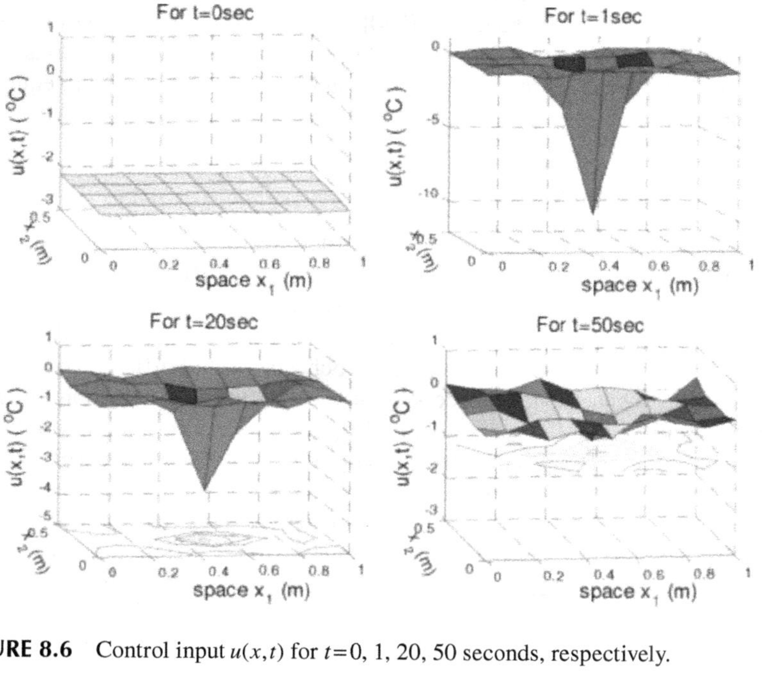

FIGURE 8.6 Control input $u(x,t)$ for $t = 0$, 1, 20, 50 seconds, respectively.

Remark 8.4.2

Compared to our study, the result in [230] has many drawbacks in the robust output feedback control of PDSs as follows: (i) Its robust control performance is not of the H_∞ stabilization performance as equations (8.28) or (8.29) in the chapter since its robust control performance is given as follows:

$$\limsup_{t\to\infty}\left|y^i\right| \le d_0, i = 1,\ldots,l \tag{8.66}$$

or

$$\limsup_{t\to\infty}\left|y^t - v_i\right| \le d_0, i = 1,\ldots,l \tag{8.67}$$

where y^i is the ith controlled output, v_i is the ith reference input, and d_0 is an uncertainty attenuation level. (ii) Some adjustable parameters in its control design need to be chosen by try and error. (iii) There is a restriction on its control design that the number of measurements is equal to the number of slow modes since static output feedback is used. (iv) The form of its inversion-based control law is too complex since the inversion-based control law is relevant to many differential operators. (v) It applies the Galerkin's method to derive an Ordinary differential equation (ODE)system with dimension equal to the number of the slow modes of the original PDS. The computation complexity will increase if the number of slow modes is large. (vi) The effect of modeling error due to the truncation of fast modes is not efficiently attenuated in [230]. In this chapter, the effect of finite difference truncation error $O(\Delta^2)$ on the stabilization can be attenuated to as small a value as possible owing to the H_∞ robust stabilization performance in equations (8.44) or (8.45).

8.5 CONCLUSION

In this chapter, we first develop the more general H_∞ robust observer-based stabilization design theory for a stochastic linear partial differential system under spatiotemporal disturbances and sensor measurement noises through a set of sensor measurements. Then in order to make the H_∞ robust observer-based stabilization design easier and implementable, the stochastic partial differential system is represented by a spatial state space system via the semi-discretization finite difference scheme so that the spatiotemporal H_∞ robust observer-based stabilization performance can be represented by the only temporal H_∞ robust observer-based stabilization performance. Therefore, based on the more suitable spatial state space model, an implementable H_∞ robust observer-based controller is designed for robust stabilization of LSPDSs with spatiotemporal system matrices, disturbances and state-dependent noises. Further, the effect of finite difference truncation error on the stabilization is also attenuated to as small a value as possible. As

$\Delta \rightarrow 0$ the proposed finite-difference-based H_∞ robust stabilization method could approach the general theoretical H_∞ robust stabilization design. The proposed method in this chapter is suitable for practical control applications in partial differential systems such as fluid and thermal dynamics in mechanical systems, reaction process systems in chemical engineering, population dynamics in ecological systems, and biodynamics in biological systems. From the simulation example of heat transfer system, the spatiotemporal H_∞ robust observer-based stabilization and state estimation performance for stochastic partial differential systems via the measurement of a set of sensors can be confirmed by the proposed method. After the introduction of robust stabilization designs of linear partial differential systems, we will introduce the more complicate robust control design of nonlinear partial differential systems in the following chapters.

9 Fuzzy State-Space Modeling and Robust Observer-Based Control Design for Nonlinear Partial Differential Systems

9.1 INTRODUCTION

In the past, most control systems are modeled by ordinary differential equations (ODEs). The conventional control theory is developed based on the models of ODEs. However, the behavior of many physical systems must depend on space, for example, some mechanical systems related with heat flows, fluid flows, elastic wave, flexible structures [268,269], chemical engineering [270,271] and biodynamics [272–274], etc. Thus, these systems can be described by partial differential equations (PDEs). In this situation, the control design problem of partial differential systems (PDSs) or distributed parameter systems is an important topic for practical application. So far some researchers have investigated the control problem of PDSs. Many related books and articles [268,269,271,275–278] have been published for solving the control design problem of linear PDSs. For flexible systems [268,269,275,276], the linear PDSs can be transferred to the equivalent infinite-dimensional ordinary differential systems based on Galerkin's method in Section 1.2.1. According to the separation of eigenvalues, the infinite-dimensional ordinary differential system can be separated into a finite-dimensional slow mode and an infinite-dimensional fast mode. A finite-dimensional controller based on the finite-dimensional mode could be designed to stabilize the linear PDS. A concept of spillovers, control spillover, and observation spillover is introduced by Balas [268]. In [269], the control and observation spillovers were treated as parametric uncertainties or unmodeled uncertainties, and the robust stability conditions were developed with respect to the two types of uncertainties. In the chemical engineering applications, Christofides [270,271] has proposed some robust control methods of linear PDSs.

In the control design problem of the nonlinear partial differential systems (NPDSs), Balas [279] and Christofides [280] proposed some control design methods.

DOI: 10.1201/9781003229230-12

However, because of the properties of nonlinearity and infinite dimension, the infinite-dimensional nonlinear ODE in [279,280] is difficult to obtain for the design of the corresponding controller. Although these NPDSs can be analyzed and designed case by case, at present, there exists no general design method for the control design problem of the nonlinear partial differential (distributed parameter) systems. For example, the stabilization of nonlinear distributed parameter systems are discussed based on the dissipative or contractive properties of systems [281]. However, how to design the corresponding controller is still obscure. To remedy this design difficulty, this chapter proposes a systematic approach based on fuzzy theory to solve the control design problem of NPDS. Recently, fuzzy control designs [282–288] have been successfully applied to the robust stabilization problem of nonlinear systems which are described by nonlinear ODEs. In [289], a new technique using the adaptive fuzzy algorithm to obtain the solutions of PDEs is proposed. According to the adaptive scheme of fuzzy logic systems, a fuzzy solution with adjustable parameters for the PDE can be obtained successfully. In [290,291], a fuzzy control method is proposed to stabilize a class of NPDS based on the infinite-dimensional nonlinear ODE represented by the Galerkin's method in Section 1.2.1. They use a finite-dimensional fuzzy model to approximate the infinite-dimensional nonlinear ODE but neglect the effect of infinite-dimensional (fast) residual mode in the control design procedure [290,291]. In general, for a complex NPDS, it is difficult to represent a NPDS by an infinite-dimensional nonlinear ODE [292]. In addition, for the control design of NPDS, the effects of control spillover and observation spillover due to the residual mode will affect the stabilization of the system [268–269], but the authors of [290,291] do not discuss this problem. Therefore, in this chapter, a new T-S fuzzy state space model via fuzzy interpolation approach is proposed to approximate the NPDS and the effects of the control spillover and observation spillover on stabilization can be attenuated by the proposed robust stabilization design.

There are many different PDE systems, such as elliptical equation, parabolic equation, and hyperbolic equation. However, many physical systems can be described as the parabolic PDEs. In this chapter, we will study the nonlinear parabolic PDE stabilization control problem in detail. However, these results can be easily extended to other nonlinear PDE stabilization control problems. Firstly, the NPDS is approximated by a so-called T-S fuzzy PDS through interpolating several linear partial differential subsystems via membership functions with some approximation errors. Then, each linear partial differential subsystem could be represented equivalently by an infinite-dimensional linear state-space system. By separating the infinite-dimensional linear ordinary differential system, we could get a finite-dimensional linear state space subsystem in controlled mode and a coupled infinite-dimensional subsystem in residual mode to represent the infinite-dimensional linear state space system. In summary, we could interpolate several finite-dimensional linear state space models to represent the NPDS via fuzzy membership functions with a coupled residual subsystem and a fuzzy approximation error. Then, a robust stabilization design is needed to control the fuzzy finite-dimensional linear state-space subsystem and to attenuate the effect of the coupled residual subsystem and fuzzy approximation error. Furthermore, the effects of the external disturbances and measurement noises can be attenuated to a prescribed attenuation level by the proposed robust H_∞ stabilization design method.

In general, it is not easy to measure the state variables of NPDS directly. In this situation, a robust fuzzy observer is proposed to estimate the state variables from the measurement output for fuzzy feedback control design of NPDSs. Based on the small gain theorem [293,294], an observer-based fuzzy control design is proposed to asymptotically stabilize the NPDSs. Furthermore, based on dissipative theory [293], a robust H_∞ fuzzy observer-based controller is proposed to stabilize the NPDSs by efficiently eliminating the effect of external disturbances, fuzzy approximation errors, and residual modes. Because of the coupling between the controlled subsystems and the residual subsystems through control spillover and observation spillover, more effort is needed for robust control design of NPDS than the conventional nonlinear systems. The LMI technique is employed to simplify the robust fuzzy stabilization design problem of NPDS.

In this chapter, the T-S fuzzy state space model is successfully applied to represent the NPDS, and a finite-dimensional H_∞ fuzzy observer-based controller is proposed to robustly stabilize the NPDS. The proposed fuzzy control design scheme is a systematic design method for the output control design problems of NPDS and it can be solved easily by LMI toolbox. The remainder of the chapter is organized as follows. A problem formulation for the control design problem of the NPDS is discussed in Section 9.2. In Section 9.3, we present a T-S fuzzy PDS to approximate the NPDS. For the convenience of control design, an infinite- dimensional T-S fuzzy state-space model is also developed to represent the NPDS by a finite-dimensional T-S fuzzy state-space subsystem in controlled mode and a coupled infinite-dimensional T-S fuzzy state-space subsystem in residual mode. We propose an H_∞ fuzzy observer-based feedback controller for the robust stabilization control problem of the NPDSs in Section 9.4. Finally, we give an example to illustrate the design procedure and to confirm the performance of the proposed method in Section 9.5.

9.2 PROBLEM FORMULATION

At present, the control design problems of the NPDSs are still very difficult to obtain a systematic design method. The first difficult problem is still lacking in a suitable state space model for NPDS for control system design. The second problem is how to develop a robust control design method based on the state space model to treat the stabilization problem of complex NPDS. Since most physical systems can be described as parabolic PDEs, we will discuss the control design problems of the nonlinear parabolic partial differential systems.

Consider the following NPDS [295]

$$\begin{aligned}\frac{\partial y(x,t)}{\partial t} &= \kappa(y(x,t))\frac{\partial^2 y(x,t)}{\partial x^2} + f(y(x,t)) + g(x)u(t)\\ z(t) &= h(x)y(x,t)\end{aligned} \tag{9.1}$$

for $0 \le x \le L$ and $t > 0$, where $y(x,t)$ is the transverse displacement; $\kappa(y(x,t)) \ge 0$ and $f(y(x,t))$ are the nonlinear functions satisfying $\kappa(0) = 0$ and $f(0) = 0$, respectively [295]; the distribution of the control force $u(t)$ is provided by p point force actuators, i.e. $u(t) \in \mathbb{R}^p$ are the applied forces to be designed as $u(t) = \left[u_1(t),...,u_p(t)\right]^T$; the influence function $g(x)$ is a $1 \times p$ vector whose elements are of Dirac delta function $\delta(x - p_i)$. For example, $g(x) = \left[g_1\delta(x - p_1),...,g_p\delta(x - p_p)\right]$ where p_i are the control force locations and $g_i\delta(x - p_i) = g_i$ for $x = p_i$ or $g_i\delta(x - p_i) = 0$ for $x \ne p_i$. The output $z(t)$ may be interpreted as observations or as parts of the system whose behavior we wish to influence, the output $z(t)$ is a vector, i.e., $z(t) = \left[z_1(t),...,z_q(t)\right]^T$, where q is the number of observations and the observation influence function $h(x)$ is defined as $\left[\delta(x - q_1),...,\delta(x - q_q)\right]^T$. Therefore, we have $z_i(t) = \delta(x - q_i)y(x,t) = y(q_i,t)$ with q_i as the sensor locations. The boundary conditions and initial-value are given by

$$y(0,t) = y(L,t) = 0,\ y_x(0,t) = y_x(L,t) = 0,\ y(x,0) = y_0(x) \tag{9.2}$$

In reality, the external disturbance and measurement noise will affect the PDS. Therefore, these influences should be considered in the design procedure, and the perturbative NPDS should be formulated as follows:

$$\begin{aligned}
\frac{\partial y(x,t)}{\partial t} &= \kappa(y(x,t))\frac{\partial^2 y(x,t)}{\partial x^2} + f(y(x,t)) + g(x)u(t) + g_d(x)d(t) \\
z(t) &= h(x)y(x,t) + D_n n(t) \\
z_o(t) &= h_o(x)y(x,t)
\end{aligned} \tag{9.3}$$

where $g_d(x)$ is an interactive location function of the external disturbance; $d(t) \in L_2(\mathbb{R}_+;\mathbb{R}^{p_d})$ is the vector of the external disturbance; $D_n \in \mathbb{R}^{q \times q_n}$ is the influence matrix of the measurement noise; and $n(t) \in L_2(\mathbb{R}_+;\mathbb{R}^{q_n})$ is the measurement noise on the location of observation. $z_o(t)$ is the controlled output. Note that $h_o(x)$ is defined similarly as $h(x)$ in equation (9.3).

For the NPDS with external disturbances and measurement noises in equation (9.3), a robust controller is necessary to attenuate the effects of external disturbances and measurement noise on the controlled output $z_o(t)$. Therefore, a prescribed disturbance attenuation level ρ is given such that the following H_∞ control performance is achieved [283,293]

$$\int_0^{t_f} z_o(t)^T z_o(t)dt \le \rho^2 \int_0^{t_f} v(t)^T v(t)dt \tag{9.4}$$

where $v(t) = [d(t), n(t)]^T$ denotes the vector of external disturbance and measurement noise, i.e., the effect of $v(t)$ on $z_o(t)$ is less than or equal to ρ^2 from the energy point of view.

For the NPDSs in equations (9.1) and (9.3), the following modeling and stabilization control problems are studied in this chapter.

1. The infinite-dimensional fuzzy state-space model will be introduced for NPDSs in equations (9.1) and (9.3). For the simplicity of controller design, the fuzzy infinite-dimensional state-space model is separated into a finite-dimensional subsystem to be controlled and an infinite-dimensional residual subsystem to be tolerated.
2. If the NPDS is free of external disturbances and measurement noises, a finite-dimensional fuzzy observer-based controller $u(t)$ is designed to stabilize the NPDS in equation (9.1). In this situation, the effects of the coupling between controlled subsystem and residual subsystem through control spillover and observation spillover [268–269] should be considered in the control design procedure.
3. Given a prescribed disturbance attenuation level ρ, a finite-dimensional robust fuzzy observer-based controller $u(t)$ is designed for the NPDS in equation (9.3) not only to tolerate the coupling of residual subsystem but also to attenuate the effects of the external disturbances and measurement noise, i.e., to achieve the H_∞ robust control performance in equation (9.4).

In the following sections, an infinite-dimensional fuzzy state space model is introduced to represent the complex NPDS at first. Then, in the disturbance free case, based on the proposed finite-dimensional fuzzy state space subsystem in controlled mode with a coupled infinite-dimensional subsystem in residual mode and fuzzy approximation error, a robust control design method is proposed to stabilize the NPDSs via a small gain theorem. Finally, based on the dissipative theory, an optimal H_∞ observer-based control design method is also proposed to efficiently attenuate the effect of residual subsystem, external disturbances, and measurement noises.

Remark 9.2.1

The following nonlinear reaction-diffusion equation is a special case of the nonlinear parabolic PDE in equation (9.1)

$$\frac{\partial y(x,t)}{\partial t} = \frac{\partial^2 y(x,t)}{\partial x^2} + f(y(x,t)) + g(x)u(t) \tag{9.5}$$

which has been widely used to model some mechanical and biochemical systems [274,278].

For the convenience of problem description and control design, we define some different separable Hilbert spaces in the following.

Definition 9.2.1

1. $L_2([0,L];\mathbb{R})$ is the space of the scalar measurable function $y(x,t)$ defined on $x \in [0,L]$ such that $\|y(x,t)\|^2_{L_2([0,L];\mathbb{R})} \triangleq \int_0^L |y(x,t)|^2 dx < \infty$.
2. The inner product is defined on Hilbert space $L_2([0,L];\mathbb{R})$ as follows:

$$\langle y_1(x,t), y_2(x,t) \rangle \triangleq \int_0^L y_1(x,t)^T y_2(x,t) dx$$

3. $L_2(\mathbb{R}_+;\mathbb{R}^n)$ is the space of the n-dimensional measurable function $z(t) \in \mathbb{R}^n$ defined on $t \in \mathbb{R}_+ = [0,\infty)$ such that $\|z(t)\|^2_{L_2(\mathbb{R}_+;\mathbb{R}^n)} \triangleq \int_0^\infty z(t)^T z(t) dt < \infty$.
4. $\|y(t)\|^2 \triangleq \Sigma_{i=1}^{\infty} |y_i(t)|^2$ where $y(t) = [y_1(t), y_2(t), \ldots]^T$.

9.3 SYSTEM REPRESENTATION BY FUZZY STATE SPACE MODEL

In the first subsection, a fuzzy PDS will be derived to approximate the NPDS. In the second subsection, a finite-dimensional fuzzy state space system to be used to design a robust fuzzy controller is proposed to separate the fuzzy PDS into a fuzzy controlled state space model and a fuzzy residual state space model.

9.3.1 Approximation of Nonlinear Partial Differential Systems by Fuzzy Partial Differential Systems

Recently, the T-S fuzzy system has been widely used to approximate a nonlinear ordinary differential system via the interpolation of several local linear ordinary differential models [282–284]. We consider the NPDSs in equations (9.1) and (9.3) without any input and disturbance, i.e., $u(t) \equiv 0$ and $d(t) \equiv 0$, as follows:

$$\frac{\partial y(x,t)}{\partial t} = \kappa(y(x,t)) \frac{\partial^2 y(x,t)}{\partial x^2} + f(y(x,t)). \tag{9.6}$$

A T-S fuzzy PDS based on fuzzy approximation method is proposed to approximate the NPDS in equation (9.6). The T-S fuzzy PDS is described by the fuzzy If-Then rules as follows:

$$\text{Rule } i : \text{If } y(x,t) \text{ is } F_i,$$
$$\text{then } \frac{\partial y(x,t)}{\partial t} = \kappa_i \frac{\partial^2 y(x,t)}{\partial x^2} + a_i y(x,t) \tag{9.7}$$

for $i = 1, 2, \ldots, M$. F_i are the fuzzy sets; $\kappa_i \geq 0$ and a_i are some known constants; M is the number of If-Then rules. The physical meaning of the fuzzy rule i is that if the premise variables $y(x,t)$ is with the fuzzy set F_i, then the NPDS in equation (9.6) could be represented by the linearized PDS in equation (9.7). The T-S fuzzy PDS in equation (9.7) is inferred as follows [282–284]:

$$\frac{\partial y(x,t)}{\partial t}=\sum_{i=1}^{M}\mu_i(y(x,t))\left[\kappa_i\frac{\partial^2 y(x,t)}{\partial x^2}+a_i y(x,t)\right] \tag{9.8}$$

where $\mu_i(y(x,t))\triangleq\frac{F_i y(x,t)}{\sum_{k=1}^{M}F_k y(x,t)}$, i.e., $F_i y(x,t)$ is the grade of the membership of $y(x,t)$ or the possibility function of $y(x,t)$. The denominator of $\mu_i(y(x,t))$ is only for normalization so that the total sum of the fuzzy bases $\sum_{i=1}^{M}\mu_i(y(x,t))=1$. Therefore, the NPDS (equation 9.1) could be approximated by the fuzzy partial differential model (equation 9.8) as follows:

$$\frac{\partial y(x,t)}{\partial t}=\sum_{i=1}^{M}\mu_i(y(x,t))\left[\kappa_i\frac{\partial^2 y(x,t)}{\partial x^2}+a_i y(x,t)\right]+\varepsilon_1(y(x,t)) \tag{9.9}$$

where the fuzzy approximation error $\varepsilon_1(y(x,t))$ is defined as

$$\begin{aligned}\varepsilon_1(y(x,t))\triangleq{}&\kappa(y(x,t))\frac{\partial^2 y(x,t)}{\partial x^2}+f(y(x,t))\\&-\sum_{i=1}^{M}\mu_i(y(x,t))\left[\kappa_i\frac{\partial^2 y(x,t)}{\partial x^2}+a_i y(x,t)\right]\end{aligned} \tag{9.10}$$

Note that, for the proposed T-S fuzzy PDS in equation (9.8), the fuzzy system only intends to approach the nonlinear terms $\kappa(y(x,t))$ and $f(y(x,t))$. Thus, the NPDS in equation (9.6) could be approximated by the proposed T-S fuzzy PDS based on the fuzzy approximation analysis in [296–298] and the bound of the fuzzy approximation error $\varepsilon_1(y(x,t))$ can be described as follows.

Assumption 9.3.1

Suppose the fuzzy approximation error $\varepsilon_1(y(x,t))$ is bounded as follows:

$$|\varepsilon_1(y(x,t))|^2\le\sigma_{1,1}^2\left|\frac{\partial^2 y(x,t)}{\partial x^2}\right|^2+\sigma_{1,2}^2|y(x,t)|^2 \tag{9.11}$$

where $\sigma_{1,1}$ and $\sigma_{1,2}$ are the positive scalar values and are dependent on the number M of the fuzzy rules.

Based on Assumption 9.3.1, the approximation theorem for T-S fuzzy PDS can be obtained in the following.

Theorem 9.3.1

Suppose $\kappa(y(x,t))$ *and* $f(y(x,t))$ *are the continuous functions defined on a compact set* $U \subset \mathbb{R}$, *i.e.,* $y(x,t) \in U$, *and* $\kappa(0) = f(0) = 0$, *then there* exist two arbitrary constants $\sigma_{1,1} > 0$ and $\sigma_{1,2} > 0$ such that

$$\left\|\varepsilon_1(y(x,t))\right\|^2_{L_2([0,L];\mathbb{R})} \leq \sigma_{1,1}^2 \left\|\frac{\partial^2 y(x,t)}{\partial x^2}\right\|^2_{L_2([0,L];\mathbb{R})} + \sigma_{1,2}^2 \left\|y(x,t)\right\|^2_{L_2([0,L];\mathbb{R})}. \quad (9.12)$$

Proof

By Assumption 9.3.1, the bound of the fuzzy approximation error $\varepsilon_1(y(x,t))$ can be estimated as follows:

$$\left\|\varepsilon_1(y(x,t))\right\|^2_{L_2([0,L];\mathbb{R})} = \int_0^L \left|\varepsilon_1(y(x,t))\right|^2 dx$$

$$\leq \int_0^L \sigma_{1,1}^2 \left|\frac{\partial^2 y(x,t)}{\partial x^2}\right|^2 + \sigma_{1,2}^2 \left|y(x,t)\right|^2 dx$$

$$\leq \sigma_{1,1}^2 \left\|\frac{\partial^2 y(x,t)}{\partial x^2}\right\|^2_{L_2([0,L];\mathbb{R})} + \sigma_{1,2}^2 \left\|y(x,t)\right\|^2_{L_2([0,L];\mathbb{R})} \quad (9.13)$$

Remark 9.3.1

Several methods have been proposed to solve the fuzzy approximation problem, for example, the least-squares estimation [299]. In this chapter, we use the least-squares estimation [300,301] to minimize the sum of the squared errors between the fuzzy function and the nonlinear function. Thus the approximation problem can be formulated as follows:

$$\kappa(y(x,t))\frac{\partial^2 y(x,t)}{\partial x^2} + f(y(x,t)) = \phi(y(x,t))^T \theta + \varepsilon_1(y(x,t)) \quad (9.14)$$

where $\phi(y(x,t))$ is a 2M-dimensional regression vector as $\left[\mu_1(y(x,t))\frac{\partial^2 y(x,t)}{\partial x^2},\ldots,\mu_M(y(x,t))\frac{\partial^2 y(x,t)}{\partial x^2},\mu_1(y(x,t))y(x,t),\ldots,\mu_M(y(x,t))y(x,t)\right]^T$, and the parameter vector θ is $[\kappa_1,\ldots,\kappa_M,a_1,\ldots,a_M]^T$. Therefore, we could use the fuzzy function to optimally approximate the nonlinear terms $\kappa(y(x,t))$ and $f(y(x,t))$ of the NPDS equation (9.6) and apply the least-square method to estimate the parameters κ_i and a_i of the fuzzy system [299].

In the state-space representation of the linear PDS, Galerkin's method [268,269,292] is a powerful method to transform a PDE into an infinite-dimensional ODE. Therefore, we will use Galerkin's method in Section 1.2.1 to obtain a fuzzy infinite-dimensional model to represent the NPDS. For the fuzzy PDS in equation (9.8), it is difficult to directly transform the T-S fuzzy PDS to an infinite-dimensional ordinary differential system. Thus, an equivalent T-S fuzzy PDS is proposed to replace the fuzzy system in equation (9.8).

From the principle of the fuzzy interpolation [302], a fuzzy function $\tilde{y}(x,t)$ could be established as follows:

$$\text{Rule } j: \text{If } x \text{ is } \bar{F}_j, \quad \text{then } \tilde{y}(x,t) = z_j(t) \tag{9.15}$$

for $j = 1,2,...,q$. $\bar{F}_j$ is the fuzzy set; $z_j(t)$ is the value of $y(x,t)$ at the location $x = q_j$ (see Figure 9.1); q is the number of If-then rules. The physical meaning of the fuzzy rule j is that if the premise variables x are with the fuzzy set $\bar{F}_j$, then $y(x,t)$ could be represented by $z_j(t)$. The fuzzy function $\tilde{y}(x,t)$ is inferred as follows:

$$\tilde{y}(x,t) = \sum_{j=1}^{q} \beta_j(x) z_j(t) \tag{9.16}$$

where $\beta_j(x) = \dfrac{\bar{F}_j(x)}{\sum_{k=1}^{q} \bar{F}_k(x)}$. $\bar{F}_j(x)$ is the grade of membership of x in $\bar{F}_j$. The denominator $\sum_{k=1}^{q} \bar{F}_k(x)$ is only for normalization so that $\sum_{j=1}^{q} \beta_j(x) = 1$. If the time t of the solution $y(x,t)$ is fixed at any time, then the solution $y(x,t)$ could be treated as a nonlinear function $y(x)$ (see Figure 9.1). Therefore, we can use the approximation principle of Mamdani fuzzy system with the fuzzy function $\tilde{y}(x,t)$ to approximate the nonlinear function $y(x,t)$. Based on the universal approximation theory for Mamdani fuzzy system in [302,303], we have the following theorem.

Theorem 9.3.2 [296–298,302,303]

Suppose $y(x,t)$ is a continuous function defined on a compact set $U \subset \mathbb{R}$, i.e., $x \in U$. Then, for an arbitrary $\sigma_2 > 0$, the fuzzy function $\tilde{y}(x,t)$ could be constructed to approximate the solution $y(x,t)$ as follows:

$$y(x,t) = \tilde{y}(x,t) + \varepsilon_2(x,t) \tag{9.17}$$

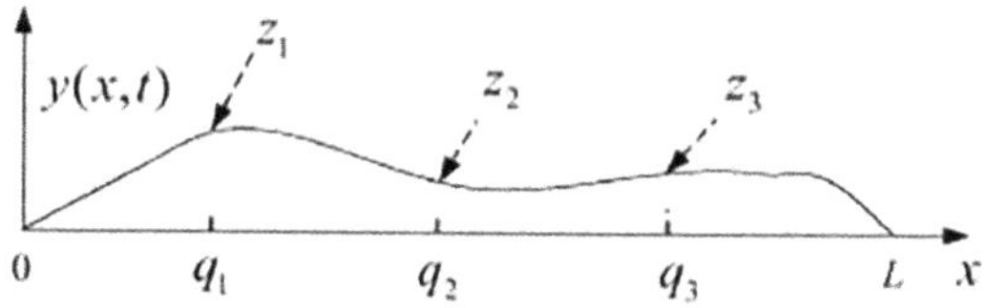

FIGURE 9.1 For a fixed time at any time t, the solution $y(x,t)$ could be treated as a nonlinear function $y(x)$.

and the approximation error is bounded by σ_2, *i.e.*,

$$\sup_{x\in U, t\in[0,\infty)} |\varepsilon_2(x,t)| \triangleq \sup_{x\in U, t\in[0,\infty)} |y(x,t) - \tilde{y}(x,t)| \le \sigma_2 \tag{9.18}$$

Remark 9.3.2

In order to apply the fuzzy function $\tilde{y}(x,t)$ to approximate the solution $y(x,t)$, enough observations $z_j(t)$ must be provided.

In order to obtain an equivalent fuzzy PDS, the fuzzy function $\tilde{y}(x,t)$ in equation (9.16) is used to approach the premise variable $y(x,t)$ of fuzzy PDS (equation 9.8). For simplicity, triangular functions are specified for the fuzzy membership functions $F_i(y(x,t))$ of the fuzzy partial differential model (equation 9.8). The results can be easily extended to other fuzzy membership functions. Thus, we have the following theorem.

Theorem 9.3.3

Suppose the triangular fuzzy membership functions $F_i(y(x,t))$ *are specified as*

$$\mathrm{F}_i\ (y(x,t)) = \begin{cases} 0, & y(x,t) < \bar{a}_i \\ \dfrac{y(x,t) - \bar{a}_i}{\bar{b}_i - \bar{a}_i}, & \bar{a}_i \le y(x,t) < \bar{b}_i \\ \dfrac{\bar{c}_i - y(x,t)}{\bar{c}_i - \bar{b}_i}, & \bar{b}_i \le y(x,t) < \bar{c}_i \\ 0, & \bar{c}_i \le y(x,t) \end{cases} \tag{9.19}$$

Given the parameters $\bar{a}_i, \bar{b}_i$ *and* $\bar{c}_i$ *with* $\bar{c}_i = \bar{b}_{i+1} = \bar{a}_{i+2}$ *(see Figure 9.2), we have*

$$\mu_i(y(x,t)) = \sum_{j=1}^{q} \beta_j(x)\mu_i(z_j(t)) + \alpha_i(\varepsilon_2(x,t)) \tag{9.20}$$

where $\alpha_i(\varepsilon_2(x,t))$ is defined as

$$\alpha_i(\varepsilon_2(x,t)) = \begin{cases} 0, & y(x,t) < \bar{a}_i \\ \dfrac{\varepsilon_2(x,t)}{\bar{b}_i - \bar{a}_i}, & \bar{a}_i \le y(x,t) < \bar{b}_i \\ \dfrac{-\varepsilon_2(x,t)}{\bar{c}_i - \bar{b}_i}, & \bar{b}_i \le y(x,t) < \bar{c}_i \\ 0, & \bar{c}_i \le y(x,t) \end{cases} \tag{9.21}$$

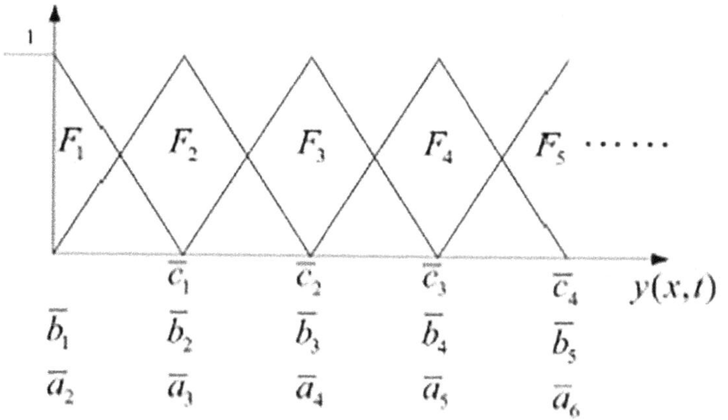

FIGURE 9.2 Fuzzy membership functions $F_i y(x,t)$.

The bound of $\alpha_i(\varepsilon_2(x,t))$ *can be obtained by* $\max\left\{\frac{\sigma_2}{\bar{b}_i-\bar{a}_i},\frac{\sigma_2}{\bar{c}_i-\bar{b}_i}\right\}$.

Proof

Firstly, for $y(x,t))\in[\bar{a}_i,\bar{b}_i]$, the ith fuzzy membership function $F_i y(x,t)$ could be represented as follows:

$$
\begin{aligned}
F_i y(x,t) &= \frac{y(x,t)-\bar{a}_i}{\bar{b}_i-\bar{a}_i} = \frac{(y(x,t)+\varepsilon_2(x,t))-\bar{a}_i}{\bar{b}_i-\bar{a}_i} \\
&= \left(\sum_{j=1}^{q}\beta_j(x)z_j(t)+\varepsilon_2(x,t)\right)\frac{1}{\bar{b}_i-\bar{a}_i}-\frac{\bar{a}_i}{\bar{b}_i-\bar{a}_i} \\
&= \sum_{j=1}^{q}\beta_j(x)\frac{z_j(t)}{\bar{b}_i-\bar{a}_i}+\frac{\varepsilon_2(x,t)}{\bar{b}_i-\bar{a}_i}-\sum_{j=1}^{q}\beta_j(x)\frac{\bar{a}_i}{\bar{b}_i-\bar{a}_i} \\
&= \sum_{j=1}^{q}\beta_j(x)\frac{z_j(t)-\bar{a}_i}{\bar{b}_i-\bar{a}_i}+\frac{\varepsilon_2(x,t)}{\bar{b}_i-\bar{a}_i} \\
&= \sum_{j=1}^{q}\beta_j(x)F_i(z_j(t))+\frac{\varepsilon_2(x,t)}{\bar{b}_i-\bar{a}_i}
\end{aligned} \tag{9.22}
$$

For $y(x,t))\in[\bar{b}_i,\bar{c}_i]$, a similar result could be obtained. Secondly, since the parameters $\bar{a}_i,\bar{b}_i$ and $\bar{c}_i$ are provided as $\bar{c}_i=\bar{b}_{i+1}=\bar{a}_{i+2}$, we have $\sum_{k=1}^{M}F_k(y(x,t))=1$ for any $y(x,t)$. Therefore, $\mu_i(y(x,t))$ could be simplified as follows:

$$\mu_i(y(x,t)) = \frac{F_i y(x,t)}{\sum_{k=1}^{M} F_k(y(x,t))} = F_i y(x,t) \tag{9.23}$$

Substituting equation (9.22) into (9.23), we can obtain equation (9.20). Further, the bound of $\alpha_i(\varepsilon_2(x,t))$ can be calculated by the bound σ_2 in equation (9.18) and the bound of the $\alpha_i(*)$ in equation (9.21), i.e. $\max\left\{\frac{\sigma_2}{\bar{b}_i - \bar{a}_i}, \frac{\sigma_2}{\bar{c}_i - \bar{b}_i}\right\}$.

By Theorem 9.3.3, an equivalent fuzzy PDS could be formulated as follows:

$$\frac{\partial y(x,t)}{\partial t} = \sum_{i=1}^{M}\sum_{j=1}^{q} \beta_j(x)\mu_i(z_j(t))\left[\kappa_i \frac{\partial^2 y(x,t)}{\partial x^2} + \alpha_i y(x,t)\right] + \varepsilon_3(y(x,t)) \tag{9.24}$$

where $\varepsilon_3(y(x,t))$ is defined as follows:

$$\varepsilon_3(y(x,t)) \triangleq \sum_{i=1}^{M} \alpha_i \varepsilon_2(x,t)\left[\kappa_i \frac{\partial^2 y(x,t)}{\partial x^2} + \alpha_i y(x,t)\right] + \varepsilon_1(y(x,t)) \tag{9.25}$$

From the $\varepsilon_3(y(x,t))$ defined in equation (9.25), we can estimate it's bound as follows:

$$\|\varepsilon_3(y(x,t))\|^2_{L_2([0,L];\mathbb{R})} = \int_0^L |\varepsilon_3(y(x,t))|^2 dx \le \int_0^L 3\left|\sum_{i=1}^{M} \alpha_i \varepsilon_2(x,t)\left[\kappa_i \frac{\partial^2 y(x,t)}{\partial x^2}\right]\right|^2 dx$$

$$+\int_0^L 3\left|\sum_{i=1}^{M} \alpha_i \varepsilon_2(x,t)[\alpha_i y(x,t)]\right|^2 dx + \int_0^L 3|\varepsilon_1(y(x,t))|^2 dx$$

$$\le \int_0^L 3\left|\sum_{i=1}^{M} \alpha_i(\varepsilon_2(x,t))\kappa_i\right|^2 \left|\left[\frac{\partial^2 y(x,t)}{\partial x^2}\right]\right|^2 dx$$

$$+\int_0^L 3\left|\sum_{i=1}^{M} \alpha_i(\varepsilon_2(x,t))\alpha_i\right|^2 |y(x,t)|^2 dx + \int_0^L 3|\varepsilon_1(y(x,t))|^2 dx$$

$$\le \int_0^L \left(\sum_{i=1}^{M} |3\alpha_i(\sigma_2)\kappa_i|^2 + 3\sigma_{1,1}^2\right)\left|\frac{\partial^2 y(x,t)}{\partial x^2}\right|^2 dx$$

$$+\int_0^L \left(\sum_{i=1}^{M} |3\alpha_i(\sigma_2)\alpha_i|^2 + 3\sigma_{1,2}^2\right)|y(x,t)|^2 dx$$

$$\le \sigma_{3,1}^2 \left\|\frac{\partial^2 y(x,t)}{\partial x^2}\right\|^2_{L_2([0,L];\mathbb{R})} + \sigma_{3,2}^2 \|y(x,t)\|^2_{L_2([0,L];\mathbb{R})} \tag{9.26}$$

where $\sigma_{3,1}^2 = \sum_{i=1}^{M} |3\alpha_i(\sigma_2)\kappa_i|^2 + 3\sigma_{1,1}^2$ and $\sigma_{3,2}^2 = \sum_{i=1}^{M} |3\alpha_i(\sigma_2)\alpha_i|^2 + 3\sigma_{1,2}^2$.

Remark 9.3.3

The bounds $\sigma_{1,1}$ and $\sigma_{1,2}$ of approximation error in equations (9.11) and (9.12) could be obtained by calculating the maximum $\sigma_{1,1}$ and $\sigma_{1,2}$ satisfying the inequality equation (9.11) for all $y(x,t) \in U$ and $\frac{\partial^2 y(x,t)}{\partial x^2} \in U$. The bound σ_2 depends on the measurement output $z(t)$ and the membership functions $\beta(x)$. Thus, σ_2 could be obtained via the error between $y(x,t)$ in equation (9.6) and $\tilde{y}(x,t)$ in equation (9.16). Finally, $\sigma_{3,1}$ and $\sigma_{3,2}$ can be obtained via equation (9.26).

For the convenience of representation, let us define the spatial differential operator A_i, for $i = 1, \ldots, M$, in $L_2([0,L];\mathbb{R})$ as in the following [277].

Definition 9.3.1

$A_i y(x,t) \triangleq \kappa_i \frac{\partial^2}{\partial x^2} y(x,t)$ where $y(x,t) \in D(A) = \{\, y(x,t) \in L_2([0,L];\mathbb{R}) \,|\, y(x,t)$, $\frac{\partial y(x,t)}{\partial x}$ are absolutely continuous, and $\frac{\partial^2 y(x,t)}{\partial x^2} \in L_2([0,L];\mathbb{R})\}$

Finally, the NPDS with external disturbances and measurement noises can be represented by

$$\frac{\partial y(x,t)}{\partial t} = \sum_{i=1}^{M}\sum_{j=1}^{q} \beta_j(x)\mu_i(z_j(t))[A_i y(x,t) + \alpha_i y(x,t)$$
$$+ g(x)u(t) + g_d(x)d(t)] + \varepsilon_3(y(x,t)) \tag{9.27}$$

$$z(t) = h(x)y(x,t) + D_n n(t) \tag{9.28}$$

Note that the terms $g(x)u(t)$ and $g_d(x)d(t)$ are independent with the nonlinear functions. Therefore, we can directly add them into the fuzzy PDS in equation (9.24).

9.3.2 Infinite-Dimensional T-S Fuzzy State-Space Model of NPDS

In this subsection, based on the Galerkin's method [268,269,292], we could transform the fuzzy partial differential model in equation (9.27) into an infinite-dimensional T-S fuzzy state space model. Since the transformation is very complicated, for the convenience, a local linear system of fuzzy partial differential in equation (9.27) is taken out for illustration of the transformation process. Consider the ith linear PDS of fuzzy PDS in equation (9.27) as

$$\frac{\partial y(x,t)}{\partial t} = A_i y(x,t) + \alpha_i y(x,t) + g(x)u(t) + g_d(x)d(t) \tag{9.29}$$

In the following, the linear partial differential model in equation (9.29) will be transformed to an equivalent infinite-dimensional state-space model.

Since A_i is a linear, symmetric, and compact operator acting on an infinite-dimensional Hilbert space $L_2([0,L];\mathbb{R})$, according to Galerkin solution in Section 1.2.1, the eigenvalue problem is defined as $A_i\phi_l(x) = \lambda_{i,l}\phi_l(x)$, for $l = 1,..., \infty$. All eigenvalues $\lambda_{i,l}$ of A_i are ordered so that $\text{Re}\{\lambda_{i,l+1}\} \le \text{Re}\{\lambda_{i,l}\}$ where $\text{Re}\{\lambda_{i,l}\}$ denotes the real part of $\lambda_{i,l}$ [268–270] and the corresponding eigenfunctions $\phi_l(x)$ are orthonormal eigenfunctions (i.e., $\langle \phi_k(x), \phi_l(x) \rangle = \delta_{i,kl}$) in $D(A_i)$, for example, $\phi_l(x) = \sin(l\pi x)$ [277]. Furthermore, the eigenfunctions $\phi_l(x)$ can be chosen so that they form an orthonormal basis for $D(A_i)$, and the action of A_i on any $y(x,t) \in L_2([0,L];\mathbb{R})$ is given by $A_i y(x,t) = \sum_{l=1}^{\infty} \lambda_{i,l} \langle y(x,t), \phi_l(x) \rangle \phi_l(x)$ [278]. For example, if the eigenfunction $\phi_l(x)$ is chosen as $\phi_l(x) = \sin(l\pi x)$, then the corresponding eigenvalue is $\lambda_{i,l} = -\kappa_i (l\pi)^2$ [277].

By the Hilbert-Schmidt theorem [278], the eigenfunctions $\phi_l(x)$ constitute a set of bases for $L_2([0,L];\mathbb{R})$. Therefore, the solution of equation (9.29) can be represented as $y(x,t) = \sum_{l=1}^{\infty} y_l(t), \phi_l(x)$ [278] where $y_l(t) = \langle y(x,t), \phi_l(x) \rangle$. From the represented solution above, we have the following properties of inner product.

$$\left\langle \frac{dy(x,t)}{dt}, \phi_l(x) \right\rangle = \frac{dy_l(t)}{dt} \text{ and } \langle A_i y(x,t), \phi_l(x) \rangle = \lambda_{i,l} y_l(t) \tag{9.30}$$

And the influence functions $g(x)$ and $g_d(x)$ have the properties as $\langle g_j(x), \phi_l(x) \rangle = \langle \delta(x - p_j), \phi_l(x) \rangle = \phi_l(p_j)$ and $\langle g_{dj}(x), \phi_l(x) \rangle = \langle \delta(x - p_{dj}), \phi_l(x) \rangle = \phi_l(p_{dj})$, respectively. So we have a set of N ODEs with $N \to \infty$, in which the lth component $y_l(t)$ is given by

$$\frac{dy_l(t)}{dt} = \lambda_{i,l} y_l(t) + a_i y_l(t) + \sum_{j=1}^{p} \phi_l(p_j) u_j(t) + \sum_{j=1}^{p_d} \phi_l(p_{dj}) d_j(t) \tag{9.31}$$

By the lth component as the ordinary differential system in (9.31), the ith linear PDS equation (9.29) can be transformed to a linear infinite-dimensional state space model as follows:

$$\dot{y}(t) = A_i y(t) + Bu(t) + B_d d(t), \quad i = 1,..., M \tag{9.32}$$

where $y(t) = [y_1(t),..., y_\infty(t)]^T \in \mathbb{R}^\infty$, $u(t) = [u_1(t),..., u_p(t)]^T$, A_i is an $\infty \times \infty$ diagonal matrix, and B is an $\infty \times p$ matrix with entries $b_{ij} = \phi_i(p_j)$, i.e.,

$$A_i = \begin{bmatrix} \lambda_{i,1} + a_i & \cdots & 0 \\ \vdots & \ddots & \vdots \\ 0 & \cdots & \lambda_{i,\infty} + a_i \end{bmatrix},$$

$$B = \begin{bmatrix} \phi_1(p_1) & \cdots & \phi_1(p_p) \\ \vdots & \phi_l(p_j) & \vdots \\ \phi_\infty(p_1) & \cdots & \phi_\infty(p_p) \end{bmatrix}$$

The matrix B_d is similar to B. Finally, the measurement output $z(t)$ in equation (9.3) can be represented as follows:

$$z(t) = Cy(t) + D_n n(t) \tag{9.33}$$

where C is an $q \times \infty$ matrix with entries $c_{ji} = \phi_i(q_j)$, i.e.,

$$C = \begin{bmatrix} \phi_1(q_1) & \cdots & \phi_\infty(q_1) \\ \vdots & \phi_l(q_j) & \vdots \\ \phi_1(q_1) & \cdots & \phi_\infty(q_q) \end{bmatrix}$$

and D_n is defined in equation (9.3).

According to the state space representation in equation (9.32) for the ith fuzzy partial differential model in equation (9.29), the state-space model of the fuzzy PDE in (equation 9.27) will be derived in the following. The difference between the fuzzy interpolation system in equation (9.27) and linear PDS in equation (9.29) is only at the fuzzy interpolation functions and the approximation error term. Thus, by the properties of inner product, an inner product for nonlinear term could be separated into two terms as follows:

$$\left\langle \kappa(y(x,t)\frac{\partial^2 y(x,t)}{\partial x^2} + f(y(x,t)), \phi_i(x) \right\rangle$$

$$= \left\langle \sum_{i=1}^{M} \sum_{j=1}^{q} \beta_j(x)\mu_i(z_j(t))[A_i y(x,t) + a_i y(x,t)], \phi_l(x) \right\rangle + \langle \varepsilon_3(y(x,t)), \phi_l(x) \rangle \tag{9.34}$$

The fuzzy approximation term in equation (9.34) could be derived as follows:

$$\left\langle \sum_{i=1}^{M}\sum_{j=1}^{q}\beta_j(x)\mu_i(z_j(t))[A_i y(x,t)+a_i y(x,t)],\phi_l(x)\right\rangle$$

$$=\left\langle \sum_{i=1}^{M}\sum_{j=1}^{q}\beta_j(x)\mu_i(z_j(t))\left[\sum_{k=1}^{\infty}\lambda_{i,k}y_k(t)\phi_k(x)+a_i\sum_{k=1}^{\infty}y_k(t)\phi_k(x)\right],\phi_l(x)\right\rangle$$

$$=\int_0^L\sum_{i=1}^{M}\sum_{j=1}^{q}\beta_j(x)\mu_i(z_j(t))\sum_{k=1}^{\infty}\lambda_{i,k}y_k(t)\phi_k(x)\phi_l(x)dx$$

$$+\sum_{i=1}^{M}\sum_{j=1}^{q}\beta_j(x)\mu_i(z_j(t))a_i\sum_{k=1}^{\infty}y_k(t)\phi_k(x)\phi_l(x)dx$$

$$=\sum_{k=1}^{\infty}\sum_{i=1}^{M}\sum_{j=1}^{q}\mu_i(z_j(t))[\lambda_{i,k}+a_i]y_k(t)\times\int_0^L\beta_j(x)\phi_k(x)\phi_l(x)dx$$

$$=\sum_{k=1}^{\infty}\sum_{i=1}^{M}\sum_{j=1}^{q}\mu_i(z_j(t))[\lambda_{i,k}+a_i]\tilde{a}_{jkl}y_k(t) \tag{9.35}$$

where $\tilde{a}_{jkl}\triangleq\int_0^L\beta_j(x)\phi_k(x)\phi_l(x)dx$. The approximation error could be decomposed as follows:

$$\varepsilon_{3l}(y(t))\triangleq\langle\varepsilon_3(y(x,t)),\phi_l(x)\rangle=\int_0^L\varepsilon_3(y(x,t))\phi_l(x)dx \tag{9.36}$$

where $y(t)$ is defined in equation (9.32). By the above analysis, the lth component $y_l(t)$ of the fuzzy PDE in equation (9.27) can be decomposed by a set of ODE as follows:

$$\frac{dy_l(t)}{dt}=\sum_{k=1}^{\infty}\sum_{i=1}^{M}\sum_{j=1}^{q}\mu_i(z_j(t))[\lambda_{i,k}+a_i]\tilde{a}_{jkl}y_k(t)+\sum_{j=1}^{p}\phi_l(p_j)u_j(t)$$

$$+\sum_{j=1}^{p_d}\phi_l(p_{dj})d_j(t)+\varepsilon_{3l}(y(t)) \tag{9.37}$$

for $l=1,2,\ldots,\infty$. From the fuzzy approximation term and approximation error term, we can know that the lth equation is dependent on the other component equations. Finally, an infinite-dimensional fuzzy state-space model for fuzzy PDS in equation (9.27) could be constructed as follows:

$$\dot{y}(t) = \sum_{i=1}^{M}\sum_{j=1}^{q}\mu_i(z_j(t))[A_{ij}y(t) + Bu(t) + B_d d(t)] + \varepsilon_3(y(t)) \tag{9.38}$$

where

$$A_{ij} = \begin{bmatrix} \tilde{\lambda}_{i,1}\tilde{a}_{j11} & \cdots & \tilde{\lambda}_{i,k}\tilde{a}_{jk1} & \cdots & \tilde{\lambda}_{i,\infty}\tilde{a}_{j\infty 1} \\ \vdots & \ddots & \vdots & \ddots & \vdots \\ \tilde{\lambda}_{i,1}\tilde{a}_{j1l} & \cdots & \tilde{\lambda}_{i,k}\tilde{a}_{jkl} & \cdots & \tilde{\lambda}_{i,\infty}\tilde{a}_{j\infty l} \\ \vdots & \ddots & \vdots & \ddots & \vdots \\ \tilde{\lambda}_{i,1}\tilde{a}_{j1\infty} & \cdots & \tilde{\lambda}_{i,k}\tilde{a}_{jk\infty} & \cdots & \tilde{\lambda}_{i,\infty}\tilde{a}_{j\infty\infty} \end{bmatrix}$$

in which $\tilde{\lambda}_{i,k} = (\lambda_{i,k} + a_i), \tilde{a}_{jkl}$ is defined in equation (9.35), and the vector $\varepsilon_3(y(t)) = [\varepsilon_{3l}(y(t)),\ldots,\varepsilon_{3\infty}(y(t))]^T$.

Remark 9.3.4

For linear infinite-dimensional state space model in equation (9.32), the system matrix A_i is a diagonal matrix. However, the system matrices A_{ij} in the infinite-dimensional fuzzy state-space model equation (9.38) are of coupling matrix due to fuzzy interpolations. The results evidence the difference between linear and NPDS.

Remark 9.3.5

In this chapter, the symbols should be differentiated clearly between $y(x,t)$ and $y(t)$. The symbol $y(x,t)$ represents the state function of NPDSs in equations (9.1) and (9.3) and the symbol $y(t)$ is used to represent the state vector of the infinite-dimensional system in equation (9.38). Similarly, the symbols $\varepsilon_3 y(x,t)$ and $\varepsilon_3 y(t)$ have different meanings.

Currently, the bound of approximate error could be proved to be less than a constant. The following lemma is necessary to prove that the approximate error $\varepsilon_3 y(x,t)$ has a same bound with infinite-dimensional-expression approximation error $\varepsilon_3 y(t)$.

Lemma 9.3.1 [304,305]

Suppose $\phi_l(x)$ is an orthonormal basis in a complete inner-product space (i.e., a Hilbert space $L_2([0,L];\mathbb{R})$) and $\frac{\partial^2 y(x,t)}{\partial x^2}, y(x,t) \in L_2([0,L];\mathbb{R})$. The lth component $y_l(t)$ is defined as $y_l(t) = \langle y(x,t), \phi_l(x) \rangle$. Then, we have

$$\|y(x,t)\|_{L_2([0,L];\mathbb{R})} = \langle y(x,t), y(x,t)\rangle = \sum_{l=1}^{\infty} |\langle y(x,t), \phi_l(x)\rangle|^2 = \sum_{l=1}^{\infty} |y_l(t)|^2 = \|y(t)\|$$

and

$$\left\|\frac{\partial^2 y(x,t)}{\partial x^2}\right\|_{L_2([0,L];\mathbb{R})} = \left\langle \frac{\partial^2 y(x,t)}{\partial x^2}, \frac{\partial^2 y(x,t)}{\partial x^2}\right\rangle$$

$$= \sum_{l=1}^{\infty} \lambda_l^2 |\langle y(x,t), \phi_l(x)\rangle|^2 = \sum_{l=1}^{\infty} \lambda_l^2 |y_l(t)|^2 = \|\Lambda y(t)\| \tag{9.39}$$

where $y(t) = [y_1(t), \dots, y_\infty(t)]^T$ and $\Lambda = diag(\lambda_1, \dots, \lambda_\infty)$.

By the above lemma, the bound for approximate error $\varepsilon_3 y(t)$ in infinite-dimensional expression could be obtained by the following theorem.

Theorem 9.3.4 (Error Estimation)

If the bound of $\varepsilon_3 y(x,t)$ is provided as follows:

$$\|\varepsilon_3 y(x,t)\|^2_{L_2([0,L];\mathbb{R})} \le \sigma_{3,1}^2 \left\|\frac{\partial^2 y(x,t)}{\partial x^2}\right\|^2_{L_2([0,L];\mathbb{R})} + \sigma_{3,2}^2 \|y(x,t)\|^2_{L_2([0,L];\mathbb{R})} \tag{9.40}$$

then the bound of $\varepsilon_3 y(t)$ in infinite-dimensional expression could also be estimated as follows:

$$\|\varepsilon_3 y(t)\|^2 \le \sigma_{3,1}^2 \|\Lambda y(t)\|^2 + \sigma_{3,2}^2 \|y(t)\|^2 \tag{9.41}$$

where the vector $\varepsilon_3 y(t) = [\varepsilon_{31}(y(t)), \dots, \varepsilon_{3\infty}(y(t))]^T$, with elements defined in equation (9.36).

Proof

By Lemma 9.3.1, we can prove

$$\|\varepsilon_3 y(t)\|^2 = \sum_{l=1}^{\infty} |\varepsilon_{3l} y(t)|^2 = \|\varepsilon_3 y(x,t)\|^2_{L_2([0,L];\mathbb{R})}$$

$$\le \sigma_{3,1}^2 \left\|\frac{\partial^2 y(x,t)}{\partial x^2}\right\|^2_{L_2([0,L];\mathbb{R})} + \sigma_{3,2}^2 \|y(x,t)\|^2_{L_2([0,L];\mathbb{R})} = \sigma_{3,1}^2 \|\Lambda y(t)\|^2 + \sigma_{3,2}^2 \|y(t)\|^2$$

In theory, for the PDE, the infinite modes have to be controlled or measured simultaneously, i.e., we need to control or measure all frequencies. Since the actuators cannot be excited to the highest frequency modes [268,269], only sufficiently large but finite N modes will be actively controlled. Therefore, for practical design, based on the eigenspectrum of spatial differential operators, the state space model of NPDS should be partitioned into a finite-dimensional slow subsystem to be controlled and an infinite-dimensional fast residual subsystem to be tolerated. Suppose the controlled modes and the residual modes are defined by $y_c(x,t)$ and $y_r(x,t)$, respectively. Let H_c and H_r be the controlled and residual modal subspaces of $L_2([0,L];\mathbb{R})$, respectively, as $L_2([0,L];\mathbb{R}) = H_c \oplus H_r$ [269,278]. The controlled mode and residual mode are defined as $y_c(x,t) \equiv P_c y(x,t)$ and $y_r(x,t) \equiv P_r y(x,t)$, where P_c and P_r are projected operators from H onto H_c and H_r, respectively. Therefore, we have $y(x,t) = y_c(x,t) + y_r(x,t)$. The controlled part $y_c(x,t)$ and the residual part $y_r(x,t)$ are given as $y_c(x,t) = \sum_{l=1}^{N} y_l(t)\phi_l(x)$ and $y_r(x,t) = \sum_{l=N+1}^{\infty} y_l(t)\phi_l(x)$, respectively.

In order to obtain a state-space representation of the system in equation (9.38) for the controlled mode and the residual mode, we define the controlled mode state vector $y_c(t)$ and the residual mode state vector $y_r(t)$ as follows:

$$\begin{aligned} y_c(t) &= [y_1(t), y_2(t), \ldots, y_N(t)]^T \\ y_r(t) &= [y_{N+1}(t), y_{N+2}(t), \ldots, y_\infty(t)]^T \end{aligned} \tag{9.42}$$

Thus, the infinite-dimensional fuzzy system in equation (9.38) for the controlled mode and the residual mode can be separated as follows:

$$\dot{y}_c(t) = \sum_{i=1}^{M}\sum_{j=1}^{q} \mu_i(z_j(t))[A_{c,ij}y_c(t) + A_{cr,ij}y_r(t) + B_c u(t) + B_{cd}d(t)] + \varepsilon_c(y_c(t), y_r(t)) \tag{9.43}$$

$$\dot{y}_r(t) = \sum_{i=1}^{M}\sum_{j=1}^{q} \mu_i(z_j(t))[A_{rc,ij}y_c(t) + A_{r,ij}y_r(t) + B_r u(t) + B_{rd}d(t)] + \varepsilon_r(y_c(t), y_r(t)) \tag{9.44}$$

where the matrices $A_{c,ij}$, $A_{cr,ij}$, $A_{rc,ij}$, $A_{r,ij}$, B_c, B_r, $\varepsilon_c(y_c(t), y_r(t))$, and $\varepsilon_r(y_c(t), y_r(t))$ have the relative relationships with A_{ij}, B and $\varepsilon_3(y(t))$ as follows:

$$A_{ij} = \begin{bmatrix} A_{c,ij} & A_{cr,ij} \\ A_{rc,ij} & A_{r,ij} \end{bmatrix}, B = \begin{bmatrix} B_c \\ B_r \end{bmatrix}, \varepsilon_3(y(t)) = \begin{bmatrix} \varepsilon_c(y_c(t), y_r(t)) \\ \varepsilon_r(y_c(t), y_r(t)) \end{bmatrix}$$

The matrices B_{cd} and B_{rd} are similar to B_c and B_r, respectively. The output equation in equation (9.33) can be represented as

$$z(t) = z_c(t) + z_r(t) = C_c y_c(t) + C_r y_r(t) + D_n n(t) \tag{9.45}$$

where C_c and C_r are related to C as $C = [C_c \;\; C_r]$.

For the convenience of control design, the infinite-dimensional fuzzy state-space system in equation (9.38) is separated into a finite-dimensional controlled subsystem $\mathfrak{C}$ and a coupled infinite-dimensional residual subsystem $\mathfrak{R}$ as follows:

$$\mathfrak{C}: \begin{cases} \dot{y}_c(t) = \sum_{i=1}^{M}\sum_{j=1}^{q} \mu_i(z_j(t))[A_{c,ij} y_c(t) + w_2(t) + B_c u(t) + B_{cd} d(t)] + \varepsilon_c(y_c(t), y_r(t)) \\ w_1(t) = y_c(t) \\ z_c(t) = C_c y_c(t) + D_n n(t) \end{cases} \tag{9.46}$$

$$\mathfrak{R}: \begin{cases} \dot{y}_r(t) = \sum_{i=1}^{M}\sum_{j=1}^{q} \mu_i(z_j(t))[A_{r,ij} y_r(t) + A_{rc,ij} w_1(t) + B_r u(t) + B_{rd} d(t)] \\ \quad + \varepsilon_r(y_c(t), y_r(t)) \\ w_2(t) = \sum_{i=1}^{M}\sum_{j=1}^{q} \mu_i(z_j(t)) A_{cr,ij} y_r(t) \\ z_r(t) = C_r y_r(t) \end{cases} \tag{9.47}$$

where $w_1(t)$ is an output of the controlled subsystem $\mathfrak{C}$ which can affect the residual subsystem $\mathfrak{R}$; $w_2(t)$ is an output of the residual subsystem $\mathfrak{R}$ which can affect the controlled subsystem $\mathfrak{C}$; $u(t)$ is a control input; $d(t)$ is an external disturbance. The block diagram of two coupled subsystems is shown in Figure 9.3a. The sensor outputs $z_r(t)$ are contaminated by the residual modes through the term $C_r y_r(t)$, which is called the observation spillover, and the feedback control excites the residual modes through the term $B_r u(t)$, which is called the control spillover [268,269]. In the fuzzy model of distributed systems in [290,291], the control and observation spillovers are not discussed. Because the controlled subsystem in equation (9.46) and the residual subsystem in equation (9.47) are coupled together through control spillover and observation spillover, the robust fuzzy control design needs more effort for NPDS than for the conventional nonlinear systems without spillover couplings.

In the rest of this section, a relationship between the approximation error $\varepsilon_3(y(t))$ and its components $\varepsilon_c(y_c(t), y_r(t))$ and $\varepsilon_r(y_c(t), y_r(t))$ will be discussed.

Therefore, the following corollary is an extension of Theorem 9.3.4.

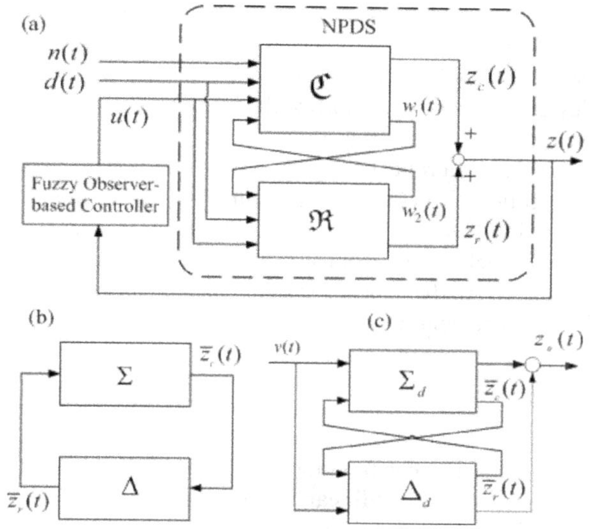

FIGURE 9.3 (a) Block diagram of the controlled subsystem $\mathfrak{C}$ in equation (9.46) and coupled residual subsystem $\mathfrak{R}$ in equation (9.47). (b) Block diagram of the two coupled subsystems without the external disturbances and the measurement noises. (c) Block diagram of the two coupled subsystems with the external disturbances and the measurement noises $\left(v(t)=[d(t),n(t)]^T\right)$.

Corollary 9.3.1

Suppose the approximation errors $\varepsilon_c(y_c(t),y_r(t))$ and $\varepsilon_r(y_c(t),y_r(t))$ are bounded as follows:

$$\|\varepsilon_c(y_c(t),y_r(t))\|^2 \le \sigma_{3,1}^2\|\Lambda_c y_c(t)\|^2+\sigma_{3,2}^2\|y_c(t)\|^2 \tag{9.48}$$

and

$$\|\varepsilon_r(y_c(t),y_r(t))\|^2 \le \sigma_{3,1}^2\|\Lambda_r y_r(t)\|^2+\sigma_{3,2}^2\|y_r(t)\|^2 \tag{9.49}$$

respectively, then we $\|\varepsilon_3(y(t))\|^2 \le \sigma_{3,1}^2\|\Lambda y(t)\|^2+\sigma_{3,2}^2\|y(t)\|^2$ where $\Lambda = diag(\Lambda_c,\Lambda_r)$.

Proof

The proof is trivial. We can use the following inequality to show.

$$\begin{aligned}\|\varepsilon_3(y(t))\|^2 &= \|\varepsilon_c(y_c(t),y_r(t))\|^2+\|\varepsilon_r(y_c(t),y_r(t))\|^2 \le \sigma_{3,1}^2\|\Lambda_c y_c(t)\|^2\\ &+\sigma_{3,1}^2\|\Lambda_r y_r(t)\|^2+\sigma_{3,2}^2\|y_c(t)\|^2+\sigma_{3,2}^2\|y_r(t)\|^2 \le \sigma_{3,1}^2\|\Lambda y(t)\|^2+\sigma_{3,2}^2\|y(t)\|^2\end{aligned} \tag{9.50}$$

In this section, a coupled T-S fuzzy controlled and residual state space model in equations (9.46) and (9.47) could successfully represent a NPDS in equation (9.3). Based on the fuzzy T-S controlled subsystem, a finite-dimensional fuzzy controller is designed to robustly stabilize the NPDS to tolerate the coupled residual subsystem and to attenuate the effect of the external disturbances and the measurement noises, i.e., a robust finite-dimensional fuzzy controller is designed based on the controlled subsystem in equation (9.46) and treats the residual subsystem $\mathfrak{R}$ in equation (9.47) as the unmodeled part or system uncertainty to be tolerated. In the sequel, an H_∞ robust fuzzy observer-based control design is employed both to tolerate the residual subsystem and to filter the external disturbances and the measurement noises, simultaneously.

Remark 9.3.6

In [270,279,280], a quasi-linear (nonlinear reaction-diffusion) equation in equation (9.5) is represented by a nonlinear infinite-dimensional ODE via Galerkin's method in Section 1.2.1. However, this nonlinear infinite-dimensional ODE is difficult to extend to a complex NPDS with nonlinear functions $\kappa(y(x, t))$ and $f(y(x, t))$ in equations (9.1) and (9.3). In [290,291], a T-S fuzzy model is proposed to approximate this nonlinear infinite-dimensional ODE by neglecting fast modes, i.e., the effects of the control and observation spillovers, which will affect the stability of NPDS, are neglected in control design procedure. In this chapter, we will propose a robust H_∞ control design method for NPDSs to remedy these shortages and to improve the performance of stabilization and disturbance attenuation.

9.4 ROBUST FUZZY OBSERVER-BASED CONTROL DESIGN OF NONLINEAR PARTIAL DIFFERENTIAL SYSTEMS

In this section, a robust fuzzy observer-based controller will be proposed to stabilize the NPDS and to guarantee the H_∞ attenuation performance of NPDS with a desired attenuation of the external disturbances and the measurement noises.

9.4.1 Robust Stabilization of Nonlinear Partial Differential Systems

Since the state information for the infinite-dimensional residual subsystem $\mathfrak{R}$ in equation (9.47) is unavailable, $\mathfrak{R}$ is considered as an uncertain part to be tolerated in the control design procedure. A fuzzy observer-based output feedback controller based on the controlled subsystem in equation (9.46) will be proposed to estimate the state of controlled mode and then to robustly stabilize the NPDS in equation (9.1) (see Figure 9.3a). The fuzzy observer-based controller has the form

$$\dot{\hat{y}}_c(t) = \sum_{i=1}^{M}\sum_{j=1}^{q} \mu_i(z_j(t))[A_{c,ij}\hat{y}_c(t) + B_c u(t) + G_{ij}(z(t) - C_c\hat{y}_c(t))] \tag{9.51}$$

$$u(t) = \sum_{i=1}^{M}\sum_{j=1}^{q}\mu_i(z_j(t))K_{ij}\hat{y}_c(t) \tag{9.52}$$

where G_{ij} and K_{ij} are the observer gains and the control gains, respectively.

The dynamic of the estimate error $e_c(t) = y_c(t) - \hat{y}_c(t)$ is given by

$$\begin{aligned}\dot{e}_c = &\sum_{i=1}^{M}\sum_{j=1}^{q}\mu_i(z_j(t))[(A_{c,ij} - G_{ij}C_c)e_c(t) + (A_{cr,ij} - G_{ij}C_r)y_r(t) \\ &+ B_{cd}d(t) - G_{ij}D_n n(t)] + \varepsilon_c(y_c(t), y_r(t))\end{aligned} \tag{9.53}$$

Combining the fuzzy system in equations (9.46) and (9.47) with the observer-based controller, the closed-loop fuzzy control system could be described as follows:

$$\begin{aligned}\begin{bmatrix}\dot{y}_c(t)\\ \dot{e}_c(t)\\ \hline \dot{y}_r(t)\end{bmatrix} = &\sum_{i=1}^{M}\sum_{j=1}^{q}\mu_i(z_j(t))\times\begin{bmatrix}A_{c,ij} - B_cK_{ij} & B_cK_{ij} & A_{cr,ij}\\ 0 & A_{c,ij} - G_{ij}C_c & A_{cr,ij} - G_{ij}C_r\\ A_{rc,ij} - B_rK_{ij} & B_rK_{ij} & A_{r,ij}\end{bmatrix}\\ &\times\begin{bmatrix}y_c(t)\\ e_c(t)\\ \hline y_r(t)\end{bmatrix} + \begin{bmatrix}B_{cd} & 0\\ B_{cd} & -G_{ij}D_n\\ \hline B_{rd} & 0\end{bmatrix}\begin{bmatrix}d(t)\\ n(t)\end{bmatrix} + \begin{bmatrix}\varepsilon_c(t)\\ \varepsilon_c(t)\\ \hline \varepsilon_r(t)\end{bmatrix}\end{aligned} \tag{9.53}$$

where $\varepsilon_c(t) \triangleq \varepsilon_c(y_c(t), y_r(t))$ and $\varepsilon_r(t) \triangleq \varepsilon_r(y_c(t), y_r(t))$.

In order to simplify the notation, the closed-loop system can be separated into a controlled part and a coupled residual part as follows:

$$\dot{\bar{y}}_c(t) = \sum_{i=1}^{M}\sum_{j=1}^{q}\mu_i(z_j(t))[A_{c,ij}\bar{y}_c(t) + \bar{B}_{c,ij}\bar{C}_{r,ij}y_r(t) + \bar{B}_{cv,ij}v(t)] + \bar{\varepsilon}_c(t) \tag{9.55}$$

$$\dot{y}_r(t) = \sum_{i=1}^{M}\sum_{j=1}^{q}\mu_i(z_j(t))[\bar{A}_{r,ij}y_r(t) + \bar{B}_{r,ij}\bar{C}_{c,ij}\bar{y}_c(t) + \bar{B}_{rv}v(t)] + \varepsilon_r(t) \tag{9.56}$$

where $\bar{y}_c = [y_c(t), e_c(t)]^T$, $v(t) = [d(t), n(t)]^T$, and the parameter matrices are defined as follows:

$$\bar{A}_{c,ij} = \begin{bmatrix} A_{c,ij} - B_c K_{ij} & B_c K_{ij} \\ 0 & A_{c,ij} - G_{ij} C_c \end{bmatrix}, \bar{B}_{c,ij} = \begin{bmatrix} I & 0 \\ I & -G_{ij} \end{bmatrix},$$

$$\bar{A}_{r,ij} = A_{r,ij}, \bar{C}_{c,ij} \begin{bmatrix} I & 0 \\ -K_{ij} & K_{ij} \end{bmatrix},$$

$$\bar{B}_{cv,ij} = \begin{bmatrix} B_{cd} & 0 \\ B_{cd} & -G_{ij} D_n \end{bmatrix}, \bar{\varepsilon}_c(t) = \begin{bmatrix} \varepsilon_c(t) \\ \varepsilon_c(t) \end{bmatrix}, \bar{B}_{r,ij} = \begin{bmatrix} A_{rc,ij} & B_r \end{bmatrix},$$

$$\bar{C}_{r,ij} = \begin{bmatrix} A_{cr,ij} \\ C_r \end{bmatrix}, \bar{B}_{rv} = \begin{bmatrix} B_{rd} & 0 \end{bmatrix}$$

In this subsection, suppose the external disturbances $d(t)$ and the measurement noises $n(t)$ are neglected in the closed-loop system equations (9.55) and (9.56), i.e., $v(t) \equiv 0$. Then the fuzzy observer-based controller should be designed to guarantee the robust stabilization of the closed-loop system to tolerate the coupled residual modes and fuzzy approximation errors, i.e., the closed-loop system must be asymptotically stable in the case $v(t) \equiv 0$. In this case, the closed-loop system could be reformulated as the following coupled systems with the block diagram shown in the Figure 9.3b.

$$\Sigma : \begin{cases} \dot{\bar{y}}_c(t) = \sum_{i=1}^{M} \sum_{j=1}^{q} \mu_i(z_j(t))[\bar{A}_{c,ij} \bar{y}_c(t) + \bar{B}_{c,ij} \bar{z}_r(t)] + \bar{\varepsilon}_c(t) \\ \bar{z}_c(t) = \sum_{i=1}^{M} \sum_{j=1}^{q} \mu_i(z_j(t)) \bar{C}_{c,ij} \bar{y}_c(t) \end{cases} \tag{9.57}$$

$$\Delta : \begin{cases} \dot{y}_r(t) = \sum_{i=1}^{M} \sum_{j=1}^{q} \mu_i(z_j(t))[\bar{A}_{r,ij} y_r(t) + \bar{B}_{r,ij} \bar{z}_c(t)] + \varepsilon_r(t) \\ \bar{z}_r(t) = \sum_{i=1}^{M} \sum_{j=1}^{q} \mu_i(z_j(t)) \bar{C}_{r,ij} y_r(t) \end{cases} \tag{9.58}$$

For the coupled systems in equations (9.57) and (9.58), the small gain theorem [293] can be used to design the control gain K_{ij} and observer gain G_{ij} such that the coupled system in equations (9.57) and (9.58) is asymptotically stable. First, we denote the storage functions [293] of Σ and Δ by $S_1(\bar{y}_c(t))$ and $S_2(y_r(t))$, respectively, resulting in the dissipation inequalities as follows:

$$S_1(\bar{y}_c(t_f)) \le S_1(\bar{y}_c(0)) + \int_0^{t_f} (\gamma_1^2 \|\bar{z}_r(t)\|^2 - \|\bar{z}_c(t)\|^2) dt \tag{9.59}$$

$$S_2(y_r(t)) \le S_2(\bar{y}_r(0)) + \int_0^{t_f} (\gamma_2^2 \|\bar{z}_c(t)\|^2 - \|\bar{z}_r(t)\|^2) dt \tag{9.60}$$

where $\gamma_1 > 0$ and $\gamma_2 > 0$. Note that the statement of the storage functions can refer to the references [293,306]. Second, we provide some assumptions on the residual subsystem Δ. Since for any choice of model reduction scheme it makes no practical sense if the residual subsystem Δ is unstable, we provide an assumption on the residual subsystem.

Assumption 9.4.1

1. *Without any inputs, the residual subsystem Δ is asymptotically stable, i.e., the real parts of the eigenvalues of $\bar{A}_{r,ij}$ are all negative.*
2. *For the residual subsystem Δ, the dissipation inequality equation (9.60) holds, i.e. γ_2 is finite.*

Remark 9.4.1

Since all eigenvalues $\lambda_{i,\,l}$ of A_i are ordered so that $\text{Re}\{\lambda_{i,\,l+1}\} \le \text{Re}\{\lambda_{i,\,l}\}$, e.g., $\lambda_{i,\,l} = -\kappa_i (l\pi)^2$, we can choose an appropriate dimension N for the controlled subsystem to include all unstable modes. Then, the residual subsystem is a collection of stable modes. Therefore, it is reasonable to assume the residual subsystem Δ without any input is asymptotically stable.

The following lemma will be applied to prove the stability of the coupled system equations (9.57) and (9.58).

Lemma 9.4.1 (Small Gain Theorem [293,294])

For the controlled subsystem Σ and the coupled residual subsystem Δ, suppose the storage functions $S_1(\bar{y}_c(t))$ and $S_2(y_r(t))$ are positive semi-definite functions and the dissipation inequalities in equations (9.59) and (9.60) hold. If $\gamma_1 \cdot \gamma_2 < 1$, then the coupled system in the Figure 9.3b is asymptotically stable.

Theorem 9.4.1

Under Assumption 9.4.1, if there exist a positive definite matrix P, the control gain K_{ij} and the observer gain G_{ij} in equations (9.51) and (9.52) such that

$$\begin{bmatrix} \bar{A}_{c,ij}^T P + P\bar{A}_{c,ij} + \bar{C}_{c,ij}^T \bar{C}_{c,ij} + \frac{1}{\xi} PP + \xi \bar{\sigma}_3 & P\bar{B}_{c,ij} \\ \bar{B}_{c,ij}^T P & -\gamma_1^2 I \end{bmatrix} < 0 \tag{9.61}$$

where $\bar{\sigma}_3 \triangleq diag\left(2\sigma_{3,1}^2 \Lambda_c + 2\sigma_{3,2}^2 I, 0\right)$ *for* $i = 1,\ldots,M$, $j = 1,\ldots,q$ *and* $\gamma_1 \cdot \gamma_2 < 1$, *then the coupled system equations (9.57) and (9.58) will be asymptotically stable, i.e.,* $\bar{y}_c(t) \to 0$ *and* $y_r(t) \to 0$ *as* $t \to \infty$. *Therefore, the fuzzy observer-based controller could guarantee the NPDS in equation (9.1) is also asymptotically stable.*

Proof

Before proving the theorem, the following fact is necessary.

Fact 9.4.1 [307]

$X^T PY + Y^T PX \le \xi X^T PX + \frac{1}{\xi} Y^T PY$ for any positive constant ξ, two vectors X, Y and a symmetric matrix $P = P^T \ge 0$ with appropriate dimensions. Let us choose the storage function for the controlled subsystem in equation (9.57) as $S_1(\bar{y}_c(t)) = \bar{y}_c(t)^T P\bar{y}_c(t)$. By differentiating the storage function, we get

$$\begin{aligned}
\dot{S}_1(\bar{y}_c(t)) &= \dot{\bar{y}}_c(t)^T P\bar{y}_c(t) + \bar{y}_c(t)^T P\dot{\bar{y}}_c(t) \\
&= \left(\sum_{i=1}^{M} \sum_{j=1}^{q} \mu_i(z_j(t))[\bar{A}_{c,ij}\bar{y}_c(t) + \bar{B}_{c,ij}\bar{z}_r(t)] + \bar{\varepsilon}_c(t) \right)^T P\bar{y}_c(t) \\
&\quad + \bar{y}_c(t)^T P \left(\sum_{i=1}^{M} \sum_{j=1}^{q} \mu_i(z_j(t))[\bar{A}_{c,ij}\bar{y}_c(t) + \bar{B}_{c,ij}\bar{z}_r(t)] + \bar{\varepsilon}_c(t) \right)
\end{aligned} \tag{9.62}$$

Adding and subtracting $\gamma_1^2 \|\bar{z}_r(t)\|^2 - \|\bar{z}_c(t)\|^2$ to equation (9.62), we obtain

$$\begin{aligned}
&\dot{S}_1(\bar{y}_c(t)) \\
&= \left(\sum_{i=1}^{M} \sum_{j=1}^{q} \mu_i(z_j(t))[\bar{A}_{c,ij}\bar{y}_c(t) + \bar{B}_{c,ij}\bar{z}_r(t)] + \bar{\varepsilon}_c(t) \right)^T P\bar{y}_c(t) \\
&\quad + \bar{y}_c(t)^T P \left(\sum_{i=1}^{M} \sum_{j=1}^{q} \mu_i(z_j(t))[\bar{A}_{c,ij}\bar{y}_c(t) + \bar{B}_{c,ij}\bar{z}_r(t)] + \bar{\varepsilon}_c(t) \right) \\
&\quad - (\gamma_1^2 \|\bar{z}_r(t)\|^2 - \|\bar{z}_c(t)\|^2) + (\gamma_1^2 \|\bar{z}_r(t)\|^2 - \|\bar{z}_c(t)\|^2)
\end{aligned} \tag{9.63}$$

By Fact 9.4.1, we have

$$\bar{\varepsilon}_c(t)^T P\bar{y}_c(t) + \bar{y}_c(t)^T P\bar{\varepsilon}_c(t) \le \xi\bar{\varepsilon}_c(t)^T\bar{\varepsilon}_c(t) + \frac{1}{\xi}\bar{y}_c(t)^T PP\bar{y}_c(t) \tag{9.64}$$

where ξ is any positive constant. According to the definition of $\bar{\varepsilon}_c(t)$ and the bound of $\varepsilon_c(t)$ in equation (9.48), we have

$$\begin{aligned}\bar{\varepsilon}_c(t)^T\bar{\varepsilon}_c(t) &= 2\varepsilon_c(t)^T\varepsilon_c(t)\\ &\le \begin{bmatrix} y_c(t) \\ e_c(t)\end{bmatrix}^T \begin{bmatrix} 2\sigma_{3,1}^2\Lambda_c + 2\sigma_{3,2}^2 I & 0 \\ 0 & 0\end{bmatrix}\begin{bmatrix} y_c(t) \\ e_c(t)\end{bmatrix}\\ &\le \bar{y}_c(t)^T\bar{\sigma}_3\bar{y}_c(t)\end{aligned} \tag{9.65}$$

where $\bar{\sigma}_3 \triangleq diag(\sigma_3, 0) = diag(2\sigma_{3,1}^2\Lambda_C + 2\sigma_{3,2}^2 I, 0)$. Based on the above two inequalities in equations (9.64) and (9.65), from equation (9.63), we can obtain

$$\begin{aligned}&\dot{S}_1(\bar{y}_c(t))\\ &\le \sum_{i=1}^{M}\sum_{j=1}^{q}\mu_i(z_j(t))\begin{bmatrix}\bar{y}_c(t)\\ \bar{z}_r(t)\end{bmatrix}^T\begin{bmatrix}\Phi & P\bar{B}_{c,ij}\\ \bar{B}_{c,ij}^T P & -\gamma_1^2 I\end{bmatrix}\begin{bmatrix}\bar{y}_c(t)\\ \bar{z}_r(t)\end{bmatrix} + \gamma_1^2\|\bar{z}_r\|^2 - \|\bar{z}_c\|^2\end{aligned} \tag{9.66}$$

where $\Phi \triangleq \bar{A}_{c,ij}^T P + P\bar{A}_{c,ij} + \bar{C}_{c,ij}^T\bar{C}_{c,ij} + \frac{1}{\xi}PP + \xi\bar{\sigma}_3$. If the inequalities in equation (9.61) hold, then we have

$$\dot{S}_1(\bar{y}_c(t)) \le \gamma_1^2\|\bar{z}_r(t)\|^2 - \|\bar{z}_c(t)\|^2 \tag{9.67}$$

Integrating equation (9.67) from $t = 0$ to $t = t_f$ yields the inequality in equation (9.59). Therefore, the dissipation inequality in equation (9.59) is guaranteed. By Assumption 9.4.1, we have the dissipation inequality equation (9.60) for the residual subsystem Δ in equation (9.58). Applying Lemma 9.4.1, if the inequality in equation (9.61) and $\gamma_1 \cdot \gamma_2 < 1$ all hold, then the coupled system equations (9.57) and (9.58) is asymptotically stable, i.e., $\bar{y}_c(t) \to 0$ and $y_r(t) \to 0$ as $t \to \infty$. Furthermore, because of $y(x,t) = \sum_{l=1}^{\infty} y_l(t)\phi_l(x)$ the fuzzy observer-based controller could guarantee the NPDS in equation (9.1) is also asymptotically stable.

Remark 9.4.2

In stability analysis of fuzzy control systems, the most results have been obtained based on a single Lyapunov function. Recently, some approaches based on different Lyapunov functions, i.e., piecewise Lyapunov function [308] and fuzzy Lyapunov function [309], have been developed to relax the conservativeness of stability and stabilization problems. Although the study of the conservativeness is important, our design purpose is to extend the fuzzy stabilization approach from the conventional nonlinear system to NPDS. Therefore, the conservative analysis of stability can refer to the related literature [308–315].

In the observer-based control design problem, the control gain K_{ij} and the observer gain G_{ij} are coupled through the Lyapunov matrix P. If we assign P as a full matrix, then the matrix inequality in equation (9.61) will be a complex bilinear matrix inequality (BMI). At present, some local optimal solvers have been proposed for BMI problem via the augmented Lagrangian method [316] or the iteration method [291,317–320]. However, because the BMI problem is nonconvex, these solvers are still inefficient to solve BMI problems with multi-variables, especially for the BMI in equation (9.61) of the observer-based control design problem of the complex coupled systems in equations (9.57) and (9.58). Therefore, in this chapter, we develop a transformed technique to transform the BMIs into LMIs which could be efficiently solved with the help of LMI toolbox in MATLAB. At first, suppose $P = diag(P_1, P_2)$ where P_1 and P_2 are positive definite matrices. The following statements will describe how to transform BMIs in equation (9.61) into equivalent linear matrix inequalities (LMIs).

Assumption 9.4.2

The matrix $B_c \in \mathbb{R}^{N\times p_p}$ *is a full-column rank, i.e.* $\text{rank}(B_c) = p_p$.

For the matrix B_c, there exist two orthogonal matrices $U \in \mathbb{R}^{N\times N}$ and $V \in \mathbb{R}^{p_p\times p_p}$, such that

$$UB_cV = \begin{bmatrix} U_1 \\ U_2 \end{bmatrix} B_cV = \begin{bmatrix} \chi \\ 0 \end{bmatrix}$$

where $U_1 \in \mathbb{R}^{p_p\times N}, U_2 \in \mathbb{R}^{(N-p_p)\times N}$ and $\chi = diag\{\chi_1,...,\chi_{p_p}\}$ in which $\chi_i, i = 1,..., p_p$, are nonzero singular values of B_c.

Suppose the structure of the P_1 in the matrix P is given in the following lemma.

Lemma 9.4.2 [321,322]

For the matrix $B_c \in \mathbb{R}^{N\times p_p}$ with full-column rank, if the matrix P_1 is of the following structures

$$P_1 = U^T \begin{bmatrix} P_{11} & 0 \\ 0 & P_{12} \end{bmatrix} U = U_1^T P_{11} U_1 + U_2^T P_{12} U_2 \tag{9.68}$$

for some P_{11} and P_{12} with $0 < P_{11} \in \mathbb{R}^{p_p \times p_p}$ and $0 < P_{12} \in \mathbb{R}^{(N-p_p)\times(N-p_p)}$, then there exists a nonsingular matrix $\bar{P}_1 \in \mathbb{R}^{p_p \times p_p}$ such that $B_c\bar{P}_1 = P_1 B_c$.

Theorem 9.4.2

For the closed-loop system in equations (9.57) and (9.58), suppose there exist three positive definite matrices P_{11}, P_{12}, and P_2, the matrices Y_{ij} and Z_{ij}, and two scalars $\gamma_1 > 0$ and $\xi > 0$ such that equation (9.69)

$$\Gamma_{ij} \triangleq \begin{bmatrix} \Omega_{11,ij} & \Omega_{12,ij} & -Y_{ij}^T & P_1 & 0 & P_1 & 0 \\ \Omega_{21,ij} & \Omega_{22,ij} & Y_{ij}^T & 0 & P_2 & P_2 & -Z_{ij} \\ -Y_{ij} & Y_{ij} & -\bar{P}_1 - \bar{P}_1^T + I & 0 & 0 & 0 & 0 \\ P_1 & 0 & 0 & -\xi I & 0 & 0 & 0 \\ 0 & P_2 & 0 & 0 & -\xi I & 0 & 0 \\ P_1 & P_2 & 0 & 0 & 0 & -\gamma_1^2 I & 0 \\ 0 & -Z_{ij}^T & 0 & 0 & 0 & 0 & -\gamma_1^2 I \end{bmatrix} < 0 \tag{9.69}$$

and

$$\gamma_1 \cdot \gamma_2 < 1 \tag{9.70}$$

where

$$\Omega_{11,ij} = A_{c,ij}^T P_1 - Y_{ij}^T B_c^T + P_1 A_{c,ij} - B_c Y_{ij} + \xi\sigma_3 + I, \Omega_{12,ij} = B_c Y_{ij}, \Omega_{21,ij} = \Omega_{12,ij}^T,$$

$$\Omega_{22,ij} = A_{c,ij}^T P_2 - C_c^T Z_{c,ij}^T + P_2 A_{c,ij} - Z_{ij} C_c, P_1 = U_1^T P_{11} U_1 + U_2^T P_{12} U_2 > 0,$$

$$\bar{P}_1 = (V^T)^{-1} \chi^{-1} P_{11} \chi V^T, \textit{ for } i = 1,\ldots,M, j = 1,\ldots,q.$$

Then the coupled system equations (9.57) and (9.58) will be asymptotically stable, i.e., the effect of the residual can be tolerated by the fuzzy observer-based controller in equations (9.51) and (9.52) with observer gains $G_{ij} = P_2^{-1} Z_{ij}$ and control gains $K_{ij} = \bar{P}_1^{-1} Y_{ij}$, $i = 1,\ldots,M$, $j = 1,\ldots,q$.

Proof

By Schur complement, the inequalities in equation (9.61) can be transformed to the following form:

$$\bar{A}_{c,ij}^T P + P\bar{A}_{c,ij} + \bar{C}_{c,ij}^T \bar{C}_{c,ij} + \frac{1}{\xi} PP + \xi\bar{\sigma}_3 + \frac{1}{\gamma_1^2} P\bar{B}_{c,ij}\bar{B}_{c,ij}^T P < 0 \tag{9.71}$$

Suppose $P = diag(P_1, P_2)$ where P_1 and P_2 are positive definite matrices, the above inequality could be simplified as

$$\Xi_{ij} \triangleq \begin{bmatrix} \Xi_{11,ij} + \frac{1}{\xi} P_1 P_1 + K_{ij}^T K_{ij} + \frac{1}{\gamma_1^2} P_1 P_1 & \Xi_{12,ij} - K_{ij}^T K_{ij} + \frac{1}{\gamma_1^2} P_1 P_2 \\ \Xi_{21,ij} - K_{ij}^T K_{ij} + \frac{1}{\gamma_1^2} P_2 P_1 & \begin{pmatrix} \Xi_{22,ij} + K_{ij}^T K_{ij} + \frac{1}{\xi} P_2 P_2 \\ + \frac{1}{\gamma_1^2} P_2 P_2 + \frac{1}{\gamma_1^2} P_2 G_{ij}^T G_{ij} P_2 \end{pmatrix} \end{bmatrix} < 0 \tag{9.72}$$

where

$$\Xi_{11,ij} = (A_{c,ij} - B_c K_{ij})^T P_1 + P_1 (A_{c,ij} - B_c K_{ij}) + \xi\sigma_3 + I,$$

$$\Xi_{12,ij} = P_1 B_c K_{ij}, \Xi_{21,ij} = \Xi_{12,ij}^T,$$

$$\Xi_{22,ij} = (A_{c,ij} - G_{ij} C_c)^T P_2 + P_2 (A_{c,ij} - G_{ij} C_c).$$

Using Schur complement and the equality, $K_{ij}^T K_{ij} = K_{ij}^T \bar{P}_1^T (\bar{P}_1^T)^{-1} \bar{P}_1^{-1} \bar{P} K_{ij}$, where $\bar{P}_1$ is a nonsingular matrix defined in Lemma 9.4.2, the inequality in equation (9.72) can be transformed to the inequality

$$\begin{bmatrix} \Xi_{11,ij} & \Xi_{12,ij} & -K_{ij}^T \bar{P}_1^T & P_1 & 0 & P_1 & 0 \\ \Xi_{21,ij} & \Xi_{22,ij} & K_{ij}^T \bar{P}_1^T & 0 & P_2 & P_2 & -P_2 G_{ij} \\ -\bar{P}_1 K_{ij} & \bar{P}_1 K_{ij} & -\bar{P}_1 - \bar{P}_1^T & 0 & 0 & 0 & 0 \\ P_1 & 0 & 0 & -\xi I & 0 & 0 & 0 \\ 0 & P_2 & 0 & 0 & -\xi I & 0 & 0 \\ P_1 & P_2 & 0 & 0 & 0 & -\gamma_1^2 I & 0 \\ 0 & -G_{ij}^T P_2 & 0 & 0 & 0 & 0 & -\gamma_1^2 I \end{bmatrix} < 0 \tag{9.73}$$

By the fact $(I - \bar{P}_1)(I - \bar{P}_1)^T \geq 0$, we have $-\bar{P}_1 \bar{P}_1^T \leq -\bar{P}_1 - \bar{P}_1^T + I$. Therefore, if the inequality

$$\begin{bmatrix} \Xi_{11,ij} & \Xi_{12,ij} & -K_{ij}^T\bar{P}_1^T & P_1 & 0 & P_1 & 0 \\ \Xi_{21,ij} & \Xi_{22,ij} & K_{ij}^T\bar{P}_1^T & 0 & P_2 & P_2 & -P_2G_{ij} \\ -\bar{P}_1K_{ij} & \bar{P}_1K_{ij} & -\bar{P}_1-\bar{P}_1^T+I & 0 & 0 & 0 & 0 \\ P_1 & 0 & 0 & -\xi I & 0 & 0 & 0 \\ 0 & P_2 & 0 & 0 & -\xi I & 0 & 0 \\ P_1 & P_2 & 0 & 0 & 0 & -\gamma_1^2 I & 0 \\ 0 & -G_{ij}^TP_2 & 0 & 0 & 0 & 0 & -\gamma_1^2 I \end{bmatrix} < 0 \tag{9.74}$$

holds, then the inequality in equation (9.73) could be guaranteed. According to Lemma 9.4.2, there exists the nonsingular matrix $\bar{P}_1 \in \mathbb{R}^{p_p \times p_p}$ such that $B_c\bar{P}_1 = P_1B_c$. The matrix $\bar{P}_1$ could be calculated from the relation $B_c\bar{P}_1 = P_1B_c$ as follows:

$$U^T\begin{bmatrix} \chi \\ 0 \end{bmatrix}V^T\bar{P}_1 = P_1U^T\begin{bmatrix} \chi \\ 0 \end{bmatrix}V^T \tag{9.75}$$

According to the structure of the matrix P_1 in equation (9.68), we have

$$U^T\begin{bmatrix} \chi \\ 0 \end{bmatrix}V^T\bar{P}_1 = U^T\begin{bmatrix} P_{11} & 0 \\ 0 & P_{12} \end{bmatrix}\begin{bmatrix} \chi \\ 0 \end{bmatrix}V^T \tag{9.76}$$

which implies that $\bar{P}_1 = \left(V^T\right)^{-1}\chi^{-1}P_{11}\chi V^T$.

Based on the above analysis, the robust fuzzy stabilization control design of Problem (2) becomes how to specify Z_{ij} and Y_{ij} to guarantee that the LMIs in equations (9.69) and (9.70) have positive definite solutions $P_{11}>0$, $P_{12}>0$, $P_2>0$. Then the control gain K_{ij} and observer gain G_{ij} of the observer-based fuzzy controller in equation (9.51) and (9.52) can be obtained by $K_{ij} = \bar{P}_1^{-1}Y_{ij}$ and $G_{ij} = P_2^{-1}Z_{ij}$, respectively. In general, NPDSs always suffer from external disturbance and measurement noise. After discussing the robust fuzzy stabilization control design of NPDS to tolerate the residual modes, the robust fuzzy H_∞ control design Problem (3) to attenuate external disturbance and measurement noise will be discussed in the next subsection.

9.4.2 Fuzzy H_∞ Observer-Based Control Design for Nonlinear Partial Differential Systems

After the residual models are tolerated, the effects of the external disturbances and measurement noises will be also attenuated to a prescribed level. First, the coupled system in equations (9.55) and (9.56) can be formulated as follows (see Figure 9.3c):

$$\Sigma_d : \begin{cases} \dot{\bar{y}}_c(t) = \sum_{i=1}^{M}\sum_{j=1}^{q}\mu_i(z_j(t))[\bar{A}_{c,ij}\bar{y}_c(t) + \bar{B}_{c,ij}\bar{z}_r(t) + \bar{B}_{cv,ij}v(t)] + \bar{\varepsilon}_c(t) \\ \bar{z}_c(t) = \sum_{i=1}^{M}\sum_{j=1}^{q}\mu_i(z_j(t))\bar{C}_{c,ij}\bar{y}_c(t) \\ z_{oc}(t) = C_{oc}y_c(t) = \bar{C}_{oc}\bar{y}_c(t) \end{cases} \tag{9.77}$$

$$\Delta_d : \begin{cases} \dot{y}_r(t) = \sum_{i=1}^{M}\sum_{j=1}^{q}\mu_i\left(z_j(t)\right)\left[\bar{A}_{r,ij}y_r(t) + \bar{B}_{r,ij}\bar{z}_c(t)\right] + \varepsilon_r(t) \\ \bar{z}_r(t) = \sum_{i=1}^{M}\sum_{j=1}^{q}\mu_i\left(z_j(t)\right)\bar{C}_{r,ij}y_r(t) \\ z_{or}(t) = C_{or}y_r(t) \end{cases} \tag{9.78}$$

$$z_o(t) = z_{oc}(t) + z_{or}(t) \tag{9.79}$$

where $z_o(t)$ is the control output; C_{oc}, C_{or} and $\bar{C}_{oc} = [C_{oc}, 0]$ are the corresponding matrices to specify the states of interest. Note that the control output $z_o(t)$ can be represented as $z_o(t) = h_0(x)y(x,t)$ in the NPDS, and $h_o(x)$ is defined similarly as $h(x)$ in equation (9.1). In this subsection, a robust H_∞ fuzzy observer-based controller is specified to guarantee that the effect of the external disturbance and measurement noise on controlled output $z_o(t)$ must be attenuated below a prescribed level ρ from the energy perspective. This design problem is called the robust H_∞ stabilization control problem to achieve the H_∞ disturbance attenuation performance in equation (9.4). If the energy of the initial condition is considered, i.e., $V(\bar{y}_c(0), y_r(0))$, the H_∞ control performance in equation (9.4) can be modified as follows

$$\int_0^{tf} z_o(t)^T z_o(t)dt \le V(\bar{y}_c(0), y_r(0)) + \rho^2 \int_0^{tf} v(t)^T v(t)dt \tag{9.80}$$

If the attenuation level ρ is to be specified as small as possible (i.e. ρ^2 is minimized), then this is an optimal H_∞ control design problem. In addition, the dissipation inequality of Δ_d in equation (9.78) could be defined as follows [293]:

$$V_2(y_r(t_f)) \leq V_2(y_r(0)) +^2 \int_0^{t_f} (\gamma_2^2 \|\bar{z}_c(t)\|^2 + \rho_2^2 v \|\bar{z}_c(t)\|^2 - (\|\bar{z}_r(t)\|^2 + \|z_{or}(t)\|^2))dt \tag{9.81}$$

Assumption 9.4.3

1. *The residual subsystem Δd is asymptotically stable in the inputs free case, i.e., $\bar{z}_r(t) \equiv 0$ and $v(t) \equiv 0$.*
2. *For the residual subsystem Δd, the dissipation inequality in equation (9.81) holds, i.e., γ_2 and ρ_2 are finite.*

Theorem 9.4.3

Under Assumption 9.4.3, for the coupled system equations (9.77)–(9.79), if the following inequalities hold

$$\begin{bmatrix} \begin{pmatrix} \bar{A}_{c,ij}^T P + P\bar{A}_{c,ij} + \frac{1}{\xi} PP + \xi\bar{\sigma}_3 \\ + \bar{C}_{c,ij}^T \bar{C}_{c,ij} + 2\bar{C}_{oc}^T \bar{C}_{oc} \end{pmatrix} & P\bar{B}_{c,ij} & P\bar{B}_{cv,ij} \\ \bar{B}_{c,ij}^T P & -\gamma_1^2 I & 0 \\ \bar{B}_{cv,ij}^T P & 0 & -\rho_1^2 I \end{bmatrix} < 0 \tag{9.82}$$

for $i = 1,\ldots, M$, $j = 1,\ldots, q$, and

$$\eta^2\gamma_2^2 - 1 < 0,\ \gamma_1^2 - \eta^2 < 0,\ 2 < \eta^2 \text{ and } \rho_1^2 + \eta^2\rho_2^2 < \rho^2, \tag{9.83}$$

then the robust H_∞ control performance in equation (9.80) is guaranteed for a prescribed disturbance attenuation level ρ_2 by the observer-based fuzzy controller in equations (9.51) and (9.52). In other words, for the NPDS in equation (9.3), the effect of the external disturbances $d(t)$ and the measurement noise $n(t)$ on the controlled output $z_o(t)$ is bounded by the H_∞ performance in equation (9.80).

Proof

Choose the storage function for the controlled system Σ_d in equation (9.77) as $V_1(\bar{y}_c(t)) = \bar{y}_c(t)^T P\bar{y}_c(t)$. By differentiating the storage function, we can get

$$\dot{V}_1(\bar{y}_c(t)) = \dot{\bar{y}}_c(t)^T P\bar{y}_c(t) + \bar{y}_c(t)^T P\dot{\bar{y}}_c(t)$$

$$= \left(\sum_{i=1}^{M}\sum_{j=1}^{q} \mu_i(z_j(t))[\bar{A}_{c,ij}\bar{y}_c(t) + \bar{B}_{c,ij}\bar{z}_r(t) + \bar{B}_{cv,ij}v(t)] + \bar{\varepsilon}_c(t) \right)^T P\bar{y}_c(t)$$

$$+ \bar{y}_c(t)^T P\left(\sum_{i=1}^{M}\sum_{j=1}^{q} \mu_i(z_j(t))[\bar{A}_{c,ij}\bar{y}_c(t) + \bar{B}_{c,ij}\bar{z}_r(t) + \bar{B}_{cv,ij}v(t)] + \bar{\varepsilon}_c(t) \right) \tag{9.84}$$

Adding and subtracting the term $\left(\gamma_1^2\|\bar{z}_r(t)\|^2 - \|\bar{z}_c(t)\|^2 + \rho_1^2\|v(t)\|^2 - \|z_o(t)\|^2\right)$ to equation (9.84), the following equation can be obtained:

$$\dot{V}_1(\bar{y}_c(t)) = \left(\sum_{i=1}^{M}\sum_{j=1}^{q} \mu_i(z_j(t))[\bar{A}_{c,ij}\bar{y}_c(t) + \bar{B}_{c,ij}\bar{z}_r(t) + \bar{B}_{cv,ij}v(t)] + \bar{\varepsilon}_c(t) \right)^T P\bar{y}_c(t)$$

$$+ \bar{y}_c(t)^T P\left(\sum_{i=1}^{M}\sum_{j=1}^{q} \mu_i(z_j(t))[\bar{A}_{c,ij}\bar{y}_c(t) + \bar{B}_{c,ij}\bar{z}_r(t) + \bar{B}_{cv,ij}v(t)] + \bar{\varepsilon}_c(t) \right)$$

$$\pm (\gamma_1^2\|\bar{z}_r(t)\|^2 - \|\bar{z}_c(t)\|^2 + \rho_1^2\|v(t)\|^2 - \|z_o(t)\|^2) \tag{9.85}$$

By Fact 9.4.1, we have

$$\|z_o(t)\|^2 = (z_{oc}(t) + z_{or}(t))^T (z_{oc}(t) + z_{or}(t)) \le 2\left(\|\bar{z}_{oc}(t)\|^2 + \|z_{or}(t)\|^2\right)$$

and

$$\bar{\varepsilon}_c(t)^T P\bar{y}_c(t) + \bar{y}_c(t)^T P\bar{\varepsilon}_c(t) \le \xi\bar{\varepsilon}_c(t)^T\bar{\varepsilon}_c(t) + \frac{1}{\xi}\bar{y}_c(t)^T PP\bar{y}_c(t)$$

where ξ is any positive constant. According to the bound of $\varepsilon_c(t)$ in equation (9.48), we have $\bar{\varepsilon}_c(t)^T\bar{\varepsilon}_c(t) \le \bar{y}_c^T(t)\bar{\sigma}_3\bar{y}_c(t)$ where $\bar{\sigma}_3$ is defined in equation (9.65). Based on three inequalities above, from equation (9.85) we can obtain

$$\dot{V}_1(\bar{y}_c(t)) \le \sum_{i=1}^{M}\sum_{j=1}^{q} \mu_i(z_j(t)) \begin{bmatrix} \bar{y}_c(t) \\ \bar{z}_r(t) \\ v(t) \end{bmatrix}^T \begin{bmatrix} \Theta_{11,ij} & P\bar{B}_{c,ij} & P\bar{B}_{cv,ij} \\ \bar{B}_{c,ij}^T P & -\gamma_1^2 I & 0 \\ \bar{B}_{cv,ij}^T P & 0 & -\rho_1^2 I \end{bmatrix} \begin{bmatrix} \bar{y}_c(t) \\ \bar{z}_r(t) \\ v(t) \end{bmatrix}$$

$$+ \gamma_1^2\|\bar{z}_r(t)\|^2 - \|\bar{z}_c(t)\|^2 + \rho_1^2\|v(t)\|^2 + 2\|z_{or}(t)\|^2 - \|z_o(t)\|^2 \tag{9.86}$$

where $\Theta_{11,ij} = \bar{A}_{c,ij}^T P + P\bar{A}_{c,ij} + \frac{1}{\xi}PP + \xi\bar{\sigma}_3 + \bar{c}_{c,ij}^T \bar{C}_{c,ij} + 2\bar{C}_{oc}^T \bar{C}_{oc}$. Suppose the inequalities in equation (9.82) hold, then we have the following inequality:

$$\dot{V}_1(\bar{y}_c(t)) \le \gamma_1^2 \|\bar{z}_r(t)\|^2 - \|\bar{z}_c(t)\|^2 + \rho_1^2 \|v(t)\|^2 + 2\|z_{or}(t)\|^2 - \|z_o(t)\|^2. \quad (9.87)$$

Integrating the above inequality equation (9.87) from $t = 0$ to $t = t_f$ yields

$$\begin{aligned} &V_1(\bar{y}_c(t_f)) \\ &\le V_1(\bar{y}_c(0)) + \int_0^{t_f} (\gamma_1^2 \|\bar{z}_r(t)\|^2 - \|\bar{z}_c(t)\|^2 + \rho_1^2 \|v(t)\|^2 + 2\|z_{or}(t)\|^2 - \|z_o(t)\|^2)\, dt \end{aligned} \quad (9.88)$$

By Assumption 9.4.3, we have the dissipation inequality in equation (9.81) for Δ_d in equation (9.78). Suppose $\gamma_1 \cdot \gamma_2 < 1$, and we can find a scalar η such that $\gamma_1 < \eta < \frac{1}{\gamma_2}$. Multiply the inequality in equation (9.81) by η^2, and then add it to the inequality in equation (9.88) to obtain

$$\begin{aligned} V_1(\bar{y}_c(t_f), y_r(t_f)) \le V_1(\bar{y}_c(0), y_r(0)) + \int_0^{t_f} &(\gamma_1^2 - \eta^2)\|\bar{z}_r(t)\|^2 + (2 - \eta^2)\|z_{or}(t)\|^2 \\ &+ (\eta^2\gamma_2^2 - 1)\|\bar{z}_c(t)\|^2 + (\rho_1^2 + \eta^2\rho_2^2)\|v(t)\|^2 - \|z_o(t)\|^2\, dt \end{aligned} \quad (9.89)$$

where $V_1(\bar{y}_c(t), y_r(t)) \triangleq V_1(\bar{y}_c(t)) + \eta^2 V_2(y_r(t))$ Since the scalar η satisfies $\gamma_1 < \eta < \frac{1}{\gamma_2}$, we have the inequalities $\eta^2\gamma_2^2 - 1 < 0$ and $\gamma_1^2 - \eta^2 < 0$. Note that the above two inequalities imply $\gamma_1 \cdot \gamma_2 < 1$. Let $2 < \eta^2$ and $\rho_1^2 + \eta^2\rho_2^2 < \rho^2$, then the H_∞ control performance inequality in equation (9.80) could be obtained. Therefore, if the inequalities in equations (9.82) and (9.83) hold, then the H_∞ control performance in equation (9.80) could be guaranteed.

Remark 9.4.3

Since the inequalities in equation (9.61) are implied by the inequalities in equation (9.82), the conditions for H_∞ control performance can also imply that the closed-loop system in equations (9.77) and (9.78) is asymptotically stable in the disturbance free case, i.e., $v(t) \equiv 0$.

Since the inequalities in equation (9.82) are the BMIs, which is still inefficient to solve by the present solvers, an LMI form for these conditions is developed to simplify the control design procedure in the following theorem.

Theorem 9.4.4

For the coupled system in equations (9.77) and (9.78), if the following linear matrix inequalities hold for the three positive definite matrices $P_{11}>0$, $P_{12}>0$, and $P_2>0$, the matrices Y_{ij} and Z_{ij} and some positive constants γ_1, ρ_1, ρ, η, and ξ

$$\begin{bmatrix} \Gamma_{ij}+\Gamma_{11} & \Gamma_{12,ij} \\ \Gamma_{21,ij} & \Gamma_{22} \end{bmatrix} < 0 \tag{9.90}$$

for $i = 1,\ldots, M$, $j = 1,\ldots, q$, and

$$\eta^2\gamma_2^2 - 1 < 0, \gamma_1^2 - \eta^2 < 0 \text{ and } \rho_1^2 + \eta^2\rho_2^2 < \rho^2 \tag{9.91}$$

where Γ_{ij} is defined in equations (9.69), and $\Gamma_{11} = diag(2\bar{C}_{oc}^T\bar{C}_{oc}, 0,0,0,0,0,0)$, $\Gamma_{21,ij} = \begin{bmatrix} B_{cd}^T P_1 & B_{cd}^T P_2 & 0 & 0 & 0 & 0 & 0 \\ 0 & -D_n^T Z_{ij}^T & 0 & 0 & 0 & 0 & 0 \end{bmatrix}$, $\Gamma_{12,ij} = \Gamma_{21,ij}^T$, $\Gamma_{22} = \begin{bmatrix} -\rho_1^2 I & 0 \\ 0 & -\rho_1^2 I \end{bmatrix}$ then the H_∞ control performance of equations (9.4) or (9.80) is guaranteed for a prescribed disturbance attenuation level ρ^2 by the fuzzy observer-based controller in equations (9.51) and (9.52) with $G_{ij} = P_2^{-1}Z_{ij}$ and $K_{ij} = \bar{P}_1^{-1}Y_{ij}$. Note that P_1 and $\bar{P}_1$ are defined in equation (9.68).

Proof

The proof procedure is similar to Theorem 9.4.2.

Therefore, the H_∞ control design problem in Problem (3) becomes how to specify the control gains K_{ij} and the observer gains G_{ij} to satisfy the LMIs in equations (9.90) and (9.91). Finally, the optimal H_∞ control design problem for NPDS equation (9.3) could be solved by the following constrained optimization problem:

$$\rho_0 = \min_{Y_{ij}, Z_{ij}} \rho$$
$$\text{subject to } P_{11} > 0, P_{12}, P_2 > 0, (9.90) - (9.91) \tag{9.92}$$

which is called an eigenvalue problem (EVP) [307] and can be easily solved by LMI toolbox in MATLAB [323].

Based on the above analysis, the robust H_∞ observer-based control design procedure for NPDS is summarized in the following.

Design Procedure

Step 1: Select the fuzzy membership functions and fuzzy rules to establish a fuzzy PDS as equation (9.27) to approximate the NPDS in equation (9.3).

Step 2: Give the orthonormal eigenfunctions $\phi_l(x)$ and the eigenvalues $\lambda_{i,1}$ to construct the infinite-dimensional fuzzy state space model as equation (9.38), and choose an appropriate controlled order N to obtain the controlled subsystem and residual subsystem as equation (9.46) and (9.47).

Step 3: Give the dissipative constants γ_2 and ρ_2 of the coupled residual subsystem in equation (9.81), and obtain the control gains K_{ij} and the observer gains G_{ij} of the fuzzy observer-based controller in equation (9.51) and (9.52) by solving the EVP in equation (9.92).

Step 4: Construct a fuzzy observer-based feedback controller as equation (9.51) and (9.52).

9.5 SIMULATION EXAMPLE

In this section, a biochemical example of the robust control design problem for NPDS is given to illustrate the design procedure and to confirm the robust H_∞ performance of the proposed observer-based controller.

Example (Inhibition Control of Monoenzyme Systems by Excess of Substrate [272–274]): This monoenzyme system is to describe the interactions between the substrate and enzyme in an artificial membrane. The thickness L of the artificial membrane is $L=1$ in this example. An enzyme (a catalyst of biochemical reaction) is employed to control the concentration of substrate and is homogeneously distributed inside an artificial membrane. The enzyme is controlled by input at different locations of membrane. Because the membrane is immersed in a solution of substrate, the reference axis is chosen perpendicular to the membrane. The enzyme system could be formulated as follows:

$$\frac{\partial y(x,t)}{\partial t} = D_s \frac{\partial^2 y(x,t)}{\partial x^2} - V_M \frac{y(x,t)}{K_M + y(x,t) + y(x,t)^2 / K_s} + g(x)u(t) + g_d(x)d(t) \tag{9.93}$$

$$z(t) = h(x)y(x,t) + D_n n(t), z_o(t) = h_o(x)y(x,t) \tag{9.94}$$

where $y(x,t)$, which is a function of space x and time t, is the concentration of substrate in the membrane and is to be controlled by adding enzyme $u(t)$ at different locations $g(x)$; D_s is a substrate diffusion coefficient; K_M is the Michaelis constant and V_M is the maximum activity per volume unit of membrane. The parameters of the enzyme system is given by $D_s=0.03$, $V_M=0.5$, $K_M=1$, and $K_s=1$. Note that the origin of the system in this example has been shifted to the desired steady state. Therefore, a regulation control to a desired steady state becomes a stabilization problem. The concentration of initial distribution of the substrate is given as $y(x, 0)=0.3\sin(\pi x)$. The boundary conditions are used to restrict the concentration to be zero at the boundary, i.e., $y(0,t)=0$, $y(1,t)=0$. A detailed discussion on enzyme system can be found in [272–274]. Suppose the enzyme system suffers the effect of external disturbances and measurement noises. The control inputs $u(t)$ are designed to control the concentration of substrate by adding enzyme at some specified locations. The fuzzy controller proposed in this chapter is applied to adjust the concentration of the enzyme based on the estimated concentration of substrate in the artificial membrane. The control influence function of enzyme is defined as $g(x)=\left[\delta\left(x-\frac{8}{30}\right),\delta\left(x-\frac{9}{30}\right),\delta\left(x-\frac{10}{30}\right),\delta\left(x-\frac{11}{30}\right),\delta\left(x-\frac{17}{30}\right),\delta\left(x-\frac{18}{30}\right),\delta\left(x-\frac{18}{30}\right),\delta\left(x-\frac{19}{30}\right),\delta\left(x-\frac{20}{30}\right)\right]$. The influence function of external disturbances is defined as $g_d(x)=\left[\delta\left(x-\frac{9}{30}\right),\delta\left(x-\frac{14}{30}\right)\right]$. The observation influence function is defined as $h(x)=\left[\delta\left(x-\frac{2}{30}\right),\delta\left(x-\frac{6}{30}\right),\delta\left(x-\frac{12}{30}\right),\delta\left(x-\frac{22}{30}\right),\delta\left(x-\frac{26}{30}\right)\right]^T$. The influence function of measurement noise is defined as $D_n=[0, 1, 0, 1, 0]^T$. The influence function of the controlled output $z_o(t)$ is defined as $h_o(x)=h(x)$. For the convenience of simulation, the external disturbance and the measurement noise are assumed as $d(t)=10\sin(t)$ and $n(t)=\sin(t)$, respectively.

The spatial-time profiles of the NPDS without control input are shown in Figure 9.4a. First, we establish a T-S fuzzy PDS as equation (9.8) with membership functions as shown in Figure 9.2. The number of fuzzy rules is $M=4$. The parameters of the fuzzy membership function F_i are defined as $\bar{b}_1=-0.5$, $\bar{b}_2=0$, $\bar{b}_3=0.5$, and $\bar{b}_4=1$ (see Figure 9.2). The parameters in the linear PDE of fuzzy system equation (9.7) or (9.8) are obtained as $a_1=-0.0690$, $a_2=-0.0507$, $a_3=-0.0278$, and $a_4=-0.0161$. The spatial-time profiles of the T-S fuzzy partial differential model in equation (9.8) without control input are shown in Figure 9.4b and we can obtain the bounds of the approximate error $\sigma_{1,1}=0$ and $\sigma_{1,2}=0.0022$ in equation (9.12). Second, we establish the equivalent T-S fuzzy partial differential model in equation (9.27) with the spatial-time profiles shown in Figure 9.4c. The bounds of approximate errors σ_2, $\sigma_{3,1}$, and $\sigma_{3,2}$ in equations (9.18) and (9.26) could be estimated as $\sigma_2=0.0471$, $\sigma_{3,1}=0$, and $\sigma_{3,2}=0.8001$. The approximation error between the NPDS in equation (9.93) and the equivalent T-S fuzzy PDS is shown on Figure 9.4d. The fuzzy approximation error is in the scale of 10^{-3}. Obviously, the proposed fuzzy model could approach the nonlinear partial system accurately. The time profiles of the output $z_o(t)$ of the NPDS without control are shown in Figure 9.5c. The H_∞ attenuation performance of uncontrolled NPDS could be computed as follows

Since the fluctuation of system output due to disturbance and measurement noise is very large, we design a fuzzy observer-based controller to attenuate the effect of external disturbances and measurement noises. Using the orthonormal eigenfunctions $\phi_n(x)=\sin(nx\pi)$ and the eigenvalues $\lambda_n(x)=-0.03(n\pi)^2$, we can

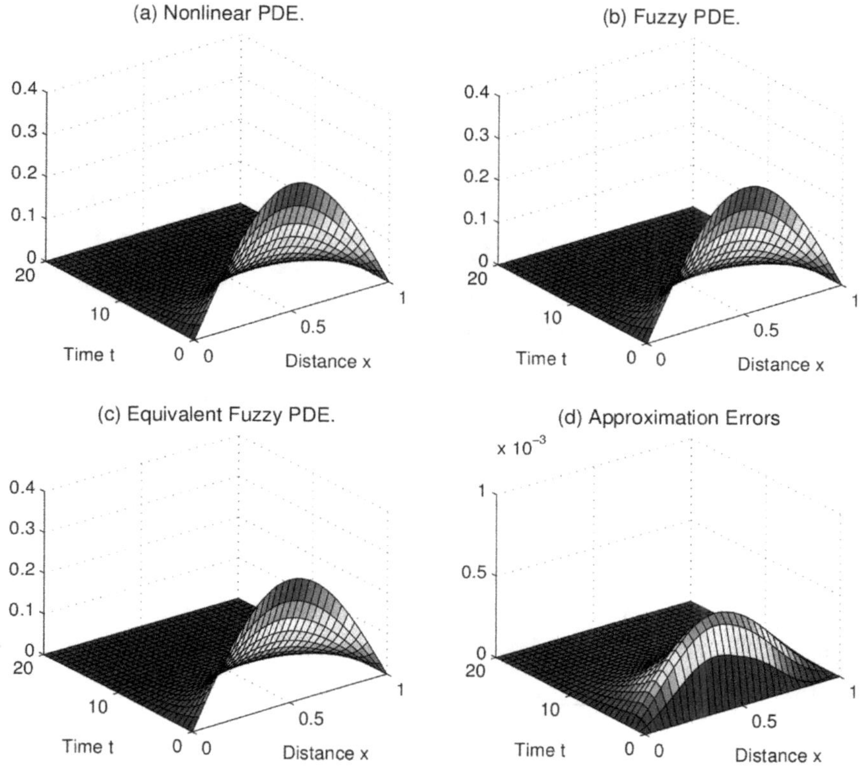

FIGURE 9.4 (a) Spatial-time profiles of the NPDS in equation (9.93). (b) Spatial-time profiles of the T-S fuzzy partial differential model in equation (9.8). (c) Spatial-time profiles of the equivalent T-S fuzzy partial differential model in equation (9.27). (d) Spatial-time profiles of the approximation error between the NPDS in equation (9.93) and the equivalent T-S fuzzy partial differential model in equation (9.27).

construct a fuzzy state space model as equation (9.38). Choose an appropriate controlled order $N=10$, then the controlled subsystem (equation 9.46) and the residual subsystem (equation 9.47) could be obtained. Following the proposed design procedure in above section, the optimal H_∞ fuzzy observer-based controller could be obtained easily with $\rho_0=0.8983$ by solving the constrained optimization problem in equation (9.92). The control gains K_{ij} and observer gains G_{ij} are omitted due to space limit. The time profiles of the estimate errors $z(t)-\hat{z}(t)=z(t)-C_c\hat{y}_c(t)$ by the proposed fuzzy observer in equation (9.51) are shown in Figure 9.5a. The control inputs $u(t)$ of enzyme in equation (9.52) are shown in Figure 9.5b. The time profiles of $z_o(t)$ of the controlled PDS in equation (9.93) are shown on Figure 9.5d. From the simulation results, the proposed H_∞ fuzzy observer-based controller obviously has improved the performance of the nonlinear enzyme system by

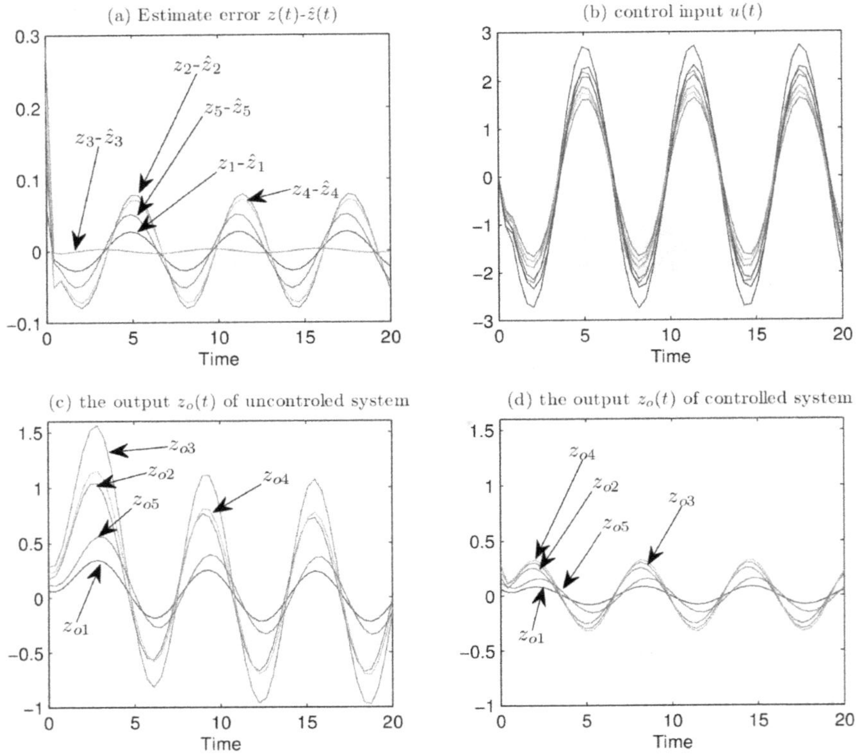

FIGURE 9.5 Simulation results. (a) Estimate error $z(t) - \hat{z}(t) = z(t) - C_c\hat{y}_c(t)$ in equation (9.51). (b) Control input $u(t)$ in equation (9.52). (c) Time profiles of the output $z_o(t)$ of the uncontrolled NPDS. (d) Time profiles of the output $z_o(t)$ of the controlled NPDS.

efficiently attenuating the external disturbances and measurement noises. The H_∞ attenuation performance could be computed as follows

$$\frac{\int_0^{20} z_o(t)^T z_o(t)\,dt}{\int_0^{20} v(t)^T v(t)\,dt} \approx 0.5602 < \rho_0 = 0.8983$$

This conservative result is due to the conservative of solving LMIs in the H_∞ control design procedure. Therefore, the simulation example has shown the feasibility of the proposed H_∞ fuzzy robust stabilization control design of the nonlinear PDS for potential practical applications. The effects of external disturbances and measurement noises in NPDSs could be efficiently attenuated by the proposed robust H_∞ observer-based control design.

9.6 CONCLUSION

In this chapter, based on the fuzzy interpolation method, NPDS can be modeled by the proposed fuzzy PDS with some approximation errors. By Galerkin's method in Section 1.2.1, the fuzzy PDS can be transformed to an equivalent fuzzy infinite-dimensional state space system. After the fuzzy infinite-dimensional system is separated into a finite- dimensional fuzzy subsystem in controlled mode and a coupled infinite-dimensional subsystem in residual mode, based on the small gain theorem, a robust fuzzy finite-dimensional observer-based controller is proposed to stabilize the fuzzy controlled subsystem and to tolerate the coupled residual subsystem. If the NPDS is free of external disturbance and measurement noise, the asymptotic stability is guaranteed to tolerate the coupled residual modes. Finally, based on the dissipative theorem, the optimal H_∞ fuzzy observer-based controller is also designed to minimize the effect of external disturbance and measurement noise on the desired control performance. The proposed optimal H_∞ fuzzy observer-based control design problem for NPDS can be transformed to how to solve the control gains K_{ij} and observer gains G_{ij} from an EVP, which can be efficiently solved by the LMI Toolbox in MATLAB. From the simulation example, the proposed optimal H_∞ fuzzy observer-based controller design method can efficiently attenuate the effect of external disturbance and measurement noise on the stabilization of NPDS. Therefore, the proposed method is potential for the robust control design of some mechanical systems related with heat flows, fluid flows, elastic wave, flexible structures, chemical engineering and biodynamic systems etc. After the introduction of robust observer-based control design based on Galerkin method and small gain theorem for NPDSs in frequency domain in this chapter, the robust control designs of NPDSs will be introduced based on finite difference method in the time domain in the following chapters.

10 Robust Tracking Control Design of Nonlinear Distributed Parameter Time-Delayed Systems

10.1 INTRODUCTION

Most physical systems are inherently distributed in space and time, e.g., chemical engineering [324]; biodynamics [325,326]; and mechanical systems related to heat flows, fluid flow, elastic wave, or flexible structure [327,328]. In the past, most physical systems were modeled by ordinary differential equation (ODE) in order to simplify and systematically solve control design problem. However, it is not sufficient to model the physical systems, if we consider the variation of the system depended on the space. In chemical engineering, many chemical processes are characterized by the presence of spatial variations and time delays [324]. Thus, the reaction-diffusion equation is introduced to represent the chemical process. In recent years, interactions between the mathematical and biological sciences have been increasing rapidly [325,326]. In biology, mathematical tools can help provide systematic analysis, e.g., the stability or the robustness of biological systems. However, the evolution of physio-logical behavior is dependent on time and space. For example, the nonlinear partial differential Hodgkin-Huxley (H-H) model has been applied to model signal transmission in a nervous system [325,329]. Therefore, the distributed parameter system (DPS), which is described by partial differential equation (PDE), is more suitable to model the spatiotemporal dynamic systems in biology.

In general, the tasks of control systems are one of two categories: stabilization and tracking. The stabilization problem is to design a controller so that the states of the closed-loop system can converge to an equilibrium point. In the tracking problem, a controller is designed to guarantee that the output of the closed-loop system can track a desired reference trajectory. Many studies have investigated the stabilization design problem of linear distributed parameter systems (LDPSs), for example, the stability analysis of the LDPSs introduced in [330,331]. Similarly, a robust stabilization of the LDPS with the external disturbances has been developed to attenuate the effect of external disturbances from the H_∞-control point of view [332]. But, the control design problem of NDPSs is more complex than the control design of the LDPSs. Based on Galerkin's method in Section 1.2.1, controller design schemes have been proposed to stabilize the

DOI: 10.1201/9781003229230-13

NDPS by a residual model filter [333] or an inertial manifold model [334]. In the past 20 years, the fuzzy approach which uses several local linear models to interpolate a nonlinear system has been widely applied to the analysis of the nonlinear systems in various fields which are described by ODE [335–352]. In Chapter 2, the fuzzy approach was already applied to the field of PDEs. A new technique using the adaptive fuzzy algorithm is proposed to obtain the solutions of PDEs [353]. According to the adaptive scheme of fuzzy logic systems, a fuzzy solution with adjustable parameters for the PDE can be obtained successfully. On the other hand, an H_∞ fuzzy observer-based control design [354] is proposed for a class of nonlinear parabolic PDE systems with control constraints. In addition, a robust stabilization problem for the NDPS with time delay is studied using a fuzzy control approach in [355]. Galerkin's method is applied to derive a set of nonlinear ODEs for the NDPS [354,355]. In Chapter 9, a new fuzzy state space model is proposed to represent the NDPSs based on Galerkin's method in Section 1.2.1, with the advantage of avoiding obtaining a complex nonlinear ODE. The robust H_∞ stabilization is developed to attenuate the effects of modeling errors, external disturbances, and measurement noises [356]. However, in a tracking design case, the asymptotic tracking cannot be achieved due to the changing reference signal via Galerkin's method, because this tracking error will affect the residual subsystem continuously. Therefore, this robust H_∞ control design method in [356] cannot be extended from the stabilization problem to the tracking problem of NDPSs.

For the reference tracking problem of distributed parameter systems, there is only a study of the linear case [357]. Byrnes et al. proposed an extended output regulation method to control the LDPS to track a reference model [357]. For the nonlinear case, to the best of our knowledge, no result of tracking design for NDPSs has been presented because of the complex nonlinearity, distributed parameter and design difficulty. However, the tracking control design is a more important control problem in practical applications, because a system needs to be controlled in order to track a reference signal. Recently, the tracking control problems of biological systems have become a very important topic in the biomedical engineering. However, biological systems are always nonlinear and compartment-dependent with process delays. For example, the control problem of the nervous system in [329,358,359] should be about how to design a controller to make the membrane voltage of a nervous system tracking a given membrane voltage reference time-course, e.g., short electrical pulses, not simply how to stabilize the nervous system. In a realistic model, a space-clamped axon was described as the nonlinear H-H model of the spatiotemporal dynamics [359]. In this situation, the tracking control design of NDPSs has the potential to create a real-time therapeutic regime for the undesired neural oscillation caused by disturbances and environmental noises [358,360–362]. In addition, the problem of time delay commonly emerges in practical systems. Therefore, this chapter studies a robust tracking control design problem for NDPSs with time delays, external disturbances, and measurement noise.

Tseng et al. studied a robust tracking problem for a nonlinear ODE system using the fuzzy tracking control scheme, where a T-S fuzzy model was initially used to represent the nonlinear ODE system [335]. Then, a fuzzy observer-based controller was developed to reduce the tracking error as much as possible via the H_∞ tracking performance. In this current study, a robust tracking control scheme for NDPSs with time delays, external disturbances and measurement noises is proposed based on a fuzzy observer-based controller. A design procedure is introduced as follows. First, a fuzzy DPS with time delay is proposed to approximate the NDPS with time delay by interpolating several linear distributed parameter time-delayed systems. Unlike using the infinite dimensional ODE system to represent the PDE system [356], for the convenience of tracking control design, the partial differential operator on space in PDE could be approximated by a finite difference operator. Then, we can obtain a set of fuzzy finite difference systems with the approximation errors and the truncation errors to represent the NDPS. When all finite difference grid points are represented by a spatial vector, a set of the fuzzy finite difference systems is represented by an equivalent fuzzy spatial state space system by the Kronecker product method. In the output feedback control design, a fuzzy observer based on fuzzy spatial state space system is developed to estimate the state of the NDPS from the output measurements at several sensor locations. Finally, a robust fuzzy observer-based tracking control scheme is proposed to control the NDPS with time delay to track a reference model. At the same time it also attenuates the effects of the time delay, the approximation errors, the truncation errors, the measurement noises, and the external disturbances on the observation and tracking error. In order to treat the robust tracking problem, a two-dimensional H_∞ tracking performance in a spatiotemporal domain is introduced for NDPSs to efficiently attenuate the effects of the time delay, the approximation error, the truncation error, the external disturbance, and the measurement noise on the tracking performance. For the convenience of tracking control design, the two-dimensional H_∞ tracking performance is transformed to an equivalent one-dimensional H_∞ tracking performance, when all finite difference grid points are represented by a spatial vector. Based on H_∞-attenuation theory [335], the proposed fuzzy observer-based tracking controller is proved to guarantee that the NDPS can robustly track a reference model by efficiently eliminating the effect of time delay, the approximation error, the truncation error, the external disturbance, and the measurement noise on the tracking error below a prescribed level. The effect of using a finite difference operator to approximate the partial differential operator on the H_∞ tracking performance is also discussed. We have found that the effect due to finite-difference approximation is of the order $O(\Delta_x^2)$, when Δ_x is the distance of the neighboring grid points. Therefore, if the grid points in a spatial domain are dense enough or the truncation error $O(\Delta_x^2)$ is small enough, the H_∞ tracking performance of the fuzzy spatial state space system will approach the H_∞ tracking performance of the NDPS.

Generally, in the observer-based control design of both ODE and PDE systems, how to solve coupling linear matrix inequalities (LMIs) for obtaining control gain and observer gain is still a difficult problem. A two-step procedure [363–365] and a transformed technique [356] have previously been proposed to conservatively solve the coupling LMIs. Although this coupling problem could be solved by a BMI optimal technique [366], the BMI problem is not a convex problem. Especially for a complex multi-variable constraint case, BMI is not an efficient method to solve the coupling LMIs of fuzzy observer- based tracking design for NDPSs. In [367], a single-step approach was proposed to solve the BMI problem for matrix decoupling of the fuzzy observer-based stabilization design of the fuzzy time-delay systems. In this chapter, the BMI problem is approximated to an LMI problem via Schur complement and several inequalities [368]. Therefore, the coupling problem can be efficiently solved by the conventional LMI technique in the design procedure. Finally, in order to emerge the importance of practical application, an example of the reference tracking control design for the H-H nervous system [325,358,369–371] in biochemical engineering is given to illustrate the design procedure and to confirm the robust tracking performance. This tracking control design of the H-H nervous system is useful to suppress oscillations and blockage of action potential transmission for patients suffering from nervous system dysfunction.

The main features of this chapter are given as follows: (i) The difficulty of the reference tracking control design for a NDPS with time delay, external disturbances, and measurement noises is overcome by the proposed fuzzy spatial state space model via a finite difference method. (ii) A two-dimensional H_∞ tracking performance in a spatiotemporal domain is proposed for robust tracking design of NDPS and then is transformed into an equivalent one-dimensional H_∞ tracking performance for a fuzzy spatial state space system. (iii) A fuzzy observer-based tracking controller is successfully developed to control the NDPSs to efficiently track a desired reference model based on the H_∞ tracking performance. This robust H_∞ reference tracking control scheme can be systematically designed through output feedback via the help of the MATLAB LMI toolbox.

Notations: For the convenience of problem description and control design, we define the separable Hilbert space in the following.

1. $\|y(t)\|^2 \triangleq \sum_{i=1}^{n}|y_i(t)|^2$ where $y(t)=[y_1(t),\dots,y_n(t)]^T$.
2. $L_2(\mathbb{R}_+;\mathbb{R}^n)$ is the space of the n-dimensional measurable functions defined $z(t)\in\mathbb{R}^n$ on $t\in\mathbb{R}_+=[0,\infty)$ such that $\|z(t)\|^2_{L_2(\mathbb{R}_+;\mathbb{R}^n)} \triangleq \int_0^\infty \|z(t)\|^2\,dt<\infty$.

10.2 REFERENCE TRACKING CONTROL PROBLEM FORMULATION FOR NONLINEAR DISTRIBUTED PARAMETER TIME-DELAYED SYSTEMS

In this section, the reference tracking control problems for the NDPS are formulated. Time delays, external disturbances, and measurement noise frequently appear in practical physical systems. Therefore, we consider the following NDPS with time delay, external disturbance, and measurement noise.

$$\frac{\partial y(x,t)}{\partial t}=Ay(x,t)+f(y(x,t),y(x,t_\tau))+g(x)u(t)+g_d(x)d(t) \tag{10.1}$$

$$z(t)=h(x)y(x,t)+D_n n(t) \tag{10.2}$$

for $x=[x_1,x_2]^T \in U \subset \mathbb{R}_+^2$ and $t>0$, where $y(x,t) \triangleq [y_1(x,t),...,y_n(x,t)]^T \in \mathbb{R}^n$ is the state variable; x and t are the space and time variables, respectively; $f(y(x,t),y(x,t_\tau)) \in \mathbb{R}^n$ is a nonlinear function satisfying $f(0,0)=0$; t_τ denotes the delayed time, i.e., $t_\tau \triangleq t-\tau$, $\tau>0$; the distribution of the control force $u(t)$ is provided by p point force actuators, i.e., $u(t) \in \mathbb{R}^p$ is the applied force to be designed as $u(t)=[u_1(t),...,u_p(t)]^T$; the influence function $g(x)$ is an $n \times p$ matrix form whose elements are of delta function $\delta(x-p_i)$. For example, $n=1$, $g(x)=[g_1\delta(x-p_1),\ldots,g_p\delta(x-p_p)]$ where p_i are control force locations and $g_i\delta(x-p_i)=g_i$ for $x=p_i$ or $g_i\delta(x-p_i)=0$ for $x=p_i$. The measured output $z(t)$ may be interpreted as observations or as parts of the system whose behavior we wish to influence, and the output $z(t) \in \mathbb{R}^q$ is a vector, i.e. $z(t)=[z_1(t),...,z_q(t)]^T$, where q is the number of observations and the observation influence function $h(x) \in \mathbb{R}^{q \times n}$ is a matrix function. For example, $n=1$, $h(x)=[\delta(x-q_1),\ldots,\delta(x-q_q)]^T$. Therefore, in the case free from measurement noise, we have $z_i(t)=\delta(x-q_i)y(x,t)=y(q_b,t)$ with q_i as the ith sensor locations. $g_d(x) \in \mathbb{R}^{n \times p_d}$ is an interactive location matrix of the external disturbance; $d(t) \in L_2(\mathbb{R}_+;\mathbb{R}^{p_d})$ is the vector of the external disturbance; $D_n \in \mathbb{R}^q$ is the influence matrix of the measurement noise and $n(t) \in L_2(\mathbb{R}_+;\mathbb{R})$ is the measurement noise at the locations of observation. The differential operator A in $L_2(U;\mathbb{R}^n)$ is defined as follows [330]:

$$\mathcal{A}y(x,t) \triangleq \kappa_1 \frac{\partial^2}{\partial x_1^2} y(x,t) + \kappa_2 \frac{\partial^2}{\partial x_2^2} y(x,t)$$

where $\frac{\partial^2}{\partial x_k^2} y(x,t) \triangleq \left[\frac{\partial^2}{\partial x_k^2} y_1(x,t),...,\frac{\partial^2}{\partial x_k^2} y_n(x,t)\right]^T \in \mathbb{R}^n$ for $k=1$, 2 in the two-dimensional case; κ_1 and κ_2 are the diagonal matrices; $y(x,t) \in D(\mathcal{A}) = \{y(x,t) \in L_2(U;\mathbb{R}^n) \,|\, y(x,t), \frac{\partial y(x,t)}{\partial x_1}, \frac{\partial y(x,t)}{\partial x_2}$ are absolutely continuous, and $\kappa_1 \frac{\partial^2 y(x,t)}{\partial x_1^2} + \kappa_2 \frac{\partial^2 y(x,t)}{\partial x_2^2} \in L_2(U;\mathbb{R}^n)\}$. The initial-value is given by $y(x,0)=y_0(x)$. The boundary condition is given by Dirichlet boundary condition, i.e. $y(x,t)=0$ on ∂U, or Neumann boundary condition, i.e., $\partial y(x,t)/\partial x=0$ on ∂U.

Remark 10.2.1

In this chapter, we address the robust tracking control design problem for NDPSs via the proposed fuzzy approach. Therefore, we study the constant delay for NDPSs. In general, the cases of time-varying delay or multiple time delays can be studied by combining the proposed fuzzy approach and the delay-dependent approach for ODE systems [344–347].

A desired reference trajectory $y_R(x,t) \in \mathbb{R}^n$ is generated by the following linear distributed parameter reference model:

$$\frac{\partial y_R(x,t)}{\partial t} = \mathcal{A}_R y_R(x,t) + g_R(x) r(t) \tag{10.3}$$

where $\mathcal{A}_R$ is a specified linear differential operator; A_R is a specified matrix; $g_R(x)$ is a specified influence function; $r(t)$ is a bounded reference input. The linear distributed parameter reference model in equation (10.3) can be designed similar to the reference model in [335,357]. Firstly, we design the matrix A_R and the reference input $r(t)$ to decide the behavior of the time evolution [357]. Then we choose the appropriate diffusion coefficients in differential operator $\mathcal{A}_R$ according to the practical application case. The task of tracking control is to make $y(x, t)$ in equations (10.1) and (10.2) track the desired trajectory $y_R(x, t)$ generated by equation (10.3).

The external disturbance $d(t)$ and the measurement noise $n(t)$ are uncertain, and the reference input $r(t)$ could be arbitrarily assigned by users, which can be all considered as disturbances of the tracking system. Therefore, the robust two-dimensional tracking control design should be specified so that the effect of the external disturbance $d(t)$, the measurement noise $n(t)$, and the reference input $r(t)$ on the tracking error in the spatiotemporal domain must be below a prescribed level ρ as follows:

$$\frac{\int_0^{t_f} \int_U \|y_R(x,t) - y(x,t)\|^2 \, dx\, dt}{\int_0^{t_f} \|v(t)\|^2 \, dt} \leq \rho^2 \tag{10.4}$$

or

$$\int_0^{t_f} \int_U \|y_R(x,t) - y(x,t)\|^2 \, dx\, dt \leq \rho^2 \int_0^{t_f} \|v(t)\|^2 \, dt \tag{10.5}$$

where $v(t) = \left[r(t)^T, d(t)^T, n(t)^T \right]^T$ is considered as a vector of disturbances. If the initial condition is considered, then

$$\int_0^{t_f} \int_U \|y_R(x,t) - y(x,t)\|^2 \, dx\, dt \leq V(y_0) + \rho^2 \int_0^{t_f} \|v(t)\|^2 \, dt \tag{10.6}$$

for some positive function $V(\cdot) > 0$. The inequality in equations (10.4) or (10.5) is called the two-dimensional H_∞ tracking performance on the spatio-temporal domain. Its physical meaning is that the effect of all disturbances on the tracking error at the total space $U \times [0, t_f]$ must be attenuated below a prescribed level ρ from the energy perspective. Note that the norm $\int_0^{t_f} \int_U \|y_R(x,t) - y(x,t)\|^2 dxdt$ is to describe the energy of tracking error for all positions in the spatio-temporal space $U \times [0, t_f]$, which is different from the conventional H_∞ tracking performance that is only in the time domain. The robust H_∞ tracking control problem is formulated as follows: Given a prescribed disturbance attenuation level ρ, a robust controller $u(t)$ is designed to attenuate the effect of the external disturbances $d(t)$ and measurement noises $n(t)$ on the tracking error $y_R(x, t)-y(x, t)$ below ρ from the view point of total energy on the space $U\times[0, t_f]$ to robustly track the desired reference trajectory $y_R(x, t)$. In other words, the H_∞ tracking performance in equation (10.5) is achieved.

Remark 10.2.2

The fuzzy tracking control design for nonlinear ODE systems in [335] has previously been used to control the states of the nonlinear dynamic system to track a desired state trajectory, but in this chapter it is more difficult to design the tracking control of the NDPS in equation (10.1) because the NDPS is a spatiotemporal dynamic system and the H_∞ tracking performance equations (10.4) or (10.6) should be achieved in the spatiotemporal domain. An output regulation for an LDPS developed previously [357] is designed to track a reference output generated by an exogenous system. This reference output is limited to some specific trajectories, e.g. stable trajectory or sinusoidal trajectory. However, in this chapter, the tracking problem of NDPSs is to track any desired spatiotemporal trajectory which could be generated in equation (10.3). Therefore, this tracking control design scheme is more general than the output regulation design in [357].

Remark 10.2.3

We previously used the Galerkin's method in Chapter 9 to represent a PDE by an infinite dimensional ODE to solve a stabilization problem for the NDPS [356]. However, the effect of residual subsystem on the tracking error cannot be neglected because the asymptotic tracking cannot be achieved in NDPS due to the change of $y_R(x, t)$. Moreover, the Galerkin's method is difficult to extend into the multi-space variable case for the state $y(x, t)$. Therefore, in this chapter, we applied the finite difference approach to solve the tracking control problem for the NDPS.

10.3 SYSTEM REPRESENTATION BY FUZZY SPATIAL STATE SPACE MODEL

In the control design problems of NDPSs, the main problem is to obtain a suitable state space model to represent the NDPSs. At first, a T-S fuzzy DPS with time delay is proposed to approximate the NDPS with time delay as follows [356]:

Rule i: If $y_1(x, t)$ is F_{1i}, and$\cdots$, and, $y_n(x, t)$ is F_{ni}

$$\text{then } \frac{\partial y(x,t)}{\partial t} = \mathcal{A}y(x,t) + A_i y(x,t) + A_{\tau,i} y(x,t_\tau) + g(x)u(t) + g_d(x)dt$$

where F_{ji} is the grade of the membership of $y_j(x, t)$; and $A_i \in \mathbb{R}^{n\times n}$ and $A_{\tau,i} \in \mathbb{R}^{n\times n}$ are the system parameters without time delay and with time delay τ in local LDPSs, respectively. The overall fuzzy DPS can be formulated as follows [335,356,372,373]:

$$\frac{\partial y(x,t)}{\partial t} = \sum_{i=1}^{M} \mu_i(y(x,t))[Ay(x,t) + A_i y(x,t) + A_{\tau,i} y(x,t_\tau)] + g(x)u(t)$$

$$+ g_d(x)d(t) + \varepsilon(x,t) \tag{10.7}$$

where $\mu_i(y(x,t)) \triangleq \frac{\prod_{j=1}^{n} F_{ji}(y_j(x,t))}{\sum_{i=1}^{M}\prod_{j=1}^{n} F_{ji}(y_j(x,t))}$. $F_{ji}\left(y_j(x,t)\right)$ is the grade of the membership of $y_j(x, t)$ or the possibility function of $y_j(x, t)$. The denominators of $\mu_i(y(x, t))$ are only for normalization so that the total sum of the fuzzy bases $\sum_{i=1}^{M} \mu_i(y(x,t)) = 1$.

In equation (10.7), we use the fuzzy interpolation via M local LDPSs to approximate the NDPS in equation (10.1). The approximation error $\varepsilon(x,t) \in \mathbb{R}^n$ is defined as follows:

$$\varepsilon(x,t) \triangleq f(y(x,t), y(x,t_\tau)) - \sum_{i=1}^{M} \mu_i(y(x,t))\left[A_i y(x,t) + A_{\tau,i} y(x,t_\tau)\right] \tag{10.8}$$

The bound of $\varepsilon(x,t)$ could be estimated according to the following theorem.

Theorem 10.3.1 [356]

Suppose $f(y(x,t), y(x,t_\tau))$ is a continuous function defined on a compact set $U \subset \mathbb{R}^n$, i.e., $y(x,t), y(x,t_\tau) \in U$. Then, for two arbitrary constants $\sigma > 0$ and $\sigma_\tau > 0$, the fuzzy function $\tilde{f}(y(x,t), y(x,t_\tau)) \triangleq \sum_{i=1}^{M} \mu_i(y(x,t))\left[A_i y(x,t) + A_{\tau,i} y(x,t_\tau)\right]$ could be constructed to approximate the nonlinear function $f(y(x,t), y(x,t_\tau))$ and the approximation error is bounded by σ and σ_τ, i.e., $\|\varepsilon(x,t)\|^2 \le \sigma^2 \|y(x,t)\|^2 + \sigma_\tau^2 \|y(x,t_\tau)\|^2$.

The finite difference scheme [374,375] has been widely applied to obtain numerical solutions of PDEs. In this chapter, we use the finite difference method to represent the NDPS. Consider a typical grid mesh as shown in Figure 10.1. The state $y(x, t)$ is represented by $y_{k,l}(t) \in \mathbb{R}^n$ at the grid node $x_{k,l}\left(x_1 = k\Delta_x, x_2 = l\Delta_x\right)$ where $k=0,\ldots, N_1+1$ and $l=0,\ldots, N_2+1$, i.e., $y(x,t)\big|_{x_{k,l}} \triangleq y_{k,l}(t)$. Note that the grid nodes $k=0$, $k=N_1+1$, $l=0$, or $l=N_2+1$ are the grid nodes at the boundary. At the interior points of grid, i.e., $0<k<N_1+1$ and $0<l<N_2+1$, the central difference approximation for the linear differential operator can be written as follows [374,375]:

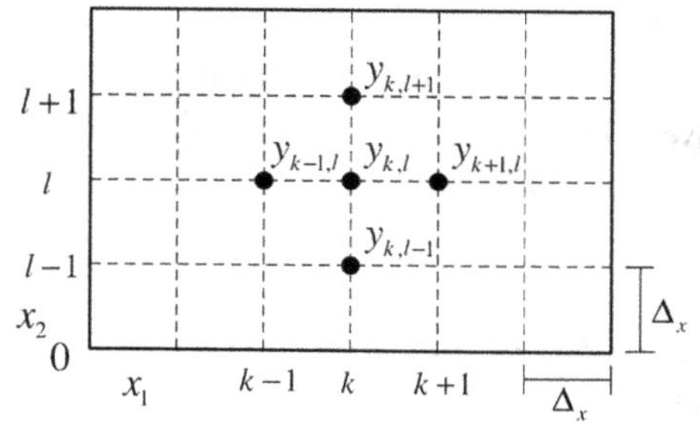

FIGURE 10.1 Finite difference grids on the spatio-domain.

$$\mathcal{A}y(x,t)_{x=x_{k,l}} = \kappa_1\left(\frac{\partial^2 y(x,t)}{\partial x_1^2}\right)_{x=x_{k,l}} + \kappa_2\left(\frac{\partial^2 y(x,t)}{\partial x_2^2}\right)_{x=x_{k,l}}$$

$$= \kappa_1 \frac{y_{k+1,l}(t) + y_{k-1,l}(t) - 2y_{k,l}(t)}{\Delta_x^2} + \kappa_2 \frac{y_{k,l+1}(t) + y_{k,l-1}(t) - 2y_{k,l}(t)}{\Delta_x^2} + O_{k,l}(\Delta_x^2) \tag{10.9}$$

The remainder term $O_{k,l}\left(\Delta_x^2\right) \in \mathbb{R}^n$ is called the local truncation error.
A fuzzy finite difference model can be constructed to represent the state $y_{k,l}(t)$ of the NDPS at $x = x_{k,l}$ in equation (10.7) as follows:

$$\dot{y}_{k,l}(t) = \sum_{i=1}^{M} \mu_i(y_{k,l})[A_i y_{k,l}(t) + A_{\tau,i} y_{k,l}(t_\tau)] + B_{k,l}u(t) + B_{d,k,l}d(t) + \varepsilon_{k,l}(t)$$

$$+ \frac{1}{\Delta_x^2}\kappa_1 y_{k+1,l}(t) + \frac{1}{\Delta_x^2}\kappa_1 y_{k-1,l}(t) - \frac{1}{\Delta_x^2}\kappa_1 y_{k,l}(t)$$

$$+ \frac{1}{\Delta_x^2}\kappa_2 y_{k,l+1}(t) + \frac{1}{\Delta_x^2}\kappa_2 y_{k,l-1}(t) - \frac{1}{\Delta_x^2}\kappa_2 y_{k,l}(t) + O_{k,l}(\Delta_x^2). \tag{10.10}$$

The elements of matrix $B_{k,l} = \left[B_{k,l,1},\ldots,B_{k,l,p}\right] \in \mathbb{R}^{n\times p}$ are defined as $B_{k,l,i} = g_i$ for $x_{k,l} = p_i$ or $B_{k,l,i} = 0$ for $x_{k,l} \neq p_i$ where g_i is the influence function at the location p_i and is defined in equation (10.1). The definition of $B_{d,k,l} = 0$ is similar to $B_{k,l}$. The approximation error $\varepsilon_{k,l}(t) \triangleq \varepsilon(x,t)\big|_{x=x_{k,l}}$.

We define a spatial state vector $y(t)$ to collect the states $y_{k,l}(t) \in \mathbb{R}^n$ at all grid nodes in Figure 10.1. For Dirichlet boundary conditions [375], the values of $y_{k,l}(t)$ at boundary are fixed, for example, $y(x, t)=0$ on ∂U. We have $y_{k,1}(t)=0$ at $k=0, N_1+1$ or $l=0, N_2+1$. Therefore, the spatial state vector $y(t) \in \mathbb{R}^{nN}$ is defined as follows:

$$y(t) = [y_{1,1}^T(t),\ldots,y_{1,l}^T(t),\ldots,y_{1,N_2}^T(t),\ldots,y_{k,l}^T(t),\ldots,y_{N_1,1}^T(t),\ldots,y_{N_1,l}^T(t),\ldots,y_{N_1,N_2}^T(t)]^T \tag{10.11}$$

where $N \triangleq N_1 \times N_2$. Note that n is the dimension of the vector $y_{k,l}(t)$ for each grid node and $N_1 \times N_2$ is the number of grid nodes. For example, let $N_1 = 2$ and $N_2 = 2$, we have $y(t) = \left[y_{1,1}^T(t), y_{1,2}^T(t), y_{2,1}^T(t), y_{2,2}^T(t)\right]^T$. For Neumann boundary conditions [375], i.e., $\partial y(x, t)/\partial x = 0$ on ∂U, the boundary condition is given as $\left(\frac{\partial y(x,t)}{\partial x}\right)_{x=x_{k,l}} = 0$ at $k=0$, N_1+1, $l=0$, N_2+1. Therefore, if the grid nodes at boundary are also considered in the spatial state vector $y(t)$, then $y(t) \in \mathbb{R}^{nN}$ in equation (10.11) should be modified as follows:

$$y(t) = \begin{bmatrix} y_{0,0}^T(t), \ldots, y_{0,l}^T(t), \ldots, y_{0,N_2+1}^T(t), \ldots, y_{k,l}^T(t), \\ \ldots, y_{N_1+1,0}^T(t), \ldots, y_{N_1+1,l}^T(t), \ldots, y_{N_1+1,N_2+1}^T(t) \end{bmatrix}^T \tag{10.12}$$

where $N \triangleq (N_1 + 1) \times (N_2 + 2)$.

In order to simplify the index of the node $y_{k,l}(t) \in \mathbb{R}^n$ in the spatial state vector $y(t) \in \mathbb{R}^{nN}$, we denote the symbol $y_j(t) \in \mathbb{R}^n$ to replace $y_{k,l}(t)$. Note that the index j is from 1 to N, i.e., $y_1(t) \triangleq y_{1,1}(t), y_2(t) \triangleq y_{1,2}(t), \ldots, y_j(t) \triangleq y_{k,l}(t), \ldots, y_N(t) \triangleq y_{N_1,N_2}(t)$ where $j = (k-1)N_1 + l$ in equation (10.11). The fuzzy finite difference model of two indices in equation (10.10) could be represented with only one index as follows:

$$\dot{y}_j(t) = \sum_{i=1}^{M} \mu_i(y_j)[A_i y_j(t) + A_{\tau,i} y_j(t_\tau)] + B_j u(t) + B_{d,j} d(t) + \varepsilon_j(t) + T_j y(t) + O_j(\Delta_x^2) \tag{10.13}$$

where finite difference matrix $T_j \in \mathbb{R}^{n \times nN}$ expresses the interaction from the other grid nodes to the node $y_j(t) \triangleq y_{k,l}(t)$ as follows:

$$T_j y(t) \triangleq \frac{\kappa_1}{\Delta_x^2} y_{k+1,l}(t) + \frac{\kappa_1}{\Delta_x^2} y_{k-1,l}(t) - \frac{2\kappa_1}{\Delta_x^2} y_{k,l}(t) + \frac{\kappa_2}{\Delta_x^2} y_{k,l+1}(t) + \frac{\kappa_2}{\Delta_x^2} y_{k,l-1}(t) - \frac{2\kappa_2}{\Delta_x^2} y_{k,l}(t)$$

The measurement output $z(t)$ in equation (10.2) can be represented as follows:

$$z(t) = \sum_{j=1}^{N} C_j y_j(t) + D_n n(t) \tag{10.14}$$

where the matrix C_j is defined as $C_j \triangleq C_{k,l} = \left[C_{k,l,1}, \ldots, C_{k,l,q}\right]^T$, in which the element is given as $C_j \triangleq C_{k,l} = \left[C_{k,l,1}, \ldots, C_{k,l,q}\right]^T$ for $x_{k,l} = q_i$ or $C_{k,l,1} = 0$ for $x_{k,l} \neq q_i$. Note that q_i is the ith sensor location as defined in equation (10.2). The matrix D_n is also defined in equation (10.2).

We collect all states $y_j(t)$ of grid nodes in equations (10.13) and (10.2) to the state $y(t)$ in equations (10.11) or (10.12). Then, a diagonal fuzzy weighting matrix $\mu_i(y) \in \mathbb{R}^{N \times N}$ is defined as $\mu_i(y) \triangleq diag(\mu_i(y_1), \ldots, \mu_i(y_N))$ and the $n \times n$ identity matrix by I_n. The Kronecker product can be used to simplify the representation. Some properties for $\mu_i(y)$ can be obtained as follows.

Lemma 10.3.1

Using the properties of Kronecker product, we have the following properties.

1. $\mu_i(y)\otimes A_i=(\mu_i(y)\otimes I_n)(I_N\otimes A_i)\in\mathbb{R}^{nN\times nN}$ where $A_i\in\mathbb{R}^{n\times n}$
2. $(\mu_i(y)\otimes I_n)(\mu_j(\hat{y})\otimes I_n)=(\mu_j(y)\mu_j(\hat{y})\otimes I_n)\in\mathbb{R}^{nN\times nN}$
3. $\Sigma_{i=1}^{M}\mu_i(y)=I_N, \Sigma_{i=1}^{M}(\mu_i(y)\otimes I_n)=I_nN$

and $\Sigma_{i=1}^{M}\Sigma_{j=1}^{M}(\mu_i(y)\mu_j(\hat{y})\otimes I_n)=I_nN$

Proof

The properties (1) and (2) are the fundamental properties for Kronecker product [376]. The property (3) can be proven via the fuzzy fundamental property, $\Sigma_{i=1}^{M}\mu_i(y_j(t))=1, j=1,\ldots,N$.

Using the Kronecker product, the systems in equations (10.13) and (10.14) can be written as the following fuzzy spatial state space system.

$$\begin{aligned}\dot{y}(t)&=\sum_{i=1}^{M}[(\mu_i(y)\otimes A_i)y(t)+(\mu_i(y)\otimes A_{\tau,i})y(t_\tau)]+Ty(t)+Bu(t)+B_d d(t)\\&\quad+\varepsilon(t)+O(\Delta_x^2)\\&=\sum_{i=1}^{M}\bar{\mu}_i(y)[(I_N\otimes A_i)y(t)+(I_N\otimes A_{\tau,i})y(t_\tau)]+Ty(t)+Bu(t)+B_d d(t)\\&\quad+\varepsilon(t)+O(\Delta_x^2)\end{aligned}\tag{10.15}$$

$$z(t)=Cy(t)+D_n n(t)\tag{10.16}$$

where we define $\bar{\mu}_i(y)\triangleq\mu_i(y)\otimes I_n$, and the spatial state vector $y(t)\in\mathbb{R}^{nN}$ is denoted in equations (10.11) or (10.12) to represent $y(x,t)$ at all finite difference grid points on the spatial domain in Figure 10.1. The corresponding matrices are defined as $T=\left[T_1^T,\ldots,T_N^T\right]^T\in\mathbb{R}^{nN\times nN}$, $B=\left[B_1^T,\ldots,B_N^T\right]^T\in\mathbb{R}^{nN\times p}$, $B_d=\left[B_{d,1}^T,\ldots,B_{d,N}^T\right]^T\in\mathbb{R}^{nN\times p_d}$, and $C=[C_1,\ldots,C_N]\in\mathbb{R}^{q\times nN}$. The matrix D_n is defined in equation (10.2). The approximation error $\varepsilon(t)$ is defined as $\varepsilon(t)=\left[\varepsilon_1(t)^T,\ldots,\varepsilon_N(t)^T\right]^T$.

The physical meaning of equations (10.15) and (10.16) is that the NDPS in equations (10.1) and (10.2) at all finite difference grid points on the spatial domain in Figure 10.1 can be represented by the fuzzy spatial state space system in equations (10.15) and (10.16). In Theorem 10.3.1, the bound of approximate error $\varepsilon(x, t)$ can be proved to be less than two arbitrary constants σ and σ_τ. The bound for approximate error $\varepsilon(t)$ can be obtained by the following corollary.

Corollary 10.3.1

If the bounds of $\varepsilon(x, t)$ are provided with σ and σ_τ, then the bound of $\varepsilon(t)$ could also be estimated as $\|\varepsilon(t)\|^2 \le \sigma^2 \|y(t)\|^2 + \sigma_\tau^2 \|y(t_\tau)\|^2$.

Proof

By Theorem 10.3.1, we have $\|\varepsilon_j(t)\|^2 \le \sigma^2 \|y_j(t)\|^2 + \sigma_\tau^2 \|y_j(t_\tau)\|^2$. Thus, $\|\varepsilon(t)\|^2 = \sum_{j=1}^{N} \|\varepsilon_j(t)\|^2 \le \sum_{j=1}^{N} \sigma^2 \|y_j(t)\|^2 + \sigma_\tau^2 \|y_j(t_\tau)\|^2 = \sigma^2 \|y(t)\|^2 + \sigma_\tau^2 \|y(t_\tau)\|^2$.

Remark 10.3.1

Similarly, the reference model in equation (10.3) could be transformed into a linear spatial state space reference model by the finite difference method, i.e., the desired reference trajectory $y_R(t)$ at all grid points can be generated by the following linear spatial state space reference model

$$\dot{y}_R(t) = (I_N \otimes A_R) y_R(t) + T_R y_R(t) + B_R r(t) + O_R(\Delta_x^2) \tag{10.17}$$

where A_R is a specified system matrix; B_R is the specified influence matrix; T_R is the finite difference matrix; and $O_R\left(\Delta_x^2\right)$ is the truncation error.

Remark 10.3.2

The state variable $y(x, t)$ in equation (10.1) depends on the space x and time t. Based on the finite difference scheme [374,375], the spatial state vector y(t) in equations (10.11) or (10.12) is used to represent the $y(x, t)$ at all grid points. Similarly, the state variables $y_R(x, t)$ at all grid points can be represented by spatial state vector $y_R(t)$. In this situation, the $\int_0^{t_f} \int_U \|y_R(x,t) - y(x,t)\|^2 \, dx \, dt$ in the two-dimensional H_∞ tracking performance equations (10.4) or (10.6) could be modified by the one-dimensional form $\int_0^{t_f} \|y_R(t) - y(t)\|^2 \Delta_x^2 \, dt$ for the finite-difference systems equations (10.15) and (10.17) to simplify the design procedure. As $\Delta_x \to 0$ the one-dimensional systems in equations (10.15) and (10.17) will approach two-dimensional systems equations (10.7) and (10.3) respectively, and the one-dimensional integration form will approach the two-dimensional integration form.

10.4 ROBUST FUZZY OBSERVER-BASED TRACKING CONTROL DESIGN

In Section 10.3, a fuzzy DPS in equation (10.7) was proposed to approximate the NDPS in equation (10.1). Then, according to the finite difference method, the fuzzy spatial state space model in equations (10.15) and (10.16) can be constructed for NDPSs. In this section, a fuzzy observer-based tracking control scheme is developed to solve the tracking control design problem of NPDSs. Then, a set of the complex matrix inequalities is derived to guarantee the robust H_∞ tracking performance. Finally, the design procedure is given to simplify the robust tracking design problem of NDPSs by solving a set of LMIs to obtain the controller gains and observer gains.

10.4.1 Robust H_∞ Fuzzy Observer-Based Tracking Control Design

A fuzzy observer-based tracking controller based on the fuzzy spatial state space system in equations (10.15) and (10.16) is proposed to estimate the state of fuzzy spatial state space system and then using a state feedback scheme to control the NDPS in equations (10.1) and (10.2) to thereby robustly track a desired trajectory $y_R(x, t)$ generated by the reference model in equation (10.3). The fuzzy observer-based tracking controller is proposed as the following form (see Figure 10.2)

$$\dot{\hat{y}}(t) = \sum_{i=1}^{M} \bar{\mu}_i(y)[(I_N \otimes A_i)\hat{y}(t) + (I_N \otimes A_{\tau,i})\hat{y}(t_\tau) + T\hat{y}(t) + Bu(t) + G_i(C\hat{y}(t) - z(t))] \tag{10.18}$$

$$u(t) = \sum_{i=1}^{M}\sum_{j=1}^{N} \mu_i(\hat{y}_j)K_{i,j}[\hat{y}_j(t) - y_{R,j}(t)] \tag{10.19}$$

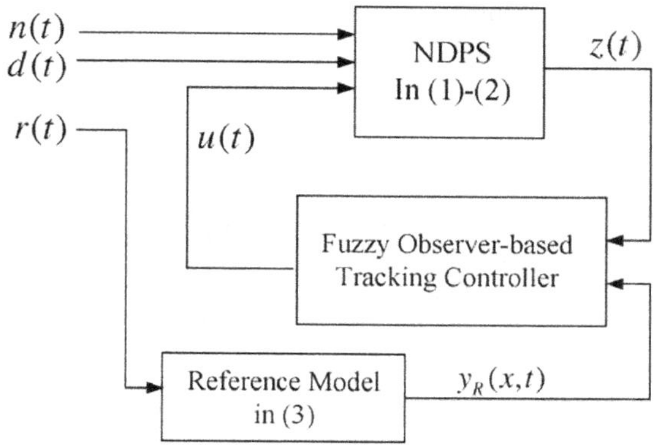

FIGURE 10.2 Block diagram of the tracking control scheme for the NDPS in equations (10.1) and (10.2).

$$= \sum_{i=1}^{M} K_i \bar{\mu}_i(\hat{y})[\hat{y}(t) - y_R(t)] \tag{10.20}$$

where $G_i \in \mathbb{R}^{nN \times q}$ and $K_i \triangleq [K_{i,1}, ..., K_{i,N}] \in \mathbb{R}^{p \times nN}$ are, respectively, the observer gains and the control gains to be designed. The fuzzy weighting matrix $\bar{\mu}_i(\hat{y})$ is defined as $\bar{\mu}_i(\hat{y}) \triangleq (\mu_i(\hat{y}) \otimes I_n)$. In the observer equation (10.18), the observer gains G_i are designed so that the spatial state estimation error $\hat{y}(t) - y(t)$ is as small as possible. In equation (10.19), the N fuzzy tracking controllers $u_j(t) = \sum_{i=1}^{M} \mu_i(\hat{y}_j) K_{i,j}[\hat{y}_j(t) - y_{R,j}(t)] \in \mathbb{R}^p$ are designed to make the tracking errors $\hat{y}_j(t) - y_{R,j}(t)$ as small as possible in spite of the fuzzy approximation error, the truncation error, the time delay, the external disturbance, and the measurement noise, i.e., to achieve the robust H_∞ tracking design. The overall controller $u(t) \triangleq \sum_{j=1}^{N} u_j(t)$ can be formulated in equation (10.20).

The dynamic of the estimated error $e(t) = y(t) - \hat{y}(t)$ is given by

$$\begin{aligned}
\dot{e}(t) &= \sum_{i=1}^{M} \bar{\mu}_i(y)[(I_N \otimes A_i)y(t) + (I_N \otimes A_{\tau,i})y(t_\tau)] + Ty(t) + B_d d(t) + \varepsilon(t) + O(\Delta_x^2) \\
&\quad - \left(\sum_{i=1}^{M} \bar{\mu}_i(\hat{y})[(I_N \otimes A_i)\hat{y}(t) + (I_N \otimes A_{\tau,i})\hat{y}(t_\tau) + T\hat{y}(t) + G_i(C\hat{y}(t) - z(t))] \right) \\
&= \sum_{i_1=1}^{M} \sum_{i_2=1}^{M} \bar{\mu}_{i_1}(y)\bar{\mu}_{i_2}(\hat{y})[((I_N \otimes A_{i_2}) + G_{i_2}C)e(t) + (I_N \otimes A_{\tau,i_2})e(t_\tau) + Te(t) \\
&\quad + (I_N \otimes (A_{i_1} - A_{i_2}))y(t) + (I_N \otimes (A_{\tau-i_1} - A_{\tau,i_2}))y(t_\tau) + B_d d(t) + G_{i_2} D_n n(t)] \\
&\quad + \varepsilon(t) + O(\Delta_x^2)
\end{aligned} \tag{10.21}$$

Combining the fuzzy system in equation (10.15) with the observer-based tracking controller, the closed-loop fuzzy tracking control system could be described by the following augmented system.

$$\dot{\bar{y}}(t) = \bar{A}(\mu, \hat{\mu})\bar{y}(t) + \bar{A}_\tau(\mu, \hat{\mu})\bar{y}(t_\tau) + \bar{B}_v(\hat{\mu})v(t) + \bar{\varepsilon}(t) + \bar{O}(\Delta_x^2) \tag{10.22}$$

where $\bar{y}(t) = \left[y_R(t)^T, y(t)^T, e(t)^T \right]^T$, $\bar{\varepsilon}(t) = \left[0, \varepsilon(t)^T, \varepsilon(t)^T \right]^T$, and $\bar{O}(\Delta_x^2) = \left[O_R(\Delta_x^2)^T, O(\Delta_x^2)^T, O(\Delta_x^2)^T \right]^T$. The other notations are defined as follows:

where $\Psi_{11} \triangleq I_N \otimes A_R + T_R$, $\Psi_{21} \triangleq -BK_{i_2}\bar{\mu}_{i_2}(\hat{y})$, $\Psi_{22} \triangleq \bar{\mu}_{i_1}(y)(I_N \otimes A_{i_1}) + T + BK_{i_2}\bar{\mu}_{i_2}(\hat{y})$, $\Psi_{23} \triangleq -BK_{i_2}\bar{\mu}_{i_2}(\hat{y})$, $\Psi_{32} \triangleq \bar{\mu}_{i_1}(y)\bar{\mu}_{i_2}(\hat{y})(I_N \otimes (A_{i_1} - A_{i_2}))$, and $\Psi_{33} \triangleq \bar{\mu}_{i_1}(y)\bar{\mu}_{i_2}(\hat{y})((I_N \otimes A_{i_2}) + T + G_{i_2}C)$. Note that the augmented system in equation (10.22) includes the reference model in equation (10.17), the fuzzy spatial state space system in equation (10.15) with fuzzy controller in equation (10.20), and the estimated error dynamic in equation (10.21).

The robust H_∞ fuzzy observer-based tracking controller is specified to guarantee that the effect of the time delay, the approximation error, the truncation error, the external disturbance, and the measurement noise on the tracking error $y_R(t) - y(t)$ and the estimation error $e(t)$ are attenuated below a prescribed level ρ from the energy perspective. This design problem is called the H_∞ observer-based tracking control design problem. Based on the augmented system in equation (10.22) and Remark 10.3.2, the H_∞ observer-based tracking control performance should be modified to include the state estimation error $e(t)$ and the effect of truncation error $\bar{O}\left(\Delta_x^2\right)$ as follows:

$$\frac{\int_0^{t_f} \left\|\bar{R}_1(y_R(t) - y(t))\right\|^2 + \left\|\bar{R}_2 e(t)\right\|^2 dt}{\int_0^{t_f} \left\|v(t)\right\|^2 + \left\|\bar{O}(\Delta_x^2)\right\|^2 dt} \le \rho^2 \tag{10.23}$$

where the weighting matrices $\bar{R}_1 \triangleq diag(\Delta_x R_1, \ldots, \Delta_x R_1) \in \mathbb{R}^{nN \times nN}$ and $\bar{R}_2 \triangleq diag(\Delta_x R_2, \ldots, \Delta_x R_2) \in \mathbb{R}^{nN \times nN}$ with $R_1 \in \mathbb{R}^{n \times n}$ and $R_2 \in \mathbb{R}^{n \times n}$, i.e., the effect of disturbances, measurement noise and truncation error on the tracking error, and the estimation error should be below the attenuation level ρ. The weighting matrices $\bar{R}_1$ and $\bar{R}_2$ denote the tradeoff between the tracking error and the estimation error. Since the truncation error due to finite difference approach appears in equation (10.22), its effect should be included in the H_∞ tracking performance in the design procedure. According to equation (10.6), the H_∞ observer-based tracking performance in equation (10.23) can be represented by the following inequality:

$$\int_0^{t_f} \left\|Ry(t)\right\|^2 dt \le V(\bar{y}(0)) + \rho^2 \int_0^{t_f} \left\|v(t)\right\|^2 + \left\|\bar{O}(\Delta_x^2)\right\|^2 dt \tag{10.24}$$

for some positive function $V(\bar{y}(0))$ when the initial condition $\bar{y}(0)$ is also considered, where

$$R = \begin{bmatrix} \bar{R}_1 & -\bar{R}_1 & 0 \\ 0 & 0 & \bar{R}_2 \end{bmatrix}$$

and $\bar{y}(t)$ is the state vector of the augmented system in equation (10.22).

Therefore, the H_∞ observer-based tracking control design for NDPSs is based on how to specify the control gains and observer gains of the fuzzy observer-based controller in equations (10.18) and (10.20) so that the H_∞ observer-based tracking performance in equations (10.23) or (10.24) could be achieved.

Remark 10.4.1

The term $\left\|\bar{R}y(t)\right\|^2$ can be represented as $\left\|\bar{R}y(t)\right\|^2 = \sum_{j=1}^{N} \left\|R_1(y_{R,j}(t) - y_j(t))\right\|^2 \Delta_x^2 + \left\|R_2(y_j(t) - y_j(t))\right\|^2 \Delta_x^2$ where Δ_x^2 is the area of the gridded rectangle; and the term $\|Ry(t)\|^2$, called the Riemann sum [54], can be used to approximate the integration of the two-dimensional H_∞ tracking performance in a spatiotemporal domain. Therefore, if the gridded spacing Δ_x of the finite difference approach is sufficiently small so that $O\left(\Delta_x^2\right) \rightarrow 0$, then the H_∞ observer-based tracking performance in equation (10.23) will approach the two-dimensional H_∞ observer-based tracking performance in spatiotemporal domain

$$\frac{\int_0^{t_f} \int_U \left\|\bar{R}_1(y_R(x,t) - y(x,t))\right\|^2 + \left\|\bar{R}_2 e(x,t)\right\|^2 dx\, dt}{\int_0^{t_f} \|v(t)\|^2 dt} \leq \rho^2 \tag{10.25}$$

where $e(x,t) \triangleq y(x,t) - \hat{y}(x,t)$.

Lemma 10.4.1 [378]

$X^T PY + Y^T PX \leq \xi X^T PX + \frac{1}{\xi} Y^T PY$ for any positive constant ξ, a symmetric matrix $P = P^T \geq 0$ and two vectors X and Y with appropriate dimensions.

For solving the H_∞ tracking problem in equation (10.25), let us choose a Lyapunov function $V(\bar{y}(t))$ for the augmented system in equation (10.22) as

$$V(\bar{y}(t)) = \bar{y}(t)^T P\bar{y}(t) + \int_{t_\tau}^{t} \bar{y}(s)^T Q\bar{y}(s)ds \tag{10.26}$$

where $P = P^T > 0$ and $Q = Q^T > 0$. Based on equation (10.26), we can obtain the following result for robust H_∞ tracking problem in equation (10.24).

Theorem 10.4.1

For the augmented system equation (10.22) with a prescribed disturbance attenuation level ρ in equation (10.24), if there exist two symmetric positive definite matrices P, Q, the control gains K_i, and observer gains G_i, for $i = 1,\ldots, M$, such that

$$\bar{\Pi}(\mu,\hat{\mu}) \triangleq \begin{bmatrix} \bar{\Theta}_{11}(\mu,\hat{\mu}) & P\bar{A}_\tau(\mu,\hat{\mu}) & P\bar{B}_v(\hat{\mu}) \\ \bar{A}_\tau^T(\mu,\hat{\mu})P & \xi\Sigma_\tau - Q & 0 \\ (\hat{\mu}) & 0 & -\rho^2 I \end{bmatrix} < 0 \tag{10.27}$$

where $\bar{\Theta}_{11}(\mu,\hat{\mu}) \triangleq \bar{A}_\tau(\mu,\hat{\mu})^T P + P\bar{A}(\mu,\hat{\mu}) + \left(\frac{1}{\rho^2} + \frac{1}{\xi}\right)PP + \xi\Sigma + R^T R + Q$, $\Sigma \triangleq$ *diag*$(0, 2\sigma^2 I, 0)$ *and* $\Sigma_\tau \triangleq$ *diag*$\left(0, 2\sigma_\tau^2 I, 0\right)$, *then the H_∞ observer-based tracking performance for NDPSs in equation (10.24) is guaranteed by the fuzzy observer-based controller in equation (10.18)–(10.20).*

Proof

Firstly, we differentiate the function $V(\bar{y}(t))$ in equation (10.26). Then adding and subtracting the term $\rho^2 v(t)^T v(t) - \bar{y}(t)^T R^T R\bar{y}(t)$, we can get

$$\begin{aligned}
\dot{V}(\bar{y}(t)) &= \dot{\bar{y}}(t)^T P\bar{y}(t) + \bar{y}(t)P\dot{\bar{y}}(t) + \bar{y}(t)^T Q\bar{y}(t) - \bar{y}(t_\tau)^T Q\bar{y}(t_\tau) \\
&= \left(\bar{A}(\mu,\hat{\mu})\bar{y}(t) + \bar{A}_\tau(\mu,\hat{\mu})\bar{y}(t_\tau) + \bar{B}_v(\hat{\mu})v(t) + \bar{\varepsilon}(t) + \bar{O}(\Delta_x^2)\right)^T P\bar{y}(t) \\
&\quad + \bar{y}(t)^T P\left(\bar{A}(\mu,\hat{\mu})\bar{y}(t) + \bar{A}_\tau(\mu,\hat{\mu})\bar{y}(t_\tau) + \bar{B}_v(\hat{\mu})v(t) + \bar{\varepsilon}(t) + \bar{O}(\Delta_x^2)\right) \\
&\quad + \bar{y}(t)^T Q\bar{y}(t) - \bar{y}(t_\tau)^T Q\bar{y}(t_\tau) - (\rho^2 v(t)^T v(t) - \bar{y}(t)^T R^T R\bar{y}(t)) \\
&\quad + (\rho^2 v(t)^T v(t) - \bar{y}(t)^T R^T R\bar{y}(t))
\end{aligned} \tag{10.28}$$

By Lemma 10.4.1, we have

$$\bar{O}(\Delta_x^2)^T P\bar{y}(t) + \bar{y}(t)^T P\bar{O}(\Delta_x^2) \le \rho^2\bar{O}(\Delta_x^2)^T\bar{O}(\Delta_x^2) + \frac{1}{\rho^2}\bar{y}(t)^T PP\bar{y}(t) \tag{10.29}$$

$$\bar{\varepsilon}(t)^T P\bar{y}(t) + \bar{y}(t)^T P\bar{\varepsilon}(t) \le \xi\bar{\varepsilon}(t)^T\bar{\varepsilon}(t) + \frac{1}{\xi}\bar{y}(t)^T PP\bar{y}(t) \tag{10.30}$$

where ξ is any positive constant. According to the bound of $\varepsilon(t)$ in Corollary 10.3.1, we have

$$\bar{\varepsilon}(t)^T\bar{\varepsilon}(t) = 2\varepsilon(t)^T\varepsilon(t) \le \bar{y}(t)^T\Sigma\bar{y}(t) + \bar{y}(t_\tau)^T\Sigma_\tau\bar{y}(t_\tau) \tag{10.31}$$

where Σ and Σ_τ are defined as $\Sigma \triangleq diag\left(0, 2\sigma^2 I, 0\right)$, and $\Sigma_\tau \triangleq diag\left(0, 2\sigma_\tau^2 I, 0\right)$. Based on the inequalities in equations (10.29)–(10.31), from equation (10.28) we can obtain

$$\dot{V}(\bar{y}(t)) \le \eta(t)^T\bar{\Pi}(\mu,\hat{\mu})\eta(t) + \rho^2 v(t)^T v(t) - \bar{y}(t)^T R^T R\bar{y}(t) + \rho^2\bar{O}(\Delta_x^2)^T\bar{O}(\Delta_x^2) \tag{10.32}$$

where $\eta(t) = [\bar{y}(t), \bar{y}(t_\tau), v(t)]$ and $\bar{\Pi}(\mu,\hat{\mu})$ is defined in equation (10.27). Suppose the inequality $\bar{\Pi}(\mu,\hat{\mu}) < 0$ holds, we have the following inequality:

$$\dot{V}_1(\bar{y}(t)) \le \rho^2 v(t)^T v(t) - \bar{y}(t)^T R^T R\bar{y}(t) + \rho^2\bar{O}(\Delta_x^2)^T\bar{O}(\Delta_x^2) \tag{10.33}$$

Integrating the above inequality equation (10.33) from $t = 0$ to $t = t_f$ yields

$$V(\bar{y}(t_f)) \le V(\bar{y}(0)) + \int_0^{t_f} \rho^2\|v(t)\|^2 - \|R\bar{y}(t)\|^2 + \rho^2\left\|\bar{O}(\Delta_x^2)\right\|^2 dt \tag{10.34}$$

Since $V(\bar{y}(t_f)) > 0$, we have the H_∞ tracking performance inequality in equation (10.24). Therefore, if the inequality in equation (10.27) holds, then the H_∞ tracking performance in equation (10.24) could be guaranteed.

Remark 10.4.2

(i) Recently, some approaches based on the fuzzy Lyapunov function [379] or piecewise Lyapunov function [380] have been developed to relax the conservativeness of stability and stabilization problems. The purpose of this chapter is to extend the fuzzy H_∞ tracking control approach to the field of NDPSs. Therefore, we use the Lyapunov function $V(\bar{y}(t))$ with common P and Q to simplify the reference tracking control design procedure of NDPSs with time delay. Therefore, the conservative analysis can refer to the related literature [340,344,345,379–382]. (ii) When $v(t) \equiv 0$ in equation (10.22), it is seen from equation (10.34) that $\bar{y}(t) \to 0$ and $R\bar{y}(t) = \left[(\bar{R}_1 y_R(t) - \bar{R}_1 y(t))^T, (\bar{R}_2 e(t))^T \right]^T \to 0$ as $t \to \infty$ [335], i.e., the inequality in equation (10.27) guarantees the asymptotical stability and asymptotical tracking of equation (10.22), simultaneously.

Since it is still very difficult to solve the matrix inequality in equation (10.27) to find control gains K_i and observer gains G_i for H_∞ observer-based tracking design of NDPSs, a simplification procedure is given below to improve the solution of matrix inequality in equation (10.27). We can define the Lyapunov function for the jth fuzzy finite difference model in equation (10.13), the jth reference finite difference model, and the jth estimated error dynamic as follows:

$$V_j(y_{R,j}(t), y_j(t), e_j(t)) = \psi_j(t)^T \begin{bmatrix} P_{11,j} & P_{12,j} & 0 \\ P_{12,j} & P_{11,j} & 0 \\ 0 & 0 & P_{33,j} \end{bmatrix} \psi_j(t) + \int_{t_\tau}^{t} \psi_j(s)^T$$

$$\begin{bmatrix} Q_{11,j} & Q_{12,j} & 0 \\ Q_{12,j} & Q_{11,j} & 0 \\ 0 & 0 & Q_{33,j} \end{bmatrix} \psi_j(s)ds \text{ where } \psi_j(t) \triangleq \left[y_{R,j}(t)^T, y_j(t)^T, e_j(t)^T \right]^T. \text{ Then,}$$

the Lyapunov function $V(\bar{y}(t))$ in equation (10.26) for the augmented system equation (10.22) can be formulated as

$$V(\bar{y}(t)) = \sum_{j=1}^{N} V_j(y_{R,j}(t), y_j(t), e_j(t))$$

Therefore, the following forms for the matrices P and Q in equation (10.26) can be easily obtained as

$$P = \begin{bmatrix} P_{11} & P_{12} & 0 \\ P_{12} & P_{11} & 0 \\ 0 & 0 & P_{33} \end{bmatrix} > 0, Q = \begin{bmatrix} Q_{11} & Q_{12} & 0 \\ Q_{12} & Q_{11} & 0 \\ 0 & 0 & Q_{33} \end{bmatrix} > 0 \quad (10.35)$$

where $P_{ij} = diag(P_{ij,1},\ldots,P_{ij,N}) \in \mathbb{R}^{nN\times nN}$ and $P_{ij} = diag(P_{ij,1},\ldots,P_{ij,N}) \in \mathbb{R}^{nN\times nN}$.

Remark 10.4.3

For simplicity, we define $\bar{P}_{(11)} \triangleq \begin{bmatrix} P_{11} & P_{12} \\ P_{12} & P_{11} \end{bmatrix}$, $\bar{Q}_{(11)} \triangleq \begin{bmatrix} Q_{11} & Q_{12} \\ Q_{12} & Q_{11} \end{bmatrix}$, $\bar{P}_{(22)} \triangleq P_{33}$, and $\bar{Q}_{(22)} \triangleq Q_{33}$ for the matrices P and Q in equation (10.35).

The following lemma gives a transformation technique from the summation of matrices into a large matrix form to simplify the design procedure of the H_∞ observer-based tracking control in Theorem 10.4.1. Thus, we can obtain the following results for $\bar{A}(\mu,\hat{\mu}), \bar{A}_\tau(\mu,\hat{\mu})$, and $\bar{B}_v(\hat{\mu})$.

Lemma 10.4.2

The system matrix $\bar{A}(\mu,\hat{\mu})$ in equation (10.22) can be represented as follows:

$$\bar{A}(\mu,\hat{\mu}) = \tilde{\mathcal{M}}^T[\tilde{A}+\tilde{\Xi}]\tilde{\mathcal{M}} \tag{10.36}$$

where

$$\tilde{\mathcal{M}} = \begin{bmatrix} \tilde{\mathcal{M}}_{(11)} & 0 \\ 0 & \tilde{\mathcal{M}}_{(22)} \end{bmatrix}, \tilde{\Xi} = \begin{bmatrix} \tilde{\Xi}_{(11)} & 0 \\ 0 & \tilde{\Xi}_{(22)} \end{bmatrix}, \tilde{A} = \begin{bmatrix} \tilde{A}_{(11)} + \tilde{B}\tilde{K}_{(11)} & \tilde{B}\tilde{K}_{(12)} \\ \tilde{A}_{(21)} & \tilde{A}_{(22)} + \tilde{G}_{(22)}\tilde{C} \end{bmatrix}.$$

Moreover, for the matrices $\bar{A}_\tau(\mu,\hat{\mu})$ and $\bar{B}_v(\hat{\mu})$, we have

$$\bar{A}_\tau(\mu,\hat{\mu}) = \tilde{\mathcal{M}}^T\tilde{A}_\tau, \bar{B}_v(\hat{\mu}) = \tilde{\mathcal{M}}^T\tilde{B}_v \tag{10.37}$$

where

$$\tilde{A}_\tau = \begin{bmatrix} \tilde{A}_{\tau(11)} & 0 \\ 0 & \tilde{A}_{\tau(22)} \end{bmatrix}, \tilde{B}_v = \begin{bmatrix} H_{(11)}^T\bar{B}_{v(11)} & 0 \\ H_{(22)}^T\bar{B}_{v(21)} & \tilde{G}_{v(22)}\bar{B}_{v(22)} \end{bmatrix},$$

$$H = \begin{bmatrix} H_{(11)} & 0 \\ 0 & H_{(22)} \end{bmatrix} \triangleq \begin{bmatrix} I_{2nN} & 0 & \cdots & 0 & 0 & 0 & \cdots & 0 \\ 0 & 0 & \cdots & 0 & I_{nN} & 0 & \cdots & 0 \end{bmatrix} \in \mathbb{R}^{3n\times 3n\tilde{N}}$$

in which the dimension number $\tilde{N}$ is defined as $\tilde{N} = N\times\tilde{M}, \tilde{M} = 1+(M\times M)$.

Before we prove Lemma 10.4.2, we first give the following arrangements. Since $\bar{\mu}_{i_1}(y)$ and $\bar{\mu}_{i_2}(\hat{y})$ are of the matrix form and have the property such as $\sum_{i_1=1}^{M}\sum_{i_2=1}^{M}\bar{\mu}_{i_1}(y)\bar{\mu}_{i_2}(\hat{y}) = I$ by Lemma 10.3.1, the term $\bar{A}(\mu,\hat{\mu})$ in equation (10.22) can be represented as follows:

$$\bar{A}(\mu,\hat{\mu}) = \begin{bmatrix} \bar{A}_0^{(11)} & 0 \\ 0 & 0 \end{bmatrix} + \sum_{i_1=1}^{M}\sum_{i_2=1}^{M} \begin{bmatrix} \bar{\mathcal{M}}_{i_1i_2}^{(11)} & 0 \\ 0 & \bar{\mathcal{M}}_{i_1i_2}^{(22)} \end{bmatrix} \begin{bmatrix} \bar{A}_{i_1i_2}^{(11)} & 0 \\ \bar{A}_{i_1i_2}^{(21)} & \bar{A}_{i_1i_2}^{(22)}\bar{G}_{i_1i_2}^{(22)}C \end{bmatrix}$$

$$+ \sum_{i_3=1}^{M}\sum_{i_4=1}^{M} \begin{bmatrix} \bar{B}\bar{K}_{i_3i_4}^{(11)} & \bar{B}\bar{K}_{i_3i_4}^{(12)} \\ 0 & 0 \end{bmatrix} \begin{bmatrix} \bar{\mathcal{M}}_{i_3i_4}^{(11)} & 0 \\ 0 & \bar{\mathcal{M}}_{i_3i_4}^{(22)} \end{bmatrix} \quad (10.38)$$

where $\bar{\mathcal{M}}_{i_1i_2}^{(11)} \triangleq diag(\bar{\mu}_{i_1}(y)\bar{\mu}_{i_2}(\hat{y}), \bar{\mu}_{i_1}(y)\bar{\mu}_{i_2}(\hat{y}))$, $\bar{\mathcal{M}}_{i_1i_2}^{(22)} \triangleq \bar{\mu}_{i_1}(y)\bar{\mu}_{i_2}(\hat{y})$, $\bar{A}_0^{(11)} \triangleq$ $\bar{A}_0^{(11)} \triangleq diag(I_N \otimes A_R + T_R, 0)$, $\bar{A}_{i_1i_2}^{(11)} \triangleq diag(0, (I_N \otimes A_{i_1}) + T)$, $\bar{A}_{i_1i_2}^{(21)} \triangleq$ $\bar{A}_{i_1i_2}^{(21)} \triangleq [0, I_N \otimes (A_{i_1} - A_{i_2})]$, $\bar{A}_{i_1i_2}^{(22)} \triangleq (I_N \otimes A_{i_2}) + T$, $\bar{B} \triangleq \left[0, B^T\right]^T$, $\bar{K}_{i_3i_4}^{(11)} \triangleq [-K_{i_4}, K_{i_4}]$, $\bar{K}_{i_3i_4}^{(12)} \triangleq -K_{i_4}$, and $\bar{G}_{i_1i_2}^{(22)} \triangleq G_{i_2}$. Similarly, for $\bar{A}_\tau(\mu,\hat{\mu})$ and $\bar{B}_v(\hat{\mu})$, we have

$$\bar{A}_\tau(\mu,\hat{\mu}) = \sum_{i_1=1}^{M}\sum_{i_2=1}^{M} \begin{bmatrix} \bar{\mathcal{M}}_{i_1i_2}^{(11)} & 0 \\ 0 & \bar{\mathcal{M}}_{i_1i_2}^{(22)} \end{bmatrix} \begin{bmatrix} \bar{A}_{\tau,i_1i_2}^{(11)} & 0 \\ 0 & \bar{A}_{\tau,i_1i_2}^{(22)} \end{bmatrix} \quad (10.39)$$

$$\bar{B}_v(\hat{\mu}) = \begin{bmatrix} \bar{B}_{v(11)} & 0 \\ \bar{B}_{v(21)} & 0 \end{bmatrix} + \sum_{i_1=1}^{M}\sum_{i_2=1}^{M} \begin{bmatrix} \bar{\mathcal{M}}_{i_1i_2}^{(11)} & 0 \\ 0 & \bar{\mathcal{M}}_{i_1i_2}^{(22)} \end{bmatrix} \begin{bmatrix} 0 & 0 \\ 0 & \bar{G}_{i_1i_2}^{(22)}\bar{B}_{v(22)} \end{bmatrix} \quad (10.40)$$

where $\bar{A}_{\tau,i_1i_2}^{(11)} \triangleq diag(0, I_N \otimes A_{\tau,i_1})$, $\bar{A}_{\tau,i_1i_2}^{(22)} \triangleq I_N \otimes A_{\tau,i_2}$, $\bar{G}_{v,i_1i_2}^{(22)} \triangleq G_{i_2}$, $\bar{B}_{v(11)} \triangleq$ $diag(B_R, B_d)$, $\bar{B}_{v(21)} \triangleq [0, B_d]$, and $\bar{B}_{v(22)} \triangleq D_n$.

Lemma 10.4.3

Given the matrices $\Xi_0, \Gamma_{i_1}, \Xi_{1,i_1}$ and Ξ_{2,i_1i_2} with appropriate dimensions for $i_1 = 1,\ldots, M$ and $i_2 = 1,\ldots, M$, we have the following equality

$$\Xi_0 + \sum_{i_1=1}^{M}\Gamma_{i_1}\Xi_{1,i_1}^T + \sum_{i_2=1}^{M}\Xi_{1,i_2}\Gamma_{i_2} + \sum_{i_1=1}^{M}\sum_{i_2=1}^{M}\Gamma_{i_1}\Xi_{2,i_1i_2}\Gamma_{i_2}$$

$$= \begin{bmatrix} I \\ \Gamma_1 \\ \vdots \\ \Gamma_M \end{bmatrix}^T \begin{bmatrix} \Xi_0 & \Xi_{1,1} & \cdots & \Xi_{1,M} \\ \Xi_{1,1}^T & \Xi_{2,11} & \cdots & \Xi_{2,1M} \\ \vdots & \vdots & \ddots & \vdots \\ \Xi_{1,M} & \Xi_{2,M1} & \cdots & \Xi_{2,MM} \end{bmatrix} \begin{bmatrix} I \\ \Gamma_1 \\ \vdots \\ \Gamma_M \end{bmatrix}$$

Before further simplification, we define some notations as $\tilde{B} \triangleq I_{\tilde{M}} \otimes \bar{B}, \tilde{C} \triangleq I_{\tilde{M}} \otimes \bar{C}$ and the following matrices

$$\tilde{\mathcal{M}}_{(k)} \triangleq \begin{bmatrix} I \\ \bar{\mathcal{M}}_{11}^{(k)} \\ \vdots \\ \bar{\mathcal{M}}_{MM}^{(k)} \end{bmatrix}, \tilde{A}_{(k)} \triangleq \begin{bmatrix} \bar{A}_0^k & 0 & \cdots & 0 \\ \bar{A}_{11}^k & 0 & \cdots & 0 \\ \vdots & \vdots & \ddots & \vdots \\ \bar{A}_{MM}^k & 0 & \cdots & 0 \end{bmatrix}, \tilde{G}_{(22)} \triangleq \begin{bmatrix} 0 & 0 & 0 & 0 \\ \bar{G}_{11}^{(22)} & 0 & \cdots & 0 \\ 0 & \vdots & \ddots & \vdots \\ \bar{G}_{MM}^{(22)} & 0 & \cdots & 0 \end{bmatrix},$$

$$\tilde{\Xi}_{(k)} \triangleq \begin{bmatrix} -\Xi_0^{(k)} - \Xi_0^{(k)T} + \Xi_3^{(k)} & \Xi_0^{(k)} - \Xi_{1,11}^{(k)} & \cdots & \Xi_0^{(k)} - \Xi_{1,MM}^{(k)} \\ \Xi_0^{(k)T} - \Xi_{2,11}^{(k)T} & \Xi_{1,11}^{(k)} + \Xi_{2,11}^{(k)T} - \Xi_3^{(k)} & \cdots & \Xi_{1,MM}^{(k)} + \Xi_{2,11}^{(k)T} \\ \vdots & \vdots & \ddots & \vdots \\ \Xi_0^{(k)T} - \Xi_{2,MM}^{(k)T} & \Xi_{1,11}^{(k)} + \Xi_{2,MM}^{(k)T} & \cdots & \Xi_{1,MM}^{(k)} + \Xi_{2,MM}^{(k)T} - \Xi_3^{(k)} \end{bmatrix},$$

$$\tilde{K}_{(k)} \triangleq \begin{bmatrix} 0 & \bar{K}_{11}^{(k)} & \cdots & \bar{K}_{MM}^{(k)} \\ 0 & 0 & \cdots & 0 \\ 0 & \vdots & \ddots & \vdots \\ 0 & 0 & \cdots & 0 \end{bmatrix}, \quad \begin{array}{l} \tilde{A}_{\tau(k)} \triangleq [0, \bar{A}_{\tau,11}^{(k)T}, \cdots, \bar{A}_{\tau,MM}^{(k)T}]^T, \\ \tilde{G}_{v,(22)} \triangleq [0, \bar{G}_{v,11}^{(22)T}, \cdots, \bar{G}_{v,MM}^{(22)T}]^T, \end{array}$$

where for $\tilde{\mathcal{M}}_{(k)}$, $\tilde{\Xi}_{(k)}$, and $\tilde{A}_{\tau(k)}$, we have k=11, 22; for $\tilde{A}_{(k)}$, k=11, 21, 22; for $\tilde{K}_{(k)}$, k=11, 12; the matrices $\Xi_0^{(11)} \in \mathbb{R}^{2nN\times 2nN}$, $\Xi_0^{(22)} \in \mathbb{R}^{nN\times nN}$, $\Xi_{1,i_1i_2}^{(11)} \in \mathbb{R}^{2nN\times 2nN}$, $\Xi_{2,i_1i_2}^{(11)} \in \mathbb{R}^{2nN\times 2nN}$, $\Xi_{1,i_1i_2}^{(22)} \in \mathbb{R}^{nN\times nN}$, $\Xi_{2,i_1i_2}^{(22)} \in \mathbb{R}^{nN\times nN}$, $\Xi_3^{(11)} \in \mathbb{R}^{2nN\times 2nN}$, and $\Xi_3^{(22)} \in \mathbb{R}^{nN\times nN}$ are some relaxed matrices applying in the following lemma and other matrices are defined in equations (10.38)–(10.40).

Remark 10.4.4

In the matrices $\tilde{A}_{(k)}, \tilde{A}_{\tau(k)}, \tilde{K}_{(k)}, \tilde{G}_{(k)}, \tilde{G}_{v(k)}$, and $\tilde{\Xi}_{(k)}$ it should be noted that the subscript index i_1i_2 of submatrices is defined as $i_1i_2 = 11, 12, \ldots, MM$ for $i_1 = 1, \ldots, M$ and $i_2 = 1, \ldots, M$, i.e., (i_1, i_2)=(1,1), (1,2), (1,3), …, (M, M).

Proof of Lemma 10.4.2

Proof

By the equality $-I + \sum_{i_1=1}^{M} \sum_{i_2=1}^{M} \bar{\mathcal{M}}_{i_1i_2}^{(k)} = 0$, for k=11, 22, we can add the following equalities to relax the equation in (10.36)

$$\left(-I+\sum_{i_1=1}^{M}\sum_{i_2=1}^{M}\bar{\mathcal{M}}_{i_1i_2}^{(k)}\right)\Xi_0^{(k)T}=0,\Xi_0^{(k)}\left(-I+\sum_{i_3=1}^{M}\sum_{i_4=1}^{M}\bar{\mathcal{M}}_{i_3i_4}^{(k)}\right)=0$$

$$\left(-I+\sum_{i_1=1}^{M}\sum_{i_2=1}^{M}\bar{\mathcal{M}}_{i_1i_2}^{(k)}\right)\left(\sum_{i_3=1}^{M}\sum_{i_4=1}^{M}\Xi_{1,i_3i_4}^{(k)}\bar{\mathcal{M}}_{i_3i_4}^{(k)}\right)=0$$

$$\left(\sum_{i_1=1}^{M}\sum_{i_2=1}^{M}\bar{\mathcal{M}}_{i_1i_2}^{(k)}\Xi_{2,i_1i_2}^{(k)T}\right)\left(-I+\sum_{i_3=1}^{M}\sum_{i_4=1}^{M}\bar{\mathcal{M}}_{i_3i_4}^{(k)}\right)=0$$

$$\left(\Xi_3^{(k)}-\sum_{i_1=1}^{M}\sum_{i_2=1}^{M}\sum_{i_3=1}^{M}\sum_{i_4=1}^{M}\bar{\mathcal{M}}_{i_1i_2}^{(k)}\Xi_3^{(k)}\bar{\mathcal{M}}_{i_3i_4}^{(k)}\right)=0$$

Adding the above three equalities into equation (10.38) and using Lemma 10.4.3, we can obtain the equation in equation (10.36). Moreover, it is easy to prove the equation in equation (10.37) by Lemma 10.4.3.

Since the matrix $\tilde{A}_{(k)}$ is singular, we add the relaxed matrices $\tilde{\Xi}_{(k)}$ to adjust the relationship between the matrices $\tilde{A}_{(k)},\tilde{K}_{(k)},\tilde{G}_{(k)}$.

Lemma 10.4.4

Since the $\bar{\mu}_{i_1}(y)\bar{\mu}_{i_2}(\hat{y})$ is the diagonal matrix and the matrix P has the form in equation (10.35), it is easy to prove the relationship $\bar{P}_{(11)}\bar{\mathcal{M}}_{i_1i_2}^{(11)}=\bar{\mathcal{M}}_{i_1i_2}^{(11)}\bar{P}_{(11)}$ and $\bar{P}_{(22)}\bar{\mathcal{M}}_{i_1i_2}^{(22)}=\bar{\mathcal{M}}_{i_1i_2}^{(22)}\bar{P}_{(22)}$. Moreover, the matrix $\tilde{\mathcal{M}}P$ can be represented as $\tilde{\mathcal{M}}P=\tilde{P}\tilde{\mathcal{M}}$ where $\tilde{P}=diag(\tilde{P}_{(11)},\tilde{P}_{(22)})\triangleq diag(I_{\tilde{M}}\otimes\bar{P}_{(11)},I_{\tilde{M}}\otimes\bar{P}_{(22)})\in\mathbb{R}^{3n\tilde{N}\times 3n\tilde{N}}$.

In Theorem 10.4.1, the sufficient condition of the H_∞ tracking control in equation (10.27) includes the fuzzy weighting matrices $\bar{\mu}(y)$ and $\bar{\mu}(\hat{y})$. Based on Lemma 10.4.2, we obtain the following main result.

Theorem 10.4.2

For the augmented system equation (10.22) with a prescribed disturbance attenuation level ρ, if there exist two symmetric positive definite matrices P, Q in equation (10.35), the relaxed matrix $\tilde{\Xi}$, the control gains K_i, and the observer gains G_i, for $i=1,\ldots,M$, in equations (10.18)–(10.20) such that

$$\tilde{\Pi}\triangleq\begin{bmatrix}\tilde{\Pi}_{11} & \tilde{P}\tilde{A}_\tau & \tilde{P}\tilde{B}_v\\ \tilde{A}_\tau^T\tilde{P} & \xi\Sigma_\tau-Q & 0\\ \tilde{B}_v^T\tilde{P} & 0 & -\rho^2 I\end{bmatrix}<0 \tag{10.41}$$

then the H_∞ tracking control performance in equation (10.24) is guaranteed for a prescribed disturbance attenuation level ρ by the fuzzy observer-based controller in equations (10.18)–(10.20). In equation (10.41),

$$\tilde{\Pi}_{11} \triangleq (\tilde{A}+\tilde{\Xi})^T \tilde{P} + \tilde{P}(\tilde{A}+\tilde{\Xi}) + H^T\left(\xi\Sigma + R^T R + Q\right)H + \left(\frac{1}{\rho^2}+\frac{1}{\xi}\right)\tilde{P}H^T H\tilde{P}$$

Proof

Using the equality $\tilde{\mathcal{M}}P = \tilde{P}\tilde{\mathcal{M}}$ in Lemma 10.4.4, we have

$$\begin{aligned}\bar{A}(\mu,\hat{\mu})^T P + P\bar{A}(\mu,\hat{\mu}) &= (\tilde{\mathcal{M}}^T \tilde{A}\tilde{\mathcal{M}})^T P + P(\tilde{\mathcal{M}}^T \tilde{A}\tilde{\mathcal{M}}) \\ &= \tilde{\mathcal{M}}^T \tilde{A}^T \tilde{P}\tilde{\mathcal{M}} + \tilde{\mathcal{M}}^T \tilde{P}\tilde{A}\tilde{\mathcal{M}}\end{aligned}$$

Based on Lemma 10.4.2, $\bar{\Pi}(\mu,\hat{\mu})$ in equation (10.27) can be reformulated as follows:

$$\bar{\Pi}(\mu,\hat{\mu}) = \begin{bmatrix} \tilde{\mathcal{M}} & 0 & 0 \\ 0 & I & 0 \\ 0 & 0 & I \end{bmatrix}^T \begin{bmatrix} \tilde{\Pi}_{11} & \tilde{P}\tilde{A}_\tau & \tilde{P}\tilde{B}_v \\ \tilde{A}_\tau^T \tilde{P} & \xi\Sigma_\tau - Q & 0 \\ \tilde{B}_v^T \tilde{P} & 0 & -\rho^2 I \end{bmatrix} \begin{bmatrix} \tilde{\mathcal{M}} & 0 & 0 \\ 0 & I & 0 \\ 0 & 0 & I \end{bmatrix} \tag{10.42}$$

Finally, if $\tilde{\Pi} < 0$ in (10.41), then $\bar{\Pi}(\mu,\hat{\mu}) < 0$.

Remark 10.4.5

The control gains K_i and the observer gains G_i, for $i = 1,\ldots, M$, are included in the matrix $\tilde{\mathrm{A}}$. The matrix $\tilde{P}$ can be obtained by the matrix P in Lemma 10.4.4.

The H_∞ tracking control design problem is to specify the observer gains G_i and the control gains K_i, for $i=1,\ldots, M$, to satisfy the inequality in equation (10.41). In the observer-based control design problem, the observer gains G_i and the control gains K_i are always coupled with the matrix P [364]. In this situation, the matrix inequality in equation (10.41) will be a complex bilinear matrix inequality (BMI). A systematic design procedure is developed to solve this problem in the next section.

10.4.2 Solving Robust H_∞ Tracking Control Problem via Linear Matrix Inequality

Some algorithms for solving local optimal BMI solutions have been proposed via the augmented Lagrangian method [366] and the iteration method [354,363–365]. However, because the BMI problem is nonconvex, these algorithms are still inefficient to solve BMI problems with multiple variables. In [367], a fuzzy observer-based H_∞ control design was studied for a T-S fuzzy time-delay system without the measurement noise. In this chapter, the measurement noise at the measured output is considered in the NDPS. For the robust H_∞ tracking problem of the NDPS, the two-dimensional H_∞ tracking performance is addressed. Therefore, we proposed a different method to overcome a more complex problem. Note that the inequality in equation (10.41) is still BMI even if the matrices P and Q were chosen as equation (10.35). The following lemma is introduced to reduce a BMI to an LMI which can be efficiently solved with the conventional LMI technique.

Lemma 10.4.5 [368, 377]

Given a positive symmetric matrix X of appropriate dimension, if the following inequality holds

$$\begin{bmatrix} \Omega_{11} - 2\varsigma X & \varsigma I \\ \varsigma I & \Omega_{22} \end{bmatrix} < 0 \tag{10.43}$$

then we have $\Omega_{11} + X\Omega_{22}X < 0$. In the other words, the inequalities, $\Omega_{22} < 0$ and $-2\varsigma X - \varsigma^2\Omega_{22}^{-1} < 0$, imply that $X\Omega_{22}X - 2\varsigma X - \varsigma^2\Omega_{22}^{-1} < 0$.

Proof

First, by Schur complement, the inequality in equation (10.43) is equivalent to $\Omega_{22} < 0$ and $\Omega_{11} - 2\varsigma X - \varsigma^2\Omega_{22}^{-1} < 0$. From $\Omega_{22} < 0$, we have $\left(X + \varsigma\,\Omega_{22}^{-1}\right)^T \Omega_{22} \left(X + \varsigma\,\Omega_{22}^{-1}\right) < 0$ which is equivalent to $X\Omega_{22}X < -2\varsigma X - \varsigma^2\Omega_{22}^{-1}$. Then, we can obtain the inequality $\Omega_{11} + X\Omega_{22}X < 0$. The lemma is completely proven.

Let us define the matrix $\tilde{X}$ as follows:

$$\tilde{X} = diag(\tilde{X}_{(11)}, \tilde{X}_{(22)}) \triangleq diag(I_{\tilde{M}} \otimes \bar{X}_{(11)}, I_{\tilde{M}} \otimes \bar{X}_{(22)}) \tag{10.44}$$

where

$$\bar{X}_{(11)} \triangleq \begin{bmatrix} X_{11} & X_{12} \\ X_{12} & X_{11} \end{bmatrix} = \begin{bmatrix} P_{11} & P_{12} \\ P_{12} & P_{11} \end{bmatrix}^{-1}$$

and $\bar{X}_{(22)} \triangleq X_{11} - X_{12}$. We also define some symbols with respect to the matrix Q in (10.35) as follows:

$$\begin{aligned} &\bar{S}_{(11)} \triangleq \bar{X}_{(11)}\bar{Q}_{(11)}\bar{X}_{(11)} \text{ and} \\ &\tilde{S}_{(11)} = \tilde{X}_{(11)}H_{(11)}^T\bar{Q}_{(11)}H_{(11)}\tilde{X}_{(11)} = H_{(11)}^T\bar{S}_{(11)}H_{(11)} \end{aligned} \tag{10.45}$$

Using the matrices P and Q in equation (10.35), the sufficient conditions for robust H_∞ tracking control in Theorem 10.4.1 can be derived as follows.

Theorem 10.4.3

For the augmented system in equation (10.22) with a prescribed disturbance attenuation level ρ, suppose there exist the symmetric matrices $\bar{X}_{(11)} > 0$, $\bar{S}_{(11)} > 0$,

$\bar{P}_{(22)} > 0$, $\bar{Q}_{(22)} > 0$, *the matrices* Y_i *and* Z_i, $i = 1,\ldots, M$, $\tilde{\Xi}_{(11)}$, $\tilde{\Xi}_{(22)}$ *and the scalars* $\xi > 0$, $\xi_2 > 0$ *such that*

$$\begin{bmatrix} \Omega_{11} & \Omega_{12} & \Omega_{13} & \Omega_{14} \\ \Omega_{21} & \Omega_{22} & \Omega_{23} & 0 \\ \Omega_{31} & \Omega_{32} & \Omega_{33} & 0 \\ \Omega_{41} & 0 & 0 & \Omega_{44} \end{bmatrix} < 0 \tag{10.46}$$

Then the H_∞ *tracking performance in equation (10.24) is guaranteed by the fuzzy observer-based tracking controller in equations (10.18)–(10.20), i.e., the approximation error and the time delay can be tolerated, and the effect of the truncation error, the external disturbance, and the measurement noise on the tracking error can be attenuated below a prescribed level* ρ *by the fuzzy observer-based controller with observer gains* $G_i = P_{33}^{-1} Z_i$ *and control gains* $K_i = Y_i (X_{11} - X_{12})^{-1}$, $i = 1,\ldots,M$. *In equation (10.46),*

$$\Omega_{11} = \begin{bmatrix} \bar{\Theta}_{11} & \tilde{B}\tilde{Y}_{(12)} & \varsigma I & 0 \\ \tilde{Y}_{(12)}^T & -2\varsigma \tilde{X}_{(22)} & 0 & \varsigma I \\ \varsigma I & 0 & -2I + \xi_2 I & \tilde{A}_{(21)}^T \tilde{P}_{(22)} \\ 0 & \varsigma I & \tilde{P}_{(22)}\tilde{A}_{(21)} & \bar{\Theta}_{22} \end{bmatrix}, \Omega_{22} = \begin{bmatrix} -\bar{S}_{(11)} & 0 \\ 0 & -\bar{Q}_{(22)} \end{bmatrix},$$

$$\Omega_{21} = \begin{bmatrix} X_{(11)}\tilde{A}_{\tau(11)}^T & 0 & 0 & 0 \\ 0 & 0 & 0 & \tilde{A}_{\tau(22)}^T \tilde{P}_{(22)} \end{bmatrix}, \Omega_{12} = \Omega_{21}^T, \Omega_{33} = -2I + \xi I,$$

$$\Omega_{31} = \begin{bmatrix} \Sigma_{(11)}^{-1/2} H_{(11)}^T \tilde{X}_{(11)}, & 0, & 0, & 0 \end{bmatrix}, \Omega_{13} = \Omega_{31}^T, \Omega_{32} = \begin{bmatrix} \Sigma_{\tau(11)}^{-1/2}, & 0 \end{bmatrix}, \Omega_{23} = \Omega_{32}^T,$$

$$\Omega_{14,1} = \begin{bmatrix} \tilde{X}_{(11)} H_{(11)}^T \bar{R}_{(11)}^T, & \tilde{X}_{(11)}, & H_{(11)}^T, & H_{(11)}^T, & H_{(11)}^T \bar{B}_{v(11)} \end{bmatrix}, \Omega_{41,1} = \Omega_{14,1}^T,$$

$$\Omega_{14,2} = \begin{bmatrix} H_{(22)}^T \bar{R}_{(22)}^T, & \tilde{P}_{(22)} H_{(22)}^T, & \tilde{P}_{(22)} H_{(22)}^T, & \tilde{P}_{(22)} H_{(22)}^T \bar{B}_{v(21)}, & \tilde{Z}_{v(22)} \bar{B}_{v(22)} \end{bmatrix},$$

$$\Omega_{41} = \begin{bmatrix} \Omega_{41,1} & 0 & 0 & 0 \\ 0 & 0 & 0 & \Omega_{41,2} \end{bmatrix}, \Omega_{14} = \Omega_{41}^T, \Omega_{44} = \begin{bmatrix} \Omega_{44,1} & 0 \\ 0 & \Omega_{44,2} \end{bmatrix}, \Omega_{41,2} = \Omega_{14,2}^T,$$

$$\Omega_{44,1} = diag(-I, -\xi_2 I, -\rho^2 I, -\xi I, -\rho^2/2I),\ \Omega_{44,2} = diag(-I, -\rho^2 I, -\xi I, -\rho^2/2I, -\rho^2 I),$$

$$\bar{\Theta}_{11} \triangleq \Theta_{11} + \tilde{S}_{(11)} - 2\varsigma \tilde{X}_{(11)}, \tilde{Y}_{(11)} \triangleq \tilde{K}_{(11)}\tilde{X}_{(11)}, \tilde{Y}_{(12)} \triangleq \tilde{K}_{(12)}\tilde{X}_{(22)}, \hat{\Xi}_{(11)} \triangleq \tilde{\Xi}_{(11)}\tilde{X}_{(11)},$$

$$\Theta_{11} \triangleq \tilde{X}_{(11)}\tilde{A}_{(11)}^T + \tilde{A}_{(11)}\tilde{X}_{(11)} + \tilde{B}\tilde{Y}_{(11)} + \tilde{Y}_{(11)}^T \tilde{B}^T + \hat{\Xi}_{(11)}^T + \hat{\Xi}_{(11)},$$

$$\bar{\Theta}_{22} \triangleq \Theta_{22} + H_{(22)}^T \bar{Q}_{(22)} H_{(22)}, \tilde{Z}_{(22)} \triangleq \tilde{P}_{(22)}\tilde{G}_{(22)}, \tilde{Z}_{v(22)} \triangleq \tilde{P}_{(22)}\tilde{G}_{v(22)}, \hat{\Xi}_{(22)} \triangleq \tilde{P}_{(22)}\tilde{\Xi}_{(22)},$$

$$\Theta_{22} \triangleq \tilde{A}_{(22)}^T \tilde{P}_{(22)} + \tilde{P}_{(22)}\tilde{A}_{(22)} + \tilde{Z}_{(22)}\tilde{C} + \tilde{C}^T \tilde{Z}_{(22)}^T + \hat{\Xi}_{(22)}^T + \hat{\Xi}_{(22)},$$

where $\bar{\Sigma}_{(11)} \triangleq diag\left(0,2\sigma^2 I\right)$, $\bar{\Sigma}_{\tau,(11)} \triangleq diag\left(0,2\sigma^2 I\right)$, $\bar{R}_{(11)} \triangleq \left[\bar{R}_1,-\bar{R}_1\right]$, and $\bar{R}_{(22)} \triangleq \bar{R}_2$.

Proof

Firstly, the fact that the inequality in equation (10.46) implies the inequality in equation (10.41) is proven in the following. Because the fact $(1-\xi)\xi^{-1}(1-\xi)\geq 0$ implies the inequality $-\xi^{-1}\leq -2+\xi$, we have $\hat{\Omega}_{33} \triangleq -\xi^{-1}I \leq \Omega_{33}$ in equation (10.46). Similarly, we have $-\xi_2^{-1} \leq -2+\xi_2$. Using the aforementioned inequality and Schur complement, the inequality in equation (10.46) implies the inequality, $\Omega_{11} - \Omega_{13}\hat{\Omega}_{33}^{-1}\Omega_{31} - \Omega_{12}\left(\Omega_{22} - \Omega_{23}\hat{\Omega}_{33}^{-1}\Omega_{32}\right)^{-1}\Omega_{21} - \Omega_{14}\Omega_{44}^{-1}\Omega_{41} < 0$. This inequality can be written as follows:

$$\begin{bmatrix} \hat{\Theta}_{11} & \tilde{B}\tilde{Y}_{(12)} & \varsigma I & 0 \\ \tilde{Y}_{(12)}^T\tilde{B}^T & -2\varsigma\tilde{X}_{(22)} & 0 & \varsigma I \\ \varsigma I & 0 & -\xi_2^{-1}I & \tilde{A}_{(21)}^T\tilde{P}_{(22)} \\ 0 & \varsigma I & \tilde{P}_{(22)}\tilde{A}_{(21)} & \Theta_{22}+\Upsilon_{22} \end{bmatrix} < 0 \tag{10.47}$$

where

$$\hat{\Theta}_{11} \triangleq \Theta_{11} + \Upsilon_{11} - 2\varsigma\tilde{X}_{(11)},$$

$$\Upsilon_{11} \triangleq \tilde{X}_{(11)}H_{(11)}^T\Phi_{(11)}H_{(11)}\tilde{X}_{(11)} + \tilde{A}_{\tau(11)}\Phi_{\tau(11)}^{-1}\tilde{A}_{\tau(11)}^T$$

$$+\frac{1}{\xi_2}\tilde{X}_{(11)}\tilde{X}_{(11)} + \left(\frac{1}{\rho^2}+\frac{1}{\xi}\right)H_{(11)}^T H_{(11)} + \frac{1}{\rho^2}2H_{(11)}^T\bar{B}_{v(11)}\bar{B}_{v(11)}^T H_{(11)},$$

$$\Upsilon_{22} \triangleq H_{(22)}^T\Phi_{(22)}H_{(22)} + \tilde{P}_{(22)}\tilde{A}_{\tau(22)}\Phi_{\tau(22)}^{-1}\tilde{A}_{\tau(22)}^T\tilde{P}_{(22)} + \left(\frac{1}{\rho^2}+\frac{1}{\xi}\right)\tilde{P}_{(22)}H_{(22)}^T H_{(22)}\tilde{P}_{(22)}$$

$$+\frac{1}{\rho^2}2\tilde{P}_{(22)}H_{(22)}^T\bar{B}_{v(21)}\bar{B}_{v(21)}^T H_{(22)}\tilde{P}_{(22)} + \frac{1}{\rho^2}\tilde{Z}_{v(22)}\bar{B}_{v(22)}\bar{B}_{v(22)}^T\tilde{Z}_{v(22)}^T,$$

in which $\Phi_{(k)} \triangleq \left(\xi\bar{\Sigma}_{(k)} + \bar{R}_{(k)}^T\bar{R}_{(k)} + \bar{Q}_{(k)}\right)$ and $\Phi_{\tau(k)} \triangleq \left(\xi\bar{\Sigma}_{\tau(k)} - \bar{Q}_{(k)}\right)$, $k = 11,22$. Note that $\bar{\Sigma}_{(22)} = 0$ and $\bar{\Sigma}_{\tau(22)} = 0$. By Lemma 10.4.5, the inequality in equation (10.47) implies the following inequality

$$\Omega \triangleq \begin{bmatrix} \tilde{\Theta}_{11} & \tilde{\Theta}_{12} \\ \tilde{\Theta}_{12}^T & \tilde{X}_{(22)}(\Theta_{22}+\Upsilon_{22})\tilde{X}_{(22)} \end{bmatrix} < 0 \tag{10.48}$$

where $\tilde{\Theta}_{11} \triangleq \Theta_{11} + \Upsilon_{11} - \xi_2^{-1}\tilde{X}_{(11)}\tilde{X}_{(11)}$ and $\tilde{\Theta}_{12} \triangleq \tilde{B}\tilde{Y}_{(12)} + \tilde{X}_{(11)}\tilde{A}_{(21)^T}\tilde{P}_{(22)}\tilde{X}_{(22)}$.

On the other hand, by Schur complement, the inequality in equation (10.41) is equivalent to the following form:

$$(\tilde{A}+\tilde{\Xi})^T\tilde{P}+\tilde{P}(\tilde{A}+\tilde{\Xi})+H^T\left(\xi\Sigma+R^TR+Q\right)H+\tilde{P}\tilde{A}_\tau(\xi\Sigma_\tau-Q)^{-1}\tilde{A}_\tau^T\tilde{P}$$
$$+\left(\frac{1}{\rho^2}+\frac{1}{\xi}\right)\tilde{P}H^TH\tilde{P}+\frac{1}{\rho^2}\tilde{P}\tilde{B}_v\tilde{B}_v^T\tilde{P}<0 \tag{10.49}$$

Note that $\Sigma = diag\left(\bar{\Sigma}_{(11)},\bar{\Sigma}_{(22)}\right)$, $R^TR = diag\left(\bar{R}_{(11)}^T\bar{R}_{(11)},\bar{R}_{(22)}^T\bar{R}_{(22)}\right)$, $Q = diag\left(\bar{Q}_{(11)},\bar{Q}_{(22)}\right)$, $\Sigma_\tau = diag\left(\bar{\Sigma}_{\tau,(11)},\bar{\Sigma}_{\tau,(22)}\right)$, $\left(\xi\Sigma_\tau - Q\right)^{-1} = diag\left(\xi\Sigma_{\tau(11)}-\bar{Q}_{(11)}\right)^{-1}$, $\left(\xi\Sigma_{\tau(22)}-\bar{Q}_{(22)}\right)^{-1}$. Pro- and post-multiplying the above inequality by the matrix $\tilde{X}$ in equation (10.44), and substituting the definitions of $\tilde{P},\tilde{A},\tilde{A}_\tau,\tilde{B}_v$, and H in equations (10.36) and (10.37) into this inequality, we can obtain

$$\bar{\Omega} \triangleq \begin{bmatrix} \tilde{\Theta}_{11} & \tilde{\Theta}_{12} \\ \tilde{\Theta}_{12}^T & \tilde{X}_{(22)}\Theta_{22}\tilde{X}_{(22)} \end{bmatrix} + \begin{bmatrix} \bar{\Psi}_{11} & 0 \\ 0 & \bar{\Psi}_{22} \end{bmatrix} + \begin{bmatrix} \bar{\Psi}_{\tau,11} & 0 \\ 0 & \bar{\Psi}_{\tau,22} \end{bmatrix}$$
$$+\left(\frac{1}{\rho^2}+\frac{1}{\xi}\right)\begin{bmatrix} H_{(11)}^TH_{(11)} & 0 \\ 0 & \tilde{X}_{(22)}\tilde{P}_{(22)}H_{(22)}^TH_{(22)}\tilde{P}_{(22)}\tilde{X}_{(22)} \end{bmatrix} \tag{10.50}$$
$$+\frac{1}{\rho^2}\begin{bmatrix} \bar{\Psi}_{v,11}^T\bar{\Psi}_{v,11} & \bar{\Psi}_{v,11}^T\bar{\Psi}_{v,22} \\ \bar{\Psi}_{v,22}^T\bar{\Psi}_{v,11} & \bar{\Psi}_{v,22}^T\bar{\Psi}_{v,22}+\bar{\Psi}_{n,22}^T\bar{\Psi}_{n,22} \end{bmatrix}<0$$

where $\tilde{Z}_{v(22)},\Theta_{11}$ and Θ_{22} are defined in equation (10.46); $\tilde{\Theta}_{12}$ is defined in equation (10.48). The other notations are defined as follows:

$$\bar{\Psi}_{11} \triangleq \tilde{X}_{(11)}H_{(11)}^T\Phi_{(11)}H_{(11)}\tilde{X}_{(11)}, \bar{\Psi}_{22} \triangleq \tilde{X}_{(22)}H_{(22)}^T\Phi_{(22)}H_{(22)}\tilde{X}_{(22)},$$
$$\bar{\Psi}_{\tau,11} \triangleq \tilde{A}_{\tau(11)}\Phi_{\tau(11)}^{-1}\tilde{A}_{\tau(11)}^T, \bar{\Psi}_{\tau,22} \triangleq \tilde{X}_{(22)}\tilde{P}_{(22)}\tilde{A}_{\tau(22)}\Phi_{\tau(22)}^{-1}\tilde{A}_{\tau(22)}^T\tilde{P}_{(22)}\tilde{X}_{(22)},$$
$$\bar{\Psi}_{v,11} \triangleq \bar{B}_{v(11)}^TH_{(11)}, \bar{\Psi}_{v,22} \triangleq \bar{B}_{v(21)}^TH_{(22)}\tilde{P}_{(22)}\tilde{X}_{(22)}, \bar{\Psi}_{n,22} \triangleq \bar{B}_{v(22)}^T\tilde{Z}_{v(22)}\tilde{P}_{(22)}\tilde{X}_{(22)}.$$

By Lemma 10.4.1, we can obtain the following inequality:

$$\begin{bmatrix} 0 & \bar{\Psi}_{v,11}^T\bar{\Psi}_{v,22} \\ \bar{\Psi}_{v,22}^T\bar{\Psi}_{v,11} & 0 \end{bmatrix} \le \begin{bmatrix} \bar{\Psi}_{v,11}^T\bar{\Psi}_{v,11} & 0 \\ 0 & \bar{\Psi}_{v,22}^T\bar{\Psi}_{v,22} \end{bmatrix} \tag{10.51}$$

Using the above inequality, we can get the inequality, $\bar{\Omega} \le \Omega$, in which Ω and $\bar{\Omega}$ are defined in equations (10.48) and (10.50), respectively. Therefore, the inequality $\Omega < 0$ implies the inequality $\bar{\Omega} < 0$. Since the inequality in equation (10.50) is equivalent to the inequality in equation (10.41) and the inequality equation (10.46) implies the inequality in equation (10.48), we know the inequality in equation (10.46) implies the inequality in equation (10.41). Finally, using the result of Theorem 10.4.1, we can prove that if the inequality in equation (10.46) holds then the H_∞ tracking control performance in equation (10.24) can be guaranteed by the fuzzy observer-based tracking controller in equations (10.18)–(10.20).

Remark 10.4.6

The control gains K_i and the observer gains G_i, for $i = 1,\ldots, M$, are included in the matrices $\tilde{K}_{(11)}, \tilde{K}_{(22)}, \tilde{G}_{(22)}$ and $\tilde{G}_{v(22)}$, respectively. We define $\bar{Y}_{i_3 i_4}^{(11)} \triangleq \bar{K}_{i_3 i_4}^{(11)} \bar{X}_{(11)} = [-Y_{i_4}, Y_{i_4}]$, $\bar{Y}_{i_3 i_4}^{(12)} \triangleq \bar{K}_{i_3 i_4}^{(12)} \bar{X}_{(22)} = -Y_{i_4}$, $\bar{Z}_{i_1 i_2}^{(22)} \triangleq \bar{P}_{(22)} \bar{G}_{i_1 i_2}^{(22)} = Z_{i_2}$, and $\bar{Z}_{v,i_1 i_2}^{(22)} \triangleq \bar{P}_{(22)} \bar{Z}_{v,i_1 i_2}^{(22)} = Z_{v,i_2}$ in which $Y_{i_4} \triangleq K_{i_4}(X_{11} - X_{12})$ and $Z_{i_2} \triangleq P_{33} G_{i_2}$. Then, we have the following forms for $\tilde{Y}_{(k)}, k = 11, 12, \tilde{Z}_{(22)}$ and $\tilde{Z}_{v(22)}$:

$$\tilde{Y}_{(k)} \triangleq \begin{bmatrix} 0 & \bar{Y}_{11}^{(k)} & \cdots & \bar{Y}_{MM}^{(k)} \\ 0 & 0 & \cdots & 0 \\ \vdots & \vdots & \ddots & \vdots \\ 0 & 0 & \cdots & 0 \end{bmatrix}, \tilde{Z}_{(22)} \triangleq \begin{bmatrix} 0 & 0 & \cdots & 0 \\ \bar{Z}_{11}^{(22)} & 0 & \cdots & 0 \\ \vdots & \vdots & \ddots & \vdots \\ \bar{Z}_{MM}^{(22)} & 0 & \cdots & 0 \end{bmatrix}$$

and $\tilde{Z}_{v(22)} \triangleq [0, \bar{Z}_{v,11}^{(22)T}, \bar{Z}_{v,12}^{(22)T}, \ldots, \bar{Z}_{v,MM}^{(22)T}]^T$.

Remark 10.4.7

In equation (10.46), we use the term $-2 + \xi$ to replace the term $-\xi^{-1}$ with the fact, $-\xi^{-1} \le -2 + \xi$. From the inequality $-2 + \xi < 0$, we know that $\xi < 2$. Alternatively, we can use the term $-2 + \xi^{-1}$ to replace the term $-\xi$ by the fact $-\xi \le -2 + \xi^{-1}$. For $-2 + \xi^{-1} < 0$ we have $\xi > 1$. Therefore, we can choose either of two constraints on ξ to obtain the suitable scalar ξ in the LMI equation (10.46). For ξ_2, we have the same result.

The H_∞ tracking control design problem reduces to how to specify the observer gains $G_i = P_{33}^{-1} Z_i$ and control gains $K_i = Y_i (X_{11} - X_{12})^{-1}, i = 1,\ldots, M$ by solving the LMI in equation (10.46) with some positive definite matrices. Finally, in order to achieve the optimal attenuation of truncation error, external disturbances and measurement noises on the tracking error and estimation error, the optimal H_∞ tracking control design problem for the NDPS equations (10.1) and (10.2) can be solved by the following constrained optimization problem:

$$\rho_0 = \min_{Y_i, Z_i, i=1,\ldots,M} \rho$$
$$\text{subject to } \bar{X}_{(11)} > 0, \bar{S}_{(11)} > 0, P_{33} > 0, Q_{33} > 0, \xi > 0, \xi_2 > 0, \text{ and (10.46)} \tag{10.52}$$

This is called an eigenvalue problem (EVP) [378] and can be easily solved by the LMI technique.

Based on the above analysis, the robust H_∞ observer-based tracking control design procedure for NDPSs is summarized in the following steps.

Design Procedure

Step 1: Given a desired reference model in equation (10.3), generate the reference output $y_R(x, t)$ and a prescribed disturbance attenuation level ρ.

Step 2: Select the fuzzy membership functions and fuzzy rules to establish a fuzzy DPS in equation (10.7) to approximate the NDPS in equation (10.1).

Step 3: Give gird size Δ_x and N to construct the fuzzy spatial state space model in equations (10.15) and (10.16).

Step 4: Solve the observer gains G_i and the tracking control gains K_i of the fuzzy observer-based tracking controller, for $i=1,\ldots, M$, in equations (10.18)–(10.20) by solving the LMI problem in equation (10.46) or solve K_i and G_i from the optimal H_∞ tracking control problem in equation (10.52).

Step 5: Construct a fuzzy observer-based tracking controller in equations (10.18)–(10.20) to control the NDPSs in equations (10.1) and (10.2) and track the reference model in equation (10.3).

Remark 10.4.8

(i) Note that the different grid size Δ_x can be chosen for the different space variables x. (ii) Note that the dimension N depends on Δ_x. In theory, the grid size Δ_x is chosen as small as possible, i.e., $\Delta_x \to 0$. However, in this situation, the dimension N will increase to infinity. The computational complexity will also increase for solving the LMI problem in equation (10.46). Therefore, how to choose the grid size is a trade-off problem. Suppose the initial value problem of the NDPS in equation (10.1) is well-posed [374] and the finite difference scheme (method of lines scheme [383]) in equation (10.9) is consistent, i.e., $O(\Delta_x^2) \to 0$ as $\Delta_x \to 0$ [374]. Applying Lax-Richtmyer Equivalence Theorem [374], the fuzzy spatial state space system with $d(t) \equiv 0$ in equation (10.15) is stable if and only if the finite difference scheme is a convergent scheme in which the solution of the fuzzy spatial state space system with $d(t) \equiv 0$ in equation (10.15) can converge to the solution of the NDPS with $d(t) \equiv 0$ in equation (10.1). Based on Remark 10.4.2 (ii), the stability of the fuzzy spatial state space system with $d(t) \equiv 0$ in equation (10.15) can be guaranteed by the solvable conditions of Theorem 10.4.3. In general, as N increases, since the stability of every finite difference dynamic equation within Δ_x should be guaranteed simultaneously, it will lead to the conservative of the stability criterion. Therefore, the grid size Δ_x is chosen such that the LMI in equation (10.46) is solvable.

10.5 AN APPLICATION TO TRACKING CONTROL OF HODGKIN-HUXLEY NERVOUS SYSTEMS

The nervous system consists of highly interconnected nerve cells, which communicate by generating and transmitting short action potential (short electrical pulse). Action potentials are stereotypical and all-or-none electrical transient deflections of the membrane voltage from its resting value at electrochemical equilibrium [358]. The phenomenological model of action potential generation in the nerve cell dynamic is described by the H-H model [329,358,359]. An action potential is generated at the initial segment of the nerve cell's axon and propagated to the synaptic contacts at the end of the axon. In nerve cells, separation of ionic charge along the cell membrane causes a difference in electrical potential across the cell membrane. From the dendrites of other nerve cells, nerve cells receive electrical input signals. Depending on the spatiotemporal distribution of the input current to depolarize the membrane voltage, the firing threshold can be reached after sufficient membrane voltage depolarization and then an action potential will be triggered. The H-H dynamic equations [359] represent a phenomenological model of action potential generation in a nerve cell as a function of a given current stimulus [325]. Electrical stimulation of a nerve cell with rectangular pulses has a range of clinical applications, for example, activation of muscles by stimulating the motor nerve cell fibers innervating muscles or activation of different sensor-motor areas in the brain or spinal cord, such as deep brain stimulation for Parkinsonism patients [358,360–362]. The H-H dynamic model is expressed by nonlinear PDEs describing the spatiotemporal evolution of the membrane voltage $y_v \triangleq y_v(x,t)$. In the H-H dynamic model, the total current across nerve cell membrane is the sum of the capacitive current, the ionic currents, and the external current. The H-H model is described by the total currents leading to an equivalent electric ionic currents as follows [325,329,359]:

$$c_m \frac{\partial y_v}{\partial t} = \kappa_m \frac{\partial^2 y_v}{\partial x^2} + f(y_v) + g(x)I_{inj}(t) + g_d(x)d(t) \tag{10.53}$$

$$z(t) = h(x)y_v + D_n n(t) \tag{10.54}$$

where $f(y_v) \triangleq g_K\phi_n^4\left(V_K - y_v\right) + g_{Na}\phi_m^3\phi_h\left(V_{Na} - y_v\right) + g_m\left(V_{\text{leak}} - y_v\right)$. The ϕ_n, ϕ_m, and ϕ_h are defined as the potassium activation, the sodium activation, and the sodium inactivation, respectively. The current generated from the flow of potassium ions is determined by a maximum potassium conductance g_K, an ionic equilibrium potential V_K expressing steady-state potassium ion separation, and potassium activation ϕ_n. Similarly, sodium ion current is modeled with a maximal sodium conductance g_N, an ionic equilibrium potential V_N, and sodium activation ϕ_m and inactivation ϕ_h. The remaining ion currents are collectively modeled by a leakage current with conductance g_m and ionic equilibrium potential V_{leak}. The constant c_m is the membrane capacity per unit area. The constant κ_m is defined as $\kappa_m \triangleq \frac{r_a}{2R_2}$, where r_a is the radius of the fiber and R_2 is the specific resistance of the axoplasm. The potassium activation ϕ_n, the sodium activation ϕ_m, and the sodium inactivation ϕ_h vary depending on the change of the membrane potential y_v and are given by the following equations:

$$\phi_n \triangleq \frac{\alpha_n(y_v)}{\alpha_n(y_v)+\beta_n(y_v)}, \phi_m \triangleq \frac{\alpha_m(y_v)}{\alpha_m(y_v)+\beta_m(y_v)}, \phi_h \triangleq \frac{\alpha_h(y_v)}{\alpha_h(y_v)+\beta_h(y_v)}$$

The specific functions $\alpha_n(y_v), \beta_n(y_v), \alpha_m(y_v), \beta_m(y_v), \alpha_h(y_v)$, and $\beta_h(y_v)$ are proposed by Hodgkin and Huxley [359] as

$$\alpha_n(y_v) = \frac{0.01(10-y_v)}{\exp\left(\frac{10-y_v}{10}\right)-1}, \beta_n(y_v) = 0.125\exp\left(\frac{-y_v}{80}\right), \alpha_m(y_v) = \frac{0.1(25-y_v)}{\exp\left(\frac{25-y_v}{10}\right)-1},$$

$$\beta_m(y_v) = 4\exp\left(\frac{-y_v}{18}\right), \alpha_h(y_v) = 0.07\exp\left(\frac{-y_v}{20}\right), \text{ and } \beta_h(y_v) = \frac{1}{\exp\left(\frac{30-y_v}{10}\right)+1}$$

The remaining constants are κ_m=0.336, c_m=1 μF/cm², g_K=36 mS/cm², g_{Na}=100 mS/cm², and g_m=0.3 mS/cm² with equilibrium potentials V_K=−12 mV, V_{Na}=115 mV, and V_{leak}=10.613 mV [329]. $I_{inj}(t)$ is an externally injected current in a spatially localized axonal compartment. The potential y_v is measured in units of mV; current density is in units of μA/cm²; the unit of time is milliseconds (msec). The initial distribution of the membrane voltage is given as $y_v(x, 0)=0$. The boundary conditions are the Neumann boundary condition, i.e., $\partial y_v(x, t)/\partial x = 0$ at $x=0$ and $x=1$.

When the H-H nervous system suffers the effect of external disturbances, the influence function of external disturbances can be defined as $g_d(x) = \left[\delta\left(x-\frac{5}{10}\right)\right]$. The observation influence function is defined as $h(x) = \left[\delta\left(x-\frac{2}{10}\right), \delta\left(x-\frac{4}{10}\right), \delta\left(x-\frac{6}{10}\right), \delta\left(x-\frac{8}{10}\right)\right]^T$, i.e., the sensors are located at $x = \frac{2}{10}, \frac{4}{10}, \frac{6}{10}, \frac{8}{10}$. The influence function of measurement noise is denoted as $D_n=[0,1,0,1]^T$. The control influence function is denoted as $g(x) = \left[\delta\left(x-\frac{1}{10}\right), \delta\left(x-\frac{3}{10}\right), \delta\left(x-\frac{5}{10}\right), \delta\left(x-\frac{7}{10}\right), \delta\left(x-\frac{9}{10}\right)\right]$. For the convenience of simulation, the measurement noise and the external disturbance are assumed as $n(t)=\sin(t)$ and $d(t)=\sin(t)$.

For communication in the nervous system, the nerve cells should transmit a desired signal. Therefore, we control the nervous system to track a reference signal. Suppose the desired response of a H-H nervous system is specified by the following reference model:

$$\frac{\partial y_R(x,t)}{\partial t} = A_R y_R(x,t) A_R y_R(x,t) g_R(x) r(t) \tag{10.55}$$

where the differential operator is defined as $A_R y_R(x,t) = 0.5\frac{\partial^2 y_R(x,t)}{\partial x^2}$, $A_R = -1$, $g_R(x) = \left[200\delta\left(x - \frac{3}{10}\right), 200\delta\left(x - \frac{7}{10}\right)\right]$. The reference input $r(t)$ is given to generate the impulse response signals to simulate a nervous system as $r(t) = \exp(-0.5(t-5)^2) + \exp(-0.5(t-15)^2)$. The spatiotemporal profile of the reference model is shown in Figure 10.3a. The control target is to design the control input $I_{inj}(t)$ in equation (10.53) so that the state $y(x, t)$ could track the desired trajectory $y_R(x, t)$ in equation (10.55) as good as possible in spite of the measurement noise $n(t)$ and the external disturbance $d(t)$, i.e., the control input $I_{inj}(t)$ is designed such that the tracking error $y_R(x, t) - y_v(x, t)$ must be as small as possible under the influence of the measurement noise $n(t)$ and the external disturbance $d(t)$.

First, we establish a T-S fuzzy DPS as equation (10.7) with the trapezoidal membership functions. The range of the state is given as $y_v(x,t) \in [-5, 30]$. The operation points of the fuzzy DPS are given at $y_v(x,t) = -5$ and $y_v(x,t) = 30$. The number of fuzzy rules is $M = 2$. The parameters in the fuzzy DPS equation (10.7) are obtained as $A_1 = -0.8089$ and $A_2 = -2.6473$. We can obtain the bounds of the approximate error $\sigma = 1.67 \times 10^{-2}$ and $\sigma_\tau = 0$ in Theorem 10.3.1. Obviously, the proposed fuzzy model can approach the nonlinear partial system accurately. We give the grid space $\Delta_x = 0.1111$ and $N = 10$, then the finite difference operator can be constructed. Following the proposed design procedure in the above section, the optimal H_∞ fuzzy observer- based tracking controller could be obtained easily with $\rho_0 = 0.0440$ by solving the constrained optimization problem in equation (10.52). The spatiotemporal profile of the controlled H-H nervous system is shown in Figure 10.3b. The spatio-time profiles of the tracking error $y_R(x, t) - y(x, t)$ between the nonlinear disturbed parameter system and the reference model are shown in Figure 10.3c. The time profiles of the estimated error $e(t)$ between the nonlinear disturbed parameter system and the fuzzy observer are shown in Figure 10.3d. The simulation results show that the proposed robust H_∞ fuzzy observer-based tracking controller obviously can control the trajectory of the H-H nervous system to track a desired trajectory by efficiently attenuating the truncation error, the external disturbances and measurement noises. The H_∞ tracking performance can be computed as follows:

$$\frac{\int_0^{30} \bar{y}(t)^T R^T R \bar{y}(t) dt}{\int_0^{30} v(t)^T v(t) dt} \approx 0.0171^2 < \rho_0^2 = 0.0440^2$$

This conservative result is due to the conservative of solving LMIs in the H_∞ control tracking design procedure. Therefore, the simulation example has shown the feasibility of the proposed robust H_∞ fuzzy tracking control design of the NDPS for potential practical applications. The effects of the external disturbance and the measurement noise in the NDPS could be efficiently attenuated by the proposed robust H_∞ observer-based tracking control design.

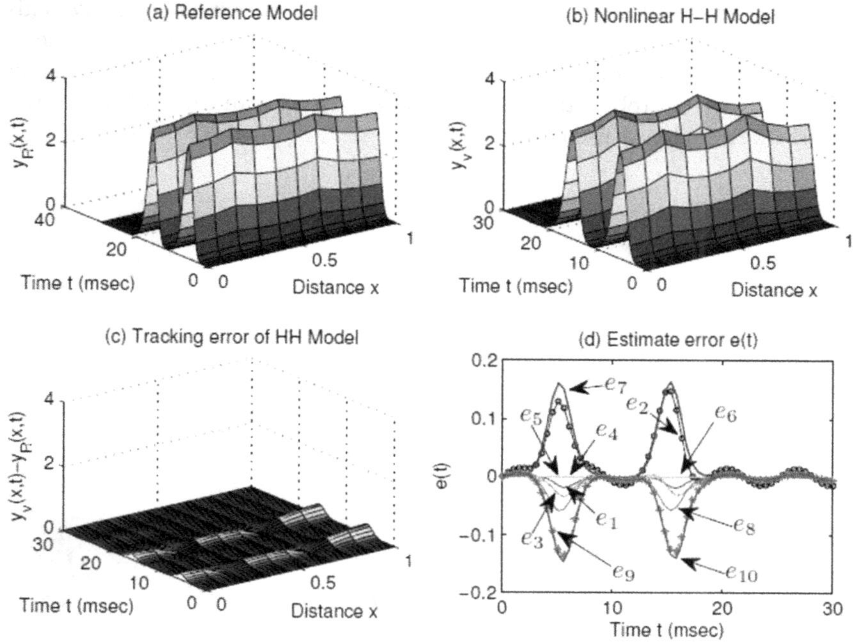

FIGURE 10.3 (a) Spatiotemporal profiles of the reference model in equation (10.55). (b) Spatiotemporal profiles of the H-H nervous system in equation (10.53). (c) Spatiotemporal profiles of the tracking error $y_R(x, t)-y_v(x, t)$. (d) Time profiles of the estimated error $e(t)$.

10.6 CONCLUSION

This chapter proposed a two-dimensional H_∞ tracking performance in spatio-temporal domain for robust model reference tracking control of NDPSs under time-delay, external disturbances, and measurement noises. The robust model reference tracking control problem for the NDPSs with time delay, external disturbances and measurement noise was successfully solved by the fuzzy spatial state space system based on finite difference model. A fuzzy observer-based tracking controller is proposed to attenuate the effect of the truncation error, external disturbances and measurement noise on the desired tracking performance below a prescribed level to achieve robust tracking control design of NDPSs. In order to simplify the design procedure, the proposed H_∞ fuzzy observer-based tracking control scheme for NDPSs can be transformed from solving a BMI problem to solving an LMI problem. Therefore, determining the observer gains and the controller gains for the optimal H_∞ observer-based tracking control becomes an LMI-based optimization problem, which can be efficiently solved by the LMI Toolbox in MATLAB. A H-H nervous tracking control problem in biology engineering is provided to illustrate the practical application of the H_∞ tracking control scheme to NDPSs and to confirm its robust tracking performance.

The proposed robust tracking design method can also be applied to many fields, e.g., heat flows, elastic wave, flexible structures, chemical engineering, biodynamic systems and so on. Therefore, the proposed design method in this chapter is potential for the robust tracking control of NDPSs with time delay, external disturbances, and measurement noises.

11 Robust Stabilization Control Design of Nonlinear Stochastic Partial Differential Systems

11.1 INTRODUCTION

In recent years, a great deal of concern has been raised regarding the study of stochastic partial differential equations (SPDEs) [384–395]. Many phenomena in science and engineering have been modeled by deterministic partial differential equations (PDEs). Examples include mechanical systems related to heat flows, fluid flows, elastic waves, flexible structures, chemical engineering [396], biology, population dynamics, neurophysiology, biodynamics [397,398], etc. Since most phenomena possess some stochastic uncertainty due to the existence of different random intrinsic fluctuations, SPDEs should be used to accurately represent their behaviors [384,385]. Some examples are turbulent flow in fluid dynamics [399], diffusion in random media, and molecule signal transduction in biology. In chemical engineering, many chemical processes are characterized by the presence of spatial variations and time delays [389,396]. In biology [397,398], mathematical tools can help provide systematic analysis, e.g., the stability or robustness of biological systems. For example, the nonlinear partial differential Hodgkin-Huxley model has been applied to model signal transmission in a nervous system [397]. Here we address the study of NSPDSs.

In the past decade, control design problems for deterministic partial differential systems have been thoroughly investigated. The stabilization design problem of linear partial differential systems (LPDSs) has been particularly widely studied and published [400]. In Chapters 6 and 7, these LPDSs can be transformed into equivalent infinite-dimensional ordinary differential systems. According to the separation of eigenvalues, the infinite-dimensional ordinary differential system can be separated into a finite-dimensional slow mode and an infinite-dimensional fast mode. A finite-dimensional controller based on the finite-dimensional mode can be designed to stabilize the LPDS. For the reference tracking problem of LPDSs, Byrnes et al. proposed an extended output regulation method to control LPDS to track a reference model [401]. In Chapter 9, the stabilization design

DOI: 10.1201/9781003229230-14

problem of NPDSs is more complex than the control design of the LPDSs. Based on Galerkin's method in Section 1.2.1, controller design schemes have been proposed to stabilize the NPDS by employing a residual model filter [402] or an inertial manifold model [403]. Due to the properties of nonlinearity and infinite dimension, the infinite-dimensional nonlinear ODE in [402,403] is difficult to obtain for the design of the corresponding controller. However, the best approach to the design of the corresponding controller is still unclear. In order to remedy this design difficulty, some researchers have provided systematic approaches based on fuzzy theory to solve the NPDSs stabilization design problem [404–408].

The fuzzy approach, which uses several local linear models to interpolate a nonlinear system, has been widely applied in the past 20 years to analyze various nonlinear systems which are described by ODE [409–415]. The fuzzy approach has recently been applied to the domain of PDEs, resulting in the proposal of a new PDE solution technique in Chapter 2 using an adaptive fuzzy algorithm. On the other hand, a fuzzy control method was also proposed to stabilize a class of NPDS based on the infinite-dimensional nonlinear ODE represented by Galerkin's method in Section 1.2.1 [404,405]. In [406], a new T-S fuzzy state space model using the fuzzy interpolation approach with Galerkin's method was proposed to approximate the NPDS. The advantage of the fuzzy state space model is that it avoids the necessity of obtaining a complex nonlinear ODE. As a result, the effects of the spillover on stabilization due to control and observation can be attenuated by the proposed robust stabilization design in Chapter 9. A robust H_∞ stabilization scheme was developed to attenuate the effects of modeling errors, external disturbances, and measurement noises. In [417], the robust H_∞ reference tracking control problem for NPDSs was solved using the fuzzy approach and the finite difference method. A robust fuzzy observer-based tracking control scheme was designed to attenuate the effects of approximation errors, truncation errors, external disturbances, and measurement noises due to the H_∞ reference tracking performance in Chapter 10. Recent investigations of SPDEs have focused on stochastic differential equation in infinite dimension [418]. In [419], the stochastic stability properties of linear stochastic second-order PDEs were introduced. The stability of NSPDSs was analyzed based on the functional Itô formula, the Lyapunov and related functions [388]. The concept of vector Lyapunov-like functional technique coupled with partial differential inequalities is described to investigate various types of stability related to NSPDSs using the comparison principle [387]. Moreover, various stability problems of NSPDSs have been presented in textbooks [384,385]. In the other words, numerical solutions for SPDEs have been studied, employing the finite difference method [386,390], the method of lines (semi-discretization) [391,392] and Taylor expansions [420]. However, few results have been presented for the control design of the SPDSs due to complex infinite-dimensional stochasticity. A predictive control model for NSPDSs with state-independent noise has been developed using Galerkin's method [389]. Since NSPDSs are the most complicated control systems, their robust stabilization design is a challenging control problem, especially in the case of random noise. We shall, therefore, discuss robust stochastic stabilization problems for NSPDSs with random external disturbances and measurement noises in the spatiotemporal domain.

In this chapter, robust stochastic stabilization control design of NSPDSs is developed by employing the Itô formula, fuzzy interpolation, and the method of lines. First, the NSPDS is approximated using a so-called FSPDS by interpolating several LSPDSs via membership functions with some approximation errors. The FSPDS can then be represented by a finite- dimensional fuzzy stochastic spatial state space model based on the Kronecker product and the method of lines (semi-discretization finite difference method). Unlike using the infinite dimensional ODE system to represent the PDE system [402,403,416], for the convenience of control design, the partial differential operator could be approximated by using a finite difference operator. The convergence of the method of lines was previously proven in [386,390–392,421]. This fuzzy stochastic spatial state space model can therefore be efficiently used to represent NSPDSs and to design the fuzzy estimator-based controller. This controller is designed to stochastically stabilize the NSPDS and to tolerate effects of the state-dependent spatiotemporal Wiener process and the fuzzy approximation error. Furthermore, the effects of random external disturbances and measurement noises in the spatiotemporal domain can also be attenuated to a prescribed attenuation level using the proposed robust stochastic H_∞ stabilization design method. In order to treat the robust stochastic H_∞ stabilization problem, we introduce two-dimensional stochastic H_∞ control performance in a spatiotemporal domain for NSPDSs. For convenience of control design, the two-dimensional stochastic H_∞ stabilization problem is transformed to an equivalent one-dimensional stochastic H_∞ stabilization problem, when all discretization grid points are represented by a spatial vector. Therefore, if grid points in a spatial domain are dense enough or the truncation error $O\left(\Delta_x^2\right)$ is small enough, the robust stochastic H_∞ stabilization problem of the fuzzy stochastic spatial state space model will approach the robust stochastic H_∞ stabilization problem of the NSPDS. Finally, a systematic approach is also developed to efficiently solve the issue of stabilization design using the conventional LMI technique.

The difficulty of stabilization control design for NSPDS with stochastic intrinsic fluctuations (state-dependent noises), external disturbances, and measurement noises is overcome in our proposed fuzzy stochastic spatial state space model by employing the method of lines. This model is successfully applied to approximate NSPDSs, and we further propose a fuzzy estimator-based controller to stochastically stabilize the NSPDS. For the free of external disturbances and measurement noises, the stochastic asymptotic stability of the NSPDS is guaranteed to tolerate state-dependent noise and fuzzy approximation error. The proposed robust stochastic H_∞ stabilization control scheme can be successfully applied to minimize the effect of random external disturbances and measurement noises on the desired stabilization performance level. The proposed fuzzy stabilization design scheme can be systematically and easily solved using the LMI toolbox in MATLAB. The features of this chapter are stated as follows: (i) The spatiotemporal Wiener process and external random disturbance are considered to represent the more general stochastic NPDSs with the stochastic parameter fluctuation and environmental disturbance in the spatiotemporal domain, respectively. When the scale of NPDSs becomes very large like ecological system or very small like molecular system in nano-scale, the stochastic spatiotemporal intrinsic fluctuations and external disturbances are unavoidable for the design of robust stabilization and state estimation of NSPDSs. (ii) The method of lines is used to replace the infinite dimensional Galerkin's method so that the fuzzy spatial state space model can be used to simplify the design procedure. (iii) The stochastic two-dimensional H_∞ stabilization performance is used to replace the one-dimensional H_∞ stabilization performance to consider the more general robust stabilization and state estimation problem of NSPDSs in the whole spatiotemporal domain. (iv) The effect of local truncation error is also considered in the H_∞ stabilization performance to attenuate the effect of local truncation error as possible in the design procedure.

Notations: For the convenience of problem description, we define some different separable Hilbert spaces in the following.

1. $\|y(t)\|^2 \triangleq \sum_{i=1}^{n}|y_i(t)|^2$ where $y(t) = [y_1(t),\dots,y_n(t)]^T$
2. $L_2(\mathbb{R}_+;\mathbb{R}^n)$ is the space of the n-dimensional measurable functions $z(t) \in \mathbb{R}^n$ defined on $t \in \mathbb{R}_+ = [0,\infty)$ such that
$$\|z(t)\|^2_{L_2(\mathbb{R}_+;\mathbb{R}^n)} \triangleq \int_0^\infty \|z(t)\|^2 \, dt < \infty$$
3. $L_2(U;\mathbb{R}^n)$ is the space of the n-dimensional measurable functions $y(x,t) \in \mathbb{R}^n$ defined on $x \in U$ for any t such that
$$\|y(x,t)\|^2_{L_2(U;\mathbb{R}^n)} \triangleq \int_U \|y(x,t)\|^2 \, dx < \infty$$
4. $L_2(U \times \mathbb{R}_+;\mathbb{R}^n)$ is the space of the n-dimensional measurable functions $y(x,t) \in \mathbb{R}^n$ defined on $(x,t) \in U \times \mathbb{R}_+$ such that
$$\|y(x,t)\|^2_{L_2(U\times\mathbb{R}_+;\mathbb{R}^n)} \triangleq \int_0^\infty \int_U \|y(x,t)\|^2 \, dx\,dt < \infty$$
5. For the matrix $A \in \mathbb{R}^{m\times n}$, the matrix norm is defined by comparing $\|A_x\|$ with $\|x\|$ as $\|A\| \triangleq \sup_{x\neq 0} \frac{\|A_x\|}{\|x\|}$ where $x \in \mathbb{R}^n$.
6. Let $(\Omega,\mathcal{F},\mathcal{P})$ denote a complete probability space. $\{W(x,t) \in \mathbb{R} : t \geq 0, x \in U\}$ is an $\mathcal{F}$-adapted Wiener process [1,2].

11.2 SYSTEM DESCRIPTION AND PROBLEM FORMULATION

In this section, we formulate the robust stochastic stabilization problem for NSPDSs. Consider the following NSPDS with random intrinsic fluctuation, external disturbance, and measurement noise [391,422].

$$\partial y(x,t) = (\kappa(y(x,t)))\frac{\partial^2 y(x,t)}{\partial x^2} + f(y(x,t)) + f_W(y(x,t))\frac{\partial^2 W(x,t)}{\partial x \partial t}$$
$$+ g(x)u(t) + g_d d(x,t))\partial t \tag{11.1}$$

$$z(t) = h(x)(y(x,t) + D_n n(x,t)) \tag{11.2}$$

for $x = [x_1, x_2]^T \in U \subset \mathbb{R}_+^2$ and $t > 0$, where $y(x,t) \triangleq [y_1(x,t),...,y_n(x,t)]^T \in \mathbb{R}^n$ is the state variable; x and t are the space and time variables, respectively. The space domain U is a bounded domain. $f(y(x,t)) \in \mathbb{R}^n$ and $f_W(y(x,t)) \in \mathbb{R}^n$ are nonlinear functions satisfying $f(0)=0$ and $f_W(0)=0$, respectively [385,391]. The nonlinear function $\kappa(y(x,t) \in \mathbb{R}^{n\times n}$ satisfies $\kappa(y(x,t) \geq 0$ and $\kappa(0) = 0$. The distribution of the control force $u(t)$ is provided by n_u point force actuators, i.e., $u(t) = [u_1(t),...,u_{n_u}(t)]^T \in \mathbb{R}^{n_u}$ is the applied force to be designed; the influence function $g(x) = [g_1\delta(x-p_1),...,g_p\delta(x-p_{n_u})]$ is an $n \times n_u$ matrix form whose elements are the delta function $g_i\delta(x-p_i)$ where p_i are control force locations and $\delta(x-p_i)$ is so-called the unit-impulse (or Dirac delta) function [444] defined as $\delta(x-p_i)=1$ for $x=p_i$ and $\delta(x-p_i)=0$ otherwise, giving $g_i\delta(x-p_i) = g_i \in \mathbb{R}^n$ for $x = p_i$ or $g_i\delta(x-p_i) = 0$ for $x \neq p_i$. The measured output $z(t)$ may be interpreted as observations or as parts of the system whose behavior we wish to influence, and the output $z(t) \in \mathbb{R}^{n_z}$ is a vector, i.e., $z(t) \triangleq [z_1(t),...,z_n(t)]^T$, where n_z is the number of observations and the observation influence function $h(x) \triangleq [h_1^T\delta(x-q_1),...,h_{n_z}^T\delta(x-q_{n_z})]^T \in \mathbb{R}^{n_z\times n}$ is a matrix function, in which $h_i \in \mathbb{R}^{1\times n}$. Therefore, in the case where there is no measurement noise, we have $z_i(t) = h_i\delta(x-q_i)y(x,t) = h_i y(q_i,t)$ with q_i as the ith sensor locations. $g_d \in \mathbb{R}^{n\times n_d}$ is the influence matrix; $d(x,t) \in L_2(U \times \mathbb{R}_+;\mathbb{R}^{n_d})$ is the vector of the random external disturbance in the spatiotemporal domain; $D_n \in \mathbb{R}^{n\times n}$ is the influence matrix and $n(x,t) \in L_2(U \times \mathbb{R}_+;\mathbb{R}^n)$ is the stochastic spatiotemporal measurement noise at the observation locations. The spatiotemporal noise $W(x,t) \in \mathbb{R}$ is a stochastic Wiener process or Brownian motion, which denotes some random fluctuation at different locations and times within the NSPDS. We recall that the Wiener process $W(x, t)$ is a zero mean Gaussian random field with covariance $EW(x,t)W(\bar{x},s) = (t \wedge s)(x \wedge \bar{x})$ [391]. We have $\frac{\partial^2}{\partial t \partial s}EW(x,t)W(\bar{x},s) = \frac{\partial^2}{\partial t \partial s}(t \wedge s)x = \delta(t-s)x$ and $\frac{\partial^2}{\partial x \partial \bar{x}}EW(x,t)W(\bar{x},t) = \frac{\partial^2}{\partial x \partial \bar{x}}(x \wedge \bar{x})t = \delta(x-\bar{x})t$. The differential operator is defined as $\frac{\partial^2}{\partial x^2}y(x,t) \triangleq \sum_{k=1}^2 \frac{\partial^2}{\partial x_k^2}y(x,t)$ where $\frac{\partial^2}{\partial x_k^2}y(x,t) \triangleq \left[\frac{\partial^2}{\partial x_k^2}y_1(x,t),...,\frac{\partial^2}{\partial x_k^2}y_n(x,t)\right]^T \in \mathbb{R}^n$ [400].

The initial-value is denoted by $y(x, 0)=y_0(x)$. The boundary condition is given by the Dirichlet boundary condition, i.e. $y(x, t)=0$ on ∂U. Note that the Neumann boundary condition can be formulated by minor modification [417].

11.2.1 STOCHASTIC STABILITY OF NSPDSs

In this section, the stochastic stability of NSPDSs is defined based on the stochastic Lyapunov theory. First, we consider an NSPDS without any input as follows:

$$\partial y(x,t) = \left(\kappa(y(x,t)) \frac{\partial^2 y(x,t)}{\partial x^2} + f(y(x,t)) + f_W(y(x,t)) \frac{\partial^2 W(x,t)}{\partial x \partial t} \right) \partial t \tag{11.3}$$

Definition 11.2.1 (Null Solution)

Suppose that $\kappa(0) = 0$, $f(0) = 0$, and $f_W(0) = 0$. In this case the system in equation (11.3) has an equilibrium solution $y(x,t) \equiv 0$. This solution is called the null solution.

Remark 11.2.1

If we are interested in the equilibrium point $y_e \neq 0$, for convenience of investigation, the origin should be shifted to y_e, i.e., the NSPDS in equation (11.3) should be modified as

$$\partial y'(x,t) = \left(\kappa(y'(x,t) + y_e) \frac{\partial^2 y'(x,t)}{\partial x^2} + f(y'(x,t) + y_e) + f_W(y'(x,t) + y_e) \frac{\partial^2 W(x,t)}{\partial x \partial t} \right) \partial t. \tag{11.4}$$

where $y'(x,t) = y(x,t) - y_e$. Therefore, instead of studying the behavior in the neighborhood of y_e, one can equivalently study the behavior in the neighborhood of $y'(x,t) = 0$ in equation (11.4).

Therefore, throughout this chapter, we shall study the stability of the null solution for the system equation (11.3). The definitions of the stability are given as follows:

Definition 11.2.2 [384,385]

(i) The null solution $y(x,t) \equiv 0$ of equation (11.3) is said to be stable in probability if for any $\epsilon_1, \epsilon_2 > 0$, there exists a $\delta = \delta(\epsilon_1, \epsilon_2) > 0$ such that if $\|y_0(x)\|_{L_2(U;\mathbb{R}^n)} < \delta$, then $P\{\|y(x,t)\|_{L_2(U;\mathbb{R}^n)} > \epsilon_1\} < \epsilon_2$ for all $t > 0$. (ii) The null solution $y(x,t) \equiv 0$

of equation (11.3) is said to be asymptotically stable in probability if it is stable in probability and for each $\epsilon > 0$, there exists a $\delta = \delta(\epsilon) > 0$ such that $\|y_0(x)\|_{L_2(U;\mathbb{R}^n)} < \delta$ guarantees $\lim_{t\to\infty} P\left\{\|y(x,t)\|_{L_2(U;\mathbb{R}^n)} > \epsilon\right\} = 0$.

Suppose that $V(y(x,t))$ is a strong Itô functional. Then we have [384,385,388]

$$V(y(x,t)) = V(y_0(x)) + \int_0^t LV(y(x,s))ds + \int_0^t \int_U \left(\frac{\partial V(y(x,s))}{\partial y}\right)^T f_W(y(x,s))dW(x,s) \tag{11.5}$$

where the infinitesimal operator $LV(y(x,s))$ is denoted as

$$LV(y(x,s)) = \int_U \left(\frac{\partial V(y(x,s))}{\partial y}\right)^T \left(\kappa(y(x,s))\frac{\partial^2 y(x,s)}{\partial x^2} + f(y(x,s))\right)dx$$
$$+ \int_U \frac{1}{2} f_W(y(x,s))^T \frac{\partial^2 V(y(x,s))}{\partial y^2} f_W(y(x,s))dx$$

Lemma 11.2.1 [384,385,388]

Let $U \subset \mathbb{R}^n$ be a neighborhood of the origin. A strong Itô functional $V(y(x,t)) : U \to \mathbb{R}$ is said to be a Lyapunov functional for the system in equation (11.3) if (i) $V(0) = 0$ and $V(y(x,t)) > 0$ for $y(x,t) \neq 0$ and (ii) $LV(y(x,t)) \leq 0$.

Theorem 11.2.1 [385,388]

Suppose the system in equation (11.3) has a strong solution $y(x,t)$ and there exists a Lyapunov functional $V(y(x,t))$. The null solution of equation (11.3) is then stable in probability. Furthermore, if $LV(y(x,t)) < 0$, then the null solution is asymptotically stable in probability.

The robust stochastic stabilization control problem is formulated as follows: A robust controller $u(t)$ is designed to stabilize the NSPDS with $d(t) \equiv 0$ in equation (11.1) and to tolerate the effect of state-dependent spatiotemporal noise $W(x, t)$.

11.2.2 Stochastic H_∞ Stabilization Control for NSPDSs

In this section, we consider the effects of random external disturbance $d(x, t)$ and measurement noise $n(x, t)$. Since the statistics of random external disturbance $d(x, t)$ and the measurement noise $n(x, t)$ are unknown and uncertain, robust stochastic H_∞ stabilization control design is required. A robust stochastic H_∞ stabilization design should be specified so that the effect of the random external disturbance $d(x, t)$ and the measurement noise $n(x, t)$ on the controlled output in the spatiotemporal domain must be below a prescribed level ρ as follows:

$$\frac{E\int_0^{t_f}\int_U \|R_1 y(x,t)\|^2\,dx\,dt}{E\int_0^{t_f}\int_U \|v(x,t)\|^2\,dx\,dt} \le \rho^2 \tag{11.6}$$

or

$$E\int_0^{t_f}\int_U \|R_1 y(x,t)\|^2\,dx\,dt \le \rho^2 E\int_0^{t_f}\int_U \|v(x,t)\|^2\,dx\,dt \tag{11.7}$$

where $v(x, t)=[d(x, t)^T, n(x, t)^T]^T$ is considered to be the vector of disturbances. $R_1 y(x, t)$ is the controlled output by the weighting matrix $R_1 \in \mathbb{R}^{n\times n}$. The inequality in equations (11.6) or (11.7) is called the two-dimensional stochastic H_∞ control performance in the spatiotemporal domain. The physical explanation of this expression is that the effect of all disturbances on the controlled output $R_1 y(x, t)$ in the total space $U\times[0, t_f]$ must be attenuated below a prescribed level ρ from the area energy perspective. Note that the integrated norm $\int_0^{t_f}\int_U \|R_1 y(x,t)\|^2\,dx$ is to describe the area energy of controlled output on the spatiotemporal space $U\times[0, t_f]$, which is different from the conventional H_∞ control performance that is only in the time domain. If the disturbance $v(x, t)$ is of a deterministic signal, then the expectation E on $v(x, t)$ in equations (11.6) and (11.7) could be neglected.

The robust stochastic H_∞ stabilization control problem is formulated as follows: given a prescribed disturbance attenuation level ρ, a robust controller $u(t)$ is designed to attenuate the effect of random external disturbances $d(x, t)$ and measurement noise $n(x, t)$ on the controlled output $R_1 y(x, t)$ below ρ from the point of view of total area energy on the space $U\times[0, t_f]$. In other words, robust H_∞ control performance, as described in equation (11.7), is achieved.

11.3 SYSTEM REPRESENTATION BY FUZZY SPATIAL STATE SPACE MODEL

Among the stabilization control design problems of NSDPSs, the foremost problem is to obtain a suitable state space model to represent the NSPDSs. First, a FSPDS is proposed to approximate the NSPDS as follows [406]

Rule i: If $y_1(x, t)$ is F_{1i}, and …, and, $y_n(x, t)$ is F_{ni} then

$$\begin{aligned}\partial y(x,t) = ((\kappa_0+\kappa_i)\frac{\partial^2 y(x,t)}{\partial x^2} + (A_0 + A_i)y(x,t)\\ +(A_{W,0}+A_{W,i})y(x,t)\frac{\partial^2 W(x,t)}{\partial x\,\partial t} + g(x)u(t) + g_d d(x,t))\partial t\end{aligned} \tag{11.8}$$

where F_{ji} is the grade of the membership of $y_j(x, t)$; $\kappa_0 \in \mathbb{R}^{n\times n}$, $\kappa_i \in \mathbb{R}^{n\times n}$, $A_0 \in \mathbb{R}^{n\times n}$, $A_i \in \mathbb{R}^{n\times n}$, $A_{W,0} \in \mathbb{R}^{n\times n}$ and $A_{W,i} \in \mathbb{R}^{n\times n}$ are the system parameters in local LSPDSs, respectively. The constant matrices κ_0 and $\kappa_i \in \mathbb{R}^{n\times n}$ satisfy $\kappa_0 \ge 0$ and $\kappa_i \ge 0$, respectively. The overall FSPDS can be formulated as follows [406,412]:

$$\partial y(x,t) = \begin{pmatrix} \sum_{i=1}^{M} \mu_i(y(x,t))[\kappa_0 + \kappa_i] \frac{\partial^2 y(x,t)}{\partial x^2} + (A_0 + A_i)y(x,t) \\ + (A_{W,0} + A_{W,i})y(x,t) \frac{\partial^2 W(x,t)}{\partial x \partial t}] + g(x)u(t) + g_d d(x,t) \\ + \varepsilon_\kappa(y(x,t)) \frac{\partial^2 y(x,t)}{\partial x^2} + \varepsilon_f(y(x,t)) + \varepsilon_{fw}(y(x,t)) \frac{\partial^2 W(x,t)}{\partial x \partial t} \end{pmatrix} \partial t \tag{11.9}$$

where $\mu_i(y(x,t)) \triangleq \frac{\prod_{j=1}^{n} F_{ji}(y_j(x,t))}{\sum_{i=1}^{M} \prod_{j=1}^{n} F_{ji}(y_j(x,t))}$. $F_{ji}(y_j(x,t))$ is the grade of the membership of $y_j(x, t)$ or the possibility function of $y_j(x, t)$. The denominators of $\mu_i(y(x, t))$ are only for normalization purposes so that the total sum of the fuzzy bases $\sum_{i=1}^{M} \mu_i(y(x,t)) = 1$.

In equation (11.9), we use the fuzzy interpolation via M local LSPDSs to approximate the NSPDS in equation (11.1). The approximation errors $\varepsilon_\kappa(y(x,t)) \in \mathbb{R}^{n \times n}$, $\varepsilon_f(y(x,t)) \in \mathbb{R}^n$, and $\varepsilon_{fw}(y(x,t)) \in \mathbb{R}^n$ are defined as follows:

$$\varepsilon_\kappa(y(\cdot,\cdot)) \triangleq \kappa(y(\cdot,\cdot)) - \sum_{i=1}^{M} \mu_i(y(\cdot,\cdot))\kappa_i + \kappa_0 \tag{11.10}$$

$$\varepsilon_f(y(\cdot,\cdot)) \triangleq f(y(\cdot,\cdot)) - \sum_{i=1}^{M} \mu_i(y(\cdot,\cdot))(A_0 + A_i)y(\cdot,\cdot) \tag{11.11}$$

$$\varepsilon_{fw}(y(\cdot,\cdot)) \triangleq f_W(y(\cdot,\cdot)) - \sum_{i=1}^{M} \mu_i(y(\cdot,\cdot))(A_{W,0} + A_{W,i})y(\cdot,\cdot) \tag{11.12}$$

where $y(\cdot,\cdot) \triangleq y(x,t)$ for any x and t. The bounds of $\varepsilon_\kappa(y(\cdot,\cdot))$, $\varepsilon_f(y(\cdot,\cdot))$, and $\varepsilon_{fw}(y(\cdot,\cdot))$ can be estimated according to the following theorem.

Theorem 11.3.1 [406]

Suppose $\kappa(y(\cdot,\cdot))$, $f(y(\cdot,\cdot))$, and $f_W(y(\cdot,\cdot))$ are the continuous nonlinear functions defined on a compact set $U \subset \mathbb{R}^n$, i.e., $y(\cdot,\cdot) \in U$. Then, for three arbitrary constants $\sigma_\kappa > 0$, $\sigma_f > 0$, and $\sigma_{fw} > 0$, the fuzzy functions in equations (11.10)–(11.12) could be constructed to approximate the nonlinear functions $\kappa(y(\cdot,\cdot))$, $f(y(\cdot,\cdot))$, and $f_W(y(\cdot,\cdot))$, and the approximation errors in equations (11.10)–(11.12) are bounded by σ_κ, σ_f, and σ_{fw}, i.e., $\|\varepsilon_\kappa(y(\cdot,\cdot))\|^2 \leq \sigma_\kappa^2$, $\|\varepsilon_f(y(\cdot,\cdot))\|^2 \leq \sigma_f^2 \|(y(\cdot,\cdot))\|^2$, and $\|\varepsilon_{fw}(y(\cdot,\cdot))\|^2 \leq \sigma_{fw}^2 \|(y(\cdot,\cdot))\|^2$.

Proof

See the proof of Theorem 1 in [406].

Remark 11.3.1

Note that the nonlinear function $\kappa(y(\cdot,\cdot)) \in \mathbb{R}^{n\times n}$ is approximated by the Mamdani fuzzy inference system [406]. The matrix norm $\|\varepsilon_\kappa(y(\cdot,\cdot))\|^2$ is defined in Notations 5) of Section 11.1 as $\|\varepsilon_\kappa(y(\cdot,\cdot))\|^2 = \sup_{y(\cdot,\cdot)\neq 0} \dfrac{\|\varepsilon_\kappa(y(\cdot,\cdot))y(\cdot,\cdot)\|}{\|(y(\cdot,\cdot))\|^2} \leq \sigma_\kappa^2$.

The finite difference scheme in Section 1.2.2 has been widely applied to obtain numerical solutions of PDEs or SPDEs. In this chapter, we use the finite difference method to semi-discretize the NSPDS. This method is the so-called method of lines [425]. Consider a typical grid mesh as shown in Figure 11.1. The state $y(x, t)$ is represented by $y_{\{k,l\}}(t) \in \mathbb{R}^n$ at the grid node $x_{k,l}(x_1 = k\Delta_x, x_2 = l\Delta_x)$ where $k = 1,\ldots,N_1$ and $l = 1,\ldots,N_2$, i.e., $y(x,t)\big|_{x=x_{\{k,l\}}} \triangleq y_{\{k,l\}}(t)$. At the grid points $x_{\{k,l\}}$, the central difference approximation for the linear differential operator can be written as follows [386,390–392,423,424]

$$\kappa_i\left(\frac{\partial^2 y(x,t)}{\partial x^2}\right)_{x=x_{\{k,l\}}} = \kappa_i \frac{y_{\{k+1,l\}}(t) + y_{\{k-1,l\}}(t) - 2y_{\{k,l\}}(t)}{\Delta_x^2}$$
$$+\kappa_i \frac{y_{\{k,l+1\}}(t) + y_{\{k,l-1\}}(t) - 2y_{\{k,l\}}(t)}{\Delta_x^2} + O_{\{k,l\}}(\Delta_x^2 \quad (11.13)$$

The Wiener process term $\dfrac{\partial W(x,t)}{\partial x}$ can be also represented by [423,424]

$$\left(\frac{\partial W(x,t)}{\partial x}\right)_{x=x_{\{k,l\}}} = \frac{W_{\{k+1,l\}}(t) - W_{\{k-1,l\}}(t)}{2\Delta_x}$$
$$+\frac{W_{\{k,l+1\}}(t) - W_{\{k,l-1\}}(t)}{2\Delta_x} + O_{W,\{k,l\}}(\Delta_x^2) \quad (11.14)$$

The remainder terms $O_{\{k,l\}}(\Delta_x^2), O_{W,\{k,l\}}(\Delta_x^2) \in^n$ are referred to as the local truncation errors, which converge to zero if the grid size Δ_x is chosen to be as small as possible by applying the **Lax-Richtmyer Equivalence Theorem** [423], i.e.,

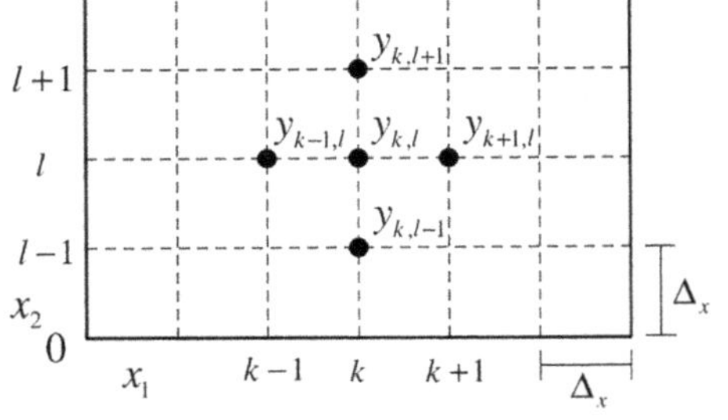

FIGURE 11.1 Finite difference grids on the spatiodomain.

$O_{\{k,l\}}(\Delta_x^2), O_{W,\{k,l\}}(\Delta_x^2) \to 0$ as $\Delta_x \to 0$. The effect of these local truncation errors on the H_∞ stabilization will be discussed in the sequel. Therefore, a fuzzy stochastic finite difference model can be constructed to represent the state $y_{\{k,l\}}(t)$ of the NSPDS at $x = x_{\{k,l\}}$ in equation (11.9) as follows:

$$dy_{\{k,l\}}(t) = \sum_{i=1}^{M} \mu_i(y_{\{k,l\}})[A_0 + A_i]y_{\{k,l\}}(t)dt + \varepsilon_{f\{k,l\}}(y_{\{k,l\}})dt + B_{\{k,l\}}u(t)dt + g_d d_{\{k,l\}}(t)dt$$

$$+\frac{1}{\Delta_x^2}(\sum_{i=1}^{M} \mu_i(y_{\{k,l\}})[\kappa_0 + \kappa_i] + \varepsilon_{\kappa\{k,l\}}(y_{\{k,l\}}))[y_{\{k+1,l\}}(t)dt + y_{\{k-1,l\}}(t)dt$$

$$-2y_{\{k,l\}}(t)dt + y_{\{k,l+1\}}(t)dt + y_{\{k,l-1\}}(t)dt - 2y_{\{k,l\}}(t)dt]$$

$$+\frac{1}{2\Delta_x}(\sum_{i=1}^{M} \mu_i(y_{\{k,l\}})[A_{W,0} + A_{W,i}]y_{\{k,l\}}(t) + \varepsilon_{fW\{k,l\}}(y_{\{k,l\}}))[dW_{\{k+1,l\}}(t)$$

$$-dW_{\{k-1,l\}}(t) + dW_{\{k,l+1\}}(t) - dW_{\{k,l-1\}}(t)] \tag{11.15}$$

The elements of matrix $B_{\{k,l\}} = [B_{\{k,l\},1},\ldots,B_{\{k,l\},p}] \in \mathbb{R}^{n\times n_u}$ are defined as $B_{\{k,l\},i} = g_i$ for $x_{\{k,l\}} = p_i$ or $B_{\{k,l\},i} = 0$ for $x_{\{k,l\}} \neq p_i$, where g_i is the influence function at location p_i and is defined in equation (11.1). $d_{\{k,l\}}(t)$ is defined as $d(x,t)\big|_{x=x_{\{k,l\}}}$. The approximation errors are defined as $\varepsilon_{\kappa\{k,l\}}(y_{\{k,l\}}) \triangleq \varepsilon_\kappa(y(x,t))\big|_{x=x_{\{k,l\}}}$, $\varepsilon_{f\{k,l\}}(y_{\{k,l\}}) \triangleq \varepsilon_f(y(x,t))\big|_{x=x_{\{k,l\}}}$, and $\varepsilon_{fW\{k,l\}}(y_{\{k,l\}}) \triangleq \varepsilon_{fW}(y(x,t))\big|_{x=x_{\{k,l\}}}$.

To simplify the design, we will define a spatial state vector $y(t)$ to collect the states $y_{\{k,l\}}(t) \in \mathbb{R}^n$ at all grid nodes in Figure 11.1. Therefore, the spatial state vector $y(t) \in \mathbb{R}^{nN}$ is defined as follows:

$$y(t) = [y_{\{1,1\}}^T(t),\ldots,y_{\{N_1,1\}}^T(t),\ldots,y_{\{k,l\}}^T(t),\ldots,y_{\{1,N_2\}}^T(t),\ldots,y_{\{N_1,N_2\}}^T(t)]^T \tag{11.16}$$

where $N \triangleq N_1 \times N_2$. Note that n is the dimension of the vector $y_{\{k,l\}}(t)$ for each grid node and $N_1 \times N_2$ is the number of grid nodes. In order to simplify the index of the node $y_{\{k,l\}}(t) \in \mathbb{R}^n$ in the spatial state vector $y(t) \in \mathbb{R}^{nN}$, we denote the symbol $y_j(t) \in \mathbb{R}^n$ to replace $y_{\{k,l\}}(t)$. Note that the index j ranges from 1 to N, i.e., $y_1(t) \triangleq y_{\{1,1\}}(t)$, $y_2(t) \triangleq y_{\{2,1\}}(t),\ldots$, $y_j(t) \triangleq y_{\{k,l\}}(t),\ldots$, $y_N(t) \triangleq y_{\{N_1,N_2\}}(t)$, where $j = (l-1)N_1 + k$ in equation (11.16). The fuzzy stochastic finite difference model of two indices in equation (11.15) could be represented with only one index as follows:

$$dy_j(t) = \sum_{i=1}^{M} \mu_i(y_j)[(\kappa_0 + \kappa_i)T_j y(t)dt + (A_0 + A_i)y_j(t)dt$$

$$+(A_{W,0} + A_{W,i})y_j(t)T_{w,j}dW(t)] + B_j u(t)dt + g_d d_j(t)dt$$

$$+\varepsilon_{\kappa,j}(y_j)T_j y(t)dt + \varepsilon_{f,j}(y_j)dt + \varepsilon_{fw,j}(y_j)T_{w,j}dW(t) \tag{11.17}$$

where $dW(t) \in \mathbb{R}^N$ is defined as $dW(t) \triangleq [dW_1(t),...,dW_N(t)]^T$. The matrix $T_j \in \mathbb{R}^{n \times nN}$ expresses the interaction from other grid nodes to the node $y_j(t) \triangleq y_{\{k,l\}}(t)$ as $T_j y(t) \triangleq [(T_{x1,j,1} + T_{x2,j,1}),...,(T_{x1,j,N} + T_{x2,j,N})]y(t) = \sum_{j2=1}^{N} (T_{x1,j,j2} + T_{x2,j,j2}) y_{j2}(t)$ in which $T_{x1,j,j2} + T_{x2,j,j2} \in \mathbb{R}^{n \times n}$ are defined as

$$T_{x1,j,j2} = \begin{cases} \frac{-2}{\Delta_x^2} I_n, \text{for } j_2 = (l-1)N_1 + k \\ \frac{1}{\Delta_x^2} I_n, \text{for } j_2 = (l-1)N_1 + k - 1 \\ \frac{1}{\Delta_x^2} I_n, \text{for } j_2 = (l-1)N_1 + k + 1 \\ 0, \text{otherwise} \end{cases} \quad \text{and}$$

$$T_{x2,j,j2} = \begin{cases} \frac{1}{\Delta_x^2} I_n, \text{for } j_2 = (l-2)N_1 + k \\ \frac{1}{\Delta_x^2} I_n, \text{for } j_2 = lN_1 + k \\ \frac{-2}{\Delta_x^2} I_n, \text{for } j_2 = (l-1)N_1 + k \\ 0, \text{otherwise} \end{cases}$$

The elements $T_{W,x1,j,j3} + T_{W,x2,j,j3}$ of the finite difference matrix $T_{W,j} = [(T_{W,x1,j,1} + T_{W,x2,j,1}),...,(T_{W,x1,j,N} + T_{W,x2,j,N})] \in \mathbb{R}^{1 \times N}$ are defined as follows:

$$T_{W,x1,j,j3} = \begin{cases} \frac{-1}{2\Delta_x}, \text{for } j_3 = (l-1)N_1 + k - 1 \\ \frac{1}{2\Delta_x}, \text{for } j_3 = (l-1)N_1 + k + 1 \\ 0, \text{otherwise} \end{cases} \quad \text{and}$$

$$T_{W,x2,j,j3} = \begin{cases} \frac{-1}{2\Delta_x}, \text{for } j_3 = (l-2)N_1 + k \\ \frac{1}{2\Delta_x}, \text{for } j_3 = lN_1 + k \\ 0, \text{otherwise} \end{cases}$$

The measurement output $z(t)$ in equation (11.2) can be represented as follows:

$$z(t)=\sum_{j=1}^{N}C_jy_j(t)+D_jn_j(t) \tag{11.18}$$

where the matrix C_j is defined as $C_j \triangleq C_{\{k,l\}}=[C_{\{k,l\},1},\ldots,C_{\{k,l\},q}]^T \in \mathbb{R}^{n_z \times n}$ where $j=(l-1)N_1+k$, in which the element is given as $C_{\{k,l\},i}=h_i$ for $x_{\{k,l\}}=q_i$ or $C_{\{k,l\},i}=0$ for $x_{\{k,l\}}\neq q_i$. The matrix D_j is defined as $D_j \triangleq D_{\{k,l\}}=[D_{\{k,l\},1},\ldots,D_{\{k,l\},q}]^T \in \mathbb{R}^{n_z \times n}$ where $j=(l-1)N_1+k$, in which the element is given as $D_{\{k,l\},i}=h_iD_n$ for $x_{\{k,l\}}=q_i$ or $C_{\{k,l\},i}=0$ for $x_{\{k,l\}}\neq q_i$. Note that q_i is the ith sensor location as defined in equation (11.2). The ith measurement noise is defined as $n_j(t)\triangleq n(x,t)\big|_{x=x\{k,l\}}$ where $j=(l-1)N_1+k$.

We collect all states $y_j(t)$ of grid nodes in equations (11.17) and (11.18) to yield $y(t)$ in equation (11.16). Then, a diagonal fuzzy weighting matrix $\mu_i(y)\in\mathbb{R}^{N\times N}$ is defined by $\mu_i(y)\triangleq diag(\mu_i(y_1),\ldots,\mu_i(y_N))$ and the $n\times n$ identity matrix by I_n. The Kronecker product can be used to simplify the representation. Some properties for $\mu_i(y)$ can be obtained as follows:

Lemma 11.3.1 [417]

Using the properties of the Kronecker product, we have the following properties:

1. $\mu_i(y)\otimes A_i=(\mu_i(y)\otimes I_n)(I_N\otimes A_i)=(I_N\otimes A_i)(\mu_i(y)\otimes I_n)\in\mathbb{R}^{nN\times nN}$ where $A_i\in\mathbb{R}^{n\times n}$.
2. $(\mu_i(y)\otimes I_n)(\mu_i(\hat{y})\otimes I_n)=(\mu_i(y)\mu_i(\hat{y})\otimes I_n)\in\mathbb{R}^{nN\times nN}$
3. $\Sigma_{i=1}^{M}\mu_i(y)=I_N$, $\Sigma_{i=1}^{M}(\mu_i(y)\otimes I_n)=I_{nN}$, and $\Sigma_{i=1}^{M}\Sigma_{i=1}^{M}(\mu_i(y)\mu_i(\hat{y})\otimes I_n)=I_{nN}$

Using the Kronecker product, the system in equations (11.17) and (11.18) can be written as the following fuzzy stochastic spatial state space model:

$$\begin{aligned}dy(t)=\sum_{i=1}^{M}\bar{\mu}_i(y)[&(I_N\otimes[\kappa_0+\kappa_i])Ty(t)dt+(I_N\otimes[A_0+A_i])y(t)dt\\&+(I_N\otimes[A_{W,0}+A_{W,i}])y(t)\circ T_w dW(t)]\\&+Bu(t)dt+B_d d(t)dt+\varepsilon(y)dt+\varepsilon_W(y)\circ T_w dW(t)\end{aligned} \tag{11.19}$$

$$z(t)=Cy(t)+Dn(t) \tag{11.20}$$

where we define $\bar{\mu}_i(y)\triangleq\mu_i(y)\otimes I_n$, and the spatial state vector $y(t)\in\mathbb{R}^{nN}$ is denoted in equation (11.16) to represent $y(x,\,t)$ at all semi-discretization grid points on the spatial domain in Figure 11.1. The corresponding matrices are defined as $T=\left[T_1^T,...,T_N^T\right]^T \in \mathbb{R}^{nN\times nN}$, $B=\left[B_1^T,...,B_N^T\right]^T \in \mathbb{R}^{nN\times n_u}$, $B_d = I_N \otimes g_d \in \mathbb{R}^{nN\times n_d N}$, $T_W = \left[T_{W,1}^T,...,T_{W,N}^T\right]^T \in \mathbb{R}^{N\times N}$, $C=[C_1,...,C_N]\in \mathbb{R}^{n_z\times nN}$, and $D=[D_1,...,D_N]\in \mathbb{R}^{n_z\times nN}$. The random external disturbance $d(t)$ is defined as $d(t)=\left[d_1(t)^T,...,d_N(t)^T\right]^T \in \mathbb{R}^{nN}$. The measurement noise $n(t)$ is defined as $n(t)=\left[n_1(t)^T,...,n_N(t)^T\right]^T \in \mathbb{R}^{nN}$. The approximation error $\varepsilon(y)$ is defined as $\varepsilon(y)=\varepsilon_\kappa(y)Ty(t)+\varepsilon_f(y)$ where $\varepsilon_f(y)=\left[\varepsilon_{f,1}(y_1)^T,...,\varepsilon_{f,N}(y_N)^T\right]^T$ and $\varepsilon_\kappa(y)=diag(\varepsilon_{\kappa,1}(y_1),...,\varepsilon_{\kappa,N}(y_N))$. The approximation error $\varepsilon_W(y)$ is defined as $\varepsilon_W(y)=\left[\varepsilon_{fw,1}(y_1)^T,...,\varepsilon_{fw,N}(y_N)^T\right]^T$.

Remark 11.3.2

(i) $T_W dW(t)$ is a zero-mean-vector-process. The covariance of $T_W dW(t)$ can be obtained as $E(T_W dW(t))(T_W dW(t))^T = I_N$. (ii) The Hadamard product of matrices (or vectors) $X=[X_{ij}]$ and $Y=[Y_{ij}]$ of the same size is the entrywise product as $X\circ Y=[[X_{ij}Y_{ij}]]$.

The physical meaning of equations (11.19) and (11.20) is that the NSPDS in equations (11.1) and (11.2) at all semi-discretization grid points on the spatial domain in Figure 11.1 can be represented by the fuzzy spatial state space model in equations (11.19) and (11.20). In Theorem 11.3.1, the bound of approximate errors $\varepsilon_\kappa(y(x,t))$, $\varepsilon_f(y(x,t))$, and $\varepsilon_{fw}(y(x,t))$ can be proved to be less than two arbitrary constants σ_κ, σ_f, and σ_{fw}. The bounds for approximate errors $\varepsilon(y)$ and $\varepsilon_W(y)$ can be obtained by the following corollary:

Corollary 11.3.1: If the bounds of $\varepsilon_\kappa(y(x,t))$, $\varepsilon_f(y(x,t))$, and $\varepsilon_{fw}(y(x,t))$ are provided with σ_κ, σ_f, and σ_{fw}, then the bounds of $\varepsilon(y)$ and $\varepsilon_W(y)$ can also be estimated as $\|\varepsilon(y)\|^2 \le \sigma_\kappa^2\|\mathrm{T}y(t)\|^2+\sigma_f^2\|y(t)\|^2$ and $\|\varepsilon_W(y)\|^2 \le \sigma_{fw}\|y(t)\|^2$.

Proof

By Theorem 11.3.1, we have $\|\varepsilon_{\kappa,j}(y)\|^2 \le \sigma_\kappa^2$, $\|\varepsilon_{f,j}(y)\|^2 \le \sigma_f^2\|y_j(t)\|^2$ and $\|\varepsilon_{fw,j}(y)\|^2 \le \sigma_{fw}^2\|y_j(t)\|^2$. Thus, $\|\varepsilon(y)\|^2 = \|\varepsilon_\kappa(y)\mathrm{T}y(t)+\varepsilon_f(y)\|^2 \le \|\varepsilon_\kappa(y)\mathrm{T}y(t)\|^2+\sum_{j=1}^N\|\varepsilon_j(y)\|^2 \le \|\varepsilon_\kappa(y)\|^2\|\mathrm{T}y(t)\|^2+\sum_{j=1}^N\|\varepsilon_j(y)\|^2 \le \sigma_\kappa^2\|\mathrm{T}y(t)\|^2+\sigma_f^2\|y(t)\|^2$. The bound of $\varepsilon_W(y)$ can be estimated in a similar manner.

The NSPDS in equation (11.1) can be represented by the fuzzy stochastic spatial state space system in equation (11.19) using the fuzzy approach and the method of lines. The following theorem provides the convergence of the solution [391].

Theorem 11.3.2

Suppose the system in equation (11.3) has a strong solution y(x, t). For $x \in [x_{\{k,l\}}, x_{\{k+1,l\}}) \times [x_{\{k,l\}}, x_{\{k,l+1\}})$*, we denote the solution* $y\Delta_x(x,t)$ *to represent the solution y(x, t) by interpolating the numerical solution* $y_{\{k,l\}}(x,t)$ *in equation (11.19) for* $0<k\le N_1$ *and* $0<l\le N_2$ *using the polygonal approximation. Then for every* $0<\alpha<\frac{1}{2}$ *and for every t > 0 there is a constant c = c(α, t) such that*

$$\sup_{x\in U} E\{\|y\Delta_x(x,t)-y(x,t)\|^2\}\le c\Delta_x^{\alpha} \tag{11.20}$$

Moreover, $y\Delta_x(x,t)$ *converges to y(x, t) almost surely, as* $\Delta_x \to 0$*, uniformly in* $t\in[0,t_f]$*,* $x\in U$ *for every* $t_f>0$*. The semi-discretization finite difference scheme is a convergence scheme.*

Remark 11.3.3

(i) Note that the different grid size Δ_x can be chosen for different space variables x. (ii) Note that the dimension of the fuzzy stochastic spatial state space model equation (11.19) N depends on the grid size Δ_x. In theory, the grid size Δ_x is chosen to be as small as possible, i.e., $\Delta_x\to 0$ to produce better local truncation errors $O_{\{k,l\}}(\Delta_x^2)$, $O_{W,\{k,l\}}(\Delta_x^2)$, i.e., $O_{\{k,l\}}(\Delta_x^2)$, $O_{W,\{k,l\}}(\Delta_x^2)\to 0$. Applying the Lax-Richtmyer Equivalence Theorem [423], if the fuzzy spatial state space model with $d(t)\equiv 0$ in equation (11.19) is stable in probability then the semi-discretization finite difference scheme is a convergent scheme in which the solution of the fuzzy spatial state space model with $d(t)\equiv 0$ in equation (11.19) can converge almost surely to the solution of the NSPDS with $d(t)\equiv 0$ in equation (11.1). (iii) In order to yield a reasonable dimension N for implementation, the grid size Δ_x should have a value depending on the different design environment. Therefore, how to choose the grid size is a trade-off problem and will be discussed in the sequel.

Remark 11.3.4

The state variable *y(x, t)* in equation (11.1) depends on the space *x* and time *t*. Based on the semi-discretization finite difference scheme [386,390–392], the spatial state vector *y(t)* in equation (11.16) was used to represent *y(x, t)* at all grid points. In this situation, the $E\int_0^{t_f}\int_U\|R_1y(x,t)\|^2\,dxdt$ and $E\int_0^{t_f}\int_U\|v(x,t)\|^2\,dxdt$ in the two-dimensional stochastic H_∞ control performance equation (11.6) could be modified by the one-dimensional form $E\int_0^{t_f}\|R_1y(t)\|^2\,\Delta_x^2\,dt$ and $E\int_0^{t_f}\|v(t)\|^2\,\Delta_x^2dt$, respectively, for the finite-difference systems equation (11.19) to simplify the design procedure. As $\Delta_x\to 0$ the one-dimensional systems in equation (11.19) will approach two-dimensional systems equation (11.9), and the one-dimensional integration form will approach the two-dimensional integration form.

11.4 ROBUST STABILIZATION DESIGN FOR NONLINEAR STOCHASTIC PARTIAL DIFFERENTIAL SYSTEMS

In Section 11.3, it was shown that a fuzzy spatial state space model, as described in equations (11.19) and (11.20), can be constructed to represent NSPDSs. In this section, a fuzzy estimator-based stabilization control scheme is developed to solve the stabilization control design problem of NSPDSs by using the measurement of a set of sensor outputs $z(t)$. Then, a set of the complex matrix inequalities is derived to guarantee robust stochastic H_∞ stabilization design. Finally, a design procedure is given to simplify the robust stabilization design of NSPDSs by solving a set of LMIs to obtain the controller and estimator gains.

11.4.1 Robust Fuzzy Estimator-Based Stabilization Control Design

We propose a fuzzy estimator-based controller based on the fuzzy spatial state space model in equations (11.19) and (11.20) to estimate the state of a fuzzy spatial state space model, and then use a state feedback scheme to stabilize the NSPDS in equations (11.1) and (11.2) or to guarantee robust stochastic H_∞ control performance in equation (11.6). The proposed fuzzy estimator-based controller has the following form (see Figure 11.2)

$$d\hat{y}(t) = \sum_{i=1}^{M} \bar{\mu}_i(\hat{y})[(I_N \otimes [\kappa_0 + \kappa_i])T\hat{y}(t)dt + (I_N \otimes [A_0 + A_i])\hat{y}(t)dt$$

$$+ Bu(t)dt + [G_0 + G_i](C\hat{y}(t) - z(t))dt] \quad (11.22)$$

$$u(t) = \sum_{i=1}^{M}\sum_{j=1}^{N} \mu_i(\hat{y}_j)[K_{0,j} + K_{i,j}]\hat{y}_j(t) = \sum_{i=1}^{M}[K_{0,j} + K_{i,j}]\bar{\mu}_i(\hat{y})\hat{y}(t) \quad (11.23)$$

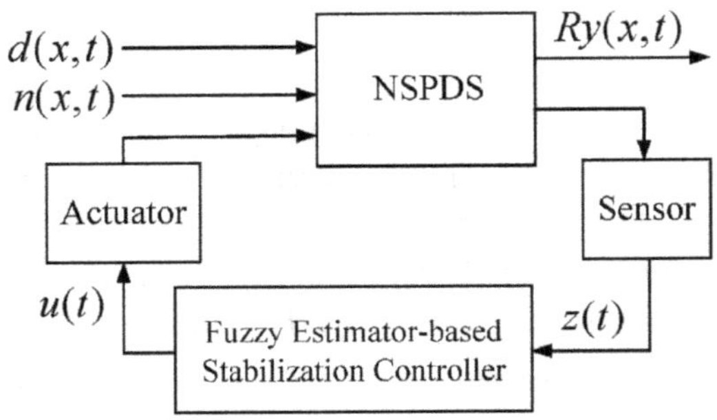

FIGURE 11.2 Block diagram of the stochastic stabilization control scheme for the NSPDS in equations (11.1) and (11.2).

where $G_i \in \mathbb{R}^{nN \times n_z}$ and $K_i \triangleq [K_{i,1},...,K_{i,N}] \in \mathbb{R}^{n_u \times nN}$ are, respectively, the estimator gains and the control gains to be designed. The fuzzy weighting matrix $\bar{\mu}_i(\hat{y})$ is defined as $\bar{\mu}_i(\hat{y}) \triangleq (\mu_i(\hat{y}) \otimes I_n)$. In equation (11.22), the estimator gains G_i are designed so that the spatial state estimation error $\hat{y}(t) - y(t)$ is as small as possible. The fuzzy controller $u(t)$ in equation (11.23) is designed to robustly stabilize the NSPDS with $d(t) \equiv 0$ and $n(t) \equiv 0$ in equations (11.1) and (11.2) in spite of the state-dependent spatiotemporal noise and the fuzzy approximation error. Furthermore, the fuzzy controller $u(t)$ can be designed to achieve robust stochastic H_∞ control performance when the NSPDS suffers from random external disturbance and measurement noise.

The dynamic of the estimated error $e(t) = y(t) - \hat{y}(t)$ is given by

$$\begin{aligned}
de(t) &= \sum_{i=1}^{M} \bar{\mu}_i(y)[(I_N \otimes [\kappa_0 + \kappa_i])Ty(t)dt + (I_N \otimes [A_0 + A_i])y(t)dt \\
&\quad + (I_N \otimes [A_{W,0} + A_{W,i}])y(t) \circ T_w dW(t)] + Bu(t)dt + B_d d(t)dt + \varepsilon(y)dt \\
&\quad + \varepsilon_W(y) \circ T_w dW(t) - (\sum_{i=1}^{M} \bar{\mu}_i(\hat{y})[(I_N \otimes [\kappa_0 + \kappa_i])T\hat{y}(t)dt \\
&\quad + (I_N \otimes [A_0 + A_i])\hat{y}(t)dt + Bu(t)dt + [G_0 + G_i](C\hat{y}(t) - z(t))dt]) \\
&= \sum_{i_1=1}^{M}\sum_{i_2=1}^{M} \bar{\mu}_{i_1}(y)\bar{\mu}_{i_2}(\hat{y})[((I_N \otimes [\kappa_0 + \kappa_{i_2}])T + (I_N \otimes [A_0 + A_{i_2}] \\
&\quad + [G_0 + G_{i_2}]C)e(t)dt + (I_N \otimes [\kappa_{i_1} + \kappa_{i_2}])Ty(t)dt + (I_N \otimes [A_{i_1} + A_{i_2}])y(t)dt \\
&\quad + (I_N \otimes [A_{W,0} + A_{W,i}])y(t) \circ T_w dW(t) + B_d d(t)dt + [G_0 + G_{i_2}]D_n(t)dt] \\
&\quad + \varepsilon(y)dt + \varepsilon y(t) \circ T_w dW(t)
\end{aligned} \tag{11.24}$$

Combining the fuzzy system in equation (11.19) with the estimator-based stabilization controller, the closed-loop fuzzy system can be described by the following augmented system:

$$\begin{aligned}
d\bar{y}(t) &= \bar{A}(\mu,\hat{\mu})\bar{y}(t)dt + \bar{A}_W(\mu)\bar{y}(t) \circ \bar{T}_W dW(t) + \bar{B}_v(\hat{\mu})v(t)dt \\
&\quad + \bar{\varepsilon}(y)dt + \bar{\varepsilon}_W(y) \circ \bar{T}_W dW(t)
\end{aligned} \tag{11.25}$$

where $\bar{y}(t) = \left[y(t)^T, e(t)^T\right]^T$, $\bar{\varepsilon}(y) = \left[\varepsilon(y)^T, \varepsilon(y)^T\right]^T$, $\bar{\varepsilon}_W(y) = \left[\varepsilon_W(y)^T, \varepsilon_W(y)^T\right]^T$, and $\bar{T}_W = \left[T_W^T, T_W^T\right]^T$. The other notations are defined as follows:

$$\bar{A}(\mu,\hat{\mu}) \triangleq \begin{bmatrix} \bar{A}_{11}(\mu) + BK(\hat{\mu}) & -BK(\hat{\mu}) \\ \bar{A}_{21}(\mu,\hat{\mu}) & \bar{A}_{22}(\hat{\mu}) + G(\hat{\mu})C \end{bmatrix}$$

$$\bar{A}_W(\mu) \triangleq \sum_{i_1-1}^{M} \begin{bmatrix} \mu_{i_1}(y)I_N \otimes [A_{W,0} + A_{W,i_1}] & 0 \\ \mu_{i_1}(y)I_N \otimes [A_{W,0} + A_{W,i_1}] & 0 \end{bmatrix}, \bar{B}_v(\hat{\mu}) \triangleq \begin{bmatrix} B_d & 0 \\ B_d & G(\hat{\mu})D \end{bmatrix}$$

where

$$\bar{A}_{11}(\mu) \triangleq \sum_{i_1=1}^{M} \bar{\mu}_{i_1}(y)\{(I_N \otimes [\kappa_0 + \kappa_{i_1}])T + (I_N \otimes [A_0 + A_i])\}$$

$$\bar{A}_{21}(\mu, \hat{\mu}) \triangleq \sum_{i_1=1}^{M}\sum_{i_2=1}^{M} \bar{\mu}_{i_1}(y)\bar{\mu}_{i_2}(\hat{y})\{(I_N \otimes [\kappa_{i_1} + \kappa_{i_2}])T + (I_N \otimes [A_{i_1} - A_{i_2}])\}$$

$$\bar{A}_{22}(\hat{\mu}) \triangleq \sum_{i_2=1}^{M} \bar{\mu}_{i_2}(\hat{y})[(I_N \otimes [\kappa_0 + \kappa_{i_2}])T + I_N \otimes A_{i_2}\}$$

$$K(\hat{\mu}) \triangleq \sum_{i_2=1}^{M} [K_0 + K_{i_2}]\bar{\mu}_{i_2}(\hat{y}), G(\hat{\mu}) \triangleq \sum_{i_2=1}^{M} \bar{\mu}_{i_2}(\hat{y})[G_0 + G_{i_2}]$$

Supposing the external disturbance $d(t)$ and the measurement noise $n(t)$ are neglected in the augmented system equation (11.25), i.e., $v(t) \equiv 0$, the fuzzy estimator-based controller should be designed to guarantee the asymptotically stochastic stability of the augmented system. This will allow it to tolerate state-dependent noise and fuzzy approximation errors.

To solve the robust asymptotically stochastic stabilization problem, let us choose a Lyapunov function $V(\bar{y})$ for the augmented system in equation (11.25) such as

$$V(\bar{y}) = \sum_{j=1}^{N} \left(y_j(t)^T P_{11,j} y_j(t) + e_j(t) P_{22,j} e_j(t)\right) = \bar{y}(t)^T \bar{P} \bar{y}(t) \quad (11.26)$$

where $\bar{P} \triangleq diag(\bar{P}_{11}, \bar{P}_{22})$ in which $\bar{P}_{11} = \bar{P}_{11}^T \triangleq diag(P_{11,1}, ..., P_{11,N}) > 0$ and $\bar{P}_{22} = \bar{P}_{22}^T \triangleq diag(P_{22,1}, ..., P_{22,N}) > 0$. The following lemma is provided for the proof of the main results.

Lemma 11.4.1 (S-Procedure [426])

Let $F_0, ..., F_p \in \mathbb{R}^{n \times n}$ be symmetric matrices. We consider the following conditions on $F_0, ..., F_p$ as $\zeta^T F_0 \zeta > 0$ for all $\zeta \neq 0$ such that $\zeta^T F_i \zeta \geq 0, i = 1, ..., p$. It is obvious that if there exist $\tau_1 \geq 0, ..., \tau_p \geq 0$ such that $F_0 - \sum_{i=1}^{p} \tau_i F_i > 0$, then $\zeta^T F_0 \zeta > 0$ for all $\zeta \neq 0$ such that $\zeta^T F_i \zeta \geq 0, i = 1, ..., p$.

Based on equation (11.26), we obtain the following result for the robust asymptotic stabilization problem.

Theorem 11.4.1

For the augmented system equation (11.25) with $v(t) \equiv 0$, suppose there exist some symmetric positive definite matrix $\bar{P}$, constants $\xi \geq 0$ and $\xi_W \geq 0$, control gains K_i, and estimator gains G_i, for $i = 0, 1,\ldots, M$, such that

$$\bar{\Theta}(\mu,\hat{\mu}) \triangleq \begin{bmatrix} \bar{\Theta}_{11}(\mu,\hat{\mu}) + \xi\Sigma + \xi_W\Sigma_W & \bar{P} & \bar{A}_W(\mu)^T\bar{P} \\ \bar{P} & -\xi I & 0 \\ \bar{P}\bar{A}_W(\mu) & 0 & \bar{P} - \xi_W I \end{bmatrix} < 0 \tag{11.27}$$

where $\bar{\Theta}_{11}(\mu,\hat{\mu}) \triangleq \bar{A}(\mu,\hat{\mu})^T\bar{P} + \bar{P}^T\bar{A}(\mu,\hat{\mu}) + \bar{A}_W(\mu)^T\bar{P}\bar{A}_W(\mu)$, $\Sigma \triangleq diag(2\sigma_\kappa^2 T^T T + 2\sigma_f^2 I, 0)$, and $\Sigma_W \triangleq diag(2\sigma_{fw}^2 I, 0)$. Then the augmented system with $v(t) \equiv 0$ is asymptotically stable in probability as per the fuzzy estimator-based controller in equations (11.22) and (11.23). Furthermore, for the case where the system is free of inputs, i.e. $d(x, t) \equiv 0$ and $n(x, t) \equiv 0$, the asymptotic stability in probability of the NSPDS in equations (11.1) and (11.2) is guaranteed.

Proof

We consider the Lyapunov function $V(\bar{y})$ in equation (11.26) for the augmented system equation (11.25). Using the Itô formula [427,428], we have

$$dV(\bar{y}) = \left(\frac{\partial V(\bar{y})}{\partial \bar{y}}\right)^T \left(\bar{A}(\mu,\hat{\mu})\bar{y}(t)dt + \bar{\varepsilon}(y)dt + \left[\bar{A}_W(\mu)\bar{y}(t) + \bar{\varepsilon}_W(y)\right] \circ \bar{T}_W dW(t)\right)$$
$$+ \frac{1}{2}\left[\bar{A}_W(\mu)\bar{y}(t) + \bar{\varepsilon}_W(y)\right]^T \frac{\partial^2 V(\bar{y})}{\partial \bar{y}^2}\left[\bar{A}_W(\mu)\bar{y}(t) + \bar{\varepsilon}_W(y)\right]dt \tag{11.28}$$

Then

$$EdV(\bar{y}) = E\left\{\begin{aligned} &\left(\frac{\partial V(\bar{y})}{\partial \bar{y}}\right)^T \left[\bar{A}(\mu,\hat{\mu})\bar{y}(t)dt + \bar{\varepsilon}(y)dt\right] \\ &+ \frac{1}{2}\left[\bar{A}_W(\mu)\bar{y}(t) + \bar{\varepsilon}_W(y)\right]^T \frac{\partial^2 V(\bar{y})}{\partial \bar{y}^2}\left[\bar{A}_W(\mu)\bar{y}(t) + \bar{\varepsilon}_W(y)\right]dt \end{aligned}\right\} \tag{11.29}$$

Since $V(\bar{y}) = \bar{y}(t)^T \bar{P}\bar{y}(t)$, we have

$$EdV(\bar{y}) = E\{\bar{y}(t)^T \bar{P}(\bar{A}(\mu,\hat{\mu})\bar{y}(t)dt + \bar{\varepsilon}(y)dt) + (\bar{A}(\mu,\hat{\mu})\bar{y}(t)dt + \bar{\varepsilon}(y)dt)^T \bar{P}\bar{y}(t)$$

$$+ [\bar{A}_W(\mu)\bar{y}(t) + \bar{\varepsilon}_W(y)]^T \bar{P}[\bar{A}_W(\mu)\bar{y}(t) + \bar{\varepsilon}_W(y)]dt\}$$

$$\leq E\{F(t)dt\} \triangleq E\left\{ \begin{bmatrix} \bar{y}(t) \\ \bar{\varepsilon}(y) \\ \bar{\varepsilon}_W(y) \end{bmatrix}^T \begin{bmatrix} \bar{\Theta}_{11}(\mu,\hat{\mu}) & \bar{P} & \bar{A}_W(\mu)^T \bar{P} \\ \bar{P} & 0 & 0 \\ \bar{P}\bar{A}_W(\mu) & 0 & \bar{P} \end{bmatrix} \begin{bmatrix} \bar{y}(t) \\ \bar{\varepsilon}(y) \\ \bar{\varepsilon}_W(y) \end{bmatrix} dt \right\}$$

$$(11.30)$$

where $\bar{\Theta}_{11}(\mu,\hat{\mu})$ is defined in equation (11.27). Therefore, if $F(t) < 0$, then the augmented system is asymptotically stable in probability [427,428]. According to the bounds of $\varepsilon(y)$ and $\varepsilon_W(y)$ in Corollary 11.3.1, we have

$$\bar{\varepsilon}(y)^T \bar{\varepsilon}(y) = 2\varepsilon(y)^T \varepsilon(y) \leq 2(\sigma_\kappa^2 y(t)^T T^T Ty(t) + \sigma_f^2 y(t)^T y(t)) = \bar{y}(t)\Sigma\bar{y}(t) \quad (11.31)$$

$$\bar{\varepsilon}_W(y)^T \bar{\varepsilon}_W(y) = 2\varepsilon_W(y)^T \varepsilon_W(y) \leq 2\sigma_{fw}^2 y(t)^T y(t) = \bar{y}(t)^T \Sigma_W \bar{y}(t) \quad (11.32)$$

where $\Sigma \triangleq diag\left(2\sigma_\kappa^2 T^T T + 2\sigma_f^2 I, 0\right)$ and $\Sigma_W \triangleq diag\left(2\sigma_{fw}^2 I, 0\right)$. In order to impose two constraints equations (11.31) and (11.32) on $F(t) < 0$, by the S-procedure in Lemma 11.4.1, if there exist $\xi > 0$ and $\xi_W > 0$ such that $\bar{\Theta}(\mu,\hat{\mu}) < 0$ in equation (11.27) holds, then not only $F(t) < 0$ but also two inequality constraints in equations (11.31) and (11.32) are satisfied. As per Theorem 11.3.2, the stochastic stability of the fuzzy stochastic spatial state space model and the consistency of the finite difference scheme imply that the semi- discretization finite difference scheme is a convergence scheme. Therefore, we can prove that the NSPDS is asymptotically stable in probability.

When we consider the effects of the random external disturbance $d(t)$ and the measurement noise $n(t)$, i.e. $d(t) \neq 0$ and $n(t) \neq 0$, the fuzzy estimator-based controller in equations (11.22) and (11.23) is specified to guarantee that the effects of the random external disturbance and the measurement noise on the controlled output $R_1 y(x, t)$ and the estimation error $e(t)$ are attenuated below a prescribed level ρ from the area energy perspective. This design problem is called the robust stochastic H_∞ estimator-based stabilization control design problem. Based on the augmented system in equation (11.25) and on Remark 11.3.4, the stochastic H_∞ estimator-based control performance should be modified to include the state estimation error $e(t)$ as follows:

$$\frac{E\int_0^{t_f} \left(\left\|\bar{R}_1 y(t)\right\|^2 + \left\|\bar{R}_2 e(t)\right\|^2\right)}{E\int_0^{t_f} \left\|\bar{R}_3 v(t)\right\|^2 dt} \leq \rho^2 \qquad (11.33)$$

where the weighting matrices $\bar{R}_1 \triangleq diag(\Delta_x R_1,...,\Delta_x R_1) \in \mathbb{R}^{nN\times nN}$, $\bar{R}_2 \triangleq diag(\Delta_x R_2,...,\Delta_x R_2) \in \mathbb{R}^{nN\times nN}$, and $\bar{R}_3 \triangleq diag(\Delta_x I,...,\Delta_x I) \in \mathbb{R}^{nN\times nN}$ with $R_1 \in \mathbb{R}^{n\times n}$ and $R_2 \in \mathbb{R}^{n\times n}$, i.e., the effect of random disturbances and measurement noise on the controlled output and the estimation error should be below the attenuation level ρ. The weighting matrices $\bar{R}_1$ and $\bar{R}_2$ denote the tradeoff between the controlled output and the estimation error. The stochastic H_∞ estimator-based control performance in equation (11.33) can be represented by the following inequality:

$$E\int_0^{tf} \left\|\bar{R}\bar{y}(t)\right\|^2 dt \le V(\bar{y}(0)) + \rho^2 E\int_0^{tf} \left\|\bar{R}_3 v(t)\right\|^2 dt \tag{11.34}$$

for some positive function $V(\bar{y}(0))$ when the initial condition $\bar{y}(0)$ is also considered, where $\bar{R} = diag\left(\bar{R}_1, \bar{R}_2\right)$, and $\bar{y}(t)$ is the state vector of the augmented system in equation (11.25). Therefore, the robust stochastic H_∞ estimator-based stabilization control design for NSPDSs is based on specification of the control gains and estimator gains of the fuzzy estimator-based controller in equations (11.22) and (11.23) so that the stochastic H_∞ estimator-based control performance as described in equations (11.33) or (11.34) could be achieved.

Remark 11.4.1

Since $\left\|\bar{R}_1 y(t)\right\|^2$ can be represented as $\left\|\bar{R}_1 y(t)\right\|^2 = \Sigma_{j=1}^{N}\left\|R_1 y_j(t)\right\|^2 \Delta_x^2$ where Δ_x^2 is the area of the gridded rectangle, the term $\left\|\bar{R}_1 y(t)\right\|^2$ can be used to approximate the integration of the two-dimensional stochastic H∞ control performance in a spatiotemporal domain. A similar result can be obtained for $\left\|\bar{R}_3 v(t)\right\|^2$. Therefore, if the gridded spacing Δ_x^2 of the finite difference approach is sufficiently small so that $O(\Delta_x^2) \to 0$, then the stochastic H_∞ estimator-based control performance in equation (11.33) will approach the two-dimensional stochastic H_∞ estimator-based control performance in the spatiotemporal domain as

$$\frac{E\int_0^{tf}\int_U \left(\left\|\bar{R}_1 y(x,t)\right\|^2 + \left\|\bar{R}_2 e(x,t)\right\|^2\right) dx\, dt}{E\int_0^{tf}\int_U \left\|\bar{R}_3 v(x,t)\right\|^2 dx\, dt} \le \rho^2 \tag{11.35}$$

where $e(x,t) = y(x,t) - \hat{y}(x,t)$ and $\hat{y}(x,t)$ is the estimation state on the spatiotemporal domain.

Theorem 11.4.2

For the augmented system equation (11.25) with a prescribed disturbance attenuation level ρ in equation (11.34), if there exist some symmetric positive definite *matrix* $\bar{P}$, *constants* $\xi \geq 0$ *and* $\xi_W \geq 0$, *control gains* K_i, *and estimator gains* G_i, *for* $i = 1,\ldots, M$, *such that*

$$\bar{\Theta}_H(\mu,\hat{\mu}) \triangleq \begin{bmatrix} \bar{\Theta}_{11}(\mu,\hat{\mu}) + \bar{\Theta}_{11,\xi} & \bar{P} & \bar{A}_W(\mu)^T \bar{P} & \bar{P}\bar{B}_v(\hat{\mu}) \\ \bar{P} & -\xi I & 0 & 0 \\ \bar{P}\bar{A}_W(\mu) & 0 & \bar{P} - \xi_W I & 0 \\ \bar{B}_v(\hat{\mu})^T \bar{P} & 0 & 0 & -\rho^2 \bar{R}_3^T \bar{R}_3 \end{bmatrix} < 0 \quad (11.36)$$

where $\bar{\Theta}_{11,\xi} \triangleq \xi\Sigma + \xi_W \Sigma_W + \bar{R}^T \bar{R}$, *then the stochastic* H_∞ *estimator-based control performance in equation (11.34) is guaranteed by the fuzzy estimator-based controller in equations (11.22) and (11.23). Furthermore, stochastic* H_∞ *estimator-based control performance in equation (11.35) for the NSPDS in equations (11.1) and (11.2) is also achieved by the fuzzy estimator-based controller described in equations (11.22) and (11.23).*

Proof

First, consider the following equivalent equation:

$$E\int_0^{t_f} \bar{y}(t)\bar{R}^T \bar{R}\bar{y}(t)dt = EV(\bar{y}(0)) - EV(\bar{y}(t_f)) + E\int_0^{t_f} (\bar{y}(t)^T \bar{R}^T \bar{R}\bar{y}(t)dt + dV(\bar{y})) \quad (11.37)$$

where $V(\bar{y}) > 0$. By applying the Itô formula, then adding and subtracting the term $\rho^2 v(t)^T \bar{R}_3^T \bar{R}_3 v(t)$, we can produce

$$\begin{aligned} E(dV(\bar{y}) + \bar{y}(t)^T \bar{R}^T \bar{R}\bar{y}(t)dt) &= E\{\bar{y}(t)^T \bar{P}(\bar{A}(\mu,\hat{\mu})\bar{y}(t)dt + \bar{B}_v(\hat{\mu})v(t)dt + \bar{\varepsilon}(y)dt) \\ &+ (\bar{A}(\mu,\hat{\mu})\bar{y}(t)dt + \bar{B}_v(\hat{\mu})v(t)dt + \bar{\varepsilon}(y)dt)\bar{P}\bar{y}(t) + [\bar{A}_W(\mu)\bar{y}(t) + \bar{\varepsilon}_W(y)]^T \bar{P} \\ &\times [\bar{A}_W(\mu)\bar{y}(t) + \bar{\varepsilon}_W(y)]dt + \bar{y}(t)^T \bar{R}^T \bar{R}\bar{y}(t)dt - \rho^2 v(t)^T \bar{R}_3^T \bar{R}_3 v(t)dt \\ &+ \rho^2 v(t)^T \bar{R}_3^T \bar{R}_3 v(t)dt\} = E\{\zeta(t)^T \hat{\Theta}_H(\mu,\hat{\mu})\zeta(t)\} + E\rho^2 v(t)^T \bar{R}_3^T \bar{R}_3 v(t) \end{aligned} \quad (11.38)$$

where $\zeta(t) = \left[\bar{y}(t)^T, \bar{\varepsilon}(y)^T, \bar{\varepsilon}_W(y)^T, v(t)^T\right]^T$ and

$$\bar{\Theta}_H(\mu,\hat{\mu}) \triangleq \begin{bmatrix} \bar{\Theta}_{11}(\mu,\hat{\mu}) + \bar{R}^T \bar{R} & \bar{P} & \bar{A}_W(\mu)^T \bar{P} & \bar{P}\bar{B}_v(\hat{\mu}) \\ \bar{P} & 0 & 0 & 0 \\ \bar{P}\bar{A}_W(\mu) & 0 & \bar{P} & 0 \\ \bar{B}_v(\hat{\mu})^T \bar{P} & 0 & 0 & -\rho^2 \bar{R}_3^T \bar{R}_3 \end{bmatrix}$$

Supposing the inequality $\zeta(t)^T \bar{\Theta}_H(\mu,\hat{\mu})\zeta(t) < 0$, holds, we have the following inequality:

$$E\int_0^{t_f} \bar{y}(t)^T \bar{R}^T \bar{R}\bar{y}(t)dt \le EV(\bar{y}(0)) - EV(\bar{y}(t_f)) + E\int_0^{t_f} \rho^2 v(t)^T \bar{R}_3^T \bar{R}_3 v(t)dt$$

Since $V(\bar{y}(t_f)) > 0$, we have the stochastic H_∞ estimator-based control performance inequality in equation (11.34). In order to impose two constraints equations (11.31) and (11.32) of error bounds on $\zeta(t)^T \bar{\Theta}_H(\mu,\hat{\mu})\zeta(t) < 0$ by the S-procedure, if there exist $\xi \ge 0$ and $\xi_W \ge 0$ satisfying $\Theta_H(\mu,\hat{\mu}) < 0$ in equation (11.36), then not only $\zeta(t)^T \bar{\Theta}_H(\mu,\hat{\mu})\zeta(t) < 0$ but also two error bounds in equations (11.31) and (11.32) hold. Therefore, if the inequality in equation (11.36) holds, then the stochastic H_∞ estimator-based control performance in equation (11.34) could be guaranteed.

Furthermore, since $\left\|\bar{R}_1 y(t)\right\|^2$ can be represented as $\left\|\bar{R}_1 y(t)\right\|^2 = \sum_{j=1}^N \left\|\bar{R}_1 y_j(t)\right\|^2 \Delta_x^2$ the term $\left\|\bar{R}_1 y(t)\right\|^2$ can be used to approximate the integration of the two-dimensional stochastic H_∞ estimator-based control performance in equation (11.35). Therefore, if the gridded spacing Δ_x^2 of the finite difference method is sufficiently small so that $O\left(\Delta_x^2\right) \to 0$, then the stochastic H_∞ estimator-based control performance in equation (11.33) will approach the two-dimensional stochastic H_∞ estimator-based control performance in equation (11.35).

Remark 11.4.2

In this chapter, the local truncation errors in equations (11.13) and (11.14) have been neglected as per Remark 11.3.3 because $O_{\{k,l\}}(\Delta_x^2), O_{W,\{k,l\}}(\Delta_x^2)$, converge to zero as $\Delta_x \to 0$. However, as the increasing of grid nodes, the dimension of the approximated fuzzy system equation (11.19) becomes very high, which will lead to an implement ability problem of controller design [394]. But, they could influence the H_∞ performance equation (11.35) if there were to be a compromise between the grid size Δ_x and the dimension N. In the consideration of the effect of truncation errors on the stabilization performance, the H_∞ performance equation (11.35) can be modified as

$$\frac{E\int_0^{t_f}\int_U (\left\|\bar{R}_1 y(t)\right\|^2 + \left\|R_2 e(x,t)\right\|^2)dxdt}{E\int_0^{t_f}\int_U (\left\|R_3 v(x,t)\right\|^2 + \left\|R_4 \bar{O}(\Delta_x^2)\right\|^2 + \left\|R_5 \bar{O}_W(\Delta_x^2)\right\|^2)dxdt} \le \rho^2 \quad (11.39)$$

where $\bar{O}\left(\Delta_x^2\right) \triangleq \left[O\left(\Delta_x^2\right)^T, O\left(\Delta_x^2\right)^T\right]^T$ and $\bar{O}_W\left(\Delta_x^2\right) \triangleq \left[O_W\left(\Delta_x^2\right)^T, O_W\left(\Delta_x^2\right)^T\right]^T$ with $O\left(\Delta_x^2\right) \triangleq \left[O_1\left(\Delta_x^2\right)^T, O_2\left(\Delta_x^2\right)^T, \ldots, O_N\left(\Delta_x^2\right)^T\right]^T$, $O_W\left(\Delta_x^2\right) \triangleq diag\left(O_{W,1}\left(\Delta_x^2\right), O_{W,2}\left(\Delta_x^2\right), \ldots, O_{W,N}\left(\Delta_x^2\right)\right)$, $R_4 \in \mathbb{R}^{n\times n}$, and $R_5 \in \mathbb{R}^{n\times n}$. Similar to the proof process of Theorem 11.4.2 described in the preceding paragraphs, let $\bar{R}_4 \triangleq diag(\Delta_x R_4, \Delta_x R_4, ..., \Delta_x R_4)$, $\bar{R}_5 \triangleq diag(\Delta_x R_5, \Delta_x R_5, ..., \Delta_x R_5)$, then we can get the following equivalent equation with the consideration of truncation errors

$$E(dV(\bar{y}) + \bar{y}(t)^T \bar{R}^T \bar{R}\bar{y}(t)dt) = E\{\zeta^T \bar{\Theta}_H(\mu, \hat{\mu})\zeta\} + E\rho^2 v(t)^T \bar{R}_3^T \bar{R}_3 v(t)dt$$

$$+E\rho^2 \bar{O}(\Delta_x^2)^T \bar{R}_4^T \bar{R}_4 \bar{O}(\Delta_x^2)dt + E\rho^2 \bar{O}_W(\Delta_x^2)^T \bar{R}_5^T \bar{R}_5 \bar{O}_W(\Delta_x^2)dt \tag{11.40}$$

where $\zeta(t) = \left[\bar{y}(t)^T, \bar{\varepsilon}(y)^T, \bar{\varepsilon}_W(y)^T, v(t)^T, \bar{O}(\Delta_x^2)^T, \bar{O}_W(\Delta_x^2)^T\right]^T$,

$$\bar{\Theta}_H(\mu, \hat{\mu})$$

$$\triangleq \begin{bmatrix} \bar{\Theta}_{11}(\mu,\hat{\mu}) + \bar{\Theta}_{11,\xi} & \bar{P} & \bar{A}_W(\mu)^T \bar{P} & \bar{P}\bar{B}_v(\hat{\mu}) & \bar{P} & \bar{A}_W(\mu)^T \bar{P} \\ \bar{P} & -\xi I & 0 & 0 & 0 & 0 \\ \bar{P}\bar{A}_W(\mu) & 0 & \bar{P} - \xi_W I & 0 & 0 & \bar{P} \\ \bar{B}_v(\hat{\mu})^T \bar{P} & 0 & 0 & -\rho^2 \bar{R}_3^T \bar{R}_3 & 0 & 0 \\ \bar{P} & 0 & 0 & 0 & -\rho^2 \bar{R}_4^T \bar{R}_4 & 0 \\ \bar{P}\bar{A}_W(\mu) & 0 & \bar{P} & 0 & 0 & -\rho^2 \bar{R}_5^T \bar{R}_5 \end{bmatrix} < 0 \tag{11.41}$$

which is similar to equations (11.38) and (11.36), respectively. By the similar procedure as in the proof of Theorem 11.4.2, it can be shown that if equation (11.41) has a symmetric positive definite solution $\bar{P}$, then H∞ performance equation (11.39) is guaranteed. Hence, the effect of the local truncation errors can be included into the stochastic H∞ estimator-based control problem as well and the H_∞ performance equation (11.39) can be achieved by solving $\bar{\Theta}_H(\mu, \hat{\mu}) < 0$ in equation (11.41) instead. Comparing equations (11.41) with (11.36), a more large ρ is needed to solve $\bar{P} > 0$ in equation (11.41) than in equation (11.36) in order to tolerate the effect of local truncation errors due to finite difference scheme, i.e. the effect of truncation errors may degrade the H_∞ performance, or even lead to the instability of the original PDE system.

Remark 11.4.3

The number of grid nodes and the implement ability are a tradeoff problem in the partial differential control design problem. The effect of the number of grid nodes (or Δ_x) on the H_∞ robust stabilization design problem is embedded in $\bar{R}_1$, $\bar{R}_2$, $\bar{R}_3$, $\bar{R}_4$ and $\bar{R}_5$ of $\bar{\Theta}_H(\mu, \hat{\mu})$, which are all functions of Δ_x (= Δ_y). From the H_∞ robust stabilization condition $\bar{\Theta}_H(\mu, \hat{\mu}) < 0$ in equations (11.36) and (11.41), which are all functions of $\bar{R}_i$, it is seen that if Δ_x becomes larger or the number of grid nodes becomes smaller, the H_∞ robust stabilization criterion $\bar{\Theta}_H(\mu,\hat{\mu}) < 0$ may be violated for a prescribed disturbance attenuation level ρ, except that we increase ρ, i.e. the truncation errors may degrade ρ. Therefore, the compromise between the number of grid nodes (or Δ_x) and the implement ability in the design procedure is that the choice of the number of grid nodes (or Δ_x) could not violate the H_∞ robust stabilization criterion $\bar{\Theta}_H(\mu,\hat{\mu}) < 0$ with a prescribed ρ.

Remark 11.4.4

Since the condition of robust H_∞ stabilization in Theorem 11.4.2 implies the condition of stochastically asymptotic stability in Theorem 11.4.1, the robust stochastic H_∞ stabilization control design implies that the system with $v(t) \equiv 0$ in equation (11.25) is asymptotically stable in probability. Therefore, the remainder of the chapter will focus on the robust stochastic H_∞ stabilization problem, because it is easy to obtain the stochastic stabilization result via the robust stochastic H_∞ stabilization results.

11.4.2 LMI Approach for Solving the Robust Stochastic H_∞ Stabilization Control Problem

Let us define $\tilde{P} \triangleq diag\left(\tilde{P}_{11},\tilde{P}_{11},\tilde{P}_{22}\right) = diag\left(I_{\tilde{M}} \otimes \bar{P}_{11}, I_{\tilde{M}} \otimes \bar{P}_{11}, I_{\tilde{M}} \otimes \bar{P}_{22}\right) \in \mathbb{R}^{3n\tilde{N}\times 3n\tilde{N}}$, $\bar{X}_{11} = \bar{P}_{11}^{-1}$ and $\tilde{X}_{11} = \tilde{P}_{11}^{-1}$, where $\bar{P}_1$ is defined in equation (11.26). The sufficient conditions for robust stochastic H_∞ stabilization control in Theorem 11.4.2 can be derived as follows:

Theorem 11.4.3

For the augmented system in equation (11.25) with a prescribed disturbance attenuation level ρ, suppose there exist the symmetric matrices $\bar{X}_{11} > 0, \bar{P}_{22} > 0$, the matrices Y_i and Z_i, $i = 1,\ldots, M$, $\hat{\Xi}_{11}$, $\hat{\Xi}_{22}$, $\hat{\Lambda}_2$ and the scalars $\xi > 0$, $\xi_W > 0$, $\eta > 0$ such that

$$\begin{bmatrix} \Omega_{11} & \Omega_{12} & \Omega_{13} & \Omega_{14} \\ \Omega_{21} & \Omega_{22} & 0 & 0 \\ \Omega_{31} & 0 & \Omega_{33} & 0 \\ \Omega_{41} & 0 & 0 & \Omega_{44} \end{bmatrix} < 0 \tag{11.42}$$

Then the stochastic H_∞ estimator-based control performance in equation (11.34) is guaranteed by the fuzzy estimator-based controller in equations (11.22) and (11.23), i.e., the state-dependent spatiotemporal noise and the approximation *error can be tolerated. Additionally, the effect of random spatiotemporal external disturbance and measurement noise on the controlled output can be attenuated below a prescribed level* ρ *by using the fuzzy estimator-based controller with estimator gains* $G_i = \bar{P}_{22}^{-1} Z_i$ *and control gains* $K_i = Y_i \bar{X}_{11}^{-1}$. *In equation (11.42),*

$$\Omega_{11} = \begin{bmatrix} \Theta_{11} - 2\varsigma\tilde{X}_{11} & -\tilde{B}\tilde{Y} & -\tilde{B}\tilde{Y} & \varsigma I & 0 & 0 \\ -\tilde{Y}^T\tilde{B}^T & -2\varsigma\tilde{X}_{11} & 0 & 0 & \varsigma I & 0 \\ -\tilde{Y}^T\tilde{B}^T & 0 & -2\varsigma\tilde{X}_{11} & 0 & 0 & \varsigma I \\ \varsigma I & 0 & 0 & -\eta I + \hat{A}_W^T\bar{P}_{22}\hat{A}_W & 0 & \tilde{A}_{21}^T\tilde{P}_{22} \\ 0 & \varsigma I & 0 & 0 & \hat{\Lambda}_2 + \hat{\Lambda}_2^T & 0 \\ 0 & 0 & \varsigma I & \tilde{P}_{22}\tilde{A}_{21} & 0 & \Theta_{22} \end{bmatrix}$$

$$\Theta_{11} \triangleq \tilde{X}_{11}\tilde{A}_{11}^T + \tilde{A}_{11}\tilde{X}_{11} + \tilde{B}\tilde{Y} + \tilde{Y}^T\tilde{B}^T + \hat{\tilde{\Xi}}_{11}^T + \hat{\tilde{\Xi}}_{11}, \tilde{Y} \triangleq \tilde{K}\tilde{X}_{11}, \hat{\tilde{\Xi}}_{11} \triangleq \tilde{\Xi}_{11}\tilde{X}_{11}, \hat{\Lambda}_2 \triangleq \tilde{\Lambda}_2\tilde{X}_{11},$$

$$\Theta_{22} \triangleq \tilde{A}_{22}^T\tilde{P}_{22} + \tilde{P}_{22}\tilde{A}_{22} + \tilde{Z}\tilde{C} + \tilde{C}^T\tilde{Z}^T + \hat{\tilde{\Xi}}_{22}^T + \hat{\tilde{\Xi}}_{22}, \tilde{Z} \triangleq \tilde{P}_{22}\tilde{G}, \hat{\tilde{\Xi}}_{22} \triangleq \tilde{P}_{22}\tilde{\Xi}_{22}, \tilde{Z}_v \triangleq \tilde{P}_{22}\tilde{G}_v,$$

$$\Omega_{21} = \begin{bmatrix} \hat{A}_W\tilde{X}_{11} & 0 & 0 & 0 & 0 & 0 \\ 0 & 0 & 0 & 0 & 0 & 0 \\ 0 & 0 & 0 & \bar{P}_{22}\hat{A}_W & 0 & 0 \end{bmatrix}, \Omega_{22} = \begin{bmatrix} -\bar{X}_{11} & I & 0 \\ I & -\xi_W I & 0 \\ 0 & 0 & \bar{P}_{22} - \xi_W I \end{bmatrix}$$

$$\Omega_{31} = \begin{bmatrix} H^T\tilde{X}_{11} & 0 & 0 & 0 & 0 & 0 \\ H^T\tilde{X}_{11} & 0 & 0 & 0 & 0 & 0 \end{bmatrix}, \Omega_{33} = diag\left(-2\Sigma_{11}^{-1/2} + \xi I, -2\Sigma_{W,11}^{-1/2} + \xi_W I\right),$$

$$\Omega_{12} = \Omega_{21}^T, \Omega_{13} = \Omega_{31}^T, \Omega_{14,1} = \left[\tilde{X}_{11}H^T\bar{R}_1^T, \tilde{X}_{11}, H^T, H^T B_d\right], \Omega_{41,1} = \Omega_{14,1}^T,$$

$$\Omega_{14,2} = \left[H^T\bar{R}_2^T, \tilde{P}_{22}H^T, \tilde{P}_{22}H^T B_d, \tilde{Z}_v D\right], \Omega_{41,2} = \Omega_{14,2}^T, \Omega_{14} = \Omega_{41}^T,$$

$$\Omega_{44,1} = diag(-I, -2I + \eta I, -\xi I, -\rho^2 / 2\bar{R}_3^T\bar{R}_3), \Omega_{41} = \begin{bmatrix} \Omega_{41,1} & 0 & 0 & 0 & 0 & 0 \\ 0 & 0 & 0 & 0 & 0 & \Omega_{41,2} \end{bmatrix},$$

$$\Omega_{44,2} = diag(-I, -\xi I, -\rho^2 / 2\bar{R}_3^T\bar{R}_3, -\rho^2\bar{R}_3^T\bar{R}_3), \Omega_{44} = \begin{bmatrix} \Omega_{44,1} & 0 \\ 0 & \Omega_{44,2} \end{bmatrix},$$

where $\Sigma_{11} \triangleq 2\sigma_\kappa^2 T^T T + 2\sigma_f^2 I$, $\Sigma_{W,11} \triangleq 2\sigma_{fW}^2 I$, and any constant ς.

Before presenting the proof of Theorem 11.4.3, some notations and definitions are clarified for convenience. Since $\bar{\mu}_{i_1}(y)$ and $\bar{\mu}_{i_2}(\hat{y})$ are in matrix form and have the property that $\sum_{i_1=1}^{M}\sum_{i_2=1}^{M}\bar{\mu}_{i_1}(y)\bar{\mu}_{i_2}(\hat{y}) = I$ by Lemma 11.3.1, the term $\bar{A}(\mu, \hat{\mu})$ in equation (11.25) can be represented as follows

$$\bar{A}(\mu,\hat{\mu}) = \begin{bmatrix} I & 0 & 0 \\ 0 & 0 & I \end{bmatrix} \begin{bmatrix} \bar{A}_0^{(11)} + B\bar{K}_0 & -B\bar{K}_0 & -B\bar{K}_0 \\ 0 & 0 & 0 \\ 0 & 0 & \bar{A}_0^{(22)} + \bar{G}_0 C \end{bmatrix} \begin{bmatrix} I & 0 \\ 0 & 0 \\ 0 & I \end{bmatrix}$$

$$+ \sum_{i_1=1}^{M} \sum_{i_2=1}^{M} \begin{bmatrix} \bar{\mu}_{i_1}(y) & \bar{\mu}_{i_1}(y) - \bar{\mu}_{i_1}(\hat{y}) & 0 \\ 0 & 0 & \bar{\mu}_{i_1}(\hat{y}) \end{bmatrix}$$

$$\times \begin{bmatrix} \bar{A}_{i_1 i_2}^{(11)} + B\bar{K}_{i_1 i_2} & -B\bar{K}_{i_1 i_2} & -B\bar{K}_{i_1 i_2} \\ 0 & 0 & 0 \\ \bar{A}_{i_1 i_2}^{(21)} & 0 & \bar{A}_{i_1 i_2}^{(22)} + \bar{G}_{i_1 i_2} C \end{bmatrix} \begin{bmatrix} \bar{\mu}_{i_2}(y) & 0 \\ \bar{\mu}_{i_2}(y) - \bar{\mu}_{i_2}(\hat{y}) & 0 \\ 0 & \bar{\mu}_{i_2}(\hat{y}) \end{bmatrix} \tag{11.43}$$

where $\bar{A}_0^{(11)} \triangleq (I_N \otimes \kappa_0)T + (I_N \otimes A_0)$, $\bar{A}_0^{(22)} \triangleq (I_N \otimes \kappa_0)T + (I_N \otimes A_0)$, $\bar{A}_0^{(21)} \triangleq 0$, $\bar{K}_0 \triangleq K_0$, $\bar{G}_0 \triangleq G_0$, $\bar{A}_{i_1 i_2}^{(11)} \triangleq (I_N \otimes \kappa_{i_1})T + (I_N \otimes A_{i_1})$, $\bar{A}_{i_1 i_2}^{(21)} \triangleq (I_N \otimes (\kappa_{i_1} \times \kappa_{i_2}))T + I_N \otimes (A_{i_1} \times A_{i_2})$, $\bar{K}_{i_1 i_2} \triangleq K_{i_2}$, $\bar{G}_{i_1 i_2} \triangleq G_{i_1}$. Similarly, for $\bar{A}_W(\mu,\hat{\mu})$ and $\bar{B}_v(\hat{\mu})$, we have

$$\bar{A}_W(\mu) = \sum_{i_2=1}^{M} \begin{bmatrix} \bar{A}_{W,i_2} & 0 & 0 \\ \bar{A}_{W,i_2} & 0 & 0 \end{bmatrix} \begin{bmatrix} \bar{\mu}_{i_2}(y) & 0 \\ \bar{\mu}_{i_2}(y) - \bar{\mu}_{i_2}(\hat{y}) & 0 \\ 0 & \bar{\mu}_{i_2}(\hat{y}) \end{bmatrix} \tag{11.44}$$

$$\bar{B}_v(\hat{\mu}) = \begin{bmatrix} B_d & 0 \\ B_d & G_0 D \end{bmatrix}$$

$$+ \sum_{i_1=1}^{M} \begin{bmatrix} \bar{\mu}_{i_1}(y) & \bar{\mu}_{i_1}(y) - \bar{\mu}_{i_1}(\hat{y}) & 0 \\ 0 & 0 & \bar{\mu}_{i_1}(\hat{y}) \end{bmatrix} \begin{bmatrix} 0 & 0 \\ 0 & 0 \\ 0 & G_{i_1}\mathrm{D} \end{bmatrix} \tag{11.45}$$

where $\bar{A}_{W,i_2} \triangleq I_N \otimes \bar{A}_{W,i_2}$. Note that the related matrices are defined in equation (11.25).

We define $\mathcal{M} \triangleq \left[I, \bar{\mu}_1(y)^T, \ldots, \bar{\mu}_M(y)^T\right]^T$, $\hat{\mathcal{M}} \triangleq \left[I, \bar{\mu}_1(\hat{y})^T, \ldots, \bar{\mu}_M(\hat{y})^T\right]^T$ and the following matrices:

$$\tilde{\Xi}_k \triangleq \begin{bmatrix} -\Xi_{0,1}^{(k)T} - \Xi_{0,2}^{(k)} + \Xi_3^{(k)} & \Xi_{0,2}^{(k)} - \Xi_{1,1}^{(k)} & \cdots & \Xi_{0,2}^{(k)} - \Xi_{1,M}^{(k)} \\ \Xi_{0,1}^{(k)T} - \Xi_{2,1}^{(k)T} & \Xi_{1,11}^{(k)} + \Xi_{2,1}^{(k)T} - \Xi_3^{(k)} & \cdots & \Xi_{1,MM}^{(k)} + \Xi_{2,1}^{(k)T} \\ \vdots & \vdots & \ddots & \vdots \\ \Xi_{0,1}^{(k)T} - \Xi_{2,M}^{(k)\,T} & \Xi_{1,1}^{(k)} + \Xi_{2,M}^{(k)\,T} & \cdots & \Xi_{1,M}^{(k)} + \Xi_{2,M}^{(k)\,T} - \Xi_3^{(k)} \end{bmatrix}$$

$$\tilde{A}_k \triangleq \begin{bmatrix} \bar{A}_0^{(k)} & 0 & \cdots & 0 \\ 0 & \bar{A}_{11}^{(k)} & \cdots & \bar{A}_{1M}^{(k)} \\ \vdots & \vdots & \ddots & \vdots \\ 0 & \bar{A}_{M1}^{(k)} & \cdots & \bar{A}_{MM}^{(k)} \end{bmatrix}, \tilde{K} \triangleq \begin{bmatrix} \bar{K}_0 & 0 & \cdots & 0 \\ 0 & \bar{K}_{11} & \cdots & \bar{K}_{1M} \\ \vdots & \vdots & \ddots & \vdots \\ 0 & \bar{K}_{M1} & \cdots & \bar{K}_{MM} \end{bmatrix},$$

$$\tilde{G} \triangleq \begin{bmatrix} \bar{G}_0 & 0 & \cdots & 0 \\ 0 & \bar{G}_{11} & \cdots & \bar{G}_{1M} \\ \vdots & \vdots & \ddots & \vdots \\ 0 & \bar{G}_{M1} & \cdots & \bar{G}_{MM} \end{bmatrix},$$

$$\tilde{A}_W \triangleq \begin{bmatrix} \hat{A}_W & 0 & 0 \\ \hat{A}_W & 0 & 0 \end{bmatrix}, \tilde{B}_v \triangleq \begin{bmatrix} H^T B_d & 0 \\ 0 & 0 \\ H^T B_d & \tilde{G}_v D \end{bmatrix}, \tilde{H} \triangleq \begin{bmatrix} H & 0 & 0 \\ 0 & 0 & H \end{bmatrix},$$

$$\hat{A}_W \triangleq [\bar{A}_{W,0}, \bar{A}_{W,1}, \ldots, \bar{A}_{W,M}],\ H \triangleq [I_{nN}, 0, \ldots, 0] \in R^{n \times n\tilde{N}},\ \tilde{G}_v \triangleq [G_0^T, G_1^T, \ldots, G_M^T]^T,$$

$$\tilde{\mathcal{M}}^T \triangleq \begin{bmatrix} \mathcal{M} & \mathcal{M} - \hat{\mathcal{M}} & 0 \\ 0 & 0 & \hat{\mathcal{M}} \end{bmatrix}, \tilde{A} \triangleq \begin{bmatrix} \tilde{A}_{11} + \tilde{B}\tilde{K} & -\tilde{B}\tilde{K} & -\tilde{B}\tilde{K} \\ 0 & \tilde{\Lambda}_2 & 0 \\ \tilde{A}_{21} & 0 & \tilde{A}_{22} + \tilde{G}\tilde{C} \end{bmatrix}.$$

$\tilde{B} \triangleq I_{\tilde{M}} \otimes B$ and $\tilde{C} \triangleq I_{\tilde{M}} \otimes C$, where $\tilde{M} \triangleq 1 + M$ and $\tilde{N} = N \times \tilde{M}$. For $\tilde{\Xi}_k$, we have k=11, 22 and for $\tilde{A}_k, k = 11, 21, 22$. The matrices $\Xi_0^{(k)} \in \mathbb{R}^{nN \times nN}$, $\Xi_{1,i_1i_2}^{(k)} \in \mathbb{R}^{nN \times nN}$, $\Xi_{2,i_1i_2}^{(k)} \in \mathbb{R}^{nN \times nN}$, and $\Xi_3^{(k)} \in \mathbb{R}^{nN \times nN}$ are relaxed matrices to be applied in Theorem 11.4.3; the other matrices are defined in equations (11.43) and (11.44).

Proof of Theorem 11.4.3

Proof

First, the fact that the inequality in equation (11.42) implies the inequality in equation (11.36) is proven in the following: Because $\left(\Sigma_{11}^{-1/2} - \xi I\right)\xi^{-1}\left(\Sigma_{11}^{-1/2} - \xi I\right) \geq 0$ and $\left(\Sigma_{W,11}^{-1/2} - \xi_W I\right)\xi^{-1}\left(\Sigma_{W,11}^{-1/2} - \xi_W I\right) \geq 0$ imply the inequalities $-\xi^{-1}\Sigma_{11}^{-1} \leq -2\Sigma_{11}^{-1/2} + \xi I$ and $-\xi_W^{-1}\Sigma_{W,11}^{-1} \leq -2\Sigma_{W,11}^{-1/2} + \xi_W I$, respectively, we have $\hat{\Omega}_{33} \triangleq diag\left(-\xi^{-1}\Sigma_{11}^{-1}, -\xi_W^{-1}\Sigma_{W,11}^{-1}\right) \leq \Omega_{33}$, in equation (11.42). Similarly, we have $-\eta^{-1} \leq -2 + \eta$. Using the aforementioned inequalities and the Schur complement, the inequality in equation (11.42) implies the inequality $\Omega_{11} - \Omega_{13}\hat{\Omega}_{33}^{-1}\Omega_{31} - \Omega_{12}\hat{\Omega}_{22}^{-1}\Omega_{21} - \Omega_{14}\hat{\Omega}_{44}^{-1}\Omega_{41} < 0$. By Lemma 5 of [417] this inequality can be written as follows:

$$\Omega \triangleq \begin{bmatrix} \Theta_{11} + \Upsilon_{11} - \eta\tilde{X}_{11}\tilde{X}_{11} + \tilde{X}_{11}\hat{\Upsilon}_{11}\tilde{X}_{11} & -\tilde{B}\tilde{Y} & -\tilde{B}\tilde{Y} + \tilde{X}_{11}\tilde{A}_{21}^T\tilde{P}_{22}\tilde{X}_{11} \\ -\tilde{Y}^T\tilde{B}^T & \hat{\Lambda}_2 + \hat{\Lambda}_2^T & 0 \\ -\tilde{Y}^T\tilde{B}^T + \tilde{X}_{11}\tilde{P}_{22}\tilde{A}_{21}\tilde{X}_{11} & 0 & \tilde{X}_{11}(\Theta_{22} + \Upsilon_{22})\tilde{X}_{11} \end{bmatrix} < 0 \tag{11.46}$$

where

$$\Upsilon_{11} \triangleq \tilde{X}_{11}H^T\Phi_{11}H\tilde{X}_{11} + \tilde{X}_{11}\hat{A}_W^T\Phi_{W,11}\hat{A}_W\tilde{X}_{11} + \eta\tilde{X}_{11}\tilde{X}_{11} + \frac{1}{\xi}H^TH + \frac{2}{\rho^2}H^TB_d(\bar{R}_3^T\bar{R}_3)^{-1}B_d^TH$$

$$\hat{\Upsilon}_{11} \triangleq \hat{A}_W^T\bar{P}_{22}\hat{A}_W - \hat{A}_W^T\bar{P}_{22}(\bar{P}_{22} - \xi_W I)^{-1}\bar{P}_{22}\hat{A}_W$$

$$\Upsilon_{22} \triangleq H^T\Phi_{22}H + \frac{1}{\xi}\tilde{P}_{22}H^TH\tilde{P}_{22} + \frac{2}{\rho^2}\tilde{P}_{22}H^TB_d(\bar{R}_3^T\bar{R}_3)^{-1}B_d^TH\tilde{P}_{22} + \frac{1}{\rho^2}\tilde{Z}_vD(\bar{R}_3^T\bar{R}_3)^{-1}D^T\tilde{Z}_v^T$$

in which $\Phi_{11} \triangleq \left(\xi\Sigma_{11} + \xi_W\Sigma_{W,11} + \bar{R}_1^T\bar{R}_1\right)$, $\Phi_{22} \triangleq \bar{R}_2^T\bar{R}_2$ and $\Phi_{W,11} = \left(\bar{X}_{11} - \frac{1}{\xi_W}I\right)^{-1}$. Note that $\Sigma = diag(\Sigma_{11}, 0)$, $\bar{R}^T\bar{R} = diag\left(\bar{R}_1^T\bar{R}_1, \bar{R}_2^T\bar{R}_2\right)$, $\Sigma_W = diag\left(\Sigma_{W,11}, 0\right)$.

On the other hand, $\bar{\Theta}_H(\mu,\hat{\mu})$ in equation (11.36) can be reformulated as follows:

$$\bar{\Theta}_H(\mu,\hat{\mu}) = \begin{bmatrix} \tilde{\mathcal{M}}^T & 0 & 0 & 0 \\ 0 & I & 0 & 0 \\ 0 & 0 & I & 0 \\ 0 & 0 & 0 & I \end{bmatrix} \begin{bmatrix} \tilde{\Theta}_{11} + \tilde{\Theta}_{11,\xi} & \tilde{H}^T\bar{P} & \tilde{A}_W^T\bar{P} & \tilde{P}\tilde{B}_v \\ \bar{P}\tilde{H} & -\xi I & 0 & 0 \\ \bar{P}\tilde{A}_W & 0 & \bar{P} - \xi_W I & 0 \\ \tilde{B}_v^T\tilde{P} & 0 & 0 & -\rho^2\bar{R}_3^T\bar{R}_3 \end{bmatrix} \begin{bmatrix} \tilde{\mathcal{M}}^T & 0 & 0 & 0 \\ 0 & I & 0 & 0 \\ 0 & 0 & I & 0 \\ 0 & 0 & 0 & I \end{bmatrix}$$

where

$$\tilde{\Theta}_{11} \triangleq (\tilde{A} + \tilde{\Xi})^T\tilde{P} + \tilde{P}(\tilde{A} + \tilde{\Xi}) + \tilde{A}_W^T\bar{P}\tilde{A}_W, \tilde{\Theta}_{11,\xi} \triangleq \tilde{H}^T(\xi\Sigma + \xi_W\Sigma_W + \bar{R}^T\bar{R})\tilde{H}.$$

Therefore, by employing the Schur complement, if the following inequality holds, then $\bar{\Theta}_H(\mu,\hat{\mu})<0$

$$(\tilde{A}+\tilde{\Xi})^T\tilde{P}+\tilde{P}(\tilde{A}+\tilde{\Xi})+\tilde{A}_W^T\bar{P}\tilde{A}_W+\tilde{H}^T(\xi\Sigma+\xi_W\Sigma_W+\bar{R}^T\bar{R})\tilde{H}$$
$$+\frac{1}{\xi}\tilde{H}^T\bar{P}\bar{P}\tilde{H}-\tilde{A}_W^T\bar{P}(\bar{P}-\xi_W I)^{-1}\bar{P}\tilde{A}_W+\frac{1}{\rho^2}\tilde{P}\tilde{B}_v(\bar{R}_3^T\bar{R}_3)^{-1}\tilde{B}_v^T\tilde{P}<0 \quad (11.47)$$

and $\bar{P}-\xi_W I<0$. From $\bar{P}=diag\left(\bar{P}_{11},\bar{P}_{22}\right)$, we have $\bar{P}-\bar{P}\left(\bar{P}-\xi_W I\right)^{-1}\bar{P}= diag\left(\bar{P}_{11}-\bar{P}_{11}(\bar{P}_{11}-\xi_W I)^{-1}\bar{P}_{11},\bar{P}_{22}-\bar{P}_{22}(\bar{P}_{22}-\xi_W I)^{-1}\bar{P}_{22}\right)$. Then, by the matrix inversion lemma, we have $\left(\bar{P}_{11}^{-1}-\xi_W^{-1}I\right)^{-1}=\bar{P}_{11}-\bar{P}_{11}\left(\bar{P}_{11}-\xi_W I\right)^{-1}\bar{P}_{11}$. As a result, we have

$$\bar{P}-\bar{P}(\bar{P}-\xi_W I)^{-1}\bar{P}=\begin{bmatrix}\Phi_{W,11} & 0\\ 0 & \Phi_{W,22}\end{bmatrix}$$

where $\Phi_{W,11}=\left(\bar{P}_{11}^{-1}-\xi_W^{-1}I\right)^{-1}$ and $\Phi_{W,22}=\bar{P}_{22}-\bar{P}_{22}\left(\bar{P}_{22}-\xi_W I\right)^{-1}\bar{P}_{22}$.

Pre- and post-multiplying the inequality in equation (11.47) by the matrix $\tilde{X}=diag\left(\tilde{X}_{11},\tilde{X}_{11},\tilde{X}_{11}\right)$ and substituting the definitions of $\tilde{P},\tilde{A},\tilde{A}_W,\tilde{B}_v$ and H into the inequality in equation (11.47), we can obtain

$$\bar{\Omega}\triangleq\begin{bmatrix}\Theta_{11} & -\tilde{B}\tilde{Y} & -\tilde{B}\tilde{Y}+\tilde{X}_{11}\tilde{A}_{21}^T\tilde{P}_{22}\tilde{X}_{11}\\ -\tilde{Y}^T\tilde{B}^T & \hat{\Lambda}_2+\hat{\Lambda}_2^T & 0\\ -\tilde{Y}^T\tilde{B}^T+\tilde{X}_{11}\tilde{P}_{22}\tilde{A}_{21}\tilde{X}_{11} & 0 & \tilde{X}_{11}\Theta_{22}\tilde{X}_{11}\end{bmatrix}$$
$$+diag(\tilde{X}_{11}H^T\Phi_{11}H\tilde{X}_{11},0,\tilde{X}_{11}H^T\Phi_{22}H\tilde{X}_{11})+diag(\tilde{X}_{11}\hat{A}_W^T(\Phi_{W,11}+\Phi_{W,22})\hat{A}_W\tilde{X}_{11},0,0)$$
$$+\frac{1}{\xi}diag(H^TH,0,\tilde{X}_{11}\tilde{P}_{22}H^TH\tilde{P}_{22}\tilde{X}_{11})$$
$$+\frac{1}{\rho^2}\begin{bmatrix}H^TB_d(\bar{R}_3^T\bar{R}_3)^{-1}B_d^TH & 0 & H^TB_d(\bar{R}_3^T\bar{R}_3)^{-1}B_d^TH\tilde{P}_{22}\tilde{X}_{11}\\ 0 & 0 & 0\\ \tilde{X}_{11}\tilde{P}_{22}H^TB_d(\bar{R}_3^T\bar{R}_3)^{-1}B_d^TH & 0 & \tilde{X}_{11}\Psi\tilde{X}_{11}\end{bmatrix}<0$$
$$(11.48)$$

where $\Psi\triangleq\tilde{P}_{22}H^TB_d\left(\bar{R}_3^T\bar{R}_3\right)^{-1}B_d^TH\tilde{P}_{22}+\tilde{Z}_vD\left(\bar{R}_3^T\bar{R}_3\right)^{-1}D^T\tilde{Z}_v^T$. The matrices $\tilde{Z}_v,\Theta_{11}$ and Θ_{22} are defined in equation (11.42). Finally, we have the following inequality:

$$\begin{bmatrix} H^T B_d(\bar{R}_3^T\bar{R}_3)^{-1}B_d^T H & 0 & H^T B_d(\bar{R}_3^T\bar{R}_3)^{-1}B_d^T H\tilde{P}_{22}\tilde{X}_{11} \\ 0 & 0 & 0 \\ \tilde{X}_{11}\tilde{P}_{22}H^T B_d(\bar{R}_3^T\bar{R}_3)^{-1}B_d^T H & 0 & \tilde{X}_{11}\tilde{P}_{22}H^T B_d(\bar{R}_3^T\bar{R}_3)^{-1}B_d^T H\tilde{P}_{22}\tilde{X}_{11} \end{bmatrix}$$

$$\leq \begin{bmatrix} 2H^T B_d(\bar{R}_3^T\bar{R}_3)^{-1}B_d^T H & 0 & 0 \\ 0 & 0 & 0 \\ 0 & 0 & 2\tilde{X}_{11}\tilde{P}_{22}H^T B_d(\bar{R}_3^T\bar{R}_3)^{-1}B_d^T H\tilde{P}_{22}\tilde{X}_{11} \end{bmatrix}$$

Using the above inequality, we can obtain the inequality, $\bar{\Omega} \leq \Omega$, in which Ω and $\bar{\Omega}$ are defined in equations (11.46) and (11.48), respectively. Therefore, the inequality $\Omega < 0$ implies the inequality $\bar{\Omega} < 0$. Since the inequality in equation (11.48) is equivalent to the inequality in equation (11.36) and the inequality in equation (11.42) implies the inequality in equation (11.46), we know the inequality in equation (11.42) implies the inequality in equation (11.36). Finally, using the result of Theorem 11.4.2, we can prove that if the inequality in equation (11.42) holds then the stochastic H_∞ estimator-based control performance in equation (11.34) can be guaranteed by the fuzzy estimator-based controller in equations (11.22) and (11.23).

Remark 11.4.5

The control gains K_i and the estimator gains G_i, for $i = 0, 1,\ldots, M$, are included in the matrices $\tilde{K}$ and $\tilde{G}$, respectively. We define $Y_{i_2} \triangleq K_{i_2}\bar{X}_{11}$, $Z_{i_1} \triangleq \bar{P}_{22}G_{i_1}$.

Based on the above analysis, the robust stochastic H_∞ stabilization design problem is reduced to the specification of the estimator gains $G_i = \bar{P}_{22}^{-1}Z_i$ and control gains $K_i = Y_i\bar{X}_{11}^{-1}, i = 0,1,...,M$ by solving the LMI in equation (11.42) with some positive definite matrices. Finally, in order to achieve the optimal attenuation of the truncation error, external disturbances and measurement noises on the controlled output and estimation error, the optimal stochastic H_∞ stabilization control design problem for the NSPDS equations (11.1) and (11.2) can be solved using the following constrained optimization problem:

$$\rho_0 = \min_{Y_i, Z_i, i=0,1,\ldots,M} \rho$$

$$\text{subject to } \bar{X}_{11} > 0, \bar{P}_{22} > 0, \xi > 0, \xi_W > 0, \eta > 0, \text{ and (11.42)} \tag{11.49}$$

This is called an eigenvalue problem [426] and can be easily solved using the LMI technique. Based on the above analysis, the robust stochastic H_∞ stabilization control design procedure for the NSPDSs is summarized in the following steps.

Design Procedure

Step 1: Give a prescribed disturbance attenuation level ρ.

Step 2: Select the fuzzy membership functions and fuzzy rules to establish a FSPDS in equation (11.9) to approximate the NSPDS in equation (11.1).

Step 3: Give grid size Δ_x and N to construct the fuzzy spatial state space model in equations (11.19) and (11.20).

Step 4: Solve the estimator gains G_i and the control gains K_i of the fuzzy estimator-based controller, for $i=0, 1,\ldots, M$, in equations (11.22) and (11.23) by solving the LMI problem in equation (11.42) or solve K_i and G_i from the optimal H_∞ stabilization control problem in equation (11.49).

Step 5: Construct a fuzzy estimator-based controller in equations (11.22) and (11.23) to stabilize the NSPDS in equations (11.1) and (11.2).

11.5 SIMULATION EXAMPLE

In nerve cells, separation of ionic charge along the cell membrane causes a difference in electrical potential across the cell membrane. Nerve cells receive electrical input signals from the dendrites of other nerve cells. Depending on the spatiotemporal distribution of the input current to depolarize the membrane voltage, the firing threshold can be reached after sufficient membrane voltage depolarization and then an action potential will be triggered. The Hodgkin-Huxley (H-H) model is a phenomenological model that describes the electrical excitations of nerve cells as a function of a given current stimulus [429]. Electrical stimulation of a nerve cell with rectangular pulses has a range of clinical applications, for example, the activation of muscles by stimulating the motor nerve cell fibers innervating muscles or activation of different sensor-motor areas in the brain or spinal cord, such as deep brain stimulation for Parkinson's patients. When the parameters of the H-H model are set to the original values in [429], the H-H model behaves as an excitable membrane. However, when parameters are appropriately changed, the H-H model can show a variety of qualitatively different behaviors [430,431]. Some abnormal neuronal disorders in clinical medicine cause many diseases such as Parkinson's disease, Alzheimer's disease, epilepsy, and arrhythmia. For example, single cell simulation is of major interest for suppression of undesired neural oscillation as it occurs in patients with Parkinson's disease or epilepsy [432]. Understanding the mechanism leading to these diseases is an important medical problem with enormous impact. During the past several decades scientists have found bifurcations due to disturbance and noise of nerve systems closely related to these diseases; they have attempted to control these bifurcations and cure these diseases [432–434]. The term "bifurcation control" denotes the design of a controller that can modify the bifurcation properties of nerve systems to avoid undesirable instability around bifurcation, i.e. to stabilize the H-H model [430,431].

The H-H model is expressed by nonlinear PDEs describing the spatiotemporal evolution of the membrane voltage $y_v \triangleq y_v(x,t)$. In the Hodgkin-Huxley dynamic model, the total current across a nerve cell membrane is the sum of the capacitive current, the ionic currents, and the external current. In [435–437], the channel noise influence on the nerve cells is studied by a stochastic version of the H-H model. A two-dimensional H-H model is developed to approximate the full H-H model [438]. Therefore, the stochastic Hodgkin-Huxley partial differential system is described by the total currents leading to equivalent electric ionic currents as follows [429–441]:

$$c_m \frac{\partial y_v}{\partial t} = \kappa_m \frac{\partial^2 y_v}{\partial x^2} + f_v(y_v, y_n) + g_v(x)u(t) + g_{d,v}(x)I_{inj}(t) \tag{11.50}$$

$$\frac{\partial y_n}{\partial t} = \left(\alpha_n(1-y_n) - \beta_n y_n\right) + \sqrt{\frac{2}{N_K}\frac{\alpha_n \beta_n}{\alpha_n + \beta_n}} \frac{\partial^2 W(x,t)}{\partial t \partial x} \tag{11.51}$$

$$z(t) = h_v(x)(y_v + D_n n(t)) \tag{11.52}$$

where $x \in [0,1]$ and $f_v(y_v, y_n) \triangleq g_K y_n^4 (V_k - y_v) + g_{Na} m^3 h (V_{Na} - y_v) + g_m (V_{leak} - y_v)$. The symbols y_n, m, and h stand for the potassium activation, the sodium activation, and the sodium inactivation, respectively. The current generated from the flow of potassium ions is determined by a maximum potassium conductance g_K, an ionic equilibrium potential V_K expressing steady-state potassium ion separation, and potassium activation y_n. Similarly, the sodium ion current is modeled with a maximal sodium conductance g_{Na}, an ionic equilibrium potential V_{Na}, sodium activation m and inactivation h. The remaining ion currents are collectively modeled by a leakage current with conductance g_m and ionic equilibrium potential V_{leak}. The constant c_m is the membrane capacity per unit area. The constant diffusion κ_m is defined as $\kappa_m \triangleq \frac{r_a}{2r_2}$, where r_a is the radius of the fiber and r_2 is the specific resistance of the axoplasm. The specific functions $\alpha_n(y_v), \beta_n(y_v), \alpha_m(y_v), \beta_m(y_v)$, $\alpha_h(y_v)$, and $\beta_h(y_v)$ as proposed by Hodgkin and Huxley [429] are

$$\alpha_n(y_v) = \frac{0.01(10-y_v)}{\exp\left(\frac{10-y_v}{10}\right)-1},\ \beta_n(y_v) = 0.125\exp\left(\frac{-y_v}{80}\right),\ \alpha_m(y_v) = \frac{0.1(25-y_v)}{\exp\left(\frac{25-y_v}{10}\right)-1},$$

$$\beta_m(y_v) = 4\exp\left(\frac{-y_v}{18}\right),\ \alpha_h(y_v) = 0.07\exp\left(\frac{-y_v}{20}\right),\ \beta_h(y_v) = \frac{1}{\exp\left(\frac{30-y_v}{10}\right)+1},$$

$$m = \frac{\alpha_m(y_v)}{\alpha_m(y_v) + \beta_m(y_v)},\ h = 0.8 - y_n$$

The remaining constants are N_K=2, κ_m=0.336, c_m=1 μF/cm², g_K=36 mS/cm², g_{Na}=120 mS/cm², and g_m=0.3 mS/cm² with equilibrium potentials V_K=−12 mV, V_{Na}=115 mV, and V_{leak}=10.613 mV. $I_{inj}(t)$ is an externally injected current in a spatially localized axonal compartment. In [442], the periodic solution bifurcates from the equilibrium point through a Hopf bifurcation when I_{inj} changes. Rinzel and Miller [443] numerically calculated the stable and the unstable periodic solution branched and parameterized by I_{inj}. Therefore, the externally injected current I_{inj} is treated as a bifurcation parameter and defined as $I_{inj}=10\sin(t)$, which is considered as an external disturbance in the case of Parkinson's or Alzheimer's disease in this study. The potassium ion channel y_n suffers from spatiotemporal Wiener noise $W(x, t)$. The influence function of I_{inj} can be defined as $g_{v,d}(x)=2\delta(x-0)$. The controller $u(t)$ is designed to stochastically stabilize the H-H model in equations (11.50) and (11.51) under Wiener noises and to attenuate the effects of externally injected current I_{inj}. The control influence function is denoted as $g_v(x)=[10\delta(x-0), 10\delta(x-1), 10\delta(x-1)]$. The potential y_v is measured in units of mV; current density is in units of μA/cm²; the unit of time is milliseconds (msec). The initial distribution of the membrane voltage is given as $y_v(x, 0)=0$. The boundary conditions are the Neumann boundary condition, i.e., $\partial y_v(x, t)/\partial x=0$ at $x=0$ and $x=1$.

When the nervous system suffers the effect of the state-dependent Wiener process and the external disturbances, the spatiotemporal profiles are depicted in Figure 11.3, which is similar to an abnormal neural disorder such as Parkinson's disease or Alzheimer's disease. From Figure 11.3, this nervous system is unable to converge to the steady state. Therefore, a fuzzy estimator-based controller is designed to robustly stabilize this nervous system. Note that the equilibrium point ye of the H-H model is at $y_e=[-0.1923; 0.3147]$. The equilibrium point of the system should be shifted to the origin.

Suppose the sensors are located at $x=0$, 0.5, and 1; the observation influence function is defined as $h_v(x)=\left[\delta(x-0),\delta\left(x-\frac{1}{2}\right),\delta(x-1)\right]^T$. The measurement noise $n(t)$ is white noise with unit variance. The matrix D_n is given as $D_n=[0.1,0.1,0.1]^T$ First, we construct the FSPDS in equation (11.9) by using trapezoidal membership functions. The region of fuzzification is chosen at $y_v\in[-10,110]$ and $y_n\in[0.01,0.485]$. For the states y_v and y_n, we choose three and two membership functions, respectively. The operation points are given as $y_v=-10,50,110$ and $y_n=0.01,0.485$. Therefore, the number of fuzzy rules is $M=3\times2=6$. The bounds on the approximation errors can be obtained as $\sigma_\kappa=0$, $\sigma_f=8.657$, and $\sigma_{fw}=0.0015$. We choose the grid spacing $\Delta_x=0.5$ and $N=3$. The finite difference operator in equations (11.11) and (11.14) can then be constructed and the fuzzy spatial state space model in equation (11.19) can be obtained.

In this robust H_∞ estimator-based control design, we give the weighting matrices $R_1=I_2, R_2=0.1I_2$ and a prescribed attenuation level $\rho=1$ in equation (11.35). Finally, by solving the LMI in equation (11.42), the controller gains K_i and estimator gains G_i in equations (11.22) and (11.23) can be obtained. The robust stochastic H_∞ control performance can be computed by Monte Carlo simulation with 20 rounds as follows:

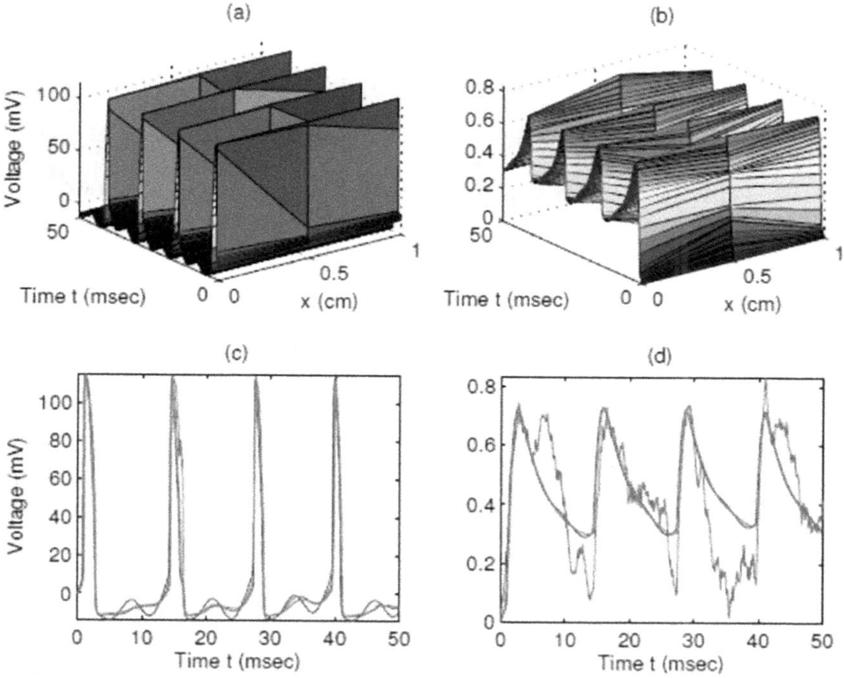

FIGURE 11.3 The stochastic H-H partial differential system with Wiener noise and external disturbances. (a) Spatiotemporal profiles of the membrane voltage y_v in equation (11.50). (b) Spatiotemporal profiles of the potassium activation y_n in equation (11.50). (c) Time profiles of the membrane voltage y_v at the grid nodes $x=0$, 0.5, 1, respectively. (d) Time profiles of the potassium activation y_n at the grid nodes $x=0$, 0.5, 1, respectively.

$$\frac{E\int_0^{50} \bar{y}(t)^T \bar{R}^T \bar{R}\bar{y}(t)\,dt}{E\int_0^{50} v(t)^T \bar{R}_3^T \bar{R}_3 v(t)\,dt} \approx 0.34136 < \rho = 1$$

The conservative of attenuation level is mainly due to the conservative in both solving LMIs equation (11.42) for H_∞ robust estimator and controller as well as the external disturbance *v(t)* in simulation being not the worst-case external disturbance in H_∞ control performance. From the simulation results at the grid nodes $x = 20$, 0.5, and 1 as shown in Figures 11.4 and 11.5, the robust stabilization performance by the proposed method is much better than the simulation results of the uncontrolled stochastic H-H model in Figure 11.3. Therefore, the simulation example has shown the robust state estimation and the robust stabilization of the proposed robust fuzzy H_∞ estimator-based stabilization design of the NSPDSs for potential practical applications to NPDSs. The effects of the external disturbance and the measurement noise in the NSPDS could be efficiently attenuated

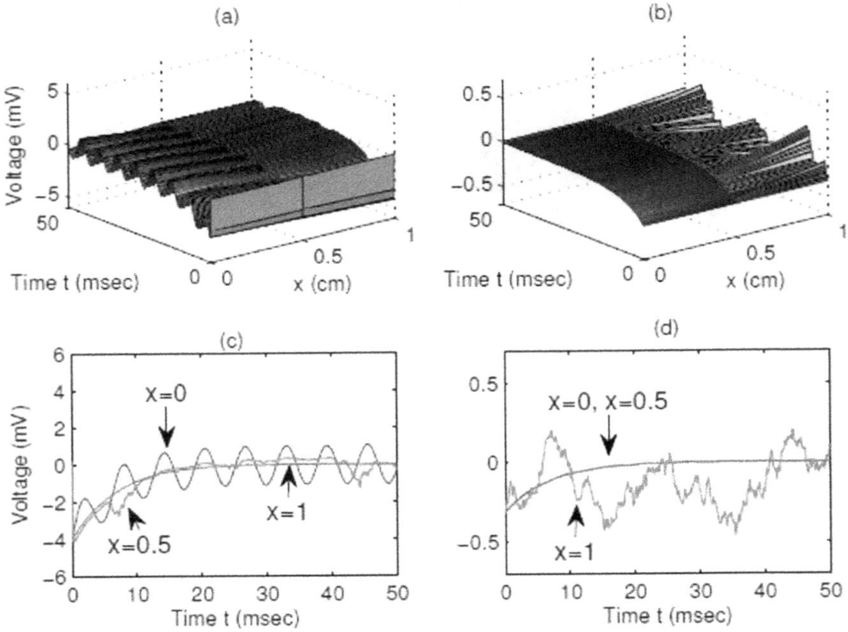

FIGURE 11.4 The controlled stochastic H-H model. (a) Spatiotemporal profiles of the membrane voltage y_v in equation (11.50). (b) Spatiotemporal profiles of the potassium activation y_n in equation (11.51). (c) Time profiles of the membrane voltage y_v at the grid nodes $x=0$, 0.5, 1, respectively. (d) Time profiles of the potassium activation y_n at the grid nodes $x=0$, 0.5, 1, respectively.

by the proposed robust H_∞ estimator-based stabilization design. From this design example, it can be observed that the proposed robust stabilization method may have potential application to the treatment of abnormal neuronal diseases such as Parkinson's disease or Alzheimer's disease, which are induced by random noise and abnormal disturbances in the neural system.

11.6 CONCLUSION

This chapter addresses the robust stabilization design problem for NSPDSs with random external disturbances and measurement noise in the spatiotemporal domain. We propose a robust stochastic stabilization design for NSPDSs based on the fuzzy approach and the method of lines. The effect of state-dependent spatiotemporal noise can be tolerated by the proposed scheme. We also successfully apply a two-dimensional H_∞ control performance of the NSPDSs under random external disturbances and measurement noise in the spatiotemporal domain. Furthermore, in order to simplify the design procedure, the two-dimensional H_∞ stabilization problem in the spatiotemporal domain is transformed to an equivalent one-dimensional H_∞ stabilization problem. A fuzzy estimator-based controller is

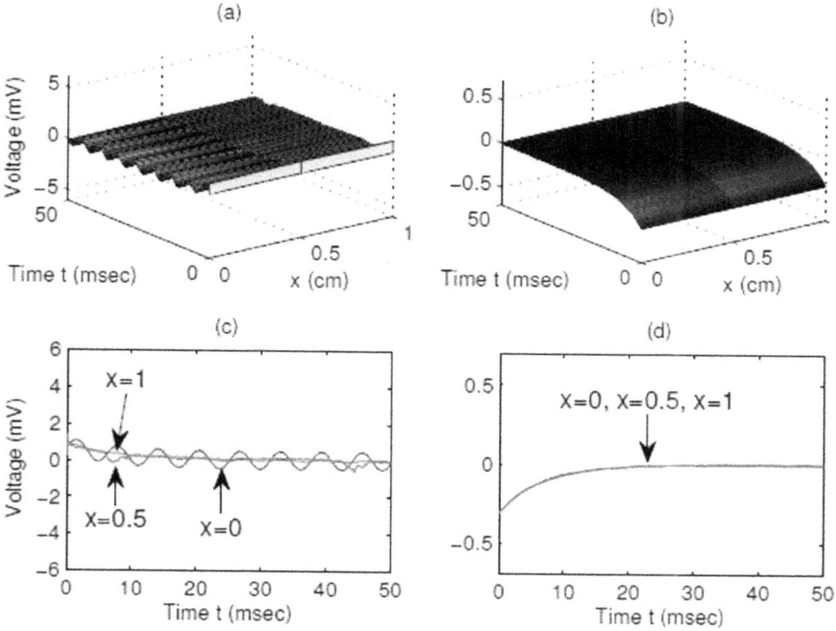

FIGURE 11.5 The results of fuzzy estimator of stochastic H-H model. (a) Spatiotemporal profiles of the membrane voltage $\hat{y}_v$ in (11.50). (b) Spatiotemporal profiles of the potassium activation $\hat{y}_n$ in equation (11.51). (c) Time profiles of the membrane voltage $\hat{y}_v$ at the grid nodes x=0, 0.5, 1, respectively. (d) Time profiles of the potassium activation $\hat{y}_n$ at the grid nodes x=0, 0.5, 1, respectively.

designed to attenuate the effect of random external disturbances and measurement noise on the desired controlled output below a prescribed level. Finally, the specification of estimator gains and controller gains for the robust stochastic H_∞ stabilization design becomes an LMI problem, which can be efficiently solved using the LMI Toolbox in MATLAB.

The proposed design method can be applied to robust stochastic stabilization design problem in many fields, e.g., heat flows, elastic waves, flexible structures, chemical engineering, biodynamic systems and so on.

12 Robust Fuzzy H_∞ Estimator-Based Stabilization Design for Nonlinear Parabolic Partial Differential Systems with Different Boundary Conditions

12.1 INTRODUCTION

System robustness is always of interest to researchers and engineers. Analysis and design for robustness is necessary to withstand potential system uncertainties and various perturbations, and can be applied in different areas, such as system control [445–457], synthetic biology [458–461], and systems biology [462,463]. In the field of system control, system robustness can be guaranteed by realizing robust control design. Consequently, many researchers have focused on resolving the robust control design problem in both the deterministic case [445–453] and the stochastic case [454–457] especially in relation to robust H_∞ control design. For example, a humanoid robot was designed to emulate human-like walking behavior [452]. The 11-degree-of-freedom robot typically encountered perturbations, including the ground reaction force, impulse disturbance, and biological electromyo noise interference. For more realistic robot walking control, a nonlinear adaptive H_∞ tracking control design was proposed. In synthetic biology [460,461], a robust H_∞ control design was proposed to systematically engineer a synthetic gene network to enable a host cell to perform new tasks in the presence of cellular perturbations from the environment. With regard to systems biology [463], a robust H_∞ observer-based tracking control was realized on a Hodgkin-Huxley nerve system, which consists of highly interconnected neurons that communicate by generating and transmitting short action potentials (short electrical pulses).

The fuzzy system [464–466] is a knowledge-based system constructed from a fuzzifier, fuzzy rule bases, a fuzzy inference engine, and a defuzzifier, and has been widely utilized in various fields. It can systematically transform knowledge

DOI: 10.1201/9781003229230-15

rule bases into a nonlinear mapping and further create a nonlinear system with an interpolation of multiple local linear systems. Mostly, the fuzzy system is employed to universally approximate a nonlinear system. Of particular note is the Takagi-Sugeno (T-S) fuzzy system [467], which has been shown to be an excellent approximation to any nonlinear system and extensively has been studied [446–448,450,452,453,461,463]. With the help of the knowledge-based fuzzy system technique [446], a fuzzy observer-based mixed H_2/H_∞ controller was developed to achieve suboptimal H_2 control performance under the H_∞ robustness constraint for T-S fuzzy systems to efficiently attenuate the effects of fuzzy approximation errors and external disturbances. In addition, based on T-S fuzzy models, the robust H_∞ observer-based control designs via the LMI scheme can be found [447,448,450]. A piecewise affine static output feedback (SOF) robust H_∞ control design was presented for a class of nonlinear systems with input constraints [9]. After constructing the T-S fuzzy affine dynamic model, the fuzzy closed-loop system was asymptotically stable with a guaranteed H_∞ performance by applying a piecewise affine SOF controller via solving a set of LMIs. However, the aforementioned works are specific only to ordinary differential systems (ODSs). Consequently, the results are not sufficient for wide application in practice because many physical systems are described by partial differential equations.

Since many real-world phenomena can be described as PDSs [468–476], studies in PDSs have become increasingly common [477–486]. In the deterministic case, robust H_∞ control designs for nonlinear parabolic PDSs with the Dirichlet boundary condition have been studied [477,478]. Using the knowledge-based fuzzy system technique and Galerkin's method [468–471], a fuzzy finite-dimensional state-space model was considered instead of a 1-dimensional nonlinear parabolic PDS. Then, the robust H_∞ stabilization control design problem can be resolved through an observed-based controller with the ability to tolerate the effect of the coupled infinite-dimensional residual subsystem and external noise [477]. An output feedback robust H_∞ reference tracking control design for a 2-dimensional nonlinear parabolic PDS with time-delay and external disturbances was developed [478] by using the finite difference method [468–471]. In this method, all states at all grid nodes of the PDS on the spatial domain are collected as a single spatial state vector; thus, the output feedback robust H_∞ reference tracking control design can be achieved by the conventional finite-dimensional T-S fuzzy control method without directly dealing with the partial derivative operator of the PDS. On the other hand, the exponential stability of the 1-dimensional nonlinear parabolic PDS with the Dirichlet boundary condition can be guaranteed [479]. By applying Lyapunov functional analysis to the parabolic PDS [479], the fuzzy controller can be constructed from the recursive LMI algorithm rather than the spatial differential linear matrix inequalities (SDLMIs). Similarly, the distributed fuzzy proportional-spatial derivative state feedback control design was proposed [480] to exponentially stabilize the 1-dimensional nonlinear parabolic PDS with the Neumann boundary condition via a recursive LMI algorithm. In addition, an optimal control design [481] and a fuzzy boundary control design [482] for 1-dimensional parabolic PDSs can be found. In relation to the stochastic case, the robust H_∞ filter design was realized for the 2-dimensional nonlinear stochastic PDS with the Dirichlet boundary condition through the finite difference method [483]. A fuzzy estimator-based controller based on the finite difference and knowledge-based fuzzy system technique was suggested to achieve a robust stochastic H_∞ estimator-based stabilization control design for 2-dimensional nonlinear stochastic PDSs [484].

However, the control designs of previous studies may be too complicated for control engineers to extensively utilize in practical applications. For example, the Galerkin's method used in [477] first leads the PDS into a fuzzy infinite-dimensional state-space model, which can be partitioned into a finite-dimensional slow controllable subsystem and a coupled infinite-dimensional residual subsystem. Then, a further processing to tolerate the effect of the infinite-dimensional fast residual subsystem is necessary in the robust control design of the finite-dimensional slow controllable subsystem, which increases the complexity of the design procedure. Taking the finite difference method [478,483,484], the complexity and H_∞ performance of robust control designs for the 2-dimensional nonlinear PDSs are related to the number of grid nodes, which induces a tradeoff problem between the grid size for the central finite difference scheme and the robust H_∞ performance. Robust control design problems, therefore, may be resolved with difficulty under some circumstances, especially when the small grid size is necessarily considered. Moreover, for stabilization control of 1-dimensional nonlinear PDSs [479,480], SDLMIs depending on both the space variables and their derivative functions are proposed. Then, recursive LMI algorithms are presented to solve the SDLMIs by using the finite difference method. Hence, a tradeoff problem between the grid size for the backward finite difference scheme and the complexity of the recursive algorithms also occurs. Besides, there is no such research that specifically proposes a general manner for the N-dimensional nonlinear parabolic PDS with either the Dirichlet or Neumann boundary conditions.

In this chapter, a new robust fuzzy H_∞ estimator-based stabilization design for a class of N-dimensional nonlinear parabolic PDSs with either the Dirichlet or Neumann boundary conditions is developed. First, an N-dimensional parabolic T-S fuzzy PDS is proposed to approximate the N-dimensional nonlinear parabolic PDS by interpolating multiple local linear systems at different operation points. Second, a robust fuzzy estimator-based controller is employed to stabilize the N-dimensional nonlinear parabolic PDS and to tolerate the effects of fuzzy approximation errors, external noise, and measurement noise. Moreover, the effects of external noise and measurement noise in the spatiotemporal domain can be attenuated below a prescribed attenuation level by the proposed robust fuzzy H_∞ estimator-based stabilization design. In order to construct the robust fuzzy estimator-based controller, the DMIs are presented in the design procedure. To solve the DMIs via the traditional algebraic matrix techniques, we utilize the divergence theorem and the Poincaré inequality [468–471] to transform the DMI problem into the BMI problem, where the Poincaré constant designed in the spatial domain of the PDS is equipped with BMIs to simplify the design procedure. Finally, the robust fuzzy H_∞ estimator-based stabilization design problem can be effectively resolved by solving a set of LMIs based on the proposed decoupled method. Additionally, the optimal robust fuzzy H_∞ estimator-based stabilization design can be realized by minimizing the noise attenuation level. A popular reaction diffusion system will be provided as a practical example to demonstrate the utility of the proposed optimal robust fuzzy H_∞ estimator-based stabilization design.

The purposes of this chapter are given as follows:

i. The chapter mainly discusses the robust fuzzy H_∞ estimator-based stabilization design for the N-dimensional nonlinear parabolic PDS and further realizes the control design without using the conventional Galerkin's method, recursive algorithms, or spatial-discretized methods, such as the finite difference method, to treat the diffusion term.
ii. With the proposed robust fuzzy H_∞ estimator-based stabilization design via the divergence theorem and the Poincaré inequality, the difficulty in treating the diffusion term of the N-dimensional nonlinear parabolic PDS is overcome with a simple and general manner; that is, we directly deal with the effect of the partial derivative on the PDS.
iii. The robust fuzzy estimator-based controller for the N-dimensional nonlinear parabolic PDS can be constructed by solving a set of LMIs instead of DMIs or BMIs by the conventional methods, and so the fuzzy controller gains and fuzzy estimator gains can be acquired simultaneously by simply using the LMI Toolbox in MATLAB. Obviously, the proposed method could significantly simplify the design procedure and save much computation time.
iv. The proposed robust fuzzy H_∞ estimator-based stabilization design can extensively serve for N-dimensional linear and nonlinear parabolic PDSs with either the Dirichlet or Neumann boundary conditions.

Notations: For convenience of analysis, we adopt the followings throughout this chapter. A symmetric matrix $P >, <, \geq,$ and ≤ 0 means that P is positive definite, negative definite, positive semidefinite, and negative semidefinite, respectively. $(\Omega;\mathbb{R}^n)$ is the Hilbert space of the n dimensional measurable function $y(x,t)\in\mathbb{R}^n$ defined on $x\in\Omega$ for any t such that $\|y(\cdot,t)\|^2_{L_2(\Omega;\mathbb{R}^n)} \triangleq \int_\Omega y(x,t)^T y(x,t)\,dx < \infty$; and $L_2(\Omega\times\mathbb{R}_+;\mathbb{R}^n)$ is the Hilbert space of the n dimensional measurable function $y(x,t)\in\mathbb{R}^n$ defined on $(x,t)\in\Omega\times\mathbb{R}_+$ such that $\|y(\cdot,t)\|^2_{L_2\left(\Omega\times\mathbb{R}_+;\mathbb{R}^n\right)} \triangleq \int_0^T\int_\Omega y(x,t)^T y(x,t)\,dx\,dt < \infty$, where T is the terminal time. Let the domain $D(\mathcal{A}) \triangleq \left\{ y(x,t)\in L_2(\Omega;\mathbb{R}^n) \middle| y(x,t), \frac{\partial y(x,t)}{\partial x}, \frac{\partial^2 y(x,t)}{\partial x^2} \in L_2\left(\Omega;\mathbb{R}^n\right)\right\}$. Throughout this chapter, π is used to denote the numerical value of a circle's circumference to its diameter; I denotes the identity matrix with appropriate dimension; and $I_n \times_n$ denotes the identity matrix of size n. Let $diag(X, Y)$ be the block diagonal matrix constructed by arbitrary matrices X and Y. For simplification, y and $\varepsilon(y)$ refer to $y(x, t)$ and $\varepsilon(y(x, t))$, respectively.

12.2 PRELIMINARIES AND PROBLEM FORMULATION

Since most phenomena can be described as PDSs [468–471], we consider a class of *N*-dimensional nonlinear parabolic PDSs in the bounded domain Ω as follows:

$$\frac{\partial y(x,t)}{\partial t} = D\Delta y(x,t) + f(y(x,t)) + Bu(x,t) + E_d d(x,t) \tag{12.1}$$

$$z(x,t) = h(y(x,t)) + E_n n(x,t) \tag{12.2}$$

where $x = [x_1, x_2, \cdots, x_N]^T \in \Omega \subset \mathbb{R}^N$ is the space variable with a smooth boundary $\partial\Omega$. $f(\cdot)$ and $h(\cdot)$ are smooth functions with $f(0)=h(0)=0$. On the spatio-temporal domain $\Omega\times[0, T]$, $y(x,t) \triangleq [y_1(x,t), y_2(x,t), \cdots, y_{n_y}(x,t)]^T \in \mathbb{R}^{n_y}$ and $y \in D(\mathcal{A})$ is the state variable; $u(x,t) \triangleq [u_1(x,t), u_2(x,t), \cdots, u_{n_u}(x,t)]^T \in \mathbb{R}^{n_u}$ is the control force to be designed with the influence matrix $B \in \mathbb{R}^{n_y \times n_u}$, and $d(x,t) \triangleq [d_1(x,t), d_2(x,t), \cdots, d_{n_d}(x,t)]^T \in L_2(\Omega \times \mathbb{R}_+; \mathbb{R}^{n_d})$ represents external noise to the PDS with the influence matrix $E_d \in \mathbb{R}^{n_y \times n_d}$. Next, $z(x,t) \triangleq [z_1(x,t), z_2(x,t), \cdots, z_{n_z}(x,t)]^T \in L_2(\Omega \times \mathbb{R}_+; \mathbb{R}^{n_n})$ represents the measurement output affected by the measurement noise $n(x,t) \triangleq [n_1(x,t), n_2(x,t), \cdots, n_{n_n}(x,t)]^T \in L_2(\Omega \times \mathbb{R}_+; \mathbb{R}^{n_n})$ with the influence matrix $E_n \in \mathbb{R}^{n_z \times n_n}$. Finally, $\Delta = \nabla^2$ is the Laplace operator for the *N*-dimensional spatial domain, and $D\Delta y(x,t)$, which represents the effect of diffusion on the system, can be written as

$$D\Delta y(x,t) = [d_{ij}]_{n_y \times n_y} \begin{bmatrix} \frac{\partial^2 y_1(x,t)}{\partial x_1^2} + \frac{\partial^2 y_1(x,t)}{\partial x_2^2} + \cdots \frac{\partial^2 y_1(x,t)}{\partial x_N^2} \\ \frac{\partial^2 y_2(x,t)}{\partial x_1^2} + \frac{\partial^2 y_2(x,t)}{\partial x_2^2} + \cdots \frac{\partial^2 y_2(x,t)}{\partial x_N^2} \\ \vdots \\ \frac{\partial^2 y_{n_y}(x,t)}{\partial x_1^2} + \frac{\partial^2 y_{n_y}(x,t)}{\partial x_2^2} + \cdots \frac{\partial^2 y_{n_y}(x,t)}{\partial x_N^2} \end{bmatrix} \tag{12.3}$$

where D is a known diffusion coefficient matrix [468–471]. For the *N*-dimensional nonlinear parabolic PDS in equations (12.1) and (12.2), it is necessary to specify both the boundary condition and the initial condition. Following physical intuition, the boundary conditions we are interested in are

i. Dirichlet boundary condition (D): $y(x,t) = 0, \forall x \in \partial\Omega$
ii. Neumann boundary condition (N): $\frac{\partial y(x,t)}{\partial \vec{n}} = 0, \forall x \in \partial\Omega$ (12.4)

where $\frac{\partial}{\partial \vec{n}}$ is the outward normal derivative, $\vec{n}$ is the normal vector to the boundary $\partial\Omega$, and the initial condition is $y(x,0) \triangleq y_0(x)$.

Using the knowledge-based fuzzy system technique, the N-dimensional nonlinear parabolic PDS in equations (12.1) and (12.2) can be efficiently approximated by interpolating multiple local linear systems at different local operation points. Therefore, the N-dimensional parabolic T-S fuzzy PDS is proposed to approximate the nonlinear parabolic PDS in equations (12.1) and (12.2) as follows:

Plant rule i:

If $\zeta_1(x,t)$ is F_{1i}, and $\zeta_2(x,t)$ is F_{2i}, and$\cdots$, and, $\zeta_{n_p}(x,t)$ is $F_{n_p i}$, then

$$\frac{\partial y(x,t)}{\partial t} = D\Delta y(x,t) + A_i y(x,t) + Bu(x,t) + E_d d(x,t) \tag{12.5}$$

$$z(x,t) = C_i y(x,t) + E_n n(x,t) \tag{12.6}$$

where $\zeta_j(x,t)$ is the jth premise variable with membership grade F_{ji} for $j = 1, 2,\ldots, n_p$, and $i = 1, 2,\ldots,$ L; $A_i \in \mathbb{R}^{n_y \times n_y}$ and $C_i \in \mathbb{R}^{n_z \times n_y}$ are the system matrices in the ith local linear PDS. Then, by using a singleton fuzzifier, product inference, and center average defuzzifier, the PDS in equations (12.1) and (12.2) can next represented by the fuzzy PDS in equations (12.5) and (12.6) in the following form [467]:

$$\frac{\partial y(x,t)}{\partial t} = \sum_{i=1}^{L} \mu_i(\zeta)\left[D\Delta y(x,t) + A_i y(x,t) + Bu(x,t) + E_d d(x,t)\right] + \varepsilon_f(y(x,t)) \tag{12.7}$$

$$z(x,t) = C_i y(x,t) + E_n n(x,t) + \varepsilon_h(y(x,t)) \tag{12.8}$$

where

$$\mu_i(\zeta) \triangleq \frac{\prod_{j=1}^{n_p} F_{ji}(\zeta_j(x,t))}{\sum_{i=1}^{L} \prod_{j=1}^{n_p} F_{ji}(\zeta_j(x,t))}$$

is the fuzzy interpolation function with the properties $0 \le \mu_i(\zeta) \le 1$ and $\sum_{i=1}^{L} \mu_i(\zeta) = 1$. The $\varepsilon_f(y(\cdot,\cdot)) \in \mathbb{R}^{n_y}$ and $\varepsilon_h(y(\cdot,\cdot)) \in \mathbb{R}^{n_h}$ are fuzzy approximation errors defined as

$$\varepsilon_f(y(\cdot,\cdot)) \triangleq f(y(\cdot,\cdot)) - \sum_{i=1}^{L} \mu_i(\zeta) A_i(y(\cdot,\cdot)) \tag{12.9}$$

$$\varepsilon_h(y(\cdot,\cdot)) \triangleq h(y(\cdot,\cdot)) - \sum_{i=1}^{L} \mu_i(\zeta) C_i(y(\cdot,\cdot)) \tag{12.10}$$

Here, $y(\cdot,\cdot)$ indicates $y(x,t)$ for any x and t. Based on the continuous functions $f(\cdot)$ and $h(\cdot)$ in equations (12.1) and (12.2) defined on the spatiotemporal domain $\Omega\times[0,T]$, the bounds of the fuzzy approximation errors $\varepsilon_f(y(\cdot,\cdot))$ and $\varepsilon_h(y(\cdot,\cdot))$ can be estimated, i.e., there exist constants σ_f and σ_h such that

$$\left\|\varepsilon_f(y(\cdot,\cdot))\right\|^2 \le \sigma_f^2\left\|y(\cdot,\cdot)\right\|^2, \text{ and } \left\|\varepsilon_h(y(\cdot,\cdot))\right\|^2 \le \sigma_h^2\left\|y(\cdot,\cdot)\right\|^2 \tag{12.11}$$

Remark 12.2.1

In general, there are two approaches to construct a fuzzy system for a nonlinear PDS in equations (12.1) and (12.2) [464–466], such as equations (12.7) and (12.8); that is, how to create the fuzzy system based on input-output data and the derivation from the known nonlinearity of the system. For the first method to create the fuzzy system for the N-dimensional nonlinear parabolic PDS in equations (12.1) and (12.2), we have the corresponding output data $f(y(x, t))$ and $h(y(x, t))$ by manufacturing input data $y(x.t)\in D(A)$. Then, after choosing appropriate premise variables and membership grades, we can obtain the system matrices A_i and C_i by employing the least-squares estimation method [472] to minimize the fuzzy approximation errors $\varepsilon_f(y(x, t))$ and $\varepsilon_h(y(x, t))$ in equations (12.9) and (12.10). As to the second method [464–466], we have the corresponding local sectors under $a_q(x) < y_q(x,t) < b_q(x), \forall t$, where $q = 1, 2,\dots, n_y$. Then, each nonlinear term of the nonlinear PDS in equations (12.1) and (12.2) can be formulated by the corresponding premise variable with the specific membership grade. Based on the knowledge-based fuzzy system technique, the system matrices A_i and C_i can be obtained. Whatever engineers choose, the bounds of the fuzzy approximation errors $\varepsilon_f(y(x, t))$ and $\varepsilon_h(y(x, t))$ in equations (12.9) and (12.10), σ_f^2 and σ_h^2, can be concluded. For the well-approximation of the nonlinear PDS in equations (12.1) and (12.2), σ_f^2 and σ_h^2 need to be as small as possible.

The robust fuzzy H_∞ stabilization design problem is to design a robust controller $u(x, t)$ to asymptotically stabilize the nonlinear parabolic PDS in equations (12.1) and (12.2) when $d(x, t)=0$ and $n(x, t)=0$ in the spatiotemporal domain, and further, to enable the N-dimensional nonlinear parabolic PDS in equations (12.1) and (12.2) to efficiently attenuate the effects of the external noise $d(x, t)$ and the measurement noise $n(x, t)$. That is, given a robust controller $u(x, t)$, the noise-free N-dimensional nonlinear parabolic PDS is asymptotically stable based on the Lyapunov stability theory in the spatiotemporal domain, and the N-dimensional nonlinear parabolic PDS equations (12.1) and (12.2) can also achieve the following robust fuzzy H_∞ noise attenuation performance in which external and measurement noises appear

$$\int_0^T \int_\Omega y^T Q_y y\, dx\, dt \le \rho^2 \int_0^T \int_\Omega v^T v\, dx\, dt \tag{12.12}$$

where $v(x,t)^T \triangleq \left[d(x,t)^T, n(x,t)^T\right]$ represents the external and measurement noises of the N-dimensional nonlinear parabolic PDS in equations (12.1) and (12.2), ρ is the noise attenuation level, and $Q_y > 0 \in \mathbb{R}^{n_y \times n_y}$ is the weighted matrix, which weights the cost of controlled output y. The physical notion of the robust fuzzy H_∞ noise attenuation performance in equation (12.12) is that the effect of the noise v on the controlled output y in the spatiotemporal domain is attenuated below a prescribed level ρ from an energy perspective, where the energy is calculated all over the spatiotemporal domain $\Omega \times [0, T]$.

12.3 ROBUST FUZZY ESTIMATOR-BASED CONTROLLER DESIGN

In this section, we propose a robust fuzzy estimator-based controller to resolve the robust fuzzy H_∞ stabilization design problem for a class of N-dimensional nonlinear parabolic PDSs in equations (12.1) and (12.2). In order to estimate the system state y via measurement output z for the robust controller u, a robust fuzzy estimator-based controller is developed. Based on the N-dimensional parabolic T-S fuzzy PDS in equations (12.7) and (12.8) with fuzzy approximation errors in equations (12.9) and (12.10), we have the following robust fuzzy estimator-based controller:

$$\frac{\partial \hat{y}(x,t)}{\partial t} = \sum_{i=1}^{L} \mu_i(\zeta)[D\Delta\hat{y}(x,t) + A_i\hat{y}(x,t) + Bu(x,t) + L_i(z(x,t) - \hat{z}(x,t))] \quad (12.13)$$

$$\hat{z}(x,t) = \sum_{j=1}^{L} \mu_j(\zeta)C_j\hat{y}(x,t) \quad (12.14)$$

$$u(x,t) = \sum_{i=1}^{L} \mu_i(\zeta)K_i\hat{y}(x,t) \quad (12.15)$$

on the spatiotemporal domain $\Omega \times [0, T]$ with boundary conditions similar to equation (12.4), so $\hat{y}(x,t) \triangleq \left[\hat{y}_1(x,t), \hat{y}_2(x,t), \ldots, \hat{y}_{n_y}(x,t)\right]^T \in \mathbb{R}^{n_y}$ vanishes on the boundary of Ω, the boundary condition (D); or the normal derivative of $\hat{y}(x,t)$ is equal to zero $\forall x \in \partial\Omega$, the boundary condition (N). Further, $K_i \in \mathbb{R}^{n_u \times n_y}$ and $L_i \in \mathbb{R}^{n_y \times n_z}$ are, respectively, the fuzzy controller gains and fuzzy estimator gains to be designed. The fuzzy estimator gains L_i, for $i = 1, 2, \ldots, L$, are designed such that the fuzzy estimation error $e(x,t) \triangleq y(x,t) - \hat{y}(x,t)$ is as small as possible. The robust fuzzy controller u in equation (12.15) with fuzzy controller gains K_i, for $i = 1, 2, \ldots, L$, is designed to asymptotically stabilize the noise-free N-dimensional nonlinear parabolic PDS and to tolerate the effects of the fuzzy approximation errors in equations (12.9) and (12.10), external noise, and measurement noise in the spatiotemporal domain.

The fuzzy estimation error PDS can be written as

$$\frac{\partial e}{\partial t}=\sum_{i=1}^{L}\sum_{j=1}^{L}\mu_i(\zeta)\mu_j(\zeta)\left[D\Delta e+(A_i-L_iC_j)e-L_iE_nn+E_dd-L_i\varepsilon_h(y)+\varepsilon_f(y)\right] \tag{12.16}$$

By combining the fuzzy estimation error PDS equation (12.16) with the fuzzy PDS in equation (12.7), the N-dimensional augmented PDS is obtained as

$$\frac{\partial \bar{y}(x,t)}{\partial t}=\sum_{i=1}^{L}\sum_{j=1}^{L}\mu_i(\zeta)\mu_j(\zeta)\left[\bar{D}\Delta\bar{y}(x,t)+\bar{A}_{ij}\bar{y}(x,t)+\bar{E}_iv(x,t)+\bar{L}_i\bar{\varepsilon}(y(x,t))\right] \tag{12.17}$$

where $\bar{y}(x,t)^T\triangleq\left[e(x,t)^T,y(x,t)^T\right]$, $\bar{\varepsilon}\left(y(x,t)\right)^T\triangleq\left[\varepsilon_f(y(x,t))^T,\varepsilon_h(y(x,t))^T\right]$, $\bar{D}\triangleq diag(D,D)$, and the remaining matrices are

$$\bar{A}_{ij}\triangleq\begin{bmatrix}A_i-L_iC_j & 0\\ -BK_i & A_i+BK_i\end{bmatrix},\ \bar{E}_i\triangleq\begin{bmatrix}E_d & -L_iE_n\\ E_d & 0\end{bmatrix},\ \bar{L}_i\triangleq\begin{bmatrix}I & -L_i\\ I & 0\end{bmatrix}$$

Accordingly, the boundary conditions for the N-dimensional augmented PDS can be easily inferred as

i. Dirichlet boundary condition (D): $\bar{y}(x,t)=0,\forall x\in\partial\Omega$

ii. Neumann boundary condition (N): $\dfrac{\partial\bar{y}(x,t)}{\partial\vec{n}_a}=0,\forall x\in\partial\Omega$ (12.18)

from the boundary conditions of the PDSs in equations (12.7) and (12.13) where the normal direction of $\bar{y}$ is denoted by $\vec{n}_a$. Considering the fuzzy estimation error e as well as the system state y, the robust fuzzy H_∞ estimator-based stabilization design is proposed. That is, by applying the robust fuzzy estimator-based controller in equations (12.13)–(12.15), the noise-free N-dimensional augmented PDS in equation (12.17) with the influences of the fuzzy approximation errors in equations (12.9) and (12.10) is asymptotically stable in the case $v=0$ based on the Lyapunov stability theory in the spatiotemporal domain, and further, the effects of the external noise d and the measurement noise n (i.e., v) on the N-dimensional augmented PDS in equation (12.17) are to be tolerated. In order to attain the asymptotical stability of the noise-free N-dimensional augmented PDS in equation (12.17), we use the Lyapunov function

$$V(\bar{y})\triangleq\int_\Omega[e^TP_ee+y^TP_yy]dx=\int_\Omega\bar{y}_e^T\bar{P}\bar{y}dx \tag{12.19}$$

where $P_e > 0, P_y > 0$, and $P \triangleq diag(P_e, P_y) > 0$. Based on the Lyapunov function in equation (12.19), the asymptotical stability of the noise-free N-dimensional augmented PDS in equation (12.17) can be guaranteed by the following theorem.

Theorem 12.3.1

Consider the N-dimensional augmented PDS in equation (12.17) with $v = 0$ and the fuzzy approximation errors equations (12.9) and (12.10). If we can find the common solutions $P > 0$, K_i, and L_i, for $i = 1, 2, \ldots, L$, from the following DMIs

$$\Delta\bar{y}^T \bar{D}^T P\bar{y} + \bar{y}^T P\bar{D}\Delta\bar{y} + \bar{y}^T \bar{A}_{ij}^T P\bar{y} + \bar{y}^T P\bar{A}_{ij}\bar{y} + \bar{\varepsilon}(y)^T \bar{L}_i^T P\bar{y} + \bar{y}^T P\bar{L}_i\varepsilon(y) < 0, \tag{12.20}$$

on the spatiotemporal domain $\Omega \times [0, T]$ for $i, j = 1, 2, \ldots, L$, then the noise-free N-dimensional augmented PDS is asymptotically stable in the spatiotemporal domain by applying the robust fuzzy estimator-based controller in equations (12.13)–(12.15).

Proof

Given the Lyapunov function in equation (12.19), we have its derivative with respect to time

$$\frac{dV(\bar{y})}{dt} = \int_\Omega \sum_{i=1}^{L}\sum_{j=1}^{L} \mu_i(\zeta)\mu_j(\zeta)[\Delta\bar{y}^T \bar{D}^T P\bar{y} + \bar{y}^T P\bar{D}\Delta\bar{y} + \bar{y}^T \bar{A}_{ij}^T P\bar{y}$$
$$+ \bar{y}^T P\bar{A}_{ij}\bar{y} + \bar{\varepsilon}(y)^T \bar{L}_i^T P\bar{y} + \bar{y}^T P\bar{L}_i\varepsilon(y)]dx \tag{12.21}$$

Hence, if we can find the common solutions $P>0$, K_i, and L_i, for $i = 1, 2, \ldots, L$, of the DMIs in equation (12.20), then the noise-free N-dimensional augmented PDS with the fuzzy approximation errors in equations (12.9) and (12.10) is asymptotically stable based on the Lyapunov stability theory in the spatiotemporal domain.

For the N-dimensional augmented PDS in equation (12.17), the robust fuzzy H_∞ noise attenuation performance in equation (12.12) should be modified as

$$\int_0^T \int_\Omega \bar{y}^T Q\bar{y}\, dx\, dt \le \rho^2 \int_0^T \int_\Omega v^T v\, dx\, dt \tag{12.22}$$

where $Q \triangleq diag(Q_e, Q_y)$ with $Q_e > 0 \in \mathbb{R}^{n_y \times n_y}$ on the spatiotemporal domain $\Omega \times [0, T]$. The weighted matrices Q_e and Q_y denote the tradeoff between the fuzzy estimation error e and the system state y. From the robust fuzzy H_∞ noise attenuation performance in equation (12.22), the effect of the noise v on the controlled output $\bar{y}$ in the spatiotemporal domain is attenuated below a prescribed level ρ from an energy perspective. That is, the effects of the external noise d and the measurement noise n on the N-dimensional augmented PDS in equation (12.17) are tolerated and attenuated under a prescribed noise attenuation level ρ on the spatiotemporal domain $\Omega \times [0,T]$. If the effect of the initial condition $\bar{y}(\cdot,0)$ of the N-dimensional augmented PDS equation (12.17) is considered, then the robust fuzzy H_∞ noise attenuation performance equation (12.22) can be modified as

$$\int_0^T \int_\Omega \bar{y}^T Q \bar{y}\, dx\, dt \le \rho^2 \int_0^T \int_\Omega v^T v\, dx\, dt + V(\bar{y}(\cdot,0)) \tag{12.23}$$

for some positive function $V(\bar{y}(\cdot,0))$. In order to achieve the robust fuzzy H_∞ noise attenuation performance in equations (12.22) or (12.23) for the N-dimensional augmented PDS in equation (12.17), we propose the following theorem to specify the fuzzy controller gains K_i and fuzzy estimator gains L_i.

Theorem 12.3.2

Consider the N-dimensional augmented PDS in (12.17) with the fuzzy approximation errors in equations (12.9) and (12.10). Given a prescribed noise attenuation level ρ*, if there exist the common solutions* $P > 0$*,* K_i*, and* L_i*, for* $i = 1, 2,\ldots, L$ *satisfying the following DMIs*

$$\begin{aligned}&\bar{y}^T Q \bar{y} + \Delta\bar{y}^T \bar{D}^T P \bar{y} + \bar{y}^T P \bar{D} \Delta \bar{y} + \bar{y}^T \bar{A}_{ij}^T P \bar{y} + \bar{y}^T P \bar{A}_{ij} \bar{y}\\ &+ v^T \bar{E}_i^T P \bar{y} + \bar{y}^T P \bar{E}_i v + \bar{\varepsilon}(y)^T \bar{L}_i^T P \bar{y} + \bar{y}^T P \bar{L}_i \varepsilon(y) - \rho^2 v^T v \le 0\end{aligned} \tag{12.24}$$

on the spatiotemporal domain $\Omega \times [0,T]$ *for* $i, j = 1, 2,\ldots, L$*, then the robust fuzzy* H_∞ *noise attenuation performance in equations (12.22) or (12.23) is guaranteed, i.e., the effects of the external noise d and the measurement noise n (i.e., v) on the N-dimensional augmented PDS in equation (12.17) can be attenuated below the noise attenuation level* ρ *in the spatiotemporal domain via the robust fuzzy estimator-based controller in equations (12.13)–(12.15).*

Proof

Given the robust fuzzy estimator-based controller u in equations (12.13)–(12.15) for the N-dimensional augmented PDS in equation (12.17), we have

$$\int_0^T \int_\Omega \bar{y}^T Q \bar{y}\, dx\, dt \le \int_0^T \int_\Omega \left[\bar{y}^T Q \bar{y} + \left(\frac{\partial \bar{y}}{\partial t}\right)^T P \bar{y} + \bar{y}^T P \left(\frac{\partial \bar{y}}{\partial t}\right) \right] dx\, dt + V(\bar{y}(\cdot,0))$$

$$= \int_0^T \int_\Omega \sum_{i=1}^{L} \sum_{j=1}^{L} \mu_i(\zeta)\mu_j(\zeta)$$

$$\times \begin{bmatrix} \bar{y}^T Q \bar{y} + \Delta\bar{y}^T \bar{D}^T P \bar{y} + \bar{y}^T P \bar{D} \Delta \bar{y} + \bar{y}^T \bar{A}_{ij}^T P \bar{y} + \bar{y}^T P \bar{A}_{ij} \bar{y} \\ + v^T \bar{E}_i^T P \bar{y} + \bar{y}^T P \bar{E}_i v + \bar{\varepsilon}(y)^T \bar{L}_i^T P \bar{y} + \bar{y}^T P \bar{L}_i \varepsilon(y) \end{bmatrix} dxdt + V(\bar{y}(\cdot,0)) \tag{12.25}$$

By adding and subtracting $\rho^2\|v(x,t)\|^2_{L_2(\Omega\times\mathbb{R}_+;\mathbb{R}^{nd+nn})}$ to equation (12.25), we obtain

$$\int_0^T\int_\Omega \bar{y}^T Q\bar{y}\,dx\,dt \le \int_0^T\int_\Omega \sum_{i=1}^{L}\sum_{j=1}^{L}\mu_i(\zeta)\mu_j(\zeta)$$

$$\times\begin{bmatrix} \bar{y}^T Q\bar{y} + \Delta\bar{y}^T\bar{D}^T P\bar{y} + \bar{y}^T P\bar{D}\Delta\bar{y} \\ +\bar{y}^T\bar{A}_{ij}^T P\bar{y} + \bar{y}^T P\bar{A}_{ij}\bar{y} + v^T\bar{E}_i^T\bar{y} + \bar{y}^T P\bar{E}_i v + \bar{\varepsilon}(y)^T\bar{L}_i^T P\bar{y} + \bar{y}^T P\bar{L}_i\bar{\varepsilon}(y) \\ -\rho^2 v^T v \end{bmatrix} dxdt$$

$$+V(\bar{y}(\cdot,0)) + \rho^2\int_0^T\int_\Omega v^T v\,dx\,dt \qquad (12.26)$$

Consequently, the robust fuzzy H_∞ noise attenuation performance in equations (12.22) or (12.23) is guaranteed if we can find the parameters $P>0$, K_i, and L_i, for $i=1, 2,\ldots, L$, such that the DMIs in equation (12.24) hold.

According to the results of Theorem 12.3.1 and 12.3.2, given the robust fuzzy estimator-based controller in equations (12.13)–(12.15) with the fuzzy controller gains K_i and fuzzy estimator gains L_i being solved from DMIs in equations (12.20) and (12.24), the fuzzy estimation error e and system state y for the noise-free N-dimensional augmented PDS in equation (12.17) converge to the equilibrium point zero while considering the influences of the fuzzy approximation errors in equations (12.9) and (12.10) and the effects of the external noise d and the measurement noise n on the N-dimensional augmented PDS in equation (12.17) can be attenuated below the noise attenuation level ρ. However, it is difficult to solve the DMIs in equations (12.20) and (12.24) owing to the effect of diffusion $D\Delta\bar{y}$ of the N-dimensional augmented PDS in equation (12.17). This effect leads to the diffusion terms $\Delta\bar{y}^T\bar{D}^T P\bar{y} + \bar{y}^T P\bar{D}\Delta\bar{y}$ in the DMIs in equations (12.20) and (12.24), which cannot be solved by using the traditional algebraic matrix techniques. In the next section, the DMI problem will be transformed into a BMI problem so that the traditional algebraic matrix techniques can be employed to realize the robust fuzzy H_∞ estimator-based stabilization design without using the conventional Galerkin's method [477], recursive algorithms [479,480], or finite difference methods [478,483,484].

12.4 ROBUST FUZZY H_∞ ESTIMATOR-BASED STABILIZATION DESIGN VIA BILINEAR MATRIX INEQUALITIES

In this section, to transform the DMI problem into a BMI problem, we directly deal with the diffusion terms $\Delta\bar{y}^T\bar{D}^T P\bar{y} + \bar{y}^T P\bar{D}\Delta\bar{y}$ in the DMIs in equations (12.20) and (12.24). Since the derivative of $\bar{y}$ with respect to x is involved, the boundary conditions in equation (12.18) of the N-dimensional augmented PDS in equation (12.17) play a crucial role. The primary idea is to reformulate the diffusion terms

$\Delta\bar{y}^T\bar{D}^TP\bar{y}+\bar{y}^TP\bar{D}\Delta\bar{y}$ in the DMIs in equations (12.20) and (12.24) into a linear combination of $\bar{y}$ with some weighted constants. Before transforming the DMI to a BMI to simplify the robust fuzzy H_∞ estimator-based stabilization design, we apply the divergence theorem and the Poincaré inequality [468–471] to deal with the partial derivative operator of PDS as follows. Initially, given $Q_d \triangleq \bar{D}^TP+P\bar{D}$, if $Q_d \geq 0$, then we have $\bar{z} \triangleq Q_d^{\frac{1}{2}}\bar{y}$. Hence, the diffusion terms can be written as

$$\Delta\bar{y}^T\bar{D}^TP\bar{y}+\bar{y}^TP\bar{D}\Delta\bar{y}=\bar{z}^T\Delta\bar{z}=\sum_{i=1}^{2n_y}\bar{z}_i\Delta\bar{z}_i \tag{12.27}$$

Next, based on the identity $\nabla\cdot(\bar{z}_i\nabla\bar{z}_i)=\bar{z}_i\Delta\bar{z}_i+|\nabla\bar{z}_i|^2$, we can obtain

$$\int_\Omega\sum_{i=1}^{2n_y}\bar{z}_i\Delta\bar{z}_i dx=\int_\Omega\sum_{i=1}^{2n_y}\nabla\cdot(\bar{z}_i\nabla\bar{z}_i)dx-\int_\Omega\sum_{i=1}^{2n_y}|\nabla\bar{z}_i|^2\,dx \tag{12.28}$$

Then, by using the divergence theorem [468–471] and the boundary conditions (D) or (N) in equations (12.18) and (12.28) can be rewritten as

$$\begin{aligned}\int_\Omega\sum_{i=1}^{2n_y}\bar{z}_i\Delta\bar{z}_i dx&=\oint_{\partial\Omega}\sum_{i=1}^{2n_y}(\bar{z}_i\nabla\bar{z}_i)\cdot\vec{n}_a ds-\int_\Omega\sum_{i=1}^{2n_y}|\nabla\bar{z}_i|^2\,dx\\&=\oint_{\partial\Omega}\sum_{i=1}^{2n_y}\bar{z}_i\frac{\partial\bar{z}_i}{\partial\vec{n}_a}ds-\int_\Omega\sum_{i=1}^{2n_y}|\nabla\bar{z}_i|^2\,dx,(by\ \nabla\bar{z}_i\cdot\partial\vec{n}_a=\frac{\partial\bar{z}_i}{\partial\vec{n}_a})\\&=-\int_\Omega\sum_{i=1}^{2n_y}|\nabla\bar{z}_i|^2\,dx\end{aligned} \tag{12.29}$$

Finally, by using the Poincaré inequality [468–471], there exists a constant $c_P>0$ such that

$$\int_\Omega\sum_{i=1}^{2n_y}\bar{z}_i\Delta\bar{z}_i dx=-\int_\Omega\sum_{i=1}^{2n_y}|\nabla\bar{z}_i|^2\,dx\leq -c_P\int_\Omega\sum_{i=1}^{2n_y}\bar{z}_i^2 dx \tag{12.30}$$

where c_P is the so-called Poincaré constant defined by the spatial domain Ω. Consequently, we have

$$\int_\Omega^T\bar{z}^T\Delta\bar{z}\,dx\leq -c_P\int_\Omega\bar{y}^TQ_d\bar{y}\,dx \tag{12.31}$$

To further this transformation, we present the following theorem.

Theorem 12.4.1

For the N-dimensional augmented PDS in (12.17) with the boundary conditions (D) or (N) in (12.18) and a prescribed noise attenuation level ρ, suppose that we can find $P>0$, K_i, L_i, for $i=1,2,\ldots,L$, and $\tau>0$ satisfying $Q_d\geq 0$ and

$$\Phi^{(ij)}(c_P)\triangleq\begin{bmatrix}\Phi_{11}(c_P)+\tau\left(\sigma_f^2\Sigma+\sigma_h^2\Sigma\right) & P\bar{L}_i & P\bar{E}_i\\ \bar{L}_i^T P & -\tau I & 0\\ \bar{E}_i^T P & 0 & -\rho^2 I\end{bmatrix}<0 \tag{12.32}$$

for $i,\ j=1,\ 2,\ldots,\ L$, where $\Phi_{11}(c_P)\triangleq Q+\bar{A}_{ij}^T P+P\bar{A}_{ij}-c_P Q_d$ with $c_P>0$ and $\Sigma\triangleq diag(0,I)$. Then, the N-dimensional augmented PDS in equation (12.17) with $v=0$ is asymptotically stable based on the Lyapunov stability theory, and further, the effects of the external noise d and the measurement noise n on the N-dimensional augmented PDS in equation (12.17) can be attenuated below the noise attenuation level ρ on the spatiotemporal domain $\Omega\times[0,T]$ by applying the robust fuzzy estimator-based controller in equations (12.13)–(12.15).

Proof

Based on the results in equations (12.25), (12.26) and (12.27)–(12.31) with $Q_d\geq 0$, we obtain

$$\begin{aligned}&\int_0^T\int_\Omega \bar{y}^T Q\bar{y}\,dx\,dt\\ &\leq\int_0^T\int_\Omega\sum_{i=1}^L\sum_{j=1}^L\mu_i(\zeta)\mu_j(\zeta)\bar{y}^T\left[Q+\bar{A}_{ij}^T P+P\bar{A}_{ij}-c_P Q_d\right]\bar{y}+v^T\bar{E}_i^T P\bar{y}+\bar{y}^T P\bar{E}_i v\\ &\quad+\bar{\varepsilon}(y)^T\bar{L}_i^T P\bar{y}+\bar{y}^T P\bar{L}_i\varepsilon(y)-\rho^2 v^T v]dxdt+V(\bar{y}(\cdot,0))+\rho^2\int_0^T\int_\Omega v^T v\,dx\,dt\end{aligned} \tag{12.33}$$

Accordingly, the robust fuzzy H_∞ noise attenuation performance in equations (12.22) or (12.23) can be guaranteed by the following inequalities:

$$\xi^T M^{(ij)}(c_P)\triangleq\xi^T\begin{bmatrix}\Phi_{11}(c_P) & P\bar{L}_i & P\bar{E}_i\\ \bar{L}_i^T P & 0 & 0\\ \bar{E}_i^T P & 0 & -\rho^2 I\end{bmatrix}\xi<0 \tag{12.34}$$

where $\xi^T \triangleq \left[\bar{y}^T, \bar{\varepsilon}(y)^T, v^T\right]$ and $\Phi_{11}(c_P) \triangleq Q + \bar{A}_{ij}^T P + P\bar{A}_{ij} - c_P Q_d$ is a function of c_P. According to the bounds of the fuzzy approximation errors in equation (12.11), we have

$$\bar{\varepsilon}(y)^T \bar{\varepsilon}(y) = \varepsilon_f(y)^T \varepsilon_f(y) + \varepsilon_h(y)^T \varepsilon_h(y) \le \sigma_f^2 y^T y + \sigma_h^2 y^T y = \sigma_f^2 \bar{y}^T \Sigma \bar{y} + \sigma_h^2 \bar{y}^T \Sigma \bar{y} \tag{12.35}$$

where $\Sigma \triangleq diag(0, I)$. In order to impose the constraint equation (12.35) on $\xi^T M^{(ij)}(c_P)\xi < 0$, by using the S-procedure [487], if there exists $\tau > 0$ such that $\Phi_{11}(c_P) < 0$ in equation (12.32) holds, then $\xi^T M^{(ij)}(c_P)\xi < 0$ and the inequality constraint in equation (12.35) are satisfied. Consequently, if we can find $P > 0$, K_i, L_i, for $i = 1, 2, \ldots, L$, and $\tau > 0$ by solving $\Phi_{11}(c_P) < 0$ in equation (12.32) and $Q_d \ge 0$, then the robust fuzzy H_∞ noise attenuation performance in equations (12.22) or (12.23) can be guaranteed. In relation to the asymptotical stability of the noise-free N-dimensional augmented PDS with the boundary conditions (D) or (N) in equation (12.18), based on the same analysis on the diffusion terms in equations (12.27)–(12.31) with $Q_d \ge 0$ and the similar results in equations (12.33)–(12.35), the common solutions $P > 0$, K_i, L_i, for $i = 1, 2, \ldots, L$, and $\tau > 0$ satisfying $Q_d \ge 0$ and

$$\Psi^{(ij)}(c_P) \triangleq \begin{bmatrix} \bar{A}_{ij}^T P + P\bar{A}_{ij} - c_P Q_d + \tau(\sigma_f^2 \Sigma + \sigma_h^2 \Sigma) & P\bar{L}_i \\ \bar{L}_i^T P & -\tau I \end{bmatrix} < 0 \tag{12.36}$$

can guarantee the asymptotical stability of the noise-free N-dimensional augmented PDS. By using the Schur complement [487], $\Phi_{11}(c_P) < 0$ and $\Psi^{(ij)}(c_P) < 0$ are equivalent to

$$\Phi_{11}(c_P) + \tau(\sigma_f^2 \Sigma + \sigma_h^2 \Sigma) + \tau^{-1} P\bar{L}_i \bar{L}_i^T P + \rho^{-2} P\bar{E}_i \bar{E}_i^T P < 0 \tag{12.37}$$

and

$$\bar{A}_{ij}^T P + P\bar{A}_{ij} - c_P Q_d + \tau(\sigma_f^2 \Sigma + \sigma_h^2 \Sigma) + \tau^{-1} P\bar{L}_i \bar{L}_i^T P < 0 \tag{12.38}$$

respectively. Obviously, the matrix inequalities in equation (12.37) imply those in equation (12.38), which mean the matrix inequalities in equation (12.37) with $Q_d \ge 0$ can also guarantee the asymptotical stability of the noise-free N-dimensional augmented PDS with the boundary conditions (D) or (N) in equation (12.18). Hence, given a prescribed noise attenuation level ρ, if we can acquire $P > 0$, fuzzy controller gains K_i, fuzzy estimator gains L_i, and $\tau > 0$ by solving the BMIs $\Phi_{11}(c_P) < 0$ in equation (12.32) with $Q_d \ge 0$, then the robust fuzzy H_∞ estimator-based stabilization design problem can be resolved easier in this way rather than that solving the DMIs equations (12.20) and (12.24).

According to the results of Theorem 12.4.1, we have successfully transformed the DMI problem into a BMI problem, which means the robust fuzzy H_∞

estimator-based stabilization design can be realized by solving the BMIs in equation (12.32) with $Q_d \geq 0$ instead of the DMIs in equations (12.20) and (12.24). However, the Poincaré constant c_P in the BMIs in equation (12.32) is defined according to the spatial domain of the corresponding PDS. That is, the constant c_P cannot be arbitrarily assigned. Moreover, the special spatial domain Ω may lead to some difficulties in obtaining the constant c_P. We therefore provide the following theorem to resolve this problem.

Theorem 12.4.2 [488–490]

Let Ω be a bounded convex set with diameter $d(\Omega)$, which is defined as $\sup\{d(x,y) | \textit{for any } x,y \in \Omega\}$. *If $Q_d \geq 0$, then the following inequality*

$$\int_{\Omega}^{T} \bar{z}^T \Delta \bar{z}\, dx \leq -\frac{\pi^2}{d(\Omega)^2} \int_{\Omega} \bar{y}^T Q_d \bar{y}\, dx \tag{12.39}$$

holds considering either the Dirichlet or Neumann boundary conditions in equation (12.18) for the Laplace operator Δ.

Proof

From [488–490], we have the lowest nonzero eigenvalue $\lambda_1^{(N)}(\Omega)$ of the Laplace operator with the boundary condition (N) in equation (12.18) satisfying

$$\lambda_1^{(N)}(\Omega) = \inf_{y \in D(\mathcal{A})\setminus\{0\}} \frac{\int_{\Omega} \|\nabla \bar{y}\|^2 dx}{\int_{\Omega} \|\bar{y}\|^2 dx}, \text{ which implies } \lambda_1^{(N)}(\Omega) \int_{\Omega} \|\bar{y}\|^2 dx \leq \int_{\Omega} \|\nabla \bar{y}\|^2 dx \tag{12.40}$$

and $\lambda_1^{(N)}(\Omega) \geq \frac{\pi^2}{d(\Omega)^2}$; thus, the inequality in equation (12.39) holds based on the similar analysis of equations (12.27)–(12.31) with $Q_d \geq 0$. Similarly, by using the results $\frac{\pi^2}{d(\Omega)^2} \leq \lambda_1^{(N)}(\Omega) \leq \lambda_1^{(D)}(\Omega)$ from studies [488–490] where $\lambda_1^{(D)}(\Omega)$ is the lowest nonzero eigenvalue for the case of the boundary condition (D) in equation (12.18), the inequality equation (12.39) holds.

Since the problem of determining the constant c_P can be resolved easily by using Theorem 12.4.2, we can give general BMIs $\Phi^{(ij)}\left(\frac{\pi^2}{d(\Omega)^2}\right) < 0$. Hence, the robust fuzzy H_∞ estimator-based stabilization design problem can be resolved by the following BMIs with $Q_d \geq 0$ instead

$$\Phi^{(ij)}\left(\frac{\pi^2}{d(\Omega)^2}\right) < 0 \tag{12.41}$$

whatever the boundary conditions in equation (12.18) are.

Remark 12.4.1

There are other methods to determine the constants c_P, i.e., to obtain different kinds of BMIs $\Phi^{(ij)}(c_P) < 0$ in equation (12.32). For example, we may consider the N-dimensional augmented PDS equation (12.17) with the boundary condition (D) in equation (12.18) and $0 \le x_i \le l_i$, where $l_i > 0$ is the length of each space variable x_i for $i = 1, 2, \ldots, N$. Then, we can get

$$\int_\Omega \sum_{i=1}^{2n_y} \bar{z}_i^2 dx \le \frac{l^2}{4N} \int_\Omega \sum_{i=1}^{2n_y} |\nabla \bar{z}_i|^2 dx \tag{12.42}$$

where $l \triangleq \max(l_1, l_2, \ldots, l_N)$ from the similar analysis in equations (12.27)–(12.31) with $Q_d \ge 0$ that follows the design procedure in [491]. Accordingly, the constant $c_P = \frac{4N}{l^2}$ is acquired via equation (12.42). Hence, based on the results in Theorem 12.4.1 and $c_P = \frac{4N}{l^2}$ in equation (12.42), the robust fuzzy H_∞ estimator-based stabilization design for the class of N-dimensional augmented PDSs in equation (12.17) with the boundary condition (D) in equation (12.18) can be realized via solving $Q_d \ge 0$ and the following BMIs:

$$\Phi^{(ij)}\left(\frac{4N}{l^2}\right) < 0 \tag{12.43}$$

According to the above analysis, $\Phi^{(ij)}(c_P)$ in equation (12.32) could be of the form $\Phi^{(ij)}\left(\frac{\pi^2}{d(\Omega)^2}\right)$ in equation (12.41) or $\Phi^{(ij)}\left(\frac{4N}{l^2}\right)$ in equation (12.43). Further, an optimal robust fuzzy H_∞ estimator-based stabilization design problem is proposed as follows

$$\begin{aligned} &\rho_0 \triangleq \min_{P>0, K_i, L_i, \tau>0} \rho, \\ &\text{subject to } \Phi^{(ij)}(c_P) < 0 \text{ in (14.32) and } Q_d \ge 0, \text{ for } i, j = 1, 2, \ldots, L. \end{aligned} \tag{12.44}$$

The minimization of the noise attenuation level ρ for the robust fuzzy H_∞ estimator-based stabilization design means the effect of the external noise v on the controlled output $\bar{y}$ within the spatiotemporal domain $\Omega \times [0, T]$ is minimized from an energy perspective.

In this section, the robust fuzzy H_∞ estimator-based stabilization design problem for the N-dimensional augmented PDS in equation (12.17) with boundary conditions (D) or (N) in equation (12.18) has been resolved by solving a BMI problem rather than a DMI problem. However, the BMIs in equation (12.32) are still very hard to solve since a BMI is not convex. Consequently, we will discuss how to transform the BMIs in equation (12.32) with $Q_d \geq 0$ into LMIs in the next section.

12.5 LMI APPROACH FOR SOLUTIONS TO ROBUST FUZZY H_∞ ESTIMATOR-BASED STABILIZATION DESIGN

According to the above, the robust fuzzy H_∞ estimator-based stabilization design for the N-dimensional augmented PDS in equation (12.17) with boundary conditions (D) or (N) in equation (12.18) can be realized by solving the BMIs $\Phi^{(ij)}(c_P) < 0$ in equation (12.32) with $Q_d \geq 0$. In this section, the robust fuzzy H_∞ estimator-based stabilization design problem is resolved by solving LMIs instead of the BMIs. To simplify the design problem, i.e., to transform the BMIs in equation (12.32) into a set of LMIs, we first give the following result.

Theorem 12.5.1

Given $X > 0$ with appropriate dimension and any constant γ, if the following inequality holds

$$\begin{bmatrix} \Omega_{11} - 2\gamma X & \gamma I \\ \gamma I & \Omega_{22} \end{bmatrix} < 0 \tag{12.45}$$

then we have $\Omega_{11} + X\Omega_{22}X < 0$.

Proof

First, by using the Schur complement, the inequality in equation (12.45) is equivalent to $\Omega_{22} < 0$ and $\Omega_{11} - 2\gamma X - \gamma^2\Omega_{22}^{-1} < 0$. As $\Omega_{22} < 0$, we have $\Omega_{11} - 2\gamma X - \gamma^2\Omega_{22}^{-1} < 0$ equivalent to $X\Omega_{22}X < -2\gamma X - \gamma^2\Omega_{22}^{-1}$. Hence, we obtain $\Omega_{11} + X\Omega_{22}X < \Omega_{11} - 2\gamma X - \gamma^2\Omega_{22}^{-1} < 0$, completing the proof.

Based on the results in Theorem 12.4.1 and 12.5.1, the robust fuzzy H_∞ estimator-based stabilization design via LMIs is provided as follows.

Theorem 12.5.2

If there exist common solutions $P_e > 0$, $X_y \triangleq P_y^{-1} > 0$, $\tau > 0$, $\eta \geq 0$, and conjunctional variables $Y_i \triangleq K_i X_y$ and $Z_i \triangleq P_e L_i$, for $i = 1, 2, \ldots, L$ satisfying the following LMIs

$$diag(D^T P_e + P_e D, X_y D^T + DX_y) \geq 0 \tag{12.46}$$

and

$$\Xi^{(ij)}(c_P) \triangleq \begin{bmatrix} \Xi_{11}(c_P) & \Xi_{12} & \Xi_{13} & \gamma I & 0 & 0 & \Xi_{17} \\ \Xi_{12}^T & -2\gamma I & 0 & 0 & \gamma I & 0 & 0 \\ \Xi_{13}^T & 0 & -2\gamma I & 0 & 0 & \gamma I & 0 \\ \gamma I & 0 & 0 & \Xi_{44}(c_P) & \Xi_{45} & \Xi_{46} & 0 \\ 0 & \gamma I & 0 & \Xi_{45}^T & -\tau I & 0 & 0 \\ 0 & 0 & \gamma I & \Xi_{46}^T & 0 & -\rho^2 I & 0 \\ \Xi_{17}^T & 0 & 0 & 0 & 0 & 0 & \Xi_{77} \end{bmatrix} < 0 \tag{12.47}$$

for $i, j = 1, 2, \ldots, L$, *where*

$$\Xi_{11}(c_P) \triangleq \begin{bmatrix} -2\gamma X_y & -Y_i^T B^T \\ -BY_i & \Theta_2(c_P) - 2\gamma X_y \end{bmatrix}, \Xi_{12} \triangleq \begin{bmatrix} 0 & 0 \\ I & 0 \end{bmatrix}, \Xi_{13} \triangleq \begin{bmatrix} 0 & 0 \\ E_d & 0 \end{bmatrix},$$

$$\Xi_{17} \triangleq \begin{bmatrix} 0 \\ X_y \end{bmatrix}, \Xi_{44}(c_P) \triangleq \begin{bmatrix} \Theta_1(c_P) & 0 \\ 0 & \Theta_3 \end{bmatrix}, \Xi_{45} \triangleq \begin{bmatrix} P_e & -Z_i \\ 0 & 0 \end{bmatrix},$$

$$\Xi_{46} \triangleq \begin{bmatrix} P_e E_d & -Z_i E_n \\ 0 & 0 \end{bmatrix}, \Xi_{77} \triangleq (-2+\eta)I,$$

with

$$\Theta_1(c_P) \triangleq A_i^T P_e - C_j^T Z_i^T + P_e A_i - Z_i C_j - c_P(D^T P_e + P_e D) + Q_e$$

$$\Theta_2(c_P) \triangleq X_y A_i^T + Y_i^T B^T + A_i X_y + BY_i - c_P(X_y D^T + DX_y)$$

$$\Theta_3 \triangleq \tau(\sigma_f^2 + \sigma_h^2)I + Q_y - \eta I$$

$c_P > 0$, *any constant* γ, *and a prescribed noise attenuation level* ρ, *then the BMIs in equation (12.32) with* $Q_d \geq 0$ *are satisfied. That is, by applying the robust fuzzy estimator- based controller in equations (12.13)–(12.15) with the fuzzy controller gains* $K_i = Y_i P_y$ *and fuzzy estimator gains* $L_i = P_e^{-1} Z_i$ *via the solutions of the LMIs in equations (12.46) and (12.47) for i, j* $= 1, 2, \ldots, L$, *the N-dimensional augmented PDS in equation (12.17) with the boundary conditions (D) or (N) in equation (12.18) and* $v = 0$ *is asymptotically stable on the spatiotemporal domain* $\Omega \times [0, T]$ *from the Lyapunov stability perspective. Furthermore, the N-dimensional* *augmented PDS in equation (12.17) with the boundary conditions (D) or (N) in equation (12.18) can achieve the robust fuzzy H_∞ noise attenuation performance in equations (12.22) or (12.23) while $v \neq 0$.*

Proof

By using the matrices defined for the N-dimensional augmented PDS in equation (12.17) as well as $P = diag(P_e, P_y)$ in equation (12.19) to expand the BMIs $\Phi^{(ij)}(c_P) < 0$ in equation (12.32) and then adding and subtracting the matrix $diag(0, \eta I, 0,0,0,0)$, we can get the following BMIs:

$$\begin{bmatrix} \Theta_1(c_P) & -K_i^T B^T P_y & P_e & -Z_i & P_e E_d & -Z_i E_n \\ P_y & \Theta_4(c_P) \pm \eta I & P_y & 0 & P_y E_d & 0 \\ P_e & P_y & -\tau I & 0 & 0 & 0 \\ -Z_i^T & 0 & 0 & -\tau I & 0 & 0 \\ E_d^T P_e & E_d^T P_y & 0 & 0 & -\rho^2 I & 0 \\ -E_n^T Z_i^T & 0 & 0 & 0 & 0 & -\rho^2 I \end{bmatrix} < 0 \qquad (12.48)$$

where

$$\Theta_4(c_P) \triangleq \left(A_i^T + K_i^T B^T\right) P_y + P_y\left(A_i + BK_i\right) - c_P\left(D^T P_y + P_y D\right) + \tau\left(\sigma_f^2 + \sigma_h^2\right) I + Q_y.$$

After pre- and post-multiplying the BMIs in equation (12.48) by the matrix $X \triangleq diag(X_y, X_y, I, I, I, I)$, we obtain

$$\Omega_{11}^{(ij)}(c_P) + X\Omega_{22}^{(ij)}(c_P)X < 0 \qquad (12.49)$$

where

$$\Omega_{11}^{(ij)}(c_P) \triangleq \begin{bmatrix} 0 & -Y_i^T B^T & 0 & 0 & 0 & 0 \\ -BY_i & \Theta_2(c_P) + X_y X_y & I & 0 & E_d & 0 \\ 0 & I & 0 & 0 & 0 & 0 \\ 0 & 0 & 0 & 0 & 0 & 0 \\ 0 & E_d^T & 0 & 0 & 0 & 0 \\ 0 & 0 & 0 & 0 & 0 & 0 \end{bmatrix} \qquad (12.50)$$

$$\Omega_{22}^{(ij)}(c_P) \triangleq \begin{bmatrix} \Theta_1(c_P) & 0 & P_e & -Z_i & P_e E_d & -Z_i E_n \\ 0 & \Theta_3 & 0 & 0 & 0 & 0 \\ P_e & 0 & -\tau I & 0 & 0 & 0 \\ -Z_i^T & 0 & 0 & -\tau I & 0 & 0 \\ E_d^T P_e & 0 & 0 & 0 & -\rho^2 I & 0 \\ -E_n^T Z_i^T & 0 & 0 & 0 & 0 & -\rho^2 I \end{bmatrix} \tag{12.51}$$

Due to the results in Theorem 12.5.1 and equations (12.50) and (12.51), if the following matrix inequalities hold

$$\begin{bmatrix} \Omega_{11}^{(ij)}(c_P) - 2\gamma X & \gamma I \\ \gamma I & \Omega_{22}^{(ij)}(c_P) \end{bmatrix} < 0 \tag{12.52}$$

then the matrix inequalities in equation (12.49) are satisfied. By using the Schur complement, the matrix inequalities in equation (12.52) can be written as

$$\begin{bmatrix} \Xi_{11}(c_P) & \Xi_{12} & \Xi_{13} & \gamma I & 0 & 0 & \Xi_{17} \\ \Xi_{12}^T & -2\gamma I & 0 & 0 & \gamma I & 0 & 0 \\ \Xi_{13}^T & 0 & -2\gamma I & 0 & 0 & \gamma I & 0 \\ \gamma I & 0 & 0 & \Xi_{44}(c_P) & \Xi_{45} & \Xi_{46} & 0 \\ 0 & \gamma I & 0 & \Xi_{45}^T & -\tau I & 0 & 0 \\ 0 & 0 & \gamma I & \Xi_{46}^T & 0 & -\rho^2 I & 0 \\ \Xi_{17}^T & 0 & 0 & 0 & 0 & 0 & -\eta^{-1} I \end{bmatrix} < 0 \tag{12.53}$$

Due to the fact that $(1-\eta)\eta^{-1}(1-\eta) \geq 0$ implies $-\eta^{-1} \leq -2+\eta$, for all $\eta \geq 0$, the LMIs in equation (12.47) imply the matrix inequalities in equation (12.53). Additionally, after pre- and postmultiplying $Q_d \geq 0$ by the matrix $diag(I, X_y)$, we obtain the LMI in equation (12.46). As a result, the BMIs in equation (12.32) with $Q_d \geq 0$ can be transformed into the LMIs in equation (12.47) with (12.46). Hence, we can construct the robust fuzzy estimator-based controller in equations (12.13)–(12.15) by solving the LMIs in equations (12.46) and (12.47) to realize the robust fuzzy H_∞ estimator-based stabilization design with the help of the LMI Toolbox in MATLAB.

Remark 12.5.1

The previous analysis can easily be developed for a class of N-dimensional linear parabolic PDSs with the boundary conditions (D) or (N). Consider the N-dimensional linear parabolic PDS

$$\frac{\partial y}{\partial t} = D\Delta y + Ay + Bu + E_d d \tag{12.54}$$

$$z = Cy + E_n n \tag{12.55}$$

where $A \in \mathbb{R}^{n_y \times n_y}$, $C \in \mathbb{R}^{n_z \times n_y}$, and the circumstances of the PDS are the same as equations (12.1) and (12.2). Following the same design methodology as earlier, the robust fuzzy H_∞ estimator-based stabilization design problem for the N-dimensional augmented PDS in the linear parabolic PDS in equations (12.54) and (12.55) can be easily resolved by finding common solutions $P_e > 0$, $X_y \triangleq P_y^{-1} > 0$, $\eta \geq 0$, and conjunctional variables $Y_i \triangleq K_i X_y$ and $Z_i \triangleq P_e L_i$, for $i = 1, 2, \ldots, L$ satisfying equation (12.46) and the following LMIs:

$$\Xi^{(i)}(c_P) \triangleq \begin{bmatrix} \Xi_{11}(c_P) & \Xi_{13} & \gamma I & 0 & \Xi_{17} \\ \Xi_{13}^T & -2\gamma I & 0 & \gamma I & 0 \\ \gamma I & 0 & \Theta_5(c_P) & \Xi_{46} & 0 \\ 0 & \gamma I & \Xi_{46}^T & -\rho^2 I & 0 \\ \Xi_{17}^T & 0 & 0 & 0 & \Xi_{77} \end{bmatrix} < 0 \tag{12.56}$$

for $i = 1, 2, \ldots, L$, with $c_P > 0$, any constant γ and a prescribed noise attenuation level ρ, where $\Theta_5(c_P) \triangleq diag(\Theta_1(c_P), (Q_y - \eta I))$, and the rest of the matrices in equation (12.56) is defined in equation (12.47) with simply adapted settings, i.e., $A_i = A$ and $C_j = C$ in matrices $\Theta_1(c_P)$ and $\Theta_2(c_P)$.

Remark 12.5.2

Many important control design problems can only be solved in terms of BMIs rather than LMIs [492–495], especially for the control design problems involving a state estimator [446,447,495–499]. However, BMIs are hard to solve because of the difficulty in solving the coupled design parameters from the BMIs, such as the coupled design parameters $P_y BK_i$ and $P_e L_i C_j$ of the BMIs in equation (12.32). For the control design purposes, in [446,496,495], the authors utilized two-step procedure methods to obtain the coupled design parameters from the corresponding BMIs in two steps. Practically, the parameters were obtained indirectly and separately by solving two sets of LMIs instead of the corresponding

BMIs. In [447], the proposed BMIs with the specific form can be transformed into the LMIs; thus, the coupled design parameters were derived via the LMI approach. However, some special matrix manipulations for the considered BMIs were necessary for the specific form. In [499], a local optimization algorithm with an iterative LMI algorithm was proposed to resolve the BMI-constrained optimization problem, such as one in equation (12.44). Basically, the suboptimal coupled design parameters in the BMI-constrained optimization problem were acquired indirectly and separately in the proposed six steps. Hence, the methods in [446,447,496–499] to solve the coupled design parameters seem conservative or not easy to use. In this chapter, we provide a simple way to deal with the difficulty in solving the coupled design parameters P_yBK_i and $P_eL_iC_j$ in equation (12.32) for the robust fuzzy H_∞ estimator-based stabilization design. By applying Theorem 12.5.1, a BMI-constrained problem like solving $\Phi^{(ij)}(c_P) < 0$ in equation (12.32) with $Q_d \geq 0$ can be easily transformed into a LMI-constrained problem like solving $\Xi^{(ij)}(c_P) < 0$ in equation (12.47) with (12.46). Thus, the conjunctional variables $Y_i \triangleq K_iX_y$ and $Z_i \triangleq P_eL_i$ can be directly and simultaneously obtained by using the LMI Toolbox in MATLAB.

Remark 12.5.3

In order to implement the robust fuzzy estimator-based controller in equations (12.13)–(12.15) in practice, large numbers of both sensors and actuators are required. Micro-electromechanical systems (MEMS) or nano-electromechanical systems (NEMS) [473–476] are therefore employed. A MEMS or NEMS device is the integration of mechanical elements, sensors, actuators, and electronics on a common silicon substrate through micro-fabrication technology. These devices or systems have the ability to sense, control, and actuate on the micro-scale or nano-scale. Additionally, the miniaturization of mass and size allows the placement of MEMS or NEMS in places where a traditional system would not fit, using physical principles that do not work at a larger scale. Furthermore, the miniaturization reduces cost by decreasing material consumption and allowing batch fabrication. Hence, MEMS or NEMS substantially increase the applicability of many control designs, including those that require large numbers of sensors and actuators. Therefore, the robust fuzzy estimator-based controller in equations (12.13)–(12.15) can be implemented in practice with the help of MEMS and NEMS.

Remark 12.5.4

The main difference in control designs between the PDSs and ODSs is the presence of the partial derivative of the system state, i.e., the Laplace operator in the spatial domain $x \in \Omega$. Hence, the effect of the partial derivative on the PDS needs to be considered, such as the effect of diffusion $D\Delta\bar{y}$ on the N-dimensional augmented PDS in equation (12.17), which leads the term $\Delta\bar{y}^T\bar{D}^TP\bar{y} + \bar{y}^TP\bar{D}\Delta\bar{y}$

in the DMIs in equations (12.20) and (12.24). Therefore, control design problems for PDSs are more difficult to resolve than those for ODSs. There are many studies that focus on control designs for parabolic PDSs [477–480,483,484]. They use different methods to deal with the effects of the partial derivative on the PDS. In [477], Galerkin's method in Section 1.2.1 has been utilized to turn the effect of diffusion on the parabolic PDSs into the infinite-dimensional expansion of the system state, e.g., the effect of diffusion $D\Delta\overline{y}$ which was turned into the infinite-dimensional expansion of the system state $\overline{y}$. It resulted in the effect of the coupled infinite-dimensional residual subsystem needed to be considered such as in Chapter 9. In relation to the methods in [34,39,40] with the help of the knowledge-based fuzzy system technique, the partial derivative of the system state of the corresponding fuzzy PDS was approximated by using the central finite difference scheme in Section 1.2.2. Next, system states at all grid nodes on spatial domain were collected as a spatial state vector and the fuzzy ODS approximation was therefore constructed such as Chapter 10. Thus, the truncation error occurs. Then, the corresponding control design in [478,483,484] was developed for the fuzzy ODS and the effect of the truncation error needed to be tolerated. For the methods in [479,480], the SDLMIs were first developed for the exponential stability of parabolic PDSs. In [479], the SDLMIs were established by mainly using the integration by parts with the Dirichlet boundary condition and Lemma 1 in [479]. Similarly, the SDLMIs in [480] were established by mainly using the integration by parts with the Neumann boundary condition and Lemma 1 in [480]. Next, for solving the SDLMIs in [479,480], the recursive LMI algorithms were proposed by employing the backward finite difference scheme without considering the truncation errors. Hence, sufficient computational resources were required. As compared with these studies [477–480,483,484], the effect of the coupled infinite-dimensional residual subsystem [477], the effects of the truncation errors [478,483,484], and the recursive LMI algorithms [479,480] can be avoided to simplify the design procedure. In our design, the robust fuzzy H∞ estimator-based stabilization design for the N-dimensional nonlinear parabolic PDS in equations (12.1) and (12.2) or the N-dimensional linear one in equation (12.54) and (12.55) can be realized by simply solving the LMIs in equations (12.46) and (12.47) or equations (12.46) and (12.56).

Finally, the optimal robust fuzzy H_∞ estimator-based stabilization design problem with the LMI constraints in equations (12.46) and (12.47) is formulated as

$$\begin{aligned}&\rho_o \triangleq \min_{P_e>0,X_y>0,Y_i,Z_i,\tau>0,\eta\geq 0} \rho \\ &\text{subject to } \Xi^{(ij)}(c_P)<0 \text{ in (12.47) and (12.46), for } i,j=1,2,\ldots,L\end{aligned} \tag{12.57}$$

Based on the above analysis, the robust fuzzy H_∞ estimator-based stabilization design for the N-dimensional nonlinear parabolic PDS with the boundary conditions (D) or (N) can be realized by applying the robust fuzzy estimator-based controller in equations (12.13)–(12.15) with the fuzzy controller gains $K_i = Y_i X^{-1}$ and fuzzy estimator gains $L_i = P^{-1}Z_i$ via solving the LMIs in equations (12.46) and (12.47). Furthermore, the optimal robust fuzzy H_∞ estimator-based stabilization design can be attained by solving the LMI-constrained optimization problem in equation (12.57). In addition, the designs for the N-dimensional linear parabolic PDS with the boundary conditions (D) or (N) in equation (12.4) can be easily concluded.

12.6 SIMULATION EXAMPLE

In this section, we present a simulation example to verify the aforementioned results. In reality, almost all phenomena are exhibited by reactions, and reaction diffusion mechanisms have been found in many areas, such as biology, geology, physics, and ecology [500–503]. Hence, the reaction diffusion system has been widely applied to model many real-world systems. In order to realize the optimal robust fuzzy H_∞ estimator-based stabilization design for a reaction diffusion system, we utilize the Belousov-Zhabotinsky (BZ) reaction system [500–503], a well-known reaction diffusion system. The BZ reaction system has been theoretically and experimentally studied, especially as a biochemical or biological reaction system. The simplified mathematical model of the BZ reaction system [500–503] can be described as

$$\frac{\partial y}{\partial t} = D\Delta y + f(y) + Bu + E_d d \tag{12.58}$$

$$z = h(y) + E_n n \tag{12.59}$$

where

$$f(y) \triangleq \begin{bmatrix} lry_2 + y_1(1 - y_1 - ry_2) \\ -my_2 - by_1y_2 \end{bmatrix}, h(y) \triangleq \begin{bmatrix} y_1 \\ y_1y_2 \end{bmatrix}, l \triangleq 10, r \triangleq 1, m \triangleq 0.25,$$

$$b \triangleq 1.25, D \triangleq I_{2\times 2}, B \triangleq I_{2\times 2}, E_d \triangleq 0.5 \times I_{2\times 2}, E_n \triangleq 0.5 \times I_{2\times 2},$$

with the system state $y \triangleq [y_1, y_2]^T$, measurement output $z \triangleq [z_1, z_2]^T$, control force $u \triangleq [u_1, u_2]^T$, measurement noise $n \triangleq [n_1, n_2]^T$, and external noise $d \triangleq [d_1, d_2]^T$. The system states y_1 and y_2 represent the chemical concentrations 1 and 2, respectively, on the spatiotemporal domain $\Omega \times [0, 10]$, where Ω is defined on $[0, 1] \times [0, 1]$ and the initial state is $y_0(x) \triangleq [0.5, 0.1]$. The measurement noises n_1 and n_2 and external noises d_1 and d_2 are the white Gaussian noises with zero mean and average power 1.5 throughout the spatiotemporal domain $\Omega \times [0, 10]$. In order to construct the 2-dimensional parabolic T-S fuzzy PDS in equations (12.5) and (12.6), we choose the premise variable $\varsigma_1 \triangleq y_1$. Moreover, we define the triangular membership functions F_{11} and F_{12} at the operating points 0.5 and 1.11493, respectively. Due to the knowledge-based fuzzy system technique [464–466], we have the fuzzy system matrices

$$A_1 \triangleq \begin{bmatrix} -0.11493 & 8.88507 \\ 0 & -1.64366 \end{bmatrix}, A_2 \triangleq \begin{bmatrix} 0.5 & 9.5 \\ 0 & -0.875 \end{bmatrix}$$

$$C_1 \triangleq \begin{bmatrix} 1 & 0 \\ 0 & 1.11493 \end{bmatrix}, C_2 \begin{bmatrix} 1 & 0 \\ 0 & 0.5 \end{bmatrix} \tag{12.60}$$

with the bounds of the fuzzy approximation errors $\sigma_f^2 \triangleq 2.25657 \times 10^{-31}$ and $\sigma_h^2 \triangleq 4.93038 \times 10^{-32}$.

After solving the optimal robust fuzzy H_∞ estimator-based stabilization design problem in equation (12.57) with $Q_e \triangleq 2 \times I_{2\times2}$, $Q_y \triangleq I_{2\times2}$, and $c_P \triangleq \dfrac{\pi^2}{d(\Omega)^2} \cong 4.93480$ chosen from the result in Theorem 12.4.2, we obtain the following results:

$$X_y \cong \begin{bmatrix} 2.7270 & -0.7970 \\ -0.7970 & 0.8356 \end{bmatrix}, P_e \cong \begin{bmatrix} 0.5721 & -0.4554 \\ -0.4554 & 3.9463 \end{bmatrix}$$

$$L_1 \cong \begin{bmatrix} 1.4466 & 0.2501 \\ 0.1993 & 0.0847 \end{bmatrix}, L_2 \cong \begin{bmatrix} 1.4586 & 0.1908 \\ 0.1997 & 0.0653 \end{bmatrix} \tag{12.61}$$

$$K_1 \cong \begin{bmatrix} -0.9159 & -0.0248 \\ -0.0909 & -1.0705 \end{bmatrix}, K_2 \cong \begin{bmatrix} -0.8726 & 0.1545 \\ 0.0086 & -0.9712 \end{bmatrix}$$

and constants $\tau \cong 289.9943$, $\eta \cong 1.7571$, and $\rho_0 \cong 0.7483$. Then, by applying the robust fuzzy estimator-based controller in equations (12.13)–(12.15) with the initial state $\hat{y}_0(x) \triangleq [0.3, 0.3]$, the simulation results for the BZ reaction system in equations (12.58) and (12.59) with the different boundary conditions (D) and (N) in equation (12.18) can be obtained. In order to clearly observe the convergence of the system trajectories on the spatial domain $\Omega = [0, 1] \times [0, 1]$, we illustrate the simulation data as samples at different time layers $t_1 = 3$, $t_2 = 6$, $t_3 = 10$, $t_4 = 0.07$, $t_5 = 0.21$, $t_6 = 10$, $t_7 = 0.29$, $t_8 = 0.93$, and $t_9 = 10$. The estimation error trajectories e_1 and e_2 for the BZ reaction system with the boundary condition (D) are shown in Figure 12.1 and those with the boundary condition (N) are in Figure 12.2. Furthermore, the states y_1 and y_2 for the system with the boundary condition (D) and those with the boundary condition (N) are shown in Figures 12.3 and 12.4, respectively. For the case of the boundary condition (D), the estimation errors e_1 and e_2, which are shown in Figure 12.1a–c and d–f, respectively, are around the equilibrium point zero at different time layers $t_1 = 3$, $t_2 = 6$, and $t_3 = 10$. That is, the states of the fuzzy estimator in equation (12.13) are sufficiently close to the states

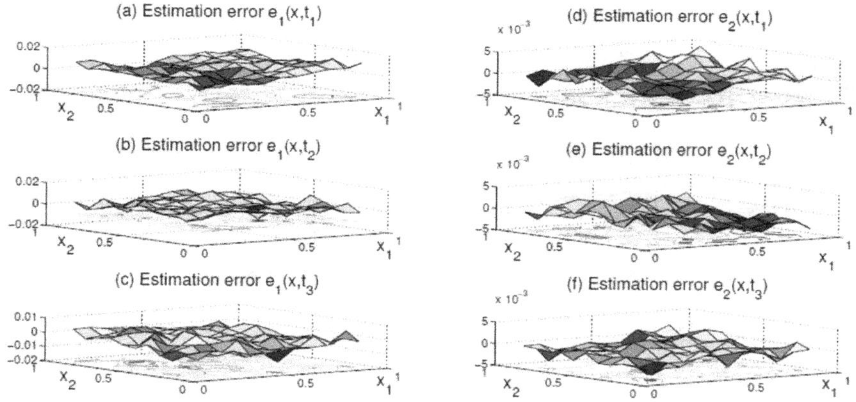

FIGURE 12.1 The estimation error trajectories $e_1(x,)$ and $e_2(x,)$ for the BZ reaction system with the boundary condition (D) under the fuzzy estimator-based controller in equations (12.13)–(12.15) on the spatial domain $\Omega=[0, 1]\times[0, 1]$: (a) spatial profile of the estimation error $e_1(x, t_1)$ at $t_1=3$, (b) spatial profile of the estimation error $e_1(x, t_2)$ at $t_2=6$, (c) spatial profile of the estimation error $e_1(x, t_3)$ at $t_3=10$, (d) spatial profile of the estimation error $e_2(x, t_1)$ at $t_1=3$, (e) spatial profile of the estimation error $e_2(x, t_2)$ at $t_2=6$, and (f) spatial profile of the estimation error $e_2(x, t_3)$ at $t_3=10$.

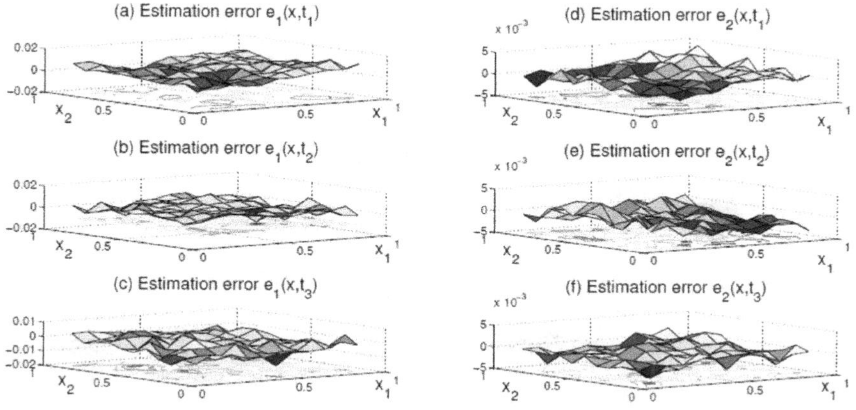

FIGURE 12.2 The estimation error trajectories $e_1(x, \cdot)$ and $e_2(x, \cdot)$ for the BZ reaction system with the boundary condition (N) under the fuzzy estimator-based controller in equations (12.13)–(12.15) on the spatial domain $\Omega=[0, 1]\times[0, 1]$: (a) spatial profile of the estimation error $e_1(x, t_1)$ at $t_1=3$, (b) spatial profile of the estimation error $e_1(x, t_2)$ at $t_2=6$, (c) spatial profile of the estimation error $e_1(x, t_3)$ at $t_3=10$, (d) spatial profile of the estimation error $e_2(x, t_1)$ at $t_1=3$, (e) spatial profile of the estimation error $e_2(x, t_2)$ at $t_2=6$, and (f) spatial profile of the estimation error $e_2(x, t_3)$ at $t_3=10$.

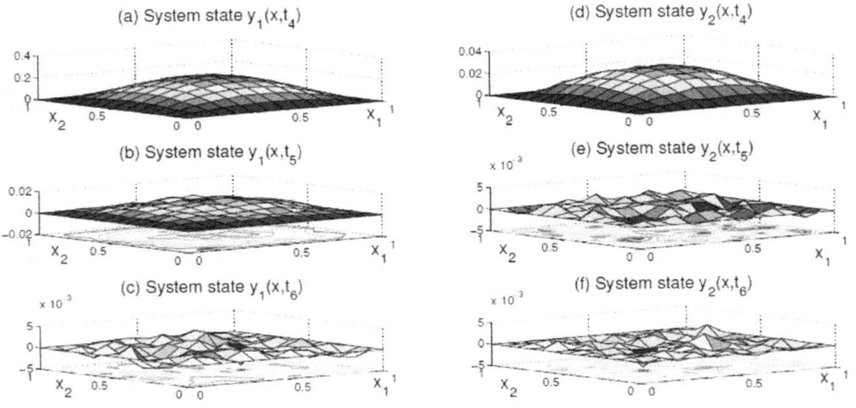

FIGURE 12.3 The system state trajectories $y_1(x, \cdot)$ and $y_2(x, \cdot)$ for the BZ reaction system with the boundary condition (D) under the fuzzy estimator-based controller in equations (12.13)–(12.15) on the spatial domain $\Omega=[0, 1]\times[0, 1]$: (a) spatial profile of the system state $y_1(x, t4)$ at $t_4=0.07$, (b) spatial profile of the system state $y_1(x, t_5)$ at $t_5=0.21$, (c) spatial profile of the system state $y_1(x, t_6)$ at $t_6=10$, (d) spatial profile of the system state $y_2(x, t_4)$ at $t_4=0.07$, (e) spatial profile of the system state $y_2(x, t_5)$ at $t_5=0.21$, and (f) spatial profile of the system state $y_2(x, t_6)$ at $t_6=10$.

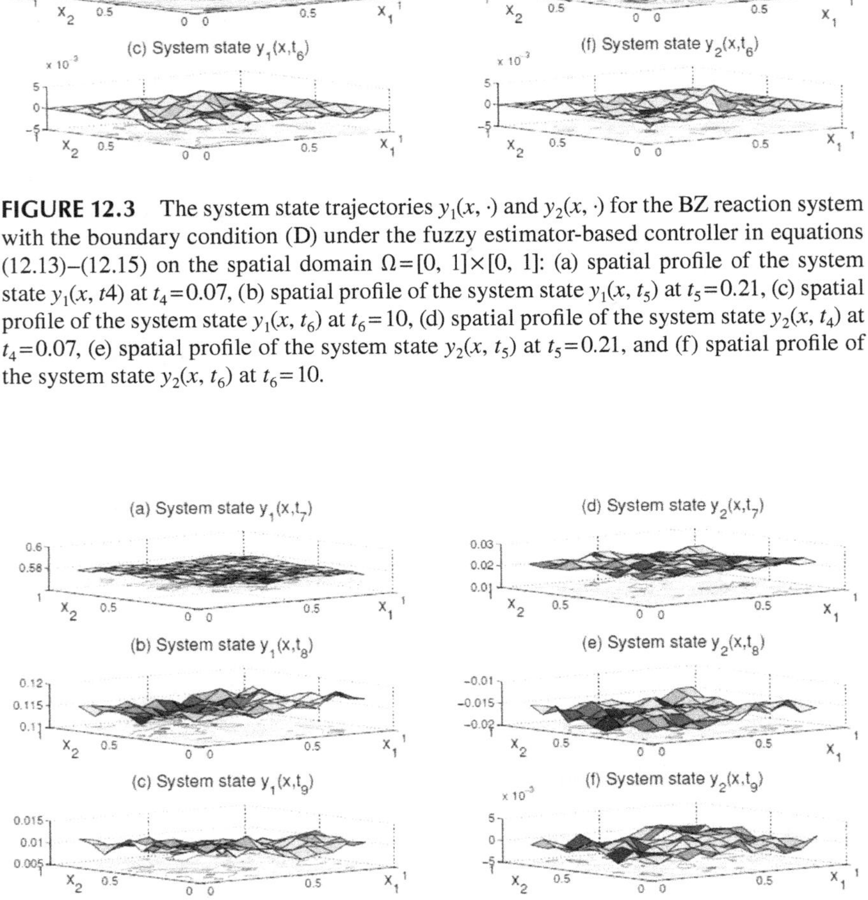

FIGURE 12.4 The system state trajectories $y_1(x, \cdot)$ and $y_2(x,)$ for the BZ reaction system with the boundary condition (N) under the fuzzy estimator-based controller in equations (12.13)–(12.15) on the spatial domain $\Omega=[0, 1]\times[0, 1]$: (a) spatial profile of the system state $y_1(x, t_7)$ at $t_7=0.29$, (b) spatial profile of the system state $y_1(x, t_8)$ at $t_8=0.93$, (c) spatial profile of the system state $y_1(x, t_9)$ at $t_9=10$, (d) spatial profile of the system state $y_2(x, t_7)$ at $t_7=0.29$, (e) spatial profile of the system state $y_2(x, t_8)$ at $t_8=0.93$, and (f) spatial profile of the system state $y_2(x, t_9)$ at $t_9=10$.

of the BZ reaction system in equation (12.58) in spite of the effects of the external noise d, the measurement noise n, and the fuzzy estimation errors on the spatio-temporal domain $\Omega\times[0, 10]$. Moreover, in Figure 12.3, the system states y_1 and y_2 converge to almost the equilibrium point zero on the spatiotemporal domain $\Omega\times[0, 10]$. In order to recognize this convergence, we plot the states y_1 and y_2 in Figure 12.3a–c and d–f, respectively, at different time layers $t_4=0.07$, $t_5=0.21$, and $t_6=10$. In the case of the boundary condition (N), the results are shown in Figures 12.2 and 12.4 in a similar way with different time layers $t_1=3$, $t_2=6$, $t_3=10$, $t_7=0.29$, $t_8=0.93$, and $t_9=10$. Based on the enormous amount of simulation data, the robust fuzzy H_∞ noise attenuation performance in equation (12.23) for the different boundary conditions (D) and (N) can be achieved as follows:

$$\begin{aligned}&(\mathrm{D}): \int_0^{10}\int_\Omega \bar{y}^T Q\bar{y}\,dx\,dt \le (1.2232\times 10^{-4})\int_0^{10}\int_\Omega v^T v\,dx\,dt + V(\bar{y}(\cdot,0))\\ &(\mathrm{N}): \int_0^{10}\int_\Omega \bar{y}^T Q\bar{y}\,dx\,dt \le (8.5\times 10^{-3})\int_0^{10}\int_\Omega v^T v\,dx\,dt + V(\bar{y}(\cdot,0))\end{aligned} \tag{12.62}$$

where $V(\bar{y}(\cdot,0)) = 49.9534$. To verify the asymptotical stability of the system without the effects of the external noise d and measurement noise n, we present Figure 12.5 for the case of the boundary condition (D) and Figure 12.6 for the case of the boundary condition (N). Figures 12.5 and 12.6 show the temporal

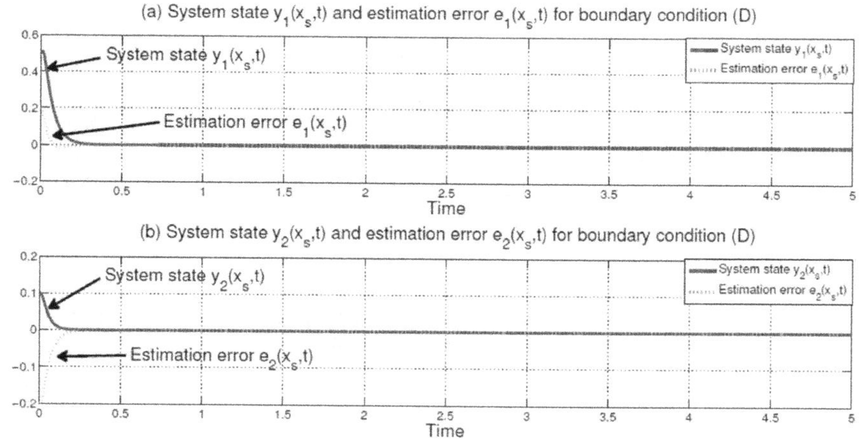

FIGURE 12.5 The trajectories of the system state $y(x_s, t)$ (solid line) and estimation error $e(x_s, t)$ (dotted line) for the BZ reaction system with the boundary condition (D) and $v=0$ under the fuzzy estimator-based controller in equations (12.13)–(12.15) where $x_s \triangleq [0.58333,0.58333]$ is randomly selected as an example: (a) temporal profile of both $y_1(x_s, t)$ and $e_1(x_s, t)$, where $t\in[0,5]$; and (b) temporal profile of both $y_2(x_s, t)$ and $e_2(x_s, t)$, where $t\in[0,5]$.

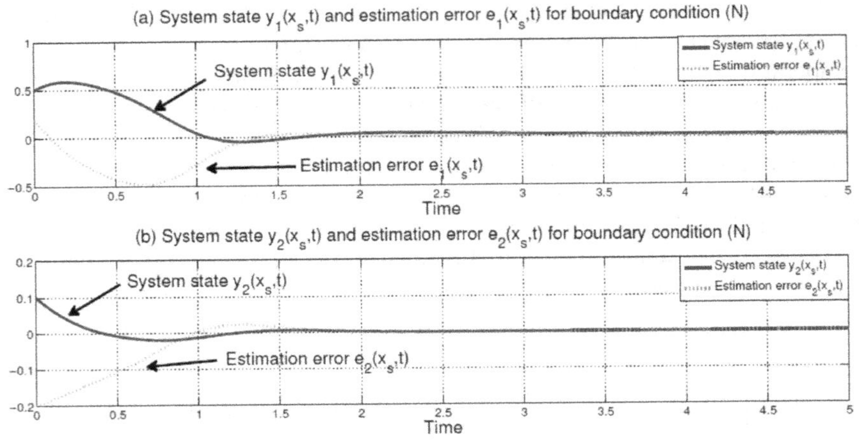

FIGURE 12.6 The trajectories of the system state $y(x_s, t)$ (solid line) and estimation error $e(x_s, t)$ (dotted line) for the BZ reaction system with the boundary condition (N) and $v=0$ under the fuzzy estimator-based controller in equations (12.13)–(12.15) where $x_s \triangleq [0.58333, 0.58333]$ is randomly selected as an example: (a) temporal profile of both $y_1(x_s, t)$ and $e_1(x_s, t)$, where $t \in [0,5]$; and (b) temporal profile of both $y_2(x_s, t)$ and $e_2(x_s, t)$, where $t \in [0,5]$.

profiles of both the system state y and estimation error e on the spatial location $x_s \triangleq [0.58333, 0.58333]$, respectively, which is randomly selected as an example. The system state y_1 and estimation error e_1 are described by Figures 12.5a and 12.6a, and the system state y_2 and estimation error e_2 are described by Figures 12.5b and 12.6b.

According to these temporal profiles, the asymptotical stability of the BZ reaction system in equations (12.58) and (12.59) and fuzzy estimation error PDS in equation (12.16) are guaranteed. As a result, both the robust fuzzy H_∞ noise attenuation performance in equation (12.62) and the asymptotical stability are satisfied for the BZ reaction system in equations (12.58) and (12.59). Thus, this simulation example demonstrates that the proposed optimal robust fuzzy H_∞ estimator-based stabilization design has potential application to the control and estimation of N-dimensional PDSs with both boundary conditions (D) and (N).

12.7 CONCLUSION

In this chapter, the robust fuzzy H_∞ estimator-based stabilization design problem for N-dimensional nonlinear parabolic PDSs with different boundary conditions (Dirichlet or Neumann boundary conditions) is resolved via a new fuzzy approach. Based on the N-dimensional parabolic Takagi-Sugeno (T-S) fuzzy PDS, the robust fuzzy estimator-based controller has been developed to stabilize the estimator-based augmented PDS, and also to attenuate the effects of external noise and measurement noise on the controlled output and estimator error in the spatiotemporal domain below a prescribed level from an energy perspective. The robust fuzzy H_∞ design specification to solve the fuzzy controller gains and fuzzy estimator gains for the robust fuzzy estimator-based controller first becomes a DMI problem. Then, the DMI problem is transformed into a BMI problem by using the divergence theorem and the Poincaré inequality via the Poincaré constant determined by the proposed computation method. Moreover, the fuzzy controller gains and fuzzy estimator gains can be obtained by solving a set of LMIs with the help of the LMI Toolbox in MATLAB. Furthermore, the minimized noise attenuation level can be attained by solving a corresponding LMI-constrained optimization problem for the optimal robust fuzzy H_∞ control design. Finally, a BZ reaction system serves as a simulation example to verify the utility of the proposed optimal robust fuzzy H_∞ estimator-based stabilization design for nonlinear parabolic PDSs. In the further work, we focus on the development of a low design cost of robust fuzzy H_∞ control for the N-dimensional parabolic PDSs; that is, the robust control design with fewer sensors and actuators.

13 Low Design-Cost Fuzzy Controllers for Robust Stabilization of Nonlinear Partial Differential Systems

13.1 INTRODUCTION

Many real phenomena are described using partial differential equations (PDEs) since PDEs express many fundamental laws of nature. Thus, PDEs frequently occur in the mathematical analysis of various problems in science and engineering [504–508]. PDSs have a wide range of applications in many branches of physics, engineering, and other sciences. Examples include PDSs of heat flows like the Fourier's equation for the diffusion of heat [508,509]; the motion of fluid substances like the Navier-Stokes equations for the motion of viscous fluids [505,508]; mechanisms with flexible structures like the Cauchy momentum equation of motion in elasticity [510–512]; electrical circuits like the Kirchhoff's equations [508,511]; quantum mechanics like the Schrödinger equation [508,513]; and chemical and biological processes [513–515]. Recently, several studies have focused on PDSs [513–528]. In general, there are three types of PDSs, parabolic, hyperbolic, and elliptic PDSs. However, several physical applications are modeled as parabolic PDSs, such as conduction of heat in bars and solids; diffusion of concentration of liquid or gaseous substances in physical chemistry or biology; diffusion of neutrons in atomic piles; telegraphic transmission in cables of low inductance or capacitance; equalization of charge in electromagnetic theory; and long wavelength electromagnetic waves in a highly conducting medium. Moreover, systems and synthetic biology, where diffusion effects should be considered, are becoming increasingly popular [529–533]. Hence, this chapter focuses on a class of N-dimensional nonlinear parabolic PDSs.

The fuzzy system [534–536], which is a knowledge-based system, uses multiple local linear systems to interpolate a nonlinear system and has been employed to analyze various kinds of nonlinear systems for many years, including both ordinary differential systems (ODSs) and PDSs. In particular, the T-S fuzzy system [537] is widely utilized and has been shown to be an excellent approximation

DOI: 10.1201/9781003229230-16

to any nonlinear system. Control designs via a fuzzy approach can be found in [531,533,538–540] for ODSs, and [516–525] for PDSs. For control design problems for ODSs, multiobjective H_2/H_∞ reference tracking control design using the fuzzy approach was proposed in order to engineer a synthetic gene circuit to be inserted into a host cell to perform new tasks in spite of the effects of external noise [531]. With the help of the knowledge-based fuzzy system technique, robust H_∞ observer-based tracking control design was realized for a Hodgkin-Huxley nerve system suffering from environmental noise [533], where the system model consists of highly interconnected neurons and each neuron communicates through generating and transmitting a short action potential (short electrical pulse). A fuzzy linear control design for nonlinear systems with optimal H_∞ robustness performance was presented in [538]. The fuzzy state-feedback and observed-based controllers were given to stabilize the nonlinear systems and to attenuate the effect of external noise on control performance to a minimum level with and without all available system states. For a class of nonlinear systems with input constraints, a piecewise affine static output feedback robust H_∞ control design was proposed to asymptotically stabilize the fuzzy closed-loop system with a guaranteed H_∞ performance based on the T-S fuzzy system technique [539]. A reliable fuzzy H_∞ controller design for uncertain active suspension systems with an actuator delay and fault based on the T-S fuzzy model approach was developed in [540]. The reliable fuzzy H_∞ controller was provided to simultaneously guarantee both the asymptotical stability and the reliable fuzzy H_∞ performance of the systems. While knowledge-based fuzzy system technologies for diverse kinds of control designs in ODSs are growing more mature, interests in fuzzy system technologies for the control designs in PDSs are raised in [516–525].

For control design problems of PDSs [516–528], various types of control designs for PDSs, with Dirichlet or Neumann boundary condition, have been investigated in recent years. The main difficulty in realizing the control designs for PDSs is that the effect of the partial derivative on PDSs needs to be considered; thus making these harder to resolve than similar issues for ODSs. Studies [516–528] use different techniques to deal with the effect of the partial derivative on the PDSs, mainly including the Galerkin's method [516,517], the central finite difference scheme [518–520], the backward finite difference scheme [521–523], and others [524–528]. In [517], the authors first acquired both the finite- and infinite-dimensional ODS approximations of the 1-dimensional parabolic PDS with the Dirichlet boundary condition by using the Galerkin's method. Then, the robust H_∞ stabilization control design was proposed for the finite-dimensional controlled subsystem to be asymptotically stable and to tolerate the effect of the coupled infinite-dimensional residual subsystem and external noise. Regarding the central finite difference scheme [518–520], the system states at all grid nodes of the T-S fuzzy parabolic PDSs with the Dirichlet boundary condition on the spatial domain were first collected as a spatial state vector and then the T-S fuzzy ODSs were constructed to approximate the PDSs with truncation errors. Next, based on the T-S fuzzy ODSs considering the truncation errors, the control designs can be achieved; that is, the output feedback robust H_∞ reference tracking control design for the 2-dimensional nonlinear parabolic PDS [518], the robust H_∞ filter design for the 2-dimensional nonlinear parabolic stochastic PDS [519], and the robust stochastic H_∞ estimator-based stabilization control design for the 2-dimensional nonlinear parabolic stochastic PDSs [520] can be realized. In [521–523], spatial differential linear matrix inequalities (SDLMIs) were individually established for the exponential stability of the 1-dimensional nonlinear hyperbolic PDS with the Dirichlet boundary condition [521], the 1-dimensional nonlinear parabolic PDS with the Dirichlet boundary condition [522], and the 1-dimensional nonlinear parabolic PDS with the Neumann boundary condition [523]. Then, a corresponding recursive LMI algorithm was developed mainly using the backward finite difference scheme to solve the SDLMI problem. Additionally, the remaining methods to deal with the effect of the partial derivative on PDSs for different kinds of control designs can be found in [524–528]. To the best of our knowledge, there is no such simple but general method which can be employed for the N-dimensional nonlinear parabolic PDSs with either the Dirichlet or Neumann boundary conditions to directly treat the effect of the partial derivative on the PDSs, i.e., the effect of diffusion on PDSs. On the other hand, for such PDSs, the design cost of any control design purpose is more expensive in general. Thus, how to reduce the design cost for wide practical use of theoretical results is an interesting topic.

In this chapter, we propose robust fuzzy H_∞ stabilization designs via the robust fuzzy full-controller, area-controller, and point-controller for N-dimensional nonlinear parabolic PDSs with either the Dirichlet or Neumann boundary conditions. The robust fuzzy H_∞ stabilization design purpose of these three robust fuzzy controllers is to stabilize the PDS and to further enable the PDS to efficiently attenuate the effect of external noise by a simple but general method utilizing the Poincaré inequality. The difference between these robust fuzzy control designs lies in their controller placement in the spatial domain. As their names suggest, the full-controller covers the entire spatial domain with the control force. With the area-controller, a specific region in the spatial domain is under control force. With the point-controller, we put control forces at only some arbitrarily assigned points. First, we establish the N-dimensional parabolic T-S fuzzy PDS with the fuzzy approximation error with the help of the knowledge-based T-S fuzzy system technique. Second, in order to realize the robust fuzzy H_∞ stabilization design via the robust fuzzy full-controller, we develop the DMI problem. Since the DMIs are too difficult to solve, they are transformed into LMIs using the Poincaré inequality, i.e., the DMI problem is transformed into a LMI problem. After resolving the LMI problem, the robust fuzzy full-controller is constructed for the robust fuzzy H_∞ stabilization design. Next, we provide two low design-cost robust fuzzy controllers, the robust fuzzy area-controller and point-controller, to realize the robust fuzzy H_∞ stabilization design via an LMI approach. Furthermore, we demonstrate the comparison of these three robust fuzzy controllers. Finally, a reaction-diffusion system is provided as a practical example to verify the performance of the proposed robust fuzzy control designs. In future, we will focus on the topic of robust control design in stochastic PDSs [541,542]. The main features of this chapter are as follows:

i. Robust fuzzy H_∞ stabilization design can be easily realized via three types of robust fuzzy controllers, including the normal design-cost robust fuzzy full-controller and two low design-cost robust fuzzy controllers called the robust fuzzy area-controller and point-controller for the N-dimensional nonlinear parabolic PDSs considering not only the Neumann boundary condition, but also the Dirichlet boundary condition.
ii. With the proposed robust fuzzy H_∞ stabilization designs of these three robust fuzzy controllers, the difficulty in treating the diffusion effect on the N-dimensional nonlinear parabolic PDSs is overcome with a simple but general method utilizing the Poincaré inequality. That is, we directly deal with the effect of the partial derivative on the PDSs without using the conventional Galerkin's method [517], spatial-discretized methods [518,519], and recursive algorithms [522,523].
iii. This chapter is the first to propose two types of low design-cost robust fuzzy controllers, the robust fuzzy area-controller and point-controller, for the robust fuzzy H_∞ stabilization design of the N-dimensional nonlinear parabolic PDSs. Furthermore, the comparison of these three robust fuzzy controllers for the robust fuzzy H_∞ stabilization designs is discussed to enable a designer to select a low cost option.

Notation: For convenience of analysis, we adopt the following notations and definitions throughout this chapter. The square of standard Euclidean norm is $\|y\|^2 \triangleq \sum_{i=1}^{n} |y_i|^2$ and the standard inner product is $\langle x, y\rangle \triangleq x^T y \triangleq \sum_{i=1}^{n} x_i y_i$ where $x = [x_1, x_2, \ldots, x_n]^T$ and $y = [y_1, y_2, \ldots, y_n]^T$. $L_2(\Omega; \mathbb{R}^n)$ is the Hilbert space of the n dimensional measurable function $y(x,t) \in \mathbb{R}^n$ defined on $x \in \Omega$ x for any t such that $\|y(x,t)\|^2_{L_2(\Omega;\mathbb{R}^n)} \triangleq \int_\Omega y^T(x,t)y(x,t)dx < \infty$; $L_2(\Omega \times \mathbb{R}_+; \mathbb{R}^n)$ is the Hilbert space of the n dimensional measurable function $y(x,t) \in \mathbb{R}^n$ defined on $(x,t) \in \Omega \times \mathbb{R}_+$ such that $\|y(x,t)\|^2_{L_2(\Omega \times \mathbb{R}_+;\mathbb{R}^n)} \triangleq \int_0^{T_f} \int_\Omega y^T(x,t)y(x,t)dxdt < \infty$, where T_f is the terminal time for the integral; and

$$D(\mathcal{A}) \triangleq \left\{ y(x,t) \in L_2(\Omega;\mathbb{R}^n) \middle| y(x,t), \frac{\partial y(x,t)}{\partial x}, \frac{\partial^2 y(x,t)}{\partial x^2} \in L_2(\Omega;\mathbb{R}^n) \right\}$$

A symmetric matrix $P >, <, \geq$, and ≤ 0 means that P is positive definite, negative definite, positive semidefinite, and negative semidefinite, respectively. Further, I is the identity matrix with appropriate dimensions, $I_n \times_n$ is the identity matrix of size n, and $diag(X, Y)$ is the block diagonal matrix constructed by arbitrary matrices X and Y. In this chapter, π is used to denote the numerical value of the ratio of a circle's circumference to its diameter; and y and $\varepsilon(y)$ refer to $y(x, t)$ and $\varepsilon(y(x, t))$, respectively, for simplification. The symbol $*$ is an ellipsis of a matrix expression, e.g.,

$$\begin{bmatrix} \Phi+(PA+A^TP) & BY \\ Y^TB^T & X \end{bmatrix} = \begin{bmatrix} \Phi+(PA+*) & BY \\ * & X \end{bmatrix}$$

where Φ, P, A, B, Y, and X are arbitrary matrices with appropriate dimensions.

13.2 SYSTEM DESCRIPTION AND PROBLEM FORMULATION

Since many physical phenomena can be formulated as parabolic PDSs [504–508], we consider a class of N-dimensional nonlinear parabolic PDSs in the bounded convex domain Ω as follows:

$$\frac{\partial y(x,t)}{\partial t} = D\Delta y(x,t) + f(y(x,t)) + Bu(x,t) \tag{13.1}$$

where $x = [x_1, x_2, ..., x_N]^T \in \Omega \subset \mathbb{R}^N$ is the space variable with a smooth boundary $\partial\Omega$ and $t \in \mathbb{R}_+$ is the time variable. On the spatiotemporal domain $\Omega \times [0, T_f]$, we have $y(x,t) \triangleq [y_1(x,t), y_2(x,t), ..., y_{n_y}(x,t)]^T \in \mathbb{R}^{n_y}$ and $y \in D(\mathcal{A})$ is the state variable; $u(x,t) \triangleq [u_1(x,t), u_2(x,t), ..., u_{n_u}(x,t)]^T \in \mathbb{R}^{n_u}$ is the control force to be designed with the influence matrix $B \in \mathbb{R}^{n_y \times n_u}$, and $f(\cdot)$ is a smooth function with $f(0) = 0$. Further, $D\Delta y(x,t)$ represents the effect of diffusion on the system where $D \in \mathbb{R}^{n_y \times n_y}$ is a known diffusion coefficient matrix and $\Delta = \nabla^2$ is the Laplace operator for the N-dimensional spatial domain [504–508]. For the N-dimensional nonlinear parabolic PDS equation (13.1), the initial condition is denoted $y(x,0) \triangleq y_0(x)$. In general, the following Dirichlet and Neumann boundary conditions are used for the PDS equation (13.1)

$$\text{(D): Dirichlet boundary condition} \quad y(x,t) = 0, \forall x \in \partial\Omega$$

$$\text{(N): Neumann boundary condition} \quad \frac{\partial y(x,t)}{\partial \vec{n}} = 0, \forall x \in \partial\Omega \tag{13.2}$$

where $\frac{\partial}{\partial \vec{n}}$ is the outward normal derivative and $\vec{n}$ is the normal vector to the boundary $\partial\Omega$.

To efficiently approximate the N-dimensional nonlinear parabolic PDS equation (13.1) by using the knowledge-based fuzzy system technique, we propose the following N-dimensional parabolic T-S fuzzy PDS

System rule i:

If $\zeta_1^S(x,t)$ is F_{1i}^S, and $\zeta_2^S(x,t)$ is F_{2i}^S, and …, and, $\zeta_{n_s}^S(x,t)$ is $F_{n_s i}^S$, then

$$\frac{\partial y(x,t)}{\partial t} = D\Delta y(x,t) + A_i y(x,t) + Bu(x,t) \tag{13.3}$$

where $\zeta_j^S(x,t)$ is the jth premise variable with membership grade F_{ji}^S for $j = 1, 2, \ldots, n_s$, and $i = 1, 2, \ldots, L_S$; $A_i \in \mathbb{R}^{n_y \times n_y}$ is the system matrix in the ith local linear PDS. The superscript and subscript S denote the fuzzy structure of the system. Then, the T-S fuzzy PDS in equation (13.3) can be represented in the following form [534–537]:

$$\frac{\partial y(x,t)}{\partial t} = \sum_{i=1}^{L_S} \mu_i^S(\zeta^S)[D\Delta y(x,t) + A_i y(x,t) + Bu(x,t)] + \varepsilon_f(y(x,t)) \tag{13.4}$$

with the fuzzy interpolation function

$$\mu_i^S(\zeta^S) \triangleq \frac{\prod_{j=1}^{n_s} F_{ji}^S(\zeta_j^S(x,t))}{\sum_{i=1}^{L_S} \prod_{j=1}^{n_s} F_{ji}^S(\zeta_j^S(x,t))} \tag{13.5}$$

Moreover, the fuzzy interpolation function has properties $0 \le \mu_i^S(\zeta^S) \le 1$ and $\sum_{i=1}^{L_S} \mu_i^S(\zeta^S) = 1$ on the spatiotemporal domain $\Omega \times [0, T_f]$. $\varepsilon_f(y(\cdot,\cdot)) \in \mathbb{R}^{n_y}$ is the fuzzy approximation error defined as

$$\varepsilon_f(y(\cdot,\cdot)) \triangleq f(y(\cdot,\cdot)) - \sum_{i=1}^{L_S} \mu_i^S(\zeta^S) A_i y(\cdot,\cdot) \tag{13.6}$$

where $y(\cdot, \cdot)$ indicates $y(x, t)$ for any x and t. Furthermore, after we construct the T-S fuzzy PDS equation (13.4), the bound of the fuzzy approximation error $\varepsilon_f(y(\cdot,\cdot))$ can be estimated, which means that there exists a constant σ_f such that

$$\|\varepsilon_f(y(\cdot,\cdot))\|^2 \le \sigma_f^2 \|(y(\cdot,\cdot))\|^2 \tag{13.7}$$

on the spatiotemporal domain $\Omega \times [0, T_f]$. Generally, the N-dimensional nonlinear parabolic PDS equation (13.1) and N-dimensional parabolic T-S fuzzy PDS equation (13.4) are called the N-dimensional noise-free nonlinear parabolic PDS and N-dimensional noise-free parabolic T-S fuzzy PDS, respectively. Additionally, when suffering from external noise, the originally noise-free PDS equation (13.1) can be adapted as follows:

$$\frac{\partial y(x,t)}{\partial t} = D\Delta y(x,t) + f(y(x,t)) + Bu(x,t) + E_n n(x,t) \tag{13.8}$$

where $n(x,t) \triangleq [n_1(x,t), n_2(x,t), \ldots, n_{n_n}(x,t)]^T \in L_2(\Omega \times \mathbb{R}_+; \mathbb{R}^{n_n})$ represents the external noise to the PDS with the influence matrix $E_n \in \mathbb{R}^{n_y \times n_n}$. Moreover, the N-dimensional parabolic T-S fuzzy PDS equation (13.8) can be represented as follows:

$$\frac{\partial y(x,t)}{\partial t}=\sum_{i=1}^{L_S}\mu_i^S(\zeta^S)[D\Delta y(x,t)+A_i y(x,t)+Bu(x,t)+E_n n(x,t)]+\varepsilon_f(y(x,t)) \tag{13.9}$$

where the circumstances of the T-S fuzzy PDS equation (13.9) are the same as equation (13.4).

Basically, the control purpose is to asymptotically stabilize the N-dimensional noise-free nonlinear parabolic PDS equation (13.1) by applying the control force u. Based on the Lyapunov stability theory [504,535,543], if there exists a Lyapunov function $V(y(x, t))>0$ and its derivative with respect to time is less than zero, then the noise-free PDS equation (13.1) is asymptotically stable. However, PDSs always suffer from external noise in practice. Since the external noise n may cause the system state y to unexpectedly oscillate or even diverge, it is important to be able to efficiently attenuate the effect of the external noise n on the N-dimensional nonlinear parabolic

PDS equation (13.8) in the spatiotemporal domain. Hence, we develop a robust fuzzy H_∞ stabilization design. The robust fuzzy H_∞ stabilization design problem is to design a robust controller u to both asymptotically stabilize the noise-free PDS equation (13.1) and enable the PDS equation (13.8) to achieve the following robust fuzzy H_∞ noise attenuation performance:

$$\int_0^{T_f}\int_\Omega y^T(x,t)Qy(x,t)\,dx\,dt\le\rho^2\int_0^{T_f}\int_\Omega n^T(x,t)n(x,t)\,dx\,dt \tag{13.10}$$

where ρ is a prescribed noise attenuation level and $Q>0\in\mathbb{R}^{n_y\times n_y}$ is the weighted matrix of the controlled output y. If the effect of the initial condition $y_0(x)$ of the PDS equation (13.8) is considered, then the robust fuzzy H_∞ noise attenuation performance equation (13.10) can be modified as

$$\int_0^{T_f}\int_\Omega y^T(x,t)Qy(x,t)\,dx\,dt\le\rho^2\int_0^{T_f}\int_\Omega n^T(x,t)n(x,t)\,dx\,dt+V(y_0(x)) \tag{13.11}$$

for some positive functions $V(y_0(x))$. The physical notion of the robust fuzzy H_∞ noise attenuation performance in equations (13.10) or (13.11) is that the effect of the external noise n on the controlled output y in the spatiotemporal domain is attenuated below a prescribed level ρ from an energy perspective, where the energy is calculated over the entire spatiotemporal domain $\Omega\times[0,T_f]$.

In order to realize the robust fuzzy H_∞ stabilization design with the robust fuzzy full-controller and further provide two low design-cost robust fuzzy controllers called the robust fuzzy area-controller and point-controller for the robust fuzzy H_∞ stabilization designs, we give these three types of robust fuzzy controllers as follows. Their sketch maps can be found in Figure 13.1. In Figure 13.1, we take the 2-dimensional square domain Ω as an example to simply illustrate how we place these three robust fuzzy controllers in practice. The

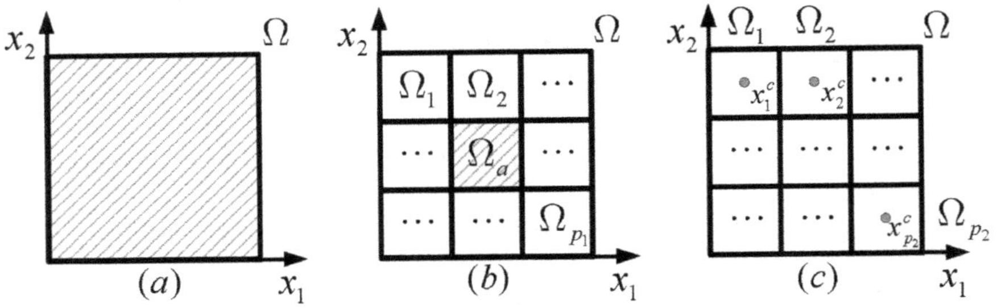

FIGURE 13.1 The sketch maps for different robust controllers $u(x, t)$ applied to the 2-dimensional bounded convex domain $[x_1,x_2] \in \Omega$ (square), for example. (a) The diagram for the robust fuzzy full-controller $u(x, t) = u_F(x, t)$. The domain with oblique line, Ω, is filled up with the control force. (b) The diagram for the robust fuzzy area-controller $u(x, t) = u_A(x, t)$. The domain with oblique line, Ω_a, is filled up with the control force. (c) The diagram for the robust fuzzy point-controller $u(x, t) = u_P(x, t)$. The points $x_k^c \subset \Omega_k$, for $k = 1, 2,\ldots, p_2$, are the spots with the control forces.

T-S fuzzy full-controller rule of the robust fuzzy full-controller $u(x, t) = u_F(x, t)$ (Figure 13.1a) is shown as follows:

Full-controller rule j:

If $\zeta_1^F(x,t)$ is F_{1j}^F, and $\zeta_2^F(x,t)$ is F_{2j}^F, and …, and, $\zeta_{nf}^F(x,t)$ is F_{nfj}^F, then

$$u_F(x,t) = K_j y(x,t) \tag{13.12}$$

where the fuzzy structure is similar to the fuzzy system rule (13.3) with simply adapted settings, i.e., the superscript and subscript S are replaced by F, which refers to the fuzzy structure for the full-controller. Accordingly, we have the following robust fuzzy full-controller:

$$u_F(x,t) = \sum_{j=1}^{L_F} \mu_j^F(\zeta^F) K_j y(x,t) \tag{13.13}$$

where $K_j \in \mathbb{R}^{n_u \times n_y}$ are the fuzzy full-controller gains, for $j = 1, 2,\ldots, L_F$, and the settings of the fuzzy interpolation function $\mu_j^F(\zeta^F)$ are similar to equation (13.5) with properties $0 \le \mu_j^F(\zeta^F) \le 1$, and $\sum_{j=1}^{L_F} \mu_j^F(\zeta^F) = 1$. In general, a large number of actuators are required to place the full-controller u_F equation (13.13) at all x on the spatial domain Ω (Figure 13.1a), which can be achieved via Micro-Electromechanical Systems (MEMS) or Nano-Electromechanical systems (NEMS) [509–512] in practice. MEMS and NEMS devices are low cost because they are manufactured with low material consumption and allow batch fabrication; even so, further reducing the cost remains an interesting topic. That is, the design cost of the robust controller u is crucial for the robust fuzzy H_∞ stabilization design of PDSs.

For the low design-cost purpose, i.e., to decrease the number of actuators, two low design-cost robust fuzzy controllers are proposed as follows, i.e., the robust fuzzy area-controller $u(x, t) = u_A(x, t)$ (Figure 13.1b), and the robust fuzzy point-controller $u(x, t) = u_P(x, t)$ (Figure 13.1c), respectively. Similar to equations (13.12) and (13.13), we have

$$u_A(x,t) = \sum_{j=1}^{L_A} \mu_j^A(\zeta^A) G(x) y(x,t) \triangleq \sum_{j=1}^{L_A} \mu_j^A(\zeta^A) G_{aj} I_{\Omega_a}(x) y(x,t) \tag{13.14}$$

$$u_P(x,t) = \sum_{j=1}^{L_P} \mu_j^P(\zeta^P) K(x) y(x,t) \triangleq \sum_{j=1}^{L_P} \mu_j^P(\zeta^P) \left[\sum_{k=1}^{p2} K_{kj} I_{x_k^c}(x) y(x,t) \right] \tag{13.15}$$

where $G(x) \triangleq G_{aj} I_{\Omega_a}(x)$ with any positive integer a, $K(x) \triangleq \sum_{k=1}^{p2} K_{kj} I_{x_k^c}(x) y(x,t)$, and the superscripts and subscripts A and P represent the fuzzy structures for the area-controller and point-controller, respectively. Additionally,

$$I_{\Omega_k}(x) \triangleq \begin{cases} 1, \text{ if } x \in \Omega_k \\ 0, \text{ if } x \notin \Omega_k \end{cases} ; \text{ and } \mathrm{I}_{x_k^c} \triangleq \begin{cases} 1, \text{ if } x = x_k^c \\ 0, \text{ if } x \neq x_k^c \end{cases} \tag{13.16}$$

are mutually distinct indicator functions of regions Ω_k, for $k = 1, 2, \ldots, a, \ldots, p_1$ and $\Omega = \Omega_1 \cup \cdots \cup \Omega_{p_1}$; and indicator functions of points x^c, for $k = 1, 2, \ldots, p_2$ and $\Omega = \Omega_1 \cup \cdots \cup \Omega_{p_2}$ with mutually distinct regions Ω_k; respectively. G_{aj} are the fuzzy area-controller gains of the ath region Ω_a, for $j = 1, 2, \ldots, L_A$ and K_{kj} are the fuzzy point-controller gains of the points x_k^c, for $k = 1, 2, \ldots, p_2$, and $j = 1, 2, \ldots, L_P$.

Remark 13.2.1

The robust fuzzy controllers u_F equation (13.13), u_A equation (13.14), and u_P equation (13.15) are developed with the help of the knowledge-based fuzzy system technique. Without loss of generality, the corresponding jth premise variable ζ_j^F, ζ_j^A, and ζ_j^P can be different for each controller. Surely, they can also be different from ζ_j^S. Further, the settings of the corresponding fuzzy systems can be varied as well. These make it possible to easily construct the three robust fuzzy controllers u_F, u_A, and u_P for the robust fuzzy H_∞ stabilization designs. Moreover, since only the information of the specific regions or spots, $I_{\Omega_a}(x)y(x,t)$ or $I_{x_k^c}(x)y(x,t)$, is needed to implement the robust fuzzy controllers u_A and u_P, the numbers of required sensors can apparently be reduced. Thus, the appropriate choices of the fuzzy structures for the system (S), area-controller (A), and point-controller (P) can reduce the amounts of required sensors, i.e., much lower design-cost robust fuzzy controllers u_A and u_P can be developed. The precise designs and discussions will be developed in future research.

For the area-controller u_A equation (13.14), the control force is applied only to the region Ω_a (Figure 13.1b), which can be arbitrarily defined by engineers based on the PDS environment. This kind of controller consumes fewer actuators, i.e., fewer MEMS or NEMS devices are required; hence, the design expense is much lower than the full-controller u_F equation (13.13). In relation to the point-controller u_P equation (13.15), we put control forces at only p_2 points, i.e., $x_k^c I \in \Omega_k$, for $k = 1, 2, \ldots, p_2$ (Figure 13.1c). Similarly, the control spots x_k^c and Ω_k can be arbitrarily assigned. Obviously, the robust fuzzy point-controller consumes the fewest actuators. Hence, in order to make these two low design-cost robust fuzzy controllers u_A and u_P realizable in practice, a sufficient diffusion effect is necessary, i.e., an adequate diffusion coefficient matrix D is required. According to these three robust fuzzy controllers, the robust fuzzy H_∞ stabilization designs and the comparison with each controller will be provided in the following sections.

13.3 ROBUST FUZZY H_∞ STABILIZATION DESIGN VIA ROBUST FUZZY FULL-CONTROLLER

In this section, we propose a robust fuzzy H_∞ stabilization design with a simple but general method for the N-dimensional parabolic T-S fuzzy PDSs equations (13.4) and (13.9) with the fuzzy approximation error equation (13.6) on the spatiotemporal domain $\Omega \times \left[0, T_f\right]$ via employing the robust fuzzy full-controller $u = u_F$ equation (13.13). To this end, we give the following theorems.

Theorem 13.3.1

Consider the N-dimensional noise-free parabolic T-S fuzzy PDS equation (13.4) with the fuzzy approximation error equation (13.6). If we can find the common solutions $P > 0$ and K_j from the following DMIs:

$$y^T PD\Delta y + y^T PA_i y + y^T PBK_j y + y^T P\varepsilon_f(y) + * < 0 \tag{13.17}$$

on the spatiotemporal domain $\Omega \times \left[0, T_f\right]$, for $i = 1, 2, \ldots, L_S$ and $j = 1, 2, \ldots, L_F$, then the noise-free T-S fuzzy PDS equation (13.4) is asymptotically stable on $\Omega \times \left[0, T_f\right]$ by applying the full-controller u_F equation (13.13).

Proof

Given the Lyapunov function

$$V(y) \triangleq \int_\Omega y^T P y dx \tag{13.18}$$

for the T-S fuzzy PDS equation (13.4) with the full-controller u_F equation (13.13), its derivative with respect to time can be rearranged as

$$\frac{dV(y)}{dt}=\int_{\Omega}\sum_{i=1}^{L_S}\sum_{j=1}^{L_F}\mu_i^S(\zeta^S)\mu_i^F(\zeta^F)$$

$$\times\left[y^T PD\Delta y+y^T PA_i y+y^T PBK_j y+y^T P\varepsilon_f(y)+*\right]dx \quad (13.19)$$

Consequently, if P and K_j, for j = 1, 2,..., L_F, satisfy the DMIs equation (13.17), then the noise-free T-S fuzzy PDS equation (13.4) with the fuzzy approximation error equation (13.6) is asymptotically stable based on the Lyapunov stability theory on the spatiotemporal domain $\Omega\times\left[0,T_f\right]$.

Theorem 13.3.2

Consider the N-dimensional parabolic T-S fuzzy PDS equation (13.9) with the fuzzy approximation error equation (13.6). Given a prescribed noise attenuation level ρ, if there exist $P>0$ and K_j satisfying the following DMIs

$$y^T Qy+\left(y^T PD\Delta y+y^T PA_i y+y^T PBK_j y+y^T P\varepsilon_f(y)+y^T PE_n n+*\right)-\rho^2 n^T n\le 0 \quad (13.20)$$

on the spatio-temporal domain $\Omega\times\left[0,T_f\right]$, for i = 1, 2,..., L_S and j = 1, 2,..., L_F, then the robust fuzzy H_∞ noise attenuation performance equations (13.10) or (13.11) is guaranteed, i.e., the effect of the external noise n on the T-S fuzzy PDS equation (13.9) can be attenuated below the noise attenuation level ρ on the spatiotemporal domain $\Omega\times\left[0,T_f\right]$ by employing the full-controller u_F equation (13.13).

Proof

Given the full-controller u_F equation (13.13) for the T-S fuzzy PDS equation (13.9) and the Lyapunov function (13.18), we have

$$\int_0^{T_f}\int_{\Omega}y^T Qy\,dx\,dt\le\int_0^{T_f}\int_{\Omega}\left[y^T Qy+\left(\frac{\partial y}{\partial t}\right)^T Py+y^T P\left(\frac{\partial y}{\partial t}\right)\right]dx\,dt+V(y_0(x))$$

$$=\int_0^{T_f}\int_{\Omega}\sum_{i=1}^{L_S}\sum_{j=1}^{L_F}\mu_i^S(\zeta^S)\mu_i^F(\zeta^F)[y^T PD\Delta y+y^T PA_i y+y^T PBK_j y$$

$$+y^T P\varepsilon_f(y)+y^T PE_n n+*]dxdt+V(y_0(x)) \quad (13.21)$$

By adding and subtracting the term $\rho^2\|n(x,t)\|^2_{L_2(\Omega\times\mathbb{R}_+;\mathbb{R}^{n_n})}$ to equation (13.21), we obtain

$$\int_0^{T_f}\int_\Omega y^T Qy\,dx\,dt \le \int_0^{T_f}\int_\Omega \sum_{i=1}^{L_S}\sum_{j=1}^{L_F}\mu_i^S(\zeta^S)\mu_i^F(\zeta^F)[y^T Qy+(y^T PD\Delta y+y^T PA_i y$$

$$+y^T PBK_j y+y^T P\varepsilon_f(y)+y^T PE_n n+*)-\rho^2 n^T n]dxdt+\rho^2\int_0^{T_f}\int_\Omega n^T n\,dx\,dt+V(y_0(x))$$

(13.22)

Accordingly, the robust fuzzy H_∞ noise attenuation performance (13.10) or (13.11) is guaranteed if there exist parameters P and K_j, for $j = 1, 2,\ldots, L_F$, such that the DMIs (13.20) hold.

However, the robust fuzzy H_∞ stabilization design problem for the T-S fuzzy PDSs equations (13.4) and (13.9) are difficult to resolve since the effect of diffusion is involved. That is, the DMIs equations (13.17) and (13.20) are difficult to solve in the presence of the diffusion terms $y^T PD\Delta y+*$. To transform the DMI problems into a LMI problem, we give the following results. First, if $Q_d \ge 0$, where $Q_d \triangleq PD+*$, we have $z \triangleq Q_d^{\frac{1}{2}} y$; thus, the diffusion terms can be written as $y^T PD\Delta y+*=z^T\Delta z=\sum_{i=1}^{n_y} z_i\Delta z_i$. Next, by using the identity $\nabla\cdot(z_i\nabla z_i)=z_i\Delta z_i+|\nabla z_i|^2$, the divergence theorem [1–5], and the boundary conditions in equation (13.2) with $\nabla z_i\cdot\vec{n}=\dfrac{\partial z_i}{\partial\vec{n}}$, we have

$$\begin{aligned}\int_\Omega\sum_{i=1}^{n_y} z_i\Delta z_i dx &= \int_\Omega\sum_{i=1}^{n_y}\nabla\cdot(z_i\nabla z_i)dx-\int_\Omega\sum_{i=1}^{n_y}|\nabla z_i|^2 dx\\ &=\oint_{\partial\Omega}\sum_{i=1}^{n_y}(z_i\nabla z_i)\cdot\vec{n}ds-\int_\Omega\sum_{i=1}^{n_y}|\nabla z_i|^2 dx\\ &=\oint_{\partial\Omega}\sum_{i=1}^{n_y} z_i\frac{\partial z_i}{\partial\vec{n}}ds-\int_\Omega\sum_{i=1}^{n_y}|\nabla z_i|^2 dx\\ &=-\int_\Omega\sum_{i=1}^{n_y}|\nabla z_i|^2 dx\end{aligned} \tag{13.23}$$

Finally, we apply the Poincaré inequality [504–508] to equation (13.23); that is, there exists a Poincaré constant $c_P>0$ defined on the spatial domain Ω such that

$$\int_\Omega\sum_{i=1}^{n_y} z_i\Delta z_i dx=-\int_\Omega\sum_{i=1}^{n_y}|\nabla z_i|^2 dx\le -c_P\int_\Omega\sum_{i=1}^{n_y} z_i^2 dx \tag{13.24}$$

Consequently, we have

$$\int_{\Omega} z^T \Delta z\, dx \leq -c_P \int_{\Omega} y^T Q_d y\, dx \tag{13.25}$$

The constant c_P can be chosen as the first nonzero eigenvalue of the Laplace operator Δ [544–551] considering the boundary conditions in equation (13.2). With the boundary condition (N), the first nonzero eigenvalue $\lambda_1^{(N)}(\Omega)$ of the Laplace operator has the property $\lambda_1^{(N)}(\Omega) \geq \frac{\pi^2}{d(\Omega)^2}$ [544–551], where $d(\Omega)$ is the diameter of Ω. Hence, we have

$$\int_{\Omega} z^T \Delta z\, dx \leq -\frac{\pi^2}{d(\Omega)^2} \int_{\Omega} y^T Q_d y\, dx \tag{13.26}$$

while $Q_d \geq 0$. In the case of the boundary condition (D), the first nonzero eigenvalue $\lambda_1^{(D)}(\Omega)$ satisfies $\lambda_1^{(D)}(\Omega) \geq \lambda_1^{(N)}(\Omega) \geq \frac{\pi^2}{d(\Omega)^2}$ [544–551]; thus, the inequality equation (13.26) holds. Accordingly, the difficulty in determining the constant c_P is easily overcome whatever the boundary conditions in equation (13.2) are. The inequality equation (13.26) makes it possible to transform the DMIs equations (13.17) and (13.20) into LMIs; hence, we give the following theorem.

Theorem 13.3.3

For the N-dimensional parabolic T-S fuzzy PDSs equations (13.4) and (13.9) with the full-controller u_F equation (13.13), the fuzzy approximation error equation (13.6), and a prescribed noise attenuation level ρ, if we can find $X \triangleq P^{-1} > 0$, Y_j, and $\tau_f > 0$ satisfying $DX + * \geq 0$ and the following LMIs:

$$\Phi_{ij}^{(F)} \triangleq \begin{bmatrix} A_i X + BY_j - \frac{\pi^2}{d(\Omega)^2} DX + * & I & E_n & X & X \\ * & -\tau_f I & 0 & 0 & 0 \\ * & * & -\rho^2 I & 0 & 0 \\ * & * & * & (-2+\tau_f)\sigma_f^{-2} I & 0 \\ * & * & * & * & -Q^{-1} \end{bmatrix} < 0 \tag{13.27}$$

for $i = 1, 2, \ldots, L_S$ and $j = 1, 2, \ldots, L_F$, where $Y_j \triangleq K_j X$, then the N-dimensional noise-free parabolic T-S fuzzy PDS equation (13.4) is asymptotically stable based on the Lyapunov stability theory, and further, the effect of the external noise n on the N-dimensional parabolic T-S fuzzy PDS equation (13.9) can be attenuated below the noise attenuation level ρ on the spatiotemporal domain $\Omega \times [0, T_f]$.

Proof

By employing the inequalities equations (13.22) and (13.26) with $Q_d \geq 0$, we get

$$\int_0^{T_f}\int_\Omega y^T Q y\,dx\,dt \leq \int_0^{T_f}\int_\Omega \sum_{i=1}^{L_S}\sum_{j=1}^{L_F}\mu_i^S(\zeta^S)\mu_i^F(\zeta^F)[y^T Q y + (y^T(PA_i + PBK_j - \frac{\pi^2}{d(\Omega)^2}PD)y + y^T P\varepsilon_f(y) + y^T PE_n n + *) - \rho^2 n^T n]dxdt + \rho^2\int_0^{T_f}\int_\Omega n^T n\,dx\,dt$$
$$+V(y_0(x)) \tag{13.28}$$

Accordingly, the robust fuzzy H_∞ noise attenuation performance equations (13.10) or (13.11) can be guaranteed by solving the common solutions $P>0$ and K_j from $Q_d \geq 0$ and the following inequalities

$$\xi^T\Phi_{ij}^{(1)}\xi \triangleq \xi^T\begin{bmatrix} Q+\left(PA_i + PBK_j - \frac{\pi^2}{d(\Omega)^2}PD + *\right) & P & PE_n \\ * & 0 & 0 \\ * & * & -\rho^2 I \end{bmatrix}\xi < 0 \tag{13.29}$$

where $\xi^T \triangleq \left[y^T, \varepsilon_f^T(y), n^T\right]$. Because the constraint of the fuzzy approximation error in equation (13.7) needs to be considered for solving $\xi^T\Phi_{ij}^{(1)}\xi < 0$ in equation (13.29), the S-procedure [552] is employed. That is, if there exist $P>0$, K_j, and $\tau_f>0$, such that the following matrix inequalities

$$\Phi_{ij}^{(2)} \triangleq \begin{bmatrix} Q+\left(PA_i + PBK_j - \frac{\pi^2}{d(\Omega)^2}PD + *\right)+\sigma_f^2\tau_f I & P & PE_n \\ * & -\tau_f I & 0 \\ * & * & -\rho^2 I \end{bmatrix} < 0 \tag{13.30}$$

hold, then $\xi^T\Phi_{ij}^{(1)}\xi < 0$ and the inequality constraint in equation (13.7) are satisfied. In addition, based on the results equations (13.19) and (13.26) with similar analysis to equations (13.28)–(13.30), the asymptotical stability of the noise-free T-S fuzzy PDS equation (13.4) can be attained by solving $Q_d \geq 0$ and the following matrix inequalities:

$$\Phi_{ij}^{(2)} \triangleq \begin{bmatrix} Q + \left(PA_i + PBK_j - \frac{\pi^2}{d(\Omega)^2} PD + * \right) + \sigma_f^2 \tau_f I & P \\ * & -\tau_f I \end{bmatrix} < 0 \quad (13.31)$$

Obviously, the matrix inequalities $\Phi_{ij}^{(2)} < 0$ in equation (13.30) imply $\Psi_{ij} < 0$ in equation (13.31), and so the asymptotical stability of the noise-free T-S fuzzy PDS equation (13.4) can also be guaranteed via solving the matrix inequalities $\Phi_{ij}^{(2)} < 0$ in equation (13.30) with $Q_d \geq 0$. To transform the matrix inequalities $\Phi_{ij}^{(2)} < 0$ in equation (13.30) into LMIs, we obtain the following via pre- and postmultiplying the matrix inequalities equation (13.30) by the matrix

$\bar{X} \triangleq diag(X, I, I)$ and using the Schur complement [49]

$$\begin{bmatrix} A_i X + BY_j - \frac{\pi^2}{d(\Omega)^2} DX + * & I & E_n & X & X \\ * & -\tau_f I & 0 & 0 & 0 \\ * & * & -\rho^2 I & 0 & 0 \\ * & * & * & -\sigma_f^{-2} \tau_f^{-1} I & 0 \\ * & * & * & * & -Q^{-1} \end{bmatrix} < 0$$

(13.32)

Similarly, we obtain DX+* ≥ 0. Due to the fact that $(1-\tau_f)\tau_f^{-1}(1-\tau_f) \geq 0$ implies $-\tau_f^{-1} \leq -2 + \tau_f, \forall \tau_f > 0$, the LMIs $\Phi_{ij}^{(F)} < 0$ in equation (13.27) imply the matrix inequalities equation (13.32). As a result, given a prescribed noise attenuation level ρ, if we can acquire $X>0$, Y_j, and $\tau_f>0$ by solving DX+* ≥ 0 and the LMIs $\Phi_{ij}^{(F)} < 0$ in equation (13.27), then the robust fuzzy H_∞ stabilization design problem via employing the full-controller u_F equation (13.13) can be resolved with the help of the LMI Toolbox in MATLAB. That is, with the full-controller u_F equation (13.13), both the asymptotical stability of the noise-free T-S fuzzy PDS equation (13.4) and the robust fuzzy H_∞ noise attenuation performance equations (13.10) or (13.11) for the T-S fuzzy PDS equation (13.9) can be achieved.

In this section, we showed that the robust fuzzy H_∞ stabilization design problem via the robust fuzzy full-controller $u = u_F$ (equation 13.13) for the N-dimensional parabolic T-S fuzzy PDSs equations (13.4) and (13.9) with the fuzzy approximation error equation (13.6) can be easily resolved via solving a LMI problem. In our design, the diffusion effect on the N-dimensional nonlinear parabolic PDSs that leads to the diffusion terms $y^T PD\Delta y + *$ can be overcome with a simple but general method through the Poincaré inequality. Therefore, for the robust fuzzy H_∞ stabilization design, the DMI problems in equations (13.17) and (13.20) are transformed into the LMI problem in equation (13.27) with $DX+* \geq 0$. Thus, the fuzzy full-controller gains $K_j = Y_jP$, for $j = 1, 2, \ldots, L_F$, can be obtained. In general, the full-controller u_F equation (13.13) is powerful, which means the control efficiency (CE) of this design is the best. The reason is that the system state $y(x, t)$ can be immediately observed and fed back all over the domain Ω for the purpose of robust control; that is, at all x on the spatial domain Ω, the state information can be utilized and the full-controller u_F equation (13.13) can promptly affect the PDSs to attain the robust fuzzy H_∞ stabilization design. However, a large number of actuators may be necessary.

13.4 LOW DESIGN-COST ROBUST FUZZY AREA-CONTROLLER FOR ROBUST FUZZY H_∞ STABILIZATION DESIGN

In order to resolve the robust fuzzy H_∞ stabilization design problem with lower design cost, we propose the robust fuzzy H_∞ stabilization design via employing the robust fuzzy area-controller $u = u_A$ equation (13.14). The concept of this design is that the control force $u_A(x,t) = \sum_{j=1}^{L_A} \mu_j^A(\zeta^A) G_{aj} y(x,t)$ on the ath region, $x \in \Omega_a$ (Figure 13.1b), contributes sufficient control effort to support the rest of the region $\Omega \backslash \Omega_a$ through an adequate diffusion to achieve the robust fuzzy H_∞ stabilization design. To solve the fuzzy area-controller gains G_{aj}, for $j = 1, 2, \ldots, L_A$, by following the analysis of the robust fuzzy H_∞ stabilization design via the full-controller u_F equation (13.13) with the adaptation $u = u_A$ equation (13.14), we have

$$\int_0^{T_f}\int_\Omega y^T Q y\,dx\,dt \leq \int_0^{T_f}\int_\Omega \sum_{i=1}^{L_S}\sum_{j=1}^{L_A} \mu_i^S(\zeta^S)\mu_i^A(\zeta^A)[y^T Q y + (y^T PA_i y + y^T PBG(x)y$$
$$-\frac{\pi^2}{d(\Omega)^2} y^T PDy + y^T P\varepsilon_f(y) + y^T PE_n n + *) - \rho^2 n^T n]dxdt + \rho^2 \int_0^{T_f}\int_\Omega n^T n\,dx\,dt$$
$$+V(y_0(x)) \tag{13.33}$$

Next, we get

$$\int_0^{T_f}\int_\Omega y^T Q y\,dx\,dt \leq \int_0^{T_f}\int_\Omega \Xi\,dx\,dt + \int_0^{T_f}\int_\Omega \sum_{j=1}^{L_A} \mu_i^A(\zeta^A)[y^T PBG_{aj} y + *]dxdt$$
$$+\rho^2 \int_0^{T_f}\int_\Omega n^T n\,dx\,dt + V(y_0(x)) \tag{13.34}$$

where

$$\Xi \triangleq \sum_{i=1}^{L_S} \mu_i^S(\zeta^S)[y^T Qy + (y^T PA_i y - \frac{\pi^2}{d(\Omega)^2} y^T PDy + y^T P\varepsilon_f(y) + y^T PE_n n + *) - \rho^2 n^T n] \tag{13.35}$$

by using the indicator function in equation (13.16) and $\Omega = \Omega_1 \cup \cdots \cup \Omega_{p_1}$. From equation (13.34), the robust fuzzy H_∞ stabilization design problem with the area-controller u_A equation (13.14) can be resolved by solving $P>0$ and the fuzzy area-controller gains G_{aj}, for $j = 1, 2,\ldots, L_A$ from the following inequality:

$$\Xi_A \triangleq \int_0^{T_f} \int_\Omega \Xi dx\, dt + \int_0^{T_f} \int_\Omega \sum_{j=1}^{L_A} \mu_i^A(\zeta^A)[y^T PBG_{aj} y + *]dxdt < 0 \tag{13.36}$$

with $Q_d \geq 0$. However, since two terms in Ξ_A cover two areas, one on Ω and the other on Ω_a, it is very difficult to obtain G_{aj} to construct the area-controller u_A via directly solving the inequality $\Xi_A < 0$ in equation (13.36). Hence, we propose a more conservative method to indirectly solve G_{aj}. To this end, we have the following results.

Theorem 13.4.1

Consider the N-dimensional parabolic T-S fuzzy PDSs equations (13.4) and (13.9) with the area-controller u_A equation (13.14), the fuzzy approximation error equation (13.6), and a prescribed noise attenuation level ρ. If there exist $X \triangleq P^{-1} > 0$, Z_{kj}, and $\tau_a > 0$ satisfying $DX+* \geq 0$ and the following LMIs:

$$\Phi_{ijk}^{(A)} \triangleq \begin{bmatrix} A_i X + BZ_{kj} - \frac{\pi^2}{d(\Omega)^2} DX + * & I & E_n & X & X \\ * & -\tau_a I & 0 & 0 & 0 \\ * & * & -\rho^2 I & 0 & 0 \\ * & * & * & (-2+\tau_a)\sigma_f^{-2} I & 0 \\ * & * & * & * & -Q^{-1} \end{bmatrix} < 0 \tag{13.37}$$

for $i = 1, 2,\ldots, L_S$, $j = 1, 2,\ldots, L_A$, and $k = 1, 2,\ldots, a, \ldots, p_1$; as well as

$$(BZ_{kj} + *) \geq 0, \; for\; j = 1,2,\ldots,L_A \text{ and for all } k \text{ except for } k = a \tag{13.38}$$

then the robust fuzzy H_∞ stabilization design for the T-S fuzzy PDSs equations (13.4) and (13.9) with the area-controller u_A equation (13.14) constructed from the fuzzy area-controller gains $G_{aj} \triangleq Z_{aj}P$, for $j = 1, 2, \ldots, L_A$, on the spatiotemporal domain $\Omega \times [0, T_f]$ can be realized.

Proof

For the robust fuzzy H_∞ stabilization design via the area-controller u_A equation (13.14), it is very difficult to obtain G_{aj} by solving the inequality $\Xi_A < 0$ in equation (13.36) with $Q_d \geq 0$. Thus, given the indicator function in equation (13.16) and $\Omega = \Omega_1 \cup \cdots \cup \Omega_{p_1}$, we present the following matrix

$$\begin{aligned}\Xi_{AS} &\triangleq \int_0^{T_f} \int_\Omega \Xi \, dx \, dt + \int_0^{T_f} \int_\Omega \sum_{j=1}^{L_A} \mu_i^A(\zeta^A)[y^T PBG_{aj} y + *]dxdt \\ &+ \int_0^{T_f} \sum_{k=1, k \neq a}^{p_1} \int_{\Omega_k} \sum_{j=1}^{L_A} \mu_i^A(\zeta^A)[y^T PBG_{kj} y + *]dxdt \\ &= \Xi_A + \int_0^{T_f} \sum_{k=1, k \neq a}^{p_1} \int_{\Omega_k} \sum_{j=1}^{L_A} \mu_i^A(\zeta^A)[y^T PBG_{kj} y + *]dxdt \end{aligned} \tag{13.39}$$

where G_{kj}, for all j and k except for $k = a$, are the parameters to be determined.

Obviously, if the inequality $\Xi_{AS} < 0$ with the constraints

$$(PBG_{kj} + *) \geq 0, \quad \text{for all } j \text{ except for } k = a \tag{13.40}$$

is true, then we can obtain the inequality $\Xi_A < 0$ in equation (13.36) due to $\Xi_A \leq \Xi_{AS}$ with the constraints equation (13.40). That is, $\Xi_{AS} < 0$ with the constraints equation (13.40) is a more conservative inequality than $\Xi_A < 0$ in equation (13.36). Moreover, $\Xi_{AS} < 0$ can be written as

$$\begin{aligned}&\int_0^{T_f} \sum_{k=1}^{p_1} \int_{\Omega_k} \sum_{i=1}^{L_S} \sum_{j=1}^{L_A} \mu_i^S(\zeta^S)\mu_i^A(\zeta^A)[y^T Q y + (y^T PA_i y + y^T PBG_{kj} y \\ &- \frac{\pi^2}{d(\Omega)^2} y^T PDy + y^T P\varepsilon_f(y) + y^T PE_n n + *) - \rho^2 n^T n]dxdt < 0\end{aligned} \tag{13.41}$$

Then, by following a similar procedure in equations (13.29)–(13.31) with a given prescribed noise attenuation level ρ, if there exist $P > 0$, the parameters G_{kj}, and $\tau_a > 0$ satisfying $Q_d \geq 0$,

$$\begin{bmatrix} Q+\left(PA_i+PBG_{kj}-\frac{\pi^2}{d(\Omega)^2}PD+*\right)+\sigma_f^2\tau_a I & P & PE_n \\ * & -\tau_a I & 0 \\ * & * & -\rho^2 I \end{bmatrix}<0 \quad (13.42)$$

for $i = 1, 2,\ldots, L_S$, $j = 1, 2,\ldots, L_A$, and $k = 1, 2,\ldots, a, \ldots, p_1$, and the constraints equation (13.40), then both the robust fuzzy H_∞ noise attenuation performance equations (13.10) or (13.11) for the T-S fuzzy PDS equation (13.9) and the asymptotical stability of the noise-free T-S fuzzy PDS equation (13.4) can be guaranteed by applying the area-controller u_A equation (13.14) with the fuzzy area-controller gains G_{aj}. To transform the matrix inequalities in equation (13.42) into the LMIs in equation (13.37) with $Z_{kj} \triangleq G_{kj}X$, we pre- and post-multiply the matrix inequalities equation (13.42) by the matrix $\bar{X} = diag(X,I,I)$ and then utilize the Schur complement [552] together with the inequality $-\tau_a^{-1} \le -2+\tau_a, \forall \tau_a > 0$. Similarly, we obtain DX+* ≥ 0 and equation (13.38). This completes the proof.

Remark 13.4.1

In the LMIs equations (13.37) and (13.38) with $DX+* \geq 0$, the parameters $G_{kj} = Z_{kj}P$, for $k = 1, 2,\ldots, p_1$, except for $k = a$ and $j = 1, 2,\ldots, L_A$, need not to be utilized in real fuzzy control implementation. Due to the difficulty in directly solving $\Xi_A < 0$ in equation (13.36) for the fuzzy area-controller gains G_{aj}, we develop the more conservative condition $\Xi_A \leq \Xi_{AS} < 0$ with the constraints equation (13.40). Hence, by solving the LMIs in equations (13.37) and (13.38) with $DX+* \geq 0$, which implies $\Xi_{AS} < 0$, not only the parameters $G_{kj} = Z_{kj}P$ but also the fuzzy area-controller gains $G_{aj} = Z_{aj}P$ can be obtained simultaneously. Consequently, only the area-controller u_A equation (13.14) with the fuzzy area-controller gains G_{aj} needs to be constructed for the robust fuzzy H_∞ stabilization design.

Remark 13.4.2

The conservative condition $\Xi_A \leq \Xi_{AS} < 0$ can be relieved via minimizing the upper bounds $\gamma_{kj} \geq 0$ satisfying

$$\gamma_{kj} I \geq (BZ_{kj}+*) > 0, \quad \textit{for} \text{ all } j \text{ and } k \text{ except for } k = a \qquad (13.43)$$

due to the quadratic form $y^T\left(PBG_{kj}+*\right)y$. The minimization of the upper bound γ_{kj} makes both values Ξ_{AS} and Ξ_A the closest. Theoretically, we have $\Xi_{AS} \to \Xi_A$ as $\gamma_{kj} \to 0$. In other words, if we solve the constrained optimization problem

$$\min_{X>0,\tau_a>0,Z_{kj}} \gamma_{kj} \geq 0,$$

$$\text{subject to } \Phi_{ijk}^{(A)} < 0, \forall i,j,k, \text{ in (13.37) with } DX + * \geq 0 \text{ and} \qquad (13.44)$$

$$\text{the constraints (13.43)}$$

then conservativity of $\Xi_A \leq \Xi_{AS} < 0$ can be reduced for the robust fuzzy H_∞ stabilization design in Theorem 13.4.1 via the area-controller u_A equation (13.14).

In summary, the robust fuzzy area-controller $u = u_A$ equation (13.14) can be easily constructed by solving the LMIs in equations (13.37) and (13.38) with $DX+* \geq 0$ or the constrained optimization problem in equation (13.44) to achieve the robust fuzzy H_∞ stabilization design for the N-dimensional parabolic T-S fuzzy PDSs equations (13.4) and (13.9). As compared with the full-controller u_F equation (13.13), only the ath region Ω_a has the control force and the others don't; thus, the design cost is much lower when we realize the robust fuzzy H_∞ stabilization design with the area-controller u_A equation (13.14) in practice. Obviously, the control efficiency of this design is lower than the full-controller u_F equation (13.13) because the control effectiveness of the region $\Omega \backslash \Omega_a$ also depends on the control force of the ath region Ω_a, i.e., the system state $y(x,t) \in \Omega \setminus \Omega_a$ is robustly stabilized through the diffusion from the control effectiveness of region Ω_a.

13.5 LOW DESIGN-COST ROBUST FUZZY POINT-CONTROLLER FOR ROBUST FUZZY H_∞ STABILIZATION DESIGN

In this section, the considerably inexpensive robust fuzzy point-controller $u = u_P$ equation (13.15) is developed to realize the robust fuzzy H_∞ stabilization design for the N-dimensional parabolic T-S fuzzy PDSs equations (13.4) and (13.9) with the fuzzy approximation error equation (13.6) on the spatio-temporal domain $\Omega \times [0, T_f]$. Basically, only p_2 points, i.e., $x_k^c \in \Omega_k$, for $k = 1, 2, \ldots, p_2$ (Figure 13.1c) have control forces to realize the control design. Each x_k^c is within each Ω_k; accordingly, we give the following Lyapunov function:

$$V(y) = \int_\Omega y^T P(x) y \, dx \triangleq \sum_{k=1}^{p_2} \int_{\Omega_k} y^T P_k y \, dx \qquad (13.45)$$

where $P(x) \triangleq P_k I_{\Omega_k}(x)$ and $P_k > 0$, for $k = 1, 2, \ldots, p_2$. The concept of this design is that each $\int_{\Omega_k} y^T P_k y \, dx$ represents the energy of each domain Ω_k and the control force $u_P(x_k^c, t) = \sum_{j=1}^{L_P} \mu_j^P(\zeta^P) K_{kj} y(x_k^c, t)$ at each control spot x_k^c provides sufficient control effort to achieve the robust fuzzy H_∞ stabilization design in Ω_k through the adequate diffusion effect with the diffusion coefficient matrix D. Furthermore, the robust fuzzy H_∞ noise attenuation performance in equations (13.10) or (13.11) can be rewritten as

$$\int_0^{T_f}\int_\Omega y^T Q(x) y\,dx\,dt \triangleq \int_0^{T_f}\sum_{k=1}^{p2}\int_{\Omega_k} y^T Q_k y\,dx\,dt \le \rho^2\int_0^{T_f} n^T n\,dx\,dt \tag{13.46}$$

or

$$\int_0^{T_f}\int_\Omega y^T Q(x) y\,dx\,dt \triangleq \int_0^{T_f}\sum_{k=1}^{p2}\int_{\Omega_k} y^T Q_k y\,dx\,dt \le \rho^2\int_0^{T_f} n^T n\,dx\,dt + V(y_0(x)), \tag{13.47}$$

with simply adapted settings $Q_k > 0 \in \mathbb{R}^{n_y \times n_y}$, for k = 1, 2,..., p_2, which are the weighted matrices of the domains Ω_k, where $Q(x) \triangleq Q_k I_{\Omega_k}(x)$ Hence, the robust fuzzy H_∞ stabilization design is now to develop the point-controller u_P equation (13.15) to both asymptotically stabilize the noise-free T-S fuzzy PDS equation (13.4) and enable the T-S fuzzy PDS equation (13.9) to achieve the robust fuzzy H_∞ noise attenuation performance equations (13.46) or (13.47) with the fuzzy approximation error equation (13.6) and a given noise attenuation level ρ. Consequently, following a similar design procedure as with the full-controller u_F equation (13.13), we have the following results.

Theorem 13.5.1

For the N-dimensional parabolic T-S fuzzy PDSs equations (13.4) and (13.9) with the point-controller u_P equation (13.15), the fuzzy approximation error equation (13.6), and a prescribed noise attenuation level ρ, if we can find $X_k \triangleq P_k^{-1} > 0$, Y_{kj}, $\tau_p > 0$, and $\tau_y > 0$ satisfying $DX_k + * \ge 0$ and the following LMIs:

$$\Phi_{ijk}^{(P)} \triangleq \begin{bmatrix} \Phi_{ijk}^{(11)} & -BY_{kj} & I & E_n X_k & X_k & X_k \\ * & -\tau_y I & 0 & 0 & 0 & 0 \\ * & * & -\tau_p I & 0 & 0 & 0 \\ * & * & * & -\rho^2 I & 0 & 0 \\ * & * & * & * & (-2+\tau_p)\sigma_f^{-2} I & 0 \\ * & * & * & * & * & -Q_k^{-1} \end{bmatrix} < 0 \tag{13.48}$$

for i = 1, 2,..., L_S, j = 1, 2,..., L_P, and k = 1, 2,..., p_2, where $\Phi_{ijk}^{(11)} \triangleq \left(A_i X_k + BY_{kj} - \frac{\pi^2}{d(\Omega)^2} DX_k + *\right) + \tau_y I$ with $Y_{kj} \triangleq K_{kj} X_k$, then the robust fuzzy H_∞ stabilization design via the point-controller u_P equation (13.15) can be achieved. That is, with the point-controller u_P equation (13.15), both the asymptotical stability of the noise-free T-S fuzzy PDS equation (13.4) and the robust *fuzzy H_∞ noise attenuation performance equations (13.46) or (13.47) with a prescribed noise attenuation level ρ of the T-S fuzzy PDS equation (13.9) can be achieved on the spatiotemporal domain $\Omega\times\left[0,T_f\right]$.*

Proof

Given the Lyapunov function (13.45) and the T-S fuzzy PDS equation (13.9), we get

$$\begin{aligned}&\int_0^{T_f}\int_\Omega y^TQ(x)y\,dx\,dt\\&\leq\int_0^{T_f}\int_\Omega[y^TQ(x)y+\left(\frac{\partial y}{\partial t}\right)^TP(x)y+y^TP(x)\left(\frac{\partial y}{\partial t}\right)]dx\,dt+V(y_0(x))\\&=\int_0^{T_f}\int_\Omega\sum_{i=1}^{L_S}\mu_i^S(\zeta^S)[y^TQ(x)y+(y^TP(x)D\Delta y+y^TP(x)A_iy+y^TP(x)Bu_P\\&+y^TP(x)\varepsilon_f(y)+y^TP(x)E_nn+*)]dxdt+V(y_0(x))\end{aligned}\tag{13.49}$$

Then, by adding and subtracting the term $\rho^2\|n(x,t)\|^2_{L_2(\Omega\times\mathbb{R}_+;\mathbb{R}^{n_n})}$ to equation (13.49), we obtain

$$\begin{aligned}&\int_0^{T_f}\int_\Omega y^TQ(x)y\,dx\,dt\\&\leq\int_0^{T_f}\int_\Omega\sum_{i=1}^{L_S}\mu_i^S(\zeta^S)[y^TQ(x)y+(y^TP(x)D\Delta y+y^TP(x)A_iy\\&+y^TP(x)Bu_P+y^TP(x)\varepsilon_f(y)+y^TP(x)E_nn+*)-\rho^2n^Tn]dxdt\\&+\rho^2\int_0^{T_f}\int_\Omega n^Tn\,dx\,dt+V(y_0(x))\end{aligned}\tag{13.50}$$

Further, similar to the inequality equation (13.26) with $Q_d\geq 0$, we can conclude the following inequality:

$$\int_\Omega(y^TP(x)D\Delta y+*)dx\leq-\frac{\pi^2}{d(\Omega^2)}\int_\Omega(y^TP(x)Dy+*)dx\tag{13.51}$$

with $P(x)D+*\geq 0$, which can be easily transformed into $P_kD+*\geq 0,\forall k$. Therefore, we have

$$\int_0^{T_f}\int_\Omega y^T Q(x) y\, dx\, dt$$

$$\leq \int_0^{T_f}\int_\Omega \sum_{i=1}^{L_S} \mu_i^S(\zeta^S)[y^T Q(x) y + (y^T P(x) A_i y + y^T P(x) B u_P$$

$$-\frac{\pi^2}{d(\Omega^2)} y^T P(x) D y + y^T P(x)\varepsilon_f(y) + y^T P(x) E_n n + *) - \rho^2 n^T n] dxdt$$

$$+\rho^2 \int_0^{T_f}\int_\Omega n^T n\, dx\, dt + V(y_0(x)) \tag{13.52}$$

Next, we substitute the point-controller u_P equation (13.15) into (13.52) with $\Omega = \Omega_1 \cup \cdots \cup \Omega_{p2}$ and $x_k^c \in \Omega_k$; thus, we have

$$\int_0^{T_f}\int_\Omega y^T Q(x) y\, dx\, dt = \int_0^{T_f}\sum_{k=1}^{p2}\int_{\Omega_k} y^T Q_k y\, dx\, dt$$

$$\leq \int_0^{T_f}\sum_{k=1}^{p2}\int_{\Omega_k}\sum_{i=1}^{L_S}\sum_{j=1}^{L_P}\mu_i^S(\zeta^S)\mu_j^P(\zeta^P)$$

$$[y^T Q_k y + (y^T P_k A_i y + y^T P_k B K_{kj} I_{x_k^c}(x) y$$

$$-\frac{\pi^2}{d(\Omega^2)} y^T P_k D y + y^T P_k \varepsilon_f(y) + y^T P_k E_n n + *)$$

$$-\rho^2 n^T n] dxdt + \rho^2 \int_0^{T_f}\int_\Omega n^T n\, dx\, dt + V(y_0(x)) \tag{13.53}$$

After giving $\bar{y}_k(x,t) \triangleq y(x,t) - I_{x_k^c}(x) y(x,t)$, we obtain

$$\int_0^{T_f}\sum_{k=1}^{p2}\int_{\Omega_k} y^T Q_k y\, dx\, dt \leq \int_0^{T_f}\sum_{k=1}^{p2}\int_{\Omega_k}\sum_{i=1}^{L_S}\sum_{j=1}^{L_P}\mu_i^S(\zeta^S)\mu_j^P(\zeta^P)$$

$$[y^T Q_k y + (y^T P_k A_i y + y^T P_k B K_{kj} I_{x_k^c}(x) y$$

$$-\frac{\pi^2}{d(\Omega^2)} y^T P_k D y + y^T P_k \varepsilon_f(y) + y^T P_k E_n n + *)$$

$$-\rho^2 n^T n] dxdt + \rho^2 \int_0^{T_f}\int_\Omega n^T n\, dx\, dt + V(y_0(x)) \tag{13.54}$$

Consequently, the robust fuzzy H_∞ stabilization design can be achieved when $P_kD + * \geq 0$ and the following inequalities are satisfied:

$$\eta^T \Theta_{ijk} \eta$$

$$\triangleq \eta^T \begin{bmatrix} Q_k + (P_kA_i + P_kBK_{kj} - \frac{\pi^2}{d(\Omega^2)} P_kD + *) & -P_kBK_{kj} & P_k & P_kE_n \\ * & 0 & 0 & 0 \\ * & * & 0 & 0 \\ * & * & * & -\rho^2 I \end{bmatrix} \eta < 0 \quad (13.55)$$

where $\eta^T \triangleq \left[y^T, \bar{y}_k^T, \varepsilon_f^T(y), n^T \right]$. Additionally, $\bar{y}_k^T P_k P_k \bar{y}_k \leq y^T P_k P_k y$ is true for $k = 1, 2, \ldots, p_2$, due to the indicator function in equation (13.16) with any $P_k > 0$. Hence, the constraints $\bar{y}_k^T P_k P_k \bar{y}_k - y^T P_k P_k y \leq 0$ and equation (13.7) are considered for solving $\eta^T \Theta_{ijk} \eta < 0$. By using the S-procedure [552], if there exist $P_k > 0$, K_{kj}, $\tau_p > 0$, and $\tau_y > 0$ such that the following matrix inequalities

$$\begin{bmatrix} \Theta_{ijk}^{(11)} & -P_kBK_{kj} & P_k & P_kE_n \\ * & -\tau_y P_k P_k & 0 & 0 \\ * & * & -\tau_p I & 0 \\ * & * & * & -\rho^2 I \end{bmatrix} < 0 \quad (13.56)$$

hold, where $\Theta_{ijk}^{(11)} \triangleq Q_k + \left(P_kA_i + P_kBK_{kj} - \frac{\pi^2}{d(\Omega^2)} P_kD + * \right) + \sigma_f^2 \tau_p I + \tau_y P_k P_k$, then $\eta^T \Theta_{ijk} \eta < 0$ and these two constraints are satisfied. Finally, to transform the matrix inequalities equation (13.56) into the LMIs in equation (13.48), we first pre- and postmultiply the matrix inequalities equation (13.56) by the matrix $\bar{X}_k = diag(X_k, X_k, I, I)$, and then apply the Schur complement [552] to them with the inequality $-\tau_p^{-1} \leq -2 + \tau_p, \forall \tau_p > 0$. Similarly, we have $DX_k + * \geq 0, \forall k$. This completes the proof.

Remark 13.5.1

In order to minimize the effect of the external noise n on the controlled output y within the spatio-temporal domain $\Omega \times (0, T_f)$ from an energy perspective, the noise attenuation level ρ needs to be minimized. Hence, we propose the optimal constrained optimization problem as follows

$$\rho_o^{(F)} \triangleq \min_{X>0,\tau_f>0,Y_j} \rho$$
$$\text{subject to } \Phi_{ij}^{(F)} < 0, \forall i,j, \text{ in (13.27) with } DX + * \geq 0 \tag{13.57}$$

for the optimal robust fuzzy H_∞ stabilization design via the full-controller u_F equation (13.13). However, all over the domain $\Omega\times\left[0,T_f\right]$, the optimal robust fuzzy H_∞ stabilization designs via the area-controller u_A equation (13.14) and the point-controller u_P equation (13.15) cannot be attained because the control efforts are not applied to whole spatial domain Ω. Thus, we propose the following sub-optimal robust fuzzy H_∞ stabilization design problems

$$(\rho_o^{(A)}, \gamma_{kj}^{(O)}) \triangleq \min_{X>0,\tau_a>0,Z_{kj}} (\rho, \gamma_{kj})$$
$$\text{subject to } \Phi_{ijk}^{(A)} < 0, \forall i,j,k, \text{ in (13.37) with } DX + * \geq 0 \text{ and} \tag{13.58}$$
$$\text{the constraints (13.43)}$$

for the area-controller u_A equation (13.14), and

$$\rho_o^{(P)} \triangleq \min_{X_k>0,\tau_p>0,\tau_y>0,Y_{kj}} \rho$$
$$\text{subject to } DX_k + * \geq 0 \text{ and } \Phi_{ijk}^{(P)} < 0, \forall i,j,k, \text{ in (13.48)} \tag{13.59}$$

for the point-controller u_P equation (13.15), respectively. In addition, the constrained optimization problem equation (13.58) is obviously a multiobjective problem, which can be properly solved via following the new design procedures in [553,554]. Otherwise, one can remove the objectives γ_{kj} from equation (13.58) to formulate a single-objective problem with the conservativity.

Remark 13.5.2

For the low design-cost robust fuzzy area-controller u_A equation (13.14) and point-controller u_P equation (13.15), the most important principle for assigning the domains Ω_k, for k = *1, 2,..., a, ...,* p_1 and k = *1, 2,...,* p_2, respectively, is that each domain Ω_k must be mutually distinct. In general, the shape of each domain Ω_k is defined as a square or a rectangle with equal area to each other for simplicity. On the other hand, for the area-controller u_A equation (13.14), the control region Ωa can be chosen as a larger area than the others for a better control performance. Similarly, there can be more control spots in each Ω_k for the point-controller u_P equation (13.15). With a simple adaptation of the LMIs equation (13.48), the robust fuzzy H_∞ or the suboptimal robust fuzzy H_∞ stabilization designs via the point-controller u_P equation (13.15) with more control spots can be realized.

From the aforementioned results, the robust fuzzy H_∞ stabilization designs via the robust fuzzy full-controller u_F equation (13.13), the area-controller u_A equation (13.14), and the point-controller u_P equation (13.15) can be accomplished for N-dimensional nonlinear parabolic PDSs. Given the N-dimensional noise-free parabolic T-S fuzzy PDS equation (13.4) and the N-dimensional parabolic T-S fuzzy PDS equation (13.9) with the sufficient small fuzzy approximation error upper bound σ_f^2 in equation (13.7), the robust fuzzy controllers u_F, u_A, and u_P can be constructed by solving the LMI problems in equation (13.27) with $DX + * \geq 0$, equations (13.37) and (13.38) with $DX + * \geq 0$ or equation (13.44) and (13.48) with $DX_k + * \geq 0$, respectively, with the help of the LMI Toolbox in MATLAB. In addition, the optimal and suboptimal robust fuzzy H_∞ stabilization design problems with these three robust fuzzy controllers can be resolved by solving the constrained optimization problems (13.57)–(13.59). Our proposed design method using the Poincaré inequality makes it possible to realize the proposed control designs for the N-dimensional nonlinear parabolic PDSs with both the boundary conditions (D) and (N) via directly dealing with the effect of the partial derivative on the PDSs. Hence, the traditional algebraic matrix techniques and robust control analysis can be simply and generally applied to the control designs of the PDSs with not only the full-controller u_F, but also the low design-cost robust fuzzy controllers u_A and u_P.

At the first sight, the full-controller u_F equation (13.13) is the most efficient robust fuzzy controller for the robust fuzzy H_∞ stabilization design. The reason is that system state at all $x \in \Omega$ has its control force. However, it is also the most expensive in terms of implementing the controller and actuator. With regard to the area-controller u_A equation (13.14) for the robust fuzzy H_∞ stabilization design, its CE is lower than that of the full-controller u_F because only the ath region $x \in \Omega_a$ has the control force. Nevertheless, its design cost is lower. Finally, with regard to the robust fuzzy H_∞ stabilization design with the point-controller u_P equation (13.15), the point-controller u_P costs the least, but its CE is also the lowest since there are only p_2 points, i.e., $x_k^c \in \Omega_k$, for $k = 1, 2, \ldots, p_2$, with the control forces. The grades of the CE and design cost (DC) are summarized in the second and third columns of Table 13.1, respectively. On the other hand, for the robust fuzzy H_∞ stabilization design, the realizability of these three robust fuzzy controllers in real applications is important. The realizability mainly relies on the adequate diffusion coefficient matrix D and the restrictivity of the placement for the three robust fuzzy controllers, which both depend on the environment of the concerned PDS in reality and cannot be arbitrarily altered by engineers. A sufficiently large diffusion effect on the N-dimensional nonlinear parabolic PDSs equations (13.1) and (13.8) is necessary for the feasibility of the LMI problems in equation (13.27) with $DX + * \geq 0$, equations (13.37) and (13.38) with $DX + * \geq 0$, equation (13.44), or (13.48) with $DX_k + * \geq 0$. According to the placement of these three robust fuzzy controllers on the spatial domain Ω (Figure 13.1), it is clear that we could apply the full-controller u_F, the area-controller u_A, and the point-controller u_P in the case of low, medium, and large diffusion effects, respectively. In other words, some PDSs with insufficient diffusion effects cannot be expected in reality to achieve the robust fuzzy H_∞ stabilization designs with the area-controller u_A or the point-controller u_P, which means the LMI problems to solve equations (13.37) and (13.38) with $DX + * \geq 0$, equation (13.44), and (13.48) with $DX_k + * \geq 0$ may be infeasible. Take the physical process of aquifer flow for example [504–512]. The diffusion coefficient matrix $D = \frac{k_s}{S_s} \geq 0$ refers to the hydraulic diffusivity, where k_s is the hydraulic conductivity and S_s is the specific storage. When k_s is much smaller than S_s in practice due to the concerned topography, i.e., D is very small due to the environment of the PDS, then the area-controller u_A or the point-controller u_P may not be applied to the PDS for the robust fuzzy H_∞ stabilization design. However, the limitation to place the three robust fuzzy controllers is reverse; that is, the restrictivities to place u_F, u_A, and u_P are considered as high, medium, and low, respectively, which can be easily observed from Figure 13.1. Thus, what kinds of the robust fuzzy controllers u_F, u_A, and u_P can be realized to achieve the robust fuzzy H_∞ stabilization design depends on the environment of the concerned PDS. The grades of the necessary diffusion effect (NDE) and restrictivity (R) are summarized in the fourth and fifth columns of Table 13.1, respectively. Additionally, the computational complexity (C) of the three robust fuzzy controllers, i.e., the numbers of the LMIs needed to solve, is summarized in the sixth column of Table 13.1. Based on Table 13.1, the benefits to employ the full-controller u_F are the excellent CE and low NDE; however, the placement of the controller is restricted the most and the design cost is relatively high. The medium choice is the area-controller u_A. Additionally, the advantages to employ the point-controller u_P are low design cost and restrictivity. For the opposite, the CE may be insufficient, especially when the diffusion effect of the environment for the concerned PDS in reality is insufficient.

Based on the above analysis, the procedure to choose and design the proposed three robust fuzzy controllers u_F, u_A, and u_P for the robust fuzzy H_∞ stabilization design of the N-dimensional nonlinear parabolic PDSs is summarized in the following.

TABLE 13.1
The Comparison of the Control Efficiency (CE), Design Cost (DC), Necessary Diffusion Effect (NDE), Restrictivity (R), and Computational Complexity (C) of Three Robust Fuzzy Controllers u_F, u_A, and u_P

Grade	(CE)	(DC)	(NDE)	(R)	(C)
u_F	High	High	Low	High	$L_S L_F + 1$
u_A	Medium	Medium	Medium	Medium	$L_S L_A p_1 + L_A(p_1 - 1) + 1$
u_P	Low	Low	High	Low	$L_S L_P p_2 + p_2$

Design Procedure

1. Give a prescribed noise attenuation level ρ.
2. Construct the T-S fuzzy PDSs equations (13.4) and (13.9) with the fuzzy approximation error upper bound σ_f^2 in equation (13.7) based on the preliminary knowledge of the N-dimensional nonlinear parabolic PDSs equations (13.1) and (13.8).
3. Based on (CE), (DC), (NDE), (R), and (C) in Table 13.1, identify what kinds of the robust fuzzy controllers u_F, u_A, and u_P can be used according to the environment of the concerned PDS; further, choose a kind of robust fuzzy controller according to the designer's requirements.
4. Define the chosen robust fuzzy controller rules, e.g., equation (13.12), and solve the corresponding LMI problem in equation (13.27) with $DX+* \geq 0$, equations (13.37) and (13.38) with $DX+*\geq 0$, equation (13.44), or (13.48) with $DX_k+* \geq 0$.
5. Construct the robust fuzzy controller in equations (13.13)–(13.15) to robustly stabilize the concerned PDS.

13.6 SIMULATION EXAMPLE

In this section, we provide a simulation example to verify the aforementioned results and further compare the robust fuzzy H_∞ stabilization designs via the robust fuzzy controllers u_F (13.13), u_A (13.14), and u_P (13.15). To simply comprehend the placement of these three controllers (Figure 13.1), and for the comparability of these three controllers, a 2-dimensional nonlinear parabolic PDS with the boundary condition (N) in equation (13.2), the Belousov-Zhabotinsky (BZ) reaction system [513–515], is chosen as an example to demonstrate our results. The BZ reaction system is a well-known reaction-diffusion system that has been theoretically and experimentally studied, especially as a biochemical or biological reaction system. The mathematical model of the BZ reaction system can be described as [513–515]

$$\frac{\partial y}{\partial t} = D\Delta y + f(y) + Bu + E_n n \tag{13.60}$$

where $f(y) \triangleq \left[k_1 y_2 + k_2 y_1 - k_3 y_1 y_2 - 2k_4 y_1^2, -k_1 y_2 - k_3 y_1 y_2 + k_6 y_3, 2k_2 y_1 - k_5 y_3\right]^T$, $k_1 \triangleq 0.0768, k_2 \triangleq 0.48, k_3 \triangleq 8\times 10^{-2}, k_4 \triangleq 2, k_5 \triangleq 2, k_6 \triangleq 3.1, D \triangleq I_{3\times 3}, E_n \triangleq I_{3\times 3}$, and $B \triangleq [1,1,1]^T$, with the system state $y \triangleq \left[y_1, y_2, y_3\right]^T$, control force $u \in \mathbb{R}$, external noise $n \triangleq \left[n_1, n_2, n_3\right]^T$ on the spatiotemporal domain $\Omega\times[0,20]$, where $\left[x_1, x_2\right]^T \in \Omega \subset \mathbb{R}^2$ and Ω is defined on $[0, 1]\times[0, 1]$; hence $\frac{\pi^2}{d(\Omega)^2} \simeq 4.93480$. The PDS equation (13.60) with the boundary condition (N) in equation (13.2) is certainly unstable. The external noises n_1, n_2, and n_3 are chosen as the white Gaussian noises with zero mean and average power 1 throughout the spatiotemporal domain $\Omega\times[0,20]$. The initial state is $y_0(x) \triangleq \left[\sin(x_1+x_2), \cos(x_1+x_2), \sin(x_1+x_2)\cos(x_1+x_2)\right]^T$.

To construct the 2-dimensional parabolic T-S fuzzy PDSs equations (13.4) and (13.9) on the spatiotemporal domain $\Omega\times[0,20]$, we choose the premise variables $\zeta_1^S \triangleq y_1$ and $\zeta_2^S \triangleq y_2$ with the triangular membership functions at the operating points 0.1659 and 0.999 and at −0.2595 and 3.6676, respectively. Based on the knowledge-based fuzzy system technique [534–536], we obtain the fuzzy system matrices as follows:

$$A_1 \simeq \begin{bmatrix} -3.8130 & 0.0768 & 0 \\ -0.2934 & -0.0768 & 3.1000 \\ 0.9600 & 0 & -2.0000 \end{bmatrix}, A_2 \simeq \begin{bmatrix} -3.4989 & 0.0768 & 0 \\ 0.0207 & -0.0768 & 3.1000 \\ 0.9600 & 0 & -2.0000 \end{bmatrix}$$

$$A_3 \simeq \begin{bmatrix} -0.4770 & 0.0768 & 0 \\ -0.2934 & -0.0768 & 3.1000 \\ 0.9600 & 0 & -2.0000 \end{bmatrix}, A_4 \simeq \begin{bmatrix} -0.1628 & 0.0768 & 0 \\ 0.0207 & -0.0768 & 3.1000 \\ 0.9600 & 0 & -2.0000 \end{bmatrix} \quad (13.61)$$

with the bound of the fuzzy approximation error $\sigma_f^2 \simeq 3.7491\times 10^{-30}$. In order to simply compare the robust fuzzy H_∞ stabilization designs via these three types of robust fuzzy controllers, the simulation environments must be the same. Thus, we give $\zeta_1^F = \zeta_1^A = \zeta_1^P = \zeta_1^S$ and $\zeta_2^F = \zeta_2^A = \zeta_2^P = \zeta_2^S$ with the same fuzzy settings; $Q \triangleq 0.2\times I_{3\times3}$, $p_1 \triangleq 9$, and $Q_k \triangleq 0.2\times I_{3\times3}$, for k = 1, 2,…, p_2 with $p_2 \triangleq 9$; and a prescribed noise attenuation level $\rho \triangleq 0.5$. The placement of the robust fuzzy controllers u_F equation (13.13), u_A equation (13.14), and u_P equation (13.15) is shown in Figure 13.1a–c, respectively. For the area-controller u_A equation (13.14), the regions Ω_k, for k = 1, 2,…, a, …, p_1, are chosen as equally-sized squares and only the central region Ω_a has the control force, where a = 5. For the point-controller u_P equation (13.15), we choose each Ω_k, for k = 1, 2,…, p_2, as an equally-sized square and each control spot $x_k^c \in \Omega_k$ is placed in the center of the corresponding region Ω_k.

After resolving the robust fuzzy H_∞ stabilization design problems for these three robust fuzzy controllers, we obtain the following results. For the robust fuzzy H_∞ stabilization designs via the full-controller u_F equation (13.13) and the area-controller u_A equation (13.14), we obtain $K_1 \simeq [-1.6037,-3.6698,0.3768]$, $K_2 \simeq [-0.8057,-3.4522,0.4675]$, $K_3 \simeq [-3.1523,-3.3481,-0.1198]$, $K_4 \simeq [-2.5659, -2.9963,0.3908]$, $\tau_f \simeq 1.9509$, $G_{51} \simeq [-0.5008,-0.2850,0.6746]$, $G_{53} \simeq [-5.7117, -1.2141,0.0578$, $G_{54} \simeq [-7.0761,-1.3987,0.0255]$, $\tau_a \simeq 1.9918$. In addition, for all j and k except for $k = a = 5$, the upper bounds γ_{kj} are around 10^{-12}, and G_{kj} are omitted since the parameters G_{kj} are not used. Finally, the simulation results for the robust fuzzy H_∞ stabilization design via the point-controller u_P equation (13.15) are $K_{11} \simeq [-0.2156, -2.4644, -2.1550]$, $K_{21} \simeq [-0.2089, -2.3939, -2.1865]$, $K_{31} \simeq [-0.2432, -2.3899, -2.0612]$, $K_{41} \simeq [-0.2252, -2.4416, -2.0685]$, $K_{51} \simeq [-0.2390, -2.4041, -2.0597]$, $K_{61} \simeq [-0.2028, -2.3948, -2.1217]$, $K_{71} \simeq [-0.1852, -2.3248, -2.0484]$, $K_{81} \simeq [-0.4466, -2.7416, -2.1764]$, $K_{91} \simeq [-0.2578, -2.4764,$

–2.1357], $K_{12} \simeq$[–0.2124, –2.2892, –2.0585], $K_{22} \simeq$[–0.2509, –2.3579, –2.1598], $K_{32} \simeq$[–0.2880,–2.4518,–2.1001],$K_{42} \simeq$[–0.2397,–2.3773,–1.9981],$K_{52} \simeq$[–0.2673, –2.4222, –2.1011], $K_{62} \simeq$[–0.2487, –2.3278, –2.0764], $K_{72} \simeq$[–0.2497, –2.4095, –2.1107], $K_{82} \simeq$[–0.4537, –2.6001, –2.0781], $K_{92} \simeq$[–0.2872, –2.4516, –2.1540], $K_{13} \simeq$[–0.8894,–2.0713,–1.8371],$K_{23} \simeq$[–0.8924,–2.0658,–1.8932],$K_{33} \simeq$[–0.9123, –2.1050, –1.8242], $K_{43} \simeq$[–0.9010, –2.1032, –1.7367], $K_{53} \simeq$[–0.8998, –2.0863, –1.7883], $K_{63} \simeq$[–0.8906, –2.0685, –1.8506], $K_{73} \simeq$[–0.8709, –2.0406, –1.8227], $K_{83} \simeq$[–1.0596,–2.3454,–1.8418],$K_{93} \simeq$[–0.9265,–2.1236,–1.8496],$K_{14} \simeq$[–1.0679, –1.9902, –1.7194], $K_{24} \simeq$[–1.0584, –1.9816, –1.6527], $K_{34} \simeq$[–1.1038, –2.0495, –1.7065], $K_{44} \simeq$[–1.0913, –2.0311, –1.6876], $K_{54} \simeq$[–1.1001, –2.0430, –1.7122], $K_{64} \simeq$[–1.0726,–1.9949,–1.7482],$K_{74} \simeq$[–1.0892,–2.0202,–1.7619],$K_{84} \simeq$[–1.2338, –2.2516,–2.7843],$K_{94} \simeq$[–1.1422,–2.1027,–2.7925], $\tau_p \simeq 1.5806$, $\tau_y \simeq 1.2732 \times 10^2$. Due to the enormous amount of simulation data, we illustrate the simulation results as samples at the different time layers $T_1 = 1$, $T_2 = 3$, and $T_f = 20$ (minutes) in Figures 13.2–13.4 to verify the robust stability of the BZ reaction system equation (13.60) with the different robust fuzzy controllers, u_F equation (13.13), u_A equation (13.14), and u_P equation (13.15). In Figure 13.2, we illustrate the states y_1 ((a)–(c)), y_2 ((d)–(f)), and y_3 ((g)–(i)) with the control force u_F equation (13.13) at the time layers T_1, T_2, and T_f. Similarly, the states y_1 ((a)–(c)), y_2 ((d)–(f)), and y_3 ((g)–(i)) with the control forces u_A equation (13.14) and u_P equation (13.15) are shown in Figure 13.3 and 13.4, respectively. In Figures 13.2–13.4, we observe that all the states y_1 ((a)–(c)), y_2 ((d)–(f)), and y_3 ((g)–(i)) with whatever control forces

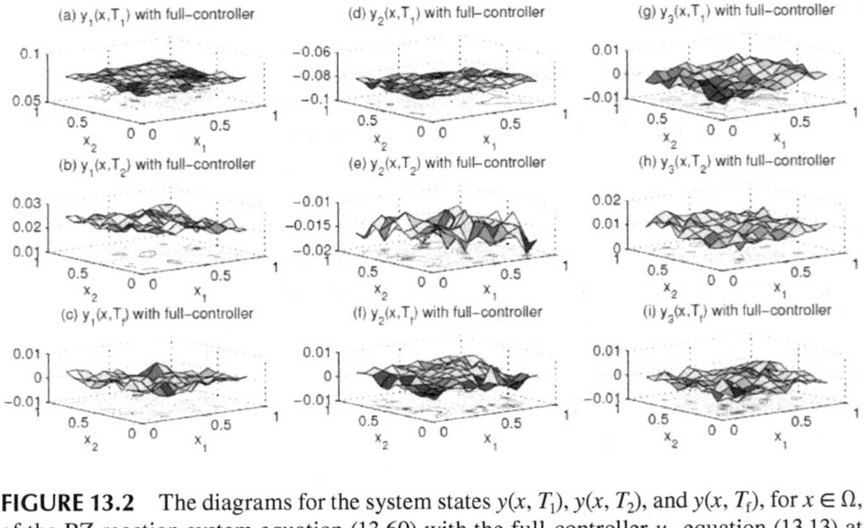

FIGURE 13.2 The diagrams for the system states $y(x, T_1)$, $y(x, T_2)$, and $y(x, T_f)$, for $x \in \Omega$, of the BZ reaction system equation (13.60) with the full-controller u_F equation (13.13) at $t = T_1$, T_2, and T_f, respectively. The graphs for the state $y_1(x, t)$, at the time layer $T_1 = 1$ (a), at the time layer $T_2 = 3$ (b), and at the time layer $T_f = 20$ (c). The graphs for the state $y_2(x, t)$, at the time layer $T_1 = 1$ (d), at the time layer $T_2 = 3$ (e), and at the time layer $T_f = 20$ (f). The graphs for the state $y_3(x, t)$, at the time layer $T_1 = 1$ (g), at the time layer $T_2 = 3$ (h), and at the time layer $T_f = 20$ (i).

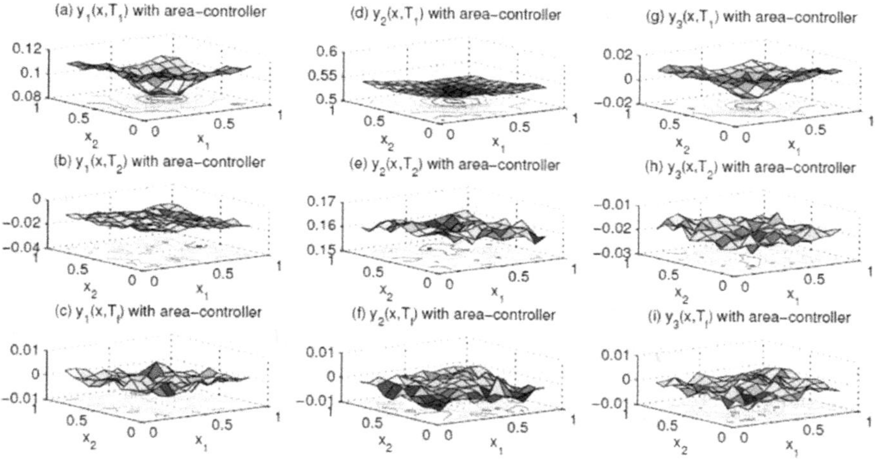

FIGURE 13.3 The diagrams for the system states $y(x, T_1)$, $y(x, T_2)$, and $y(x, T_f)$, for $x \in \Omega$, of the BZ reaction system equation (13.60) with the area-controller u_A equation (13.14) at $t = T_1$, T_2, and T_f, respectively. The graphs for the state $y_1(x, t)$, at the time layer $T_1 = 1$ (a), at the time layer $T_2 = 3$ (b), and at the time layer $T_f = 20$ (c). The graphs for the state $y_2(x, t)$, at the time layer $T_1 = 1$ (d), at the time layer $T_2 = 3$ (e), and at the time layer $T_f = 20$ (f). The graphs for the state $y_3(x, t)$, at the time layer $T_1 = 1$ (g), at the time layer $T_2 = 3$ (h), and at the time layer $T_f = 20$ (i).

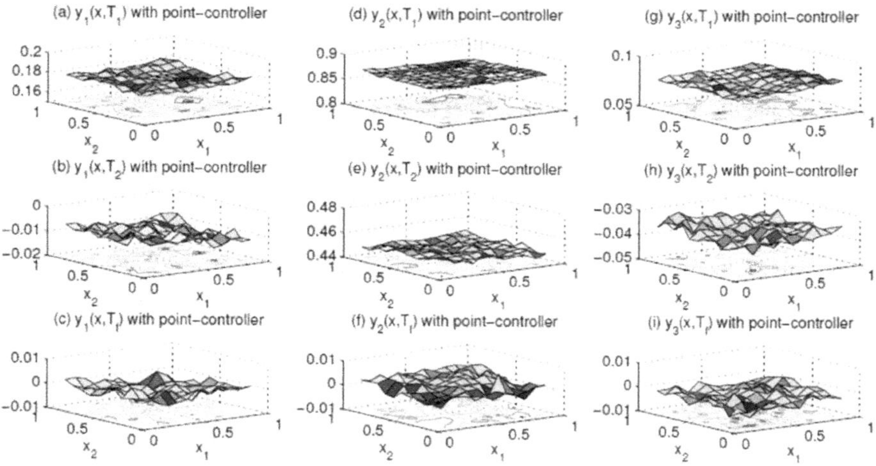

FIGURE 13.4 The diagrams for the system states $y(x, T_1)$, $y(x, T_2)$, and $y(x, T_f)$, for $x \in \Omega$, of the BZ reaction system equation (13.60) with the point-controller u_P equation (13.15) at $t = T_1$, T_2, and T_f, respectively. The graphs for the state $y_1(x, t)$, at the time layer $T_1 = 1$ (a), at the time layer $T_2 = 3$ (b), and at the time layer $T_f = 20$ (c). The graphs for the state $y_2(x, t)$, at the time layer $T_1 = 1$ (d), at the time layer $T_2 = 3$ (e), and at the time layer $T_f = 20$ (f). The graphs for the state $y_3(x, t)$, at the time layer $T_1 = 1$ (g), at the time layer $T_2 = 3$ (h), and at the time layer $T_f = 20$ (i).

we employ gradually converge almost to zero in the time layer presentations. In addition, Figures 13.5–13.7 show the temporal profiles of the states $y((0.58, 0.58), t)$ with $n((0.58, 0.58), t) = 0$, for $t \in [0,15]$, where the spatial location $x = (0.58, 0.58)$ is randomly selected as an example. In Figures 13.5–13.7, all the system states with and without the control force are displayed, including the system states

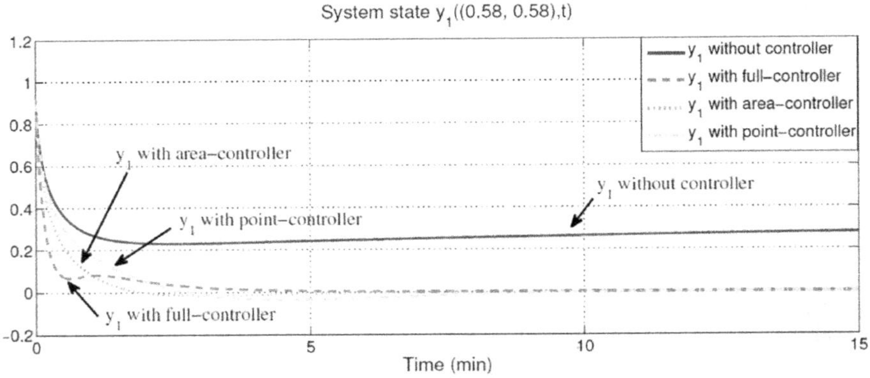

FIGURE 13.5 The temporal profiles of the system states $y_1((0.58, 0.58), t)$ with $n((0.58, 0.58), t) = 0$, for $t \in [0, 15]$, of the BZ reaction system (13.60) with and without the control forces. The trajectory of the state $y_1((0.58, 0.58), t)$ without any control force (solid line). The trajectory of the state $y_1((0.58, 0.58), t)$ with the control force u_F (dashed line). The trajectory of the state $y_1((0.58,0.58), t)$ with the control force u_A (dotted line). The trajectory of the state $y_1((0.58, 0.58), t)$ with the control force u_P (dash dot line).

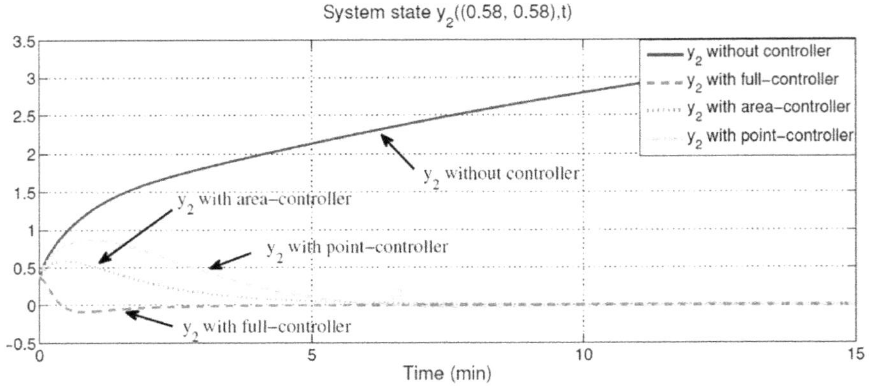

FIGURE 13.6 The temporal profiles of the system states $y_2((0.58, 0.58), t)$ with $n((0.58, 0.58), t) = 0$, for $t \in [0, 15]$, of the BZ reaction system equation (13.60) with and without the control forces. The trajectory of the state $y_2((0.58, 0.58), t)$ without any control force (solid line). The trajectory of the state $y_2((0.58, 0.58), t)$ with the control force u_F (dashed line). The trajectory of the state $y_2((0.58, 0.58), t)$ with the control force u_A (dotted line). The trajectory of the state $y_2((0.58, 0.58), t)$ with the control force u_P (dash dot line).

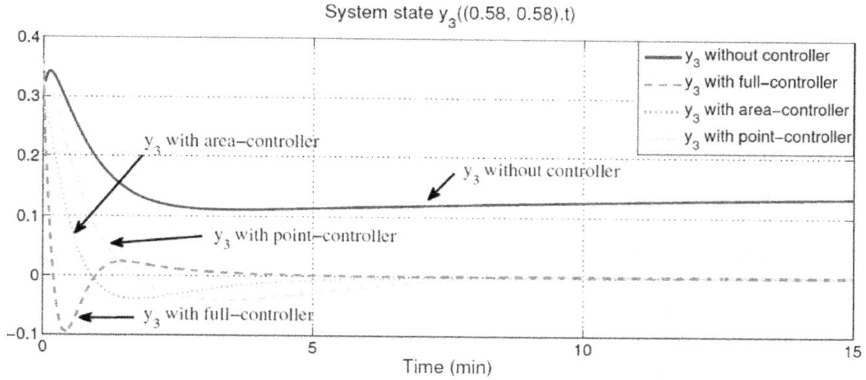

FIGURE 13.7 The temporal profiles of the system states $y_3((0.58, 0.58), t)$ with $n((0.58, 0.58), t) = 0$, for $t \in [0, 15]$, of the BZ reaction system equation (13.60) with and without the control forces. The trajectory of the state $y_3((0.58, 0.58), t)$ without any control force (solid line). The trajectory of the state $y_3((0.58, 0.58)$, t) with the control force u_F (dashed line). The trajectory of the state $y_3((0.58,0.58), t)$ with the control force u_A (dotted line). The trajectory of the state $y_3((0.58, 0.58), t)$ with the control force u_P (dash dot line).

without any control force (solid line), those with the control force u_F (dashed line), those with the control force u_A (dotted line), and those with the control force u_P (dash dot line). According to the simulation results in Figures 13.5–13.7, all the states y_1, y_2, and y_3 converge to zero with the respective control efforts of u_F, u_A, and u_P; that is, the asymptotical stability of the noise-free BZ reaction system with these three control forces can be guaranteed.

Further, based on the enormous amount of simulation data from 100 Monte Carlo simulation runs, the CEs of these three robust fuzzy H_∞ stabilization designs for the BZ reaction system equation (13.60) can be confirmed as follows. The mean robust fuzzy H_∞ noise attenuation performance equation (13.11) for the robust fuzzy H_∞ stabilization design via the full-controller u_F equation (13.13) and via the area-controller u_A equation (13.14) can be achieved by

$$E\left\{\int_0^{20}\int_\Omega y^T Q y\,dx\,dt\right\} \le (0.0194)^2 E\left\{\int_0^{20}\int_\Omega n^T n\,dx\,dt\right\} + E\{V^{(F)}(y_0(x))\} \quad (13.62)$$

and

$$E\left\{\int_0^{20}\int_\Omega y^T Q y\,dx\,dt\right\} \le (0.0505)^2 E\left\{\int_0^{20}\int_\Omega n^T n\,dx\,dt\right\} + E\{V^{(A)}(y_0(x))\} \quad (13.63)$$

where $E\{V^{(F)}(y_0(x))\} \simeq 17.3734$, and $E\{V^{(A)}(y_0(x))\} \simeq 36.8743$, respectively. In relation to the mean robust fuzzy H_∞ noise attenuation performance equation (13.47) for the robust fuzzy H_∞ stabilization design via the point-controller u_P equation (13.15), we obtain

$$E\left\{\int_0^{20}\sum_{k=1}^{9}\int_{\Omega_k} y^T Q_k y\,dx\,dt\right\} \le (0.0805)^2 E\left\{\int_0^{20}\int_{\Omega} n^T n\,dx\,dt\right\} + E\{V^{(P)}(y_0(x))\} \tag{13.64}$$

where $E\{V^{(P)}(y_0(x))\} \simeq 6.1433$. From equations (13.62)–(13.64), the mean values of the real noise attenuation levels via the simulation data in order are 0.0194 (for u_F) < 0.0505 (for u_A) < 0.0805 (for u_P) < 0.5 = ρ. In addition, at the spatial location $x = (0.58, 0.58)$, the convergence times of each state, y_1, y_2, and y_3, with these robust fuzzy controllers but without the noises n_1, n_2, and n_3 are shown in Table 13.2. That is, the times that the system states converge to and then stay on the equilibrium point zero (precision $\pm 10^{-7}$) can be found in Table 13.2. Based on Table 13.2, the states with the control force u_F converge at the fastest rate, followed by the states with the control force u_A. The states with the control force u_P converge at the slowest rate. Obviously, in Table 13.2, u_F with less ρ in equation (13.62) needs more control effort to make the BZ reaction system converge more quickly at the noise-free case. u_P with larger ρ in equation (13.64) makes the system converge more slowly.

Hence, the CEs (Table 13.1) on both the noise attenuation ability and the asymptotical stability for these three robust fuzzy H_∞ stabilization designs are confirmed as expected. In relation to the NDE (Table 13.1) for the feasibility of the LMI problems in equation (13.27) with DX+* ≥ 0, equation (13.44), and (13.48) with DX_k+* ≥ 0, the feasible regions of the diffusion coefficient matrix D are $\|D\|_2 > 0.08$ (for u_F), $\|D\|_2 > 0.318$ (for u_A), and $\|D\|_2 > 0.342$ (for u_P), respectively, where $\|D\|_2$ denotes the square root of the largest eigenvalue of D^TD. As a result, the robust fuzzy H_∞ stabilization designs via these three robust fuzzy controllers u_F equation (13.13), u_A equation (13.14), and u_P equation (13.15) for the BZ reaction system equation (13.60) are achieved. Furthermore, the comparison of the robust fuzzy H_∞ stabilization designs via these three robust fuzzy controllers has been demonstrated. Thus, this simulation example demonstrates that the proposed robust fuzzy H_∞ stabilization designs have potential application to the control of N-dimensional nonlinear parabolic PDSs with the boundary condition (N) in equation (13.2). Further, the robust fuzzy H_∞ stabilization designs for

TABLE 13.2
The Convergence Times of y_1, y_2, and y_3 Using the Three Robust Fuzzy Controllers u_F, u_A, and u_P in the Noise-Free Cases

Convergence Time	y_1	y_2	y_3
u_F	8.3486	7.9257	7.0971
u_A	11.8371	15.3257	11.1800
u_P	16.5271	19.8871	15.2086

the N-dimensional nonlinear parabolic PDSs with the boundary condition (D) in equation (13.2) can be also expected to achieve by the proposed robust fuzzy controllers.

13.7 CONCLUSION

In this chapter, the robust fuzzy H_∞ stabilization designs via not only the normal design-cost robust fuzzy full-controller, but also low design-cost robust fuzzy area-controller and point-controller for the N-dimensional nonlinear parabolic PDSs with either the Dirichlet or Neumann boundary conditions are developed. Additionally, the comparison of the three controllers is discussed. Based on the knowledge-based fuzzy system technique, these three robust fuzzy controllers are successfully utilized to both asymptotically stabilize the N-dimensional noise-free parabolic T-S PDS and to enable the N-dimensional parabolic T-S fuzzy PDS to efficiently attenuate the effect of external noise in the spatio-temporal domain from the H_∞ filtering perspective. To specify the fuzzy full-controller, area-controller, and point-controller gains for the corresponding robust fuzzy H_∞ stabilization design problem, the Poincaré inequality is utilized allowing the LMI problem for each design to be easily obtained instead of solving the complex DMI problem. By solving three sets of LMIs with the help of the LMI Toolbox in Mat- lab, these three fuzzy control gains can be derived to construct the three robust fuzzy controllers for the robust fuzzy H_∞ stabilization designs of the non-linear parabolic PDSs. The benefits and drawbacks of the three controllers are also given in Table 13.1 so that control engineers could select an adequate one according to the implementation feasibility, computational complexity, and environment restriction of the concerned PDS. Finally, a BZ reaction system serves as a simulation example to verify the proposed robust fuzzy H_∞ stabilization designs for the nonlinear parabolic PDSs and to demonstrate the comparison of the three controllers.

References

1. P.D. Christofides, *Nonlinear and Robust Control of PDE Systems: Methods and Applications to Transport-Reaction Process*, Springer, Berlin, 2001.
2. U.T. Myint and L. Debnath, *Partial Differential Equations for Scientists and Engineers*, 3rd ed, Dover Publications Inc., New York, 1987.
3. U.T. Myint and L. Debnath, *Linear Partial Differential Equations for Scientist and Engineers*, 4th ed, Springer, Berlin, 2007.
4. B.S. Chen and Y.T. Chang, "Fuzzy state-space modeling and robust observer-based control design for nonlinear partial differential systems," *IEEE Trans. Fuzzy Syst.*, vol. 17, pp. 1025–1043, 2009.
5. Y.T. Chang and B.S. Chen, "A fuzzy approach for robust reference-tracking-control design of nonlinear distributed parameter time-delayed systems and its application," *IEEE Trans. Fuzzy Syst.*, vol. 18, pp. 1041–1057, 2010.
6. C.L. Lin, B.S. Chen, and F.B. Hsiao, "On stability and stabilization of mechanical structures under nonlinear time-varying perturbation," *ASME J. Appl. Mech.*, vol. 58, no. 2, pp. 527–535, 1991.
7. C.L. Lin and B.S. Chen, "Robust observer-based control of large flexible structures," *ASME J. Dyn. Syst. Meas. Control*, vol. 116, no. 4, pp. 713–722, 1994.
8. W.H. Chen and B.S. Chen, "Robust filter for linear stochastic partial systems via a set of sensor measurement," *IEEE Trans. Circuits Syst. I*, vol. 59, no. 6, pp. 1251–1264, 2012.
9. B.S. Chen, W.H. Chen, and W. Zhang, "Robust filter for nonlinear partial differential systems in sensor signal processing: Fuzzy approach," *IEEE Trans. Fuzzy Syst.*, vol. 20, no. 5, pp. 957–970, 2012.
10. Y.Y. Chen, Y.T. Chang, and B.S. Chen, "Fuzzy solution to partial differential equation: Adaptive approach," *IEEE Trans. Fuzzy Syst.*, vol. 17, no. 1, pp. 116–127, 2009.
11. W.H. Chen and B.S. Chen, "Robust stabilization for stochastic partial differential systems under spatio-temporal disturbance and sensor measurement noises," *IEEE Trans. Circuits Syst., I: Reg. Pap.*, vol. 60, no. 4, pp. 1013–1026, 2013.
12. K. Wu and B.S. Chen, "Synchronization of partial differential systems via diffusion coupling," *IEEE Trans. Circuits Syst., I*, vol. 59, no. 11, pp. 2655–2668, 2012.
13. B.S. Chen and S.J. Ho, "Robust fuzzy H_∞ estimator-based stabilizing design for nonlinear parabolic partial differential systems with different boundary conditions," *IEEE Trans. Fuzzy Syst.*, vol. 24, no. 1, pp. 208–222, 2016.
14. S.J. Ho and B.S. Chen, "Low design-cost fuzzy controllers for robust stabilization of nonlinear partial differential systems," *IEEE Trans. Fuzzy Syst.*, vol. 24, no. 1, pp. 99–113, 2016.
15 B.S. Chen, C.Y. Hsieh, and S.J. Ho, "System entropy measurement of stochastic partial differential systems," *Entropy*, vol. 18, no. 3, p. 99, 2016.
16. Y.P. Lin and B.S. Chen, "Natural resource management for nonlinear stochastic biotic-abiotic ecosystems: Robust reference tracking control strategy using limit set of controllers," *J. Environ. Inform.*, vol. 27, no. 1, pp. 14–30, 2016.
17. A. Kharab and R. Kharab, "Spreadsheet solution of hyperbolic partial differential equations," *IEEE Trans. Educ.*, vol. 40, pp. 103–110, 1997.
18. D. Kincaid and W. Cheney, *Numerical Analysis: Mathematics of Scientific Computing*, American Mathematical Society, Providence, RI, 1991.

19. O. Bíró and K. Preis, "Finite element analysis of 3-D eddy currents," *IEEE Trans. Magn.*, vol. 26, no. 2, pp. 418–423, 1990.
20 B.H. McDonald and A. Wexler, "Finite-element solution of unbounded filed problems," *IEEE Trans. Microwave Theory Tech.*, vol. MT20, pp. 841–847, 1972.
21. S. Gratkowski, L. Pichon, and A. Razek, "New infinite elements for a finite element analysis of 2-D scattering problems," *IEEE Trans. Magn.*, vol. 32, no. 3, pp. 882–885, 1996.
22. O.C. Zienkiewicz and R.L. Taylor, *The Finite Element Method*, Dover Publications Inc., New York, vol. 1, 1989.
23. D.H. Norrie and G. de Vries, *An Introduction to Finite Element Analysis*, Academic, London, 1978.
24. A. Shmilovici and O. Maimon, "On the solution of differential equations with fuzzy spline wavelets," *Fuzzy Sets Syst.*, vol. 96, pp. 77–99, 1998.
25. I. Perfilieva, "Fuzzy transforms: Theory and applications," *Fuzzy Sets Syst.*, vol. 157, pp. 993–1023, 2006.
26. M. Stepnicka and R. Valasek, "Numerical Solution of Partial Differential Equations with Help of Fuzzy Transform," Proc. 14th IEEE International Conference on Fuzzy Systems, May 25, 2005, pp. 1104–1109.
27. H.O. Wang, K. Tanaka, and M.F. Griffin, "An approach to fuzzy control of nonlinear systems: Stability and design issues," *IEEE Trans. Fuzzy Syst.*, vol. 4, pp. 14–23, 1996.
28. T. Takagi and M. Sugeno, "Fuzzy identification of systems and its applications to modeling and control," *IEEE Trans. Syst., Man Cybern., B: Cybern.*, vol. SMC-15, no. 1, pp. 116–132, 1985.
29. K. Tanaka, T. Ikeda, and H.O. Wang, "Robust stabilization of a class of uncertain nonlinear systems via fuzzy control: Quadratic stabilizability, H-infinity control theory, and linear matrix inequalities," *IEEE Trans. Fuzzy Syst.*, vol. 4, pp. 1–13, 1996.
30. B.S. Chen, C.T. Tseng, and H.J. Uang, "Robustness design of nonlinear dynamic systems via fuzzy linear control," *IEEE Trans. Fuzzy Syst.*, vol. 7, no. 5, pp. 571–585, 1999.
31. B.S. Chen, C.H. Lee, and Y.C. Chang, "H_∞ tracking design of uncertain nonlinear SISO systems: Adaptive fuzzy approach," *IEEE Trans. Fuzzy Syst.*, vol. 4, no. 1, pp. 32–43, 1996.
32. S.H. Zak, *Systems and Control*, Oxford University Press, Oxford, 2003.
33. K. Tanaka and H.O. Wang, *Fuzzy Control Systems Design and Analysis: A Linear Matrix Inequality Approach*, Wiley, 2001.
34. T.H.S. Li and K.J. Lin, "Composite fuzzy control of nonlinear singularly perturbed systems," *IEEE Trans. Fuzzy Syst.*, vol. 15, pp. 176–187, 2007.
35. C.S. Tseng, "Robust fuzzy filter design for a class of nonlinear stochastic systems," *IEEE Trans. Fuzzy Syst.*, vol. 15, pp. 261–274, 2007.
36. J.S.R. Jang, C.T. Sun, and E. Mizutani, *Neuro-Fuzzy and Soft Computing: A Computational Approach to Learning and Machine Intelligence*, Prentice-Hall, Englewood Cliffs, NJ, 1997.
37. B. Kosko, *Fuzzy Engineering*, Prentice-Hall, Englewood Cliffs, NJ, 1997.
38. S.P. Banks, *Control Systems Engineering: Modelling and Simulation, Control Theory, and Microprocessor Implementation*, Prentice-Hall, Englewood Cliffs, NJ, 1986.
39. T. Leephakpreeda, "Novel determination of differential-equation solutions: Universal approximation method," *J. Comput. Appl. Math.*, vol. 146, pp. 443–457, 2002.

40. T.M. Apostol, *Mathematical Analysis*, Addison-Wesley, Boston, MA, 1974.
41. R. Fletcher, *Practical Methods of Optimization*, Wiley, New York, 1987.
42. S. Saatry and M. Bodson, *Adaptive Control: Stability, Convergence, and Robustness*, Prentice-Hall, Englewood Cliffs, NJ, 1989.
43. J.J.E. Slotine and W. Li, *Applied Nonlinear Control*, Prentice-Hall, Englewood Cliffs, NJ, 1991.
44. G. Rudolph, "Convergence analysis of canonical genetic algorithms," *IEEE Trans. Neural Netw.*, vol. 5, pp. 96–101, 1994.
45. H.G. Kaper and M. Garbey, *Asymptotic Analysis and the Numerical Solution of Partial Differential Equations*, Marcel Dekker, New York, 1991.
46. S.K. Nguang and W. Assawinchaichote, "H_∞ filtering for fuzzy dynamical systems with pole placement constraints," *IEEE Trans. Circuits Syst., I: Fund. Theory Appl.*, pp. 1503–1508, 2003.
47. W. Assawinchaichote and S.K. Nguang, "H_∞ filtering for nonlinear singularly perturbed systems with pole placement constraints: An LMI approach," *IEEE Trans. Signal Process.*, vol. 52, pp. 1659–1667, 2004.
48. H. Du, L. Xie, and Y.C. Soh, "H_∞ reduced-order approximation of 2-D digital filters," *IEEE Trans. Circuits Syst., I: Fund. Theory Appl.*, vol. 48, pp. 688–698, 2001.
49. Z. Tan, Y.C. Soh, and L. Xie, "Envelope-constrained H_∞ FIR filter design," *IEEE Trans. Circuits Syst., II: Analog Digit. Signal Process.*, vol. 47, pp. 79–82, 2000.
50. H. Gao, J. Lam, L. Xie, and C. Wang, "New approach to mixed H_2/H_∞ filtering for polytopic discrete-time systems," *IEEE Trans. Signal Process.*, vol. 53, pp. 3183–3192, 2005.
51. C. Du, L. Xie, and Y.C. Soh, "H_∞ filtering of 2-D discrete systems," *IEEE Trans. Signal Process.*, vol. 48, pp. 1760–1768, 2000.
52. M.J. Grimble and A.E. Sayed, "Solution of the H_∞ optimal linear filtering problem for discrete-time systems," *IEEE Trans. Acoust., Speech, Signal Process.*, vol. 38, no. 7, pp. 1092–1104, 1990.
53. S. Xu and T. Chen, "An LMI approach to the H_∞ filter design for uncertain systems with distributed delays," *IEEE Trans. Circuits Syst., II: Exp. Briefs*, vol. 51, p. 195201, 2004.
54. H. Shu, T. Chen, and B.A. Francis, "Minaimax design of hybrid multirate filter banks," *IEEE Trans. Circuits Syst., II: Analog Digit. Signal Process*, vol. 44, p. 120128, 1997.
55. A. ElSayed, M.J. Grimble, and D. Ho, "On H_∞ filtering in multivariable discretetime systems," *Proc. IEEE Int. Symp. Circuits Syst.*, vol. 2, pp. 1573–1576, 1988.
56. M.J. Grimble, "Polynomial matrix solution of the H_∞ filtering problem and the relationship to riccati equation state-space results," *IEEE Trans. Signal Process.*, vol. 41, pp. 67–81, 1993.
57. B.S. Chen, W.H. Chen, and H.L. Wu, "Robust H_2/H_∞ global linearization filter design for nonlinear stochastic systems," *IEEE Trans. Circuits Syst., I: Reg. Pap.*, vol. 56, no. 7, pp. 1441–1454, 2009.
58. B.S. Chen, C.L. Tsai, and Y.F. Chen, "Mixed H_2/H_∞ filtering design in multirate transmultiplexer systems: LMI approach," *IEEE Trans. Signal Process.*, vol. 49, pp. 2693–2701, 2001.
59. D. Hinrichsen and A.J. Pritchard, "Stochastic H_∞," *SIAM J. Control Optim.*, vol. 36, pp. 1504–1538, 1998.
60. B.S. Chen and W. Zhang, "Stochastic H_2/H_∞ control with state-dependent noise," *IEEE Trans. Autom. Control*, vol. 49, no. 1, pp. 45–57, 2004.
61. W. Zhang and B.S. Chen, "State feedback H_∞ control for a class of nonlinear stochastic systems," *SIAM J. Control Optim.*, vol. 44, no. 6, pp. 1973–1991, 2006.

62. W. Zhang, B.S. Chen, and C.S. Tseng, "Robust H_∞ filtering for nonlinear stochastic systems," *IEEE Trans. Signal Process.*, vol. 53, pp. 589–598, 2005.
63. B.S. Chen, C.H. Chiang, and S.K. Nguang, "Robust H_∞ synchronization design of nonlinear coupled network via fuzzy interpolation method," *IEEE Trans. Circuits Syst., I: Reg. Pap.*, vol. 58, pp. 349–362, 2011.
64. F. Yang, Z. Wang, G. Feng, and X. Liu, "Robust filtering with randomly varying sensor delay: The finite-horizon case," *IEEE Trans. Circuits Syst., I: Reg. Pap.*, vol. 56, pp. 664–672, 2009.
65. L. Ma, F. Da, and K.J. Zhang, "Exponential H_∞ filter design for discrete time-delay stochastic systems with Markovian jump parameters and missing measurements," *IEEE Trans. Circuits Syst., I: Reg. Pap.*, vol. 58, pp. 994–1007, 2011.
66. B. Shen, Z. Wang, and X. Liu, "A stochastic sampled-data approach to distributed H_∞ filtering in sensor networks," *IEEE Trans. Circuits Syst., I: Reg. Pap.*, vol. 58, pp. 2237–2246, 2011.
67. J. Zhang and Y. Xia, "New LMI approach to fuzzy H_∞ filter designs," *IEEE Trans. Circuits Syst., II: Exp. Briefs*, vol. 56, pp. 739–743, 2009.
68. E. Gershon, D.J.N. Limebeer, U. Shaked, and I. Yaesh, "Robust H_∞ filtering of stationary continuous-time linear systems with stochastic uncertainties," *IEEE Trans. Autom. Control*, vol. 46, pp. 1788–1793, 2001.
69. S. Xu and T. Chen, "Reduced-order H_∞ filtering for stochastic systems," *IEEE Trans. Signal Process.*, vol. 50, no. 12, pp. 2998–3007, 2002.
70. W.Y. Chiu and B.S. Chen, "Multi-source prediction under nonlinear dynamics in WSNs using a robust fuzzy approach," *IEEE Trans. Circuits Syst., I: Reg. Pap.*, vol. 58, pp. 137–149, 2011.
71. H.R. Karimi, "Robust delay-dependent H_∞ control of uncertain time-delay systems with mixed neutral, discrete, and distributed time-delays and Markovian switching parameters," *IEEE Trans. Circuits Syst., I: Reg. Pap.*, vol. 58, pp. 1910–1923, 2011.
72. K. Liu, *Stability of Infinite Dimensional Stochastic Differential Equations with Applications*, Chapman & Hall/CRC, Boca Raton, FL, 2006.
73. P.L. Chow, *Stochastic Partial Differential Equations*, Chapman & Hall/CRC, Boca Raton, FL, 2007.
74. Y. Lou, G. Hu, and P.D. Christofides, "Model predictive control of nonlinear stochastic partial differential equations with application to a sputtering process," *Aiche J.*, vol. 54, pp. 2065–2081, 2008.
75. C.L. Lin and B.S. Chen, "Robust observer-based control of large flexible structures," *J. Dyn. Syst. Meas. Control*, vol. 116, pp. 713–722, 1994.
76. P.D. Christofides, *Nonlinear and Robust Control of PDE Systems: Methods and Applications to Transport-Reaction Processes*, Boston MA, Birkhäuser, 2001.
77. J. Keener and J. Sneyd, *Mathematical Physiology*, Springer-Verlag, Berlin, 1998.
78. H.T. Banks, *Modeling and Control in the Biomedical Sciences*, Springer-Verlag, Berlin, 1975.
79. C.V. Pao, *Nonlinear Parabolic and Elliptic Equations*, Plenum, New York, 1992.
80. P.L. Chow, *Probabilistic Analysis and Related Topics*, Academic, vol. I, pp. 1–43, London, 1978.
81. D. Dawson, "Stochastic evolution equation," *Math. Biosci.*, vol. 15, pp. 287–316, 1972.
82. W. Fleming, *Distributed Parameter Stochastic Systems in Population Biology*, Springer, vol. 107, Berlin, 1975.
83. P.L. Chow, "Stability of nonlinear stochastic-evolution equations," *J. Math. Anal. Appl.*, vol. 89, pp. 400–419, 1982.
84. K. Zhou and P.P. Khargonekar, "Robust stabilization of linear systems with norm bounded time varying uncertainty," *Syst. Control Lett.*, vol. 10, pp. 17–20, 1998.

85. M. Kamrani and S.M. Hosseini, "The role of coefficients of a general SPDE on the stability and convergence of a finite difference method," *J. Comput. Appl. Math.*, vol. 234, pp. 1426–1434, 2010.
86. A.M. Davie and J.G. Gaines, "Convergence of numerical schemes for the solution of parabolic stochastic partial differential equations," *Math. Comput.*, vol. 70, no. 233, pp. 121–134, 2001.
87. J.C. Strikwerda, *Finite Difference Schemes and Partial Differential Equations*, 2nd ed, SIAM, Philadelphia, PA, 2004.
88. G. Evans, J. Blackledge, and P. Yardley, *Numerical Methods for Partial Differential Equations*, Springer-Verlag, Berlin, 2000.
89. H. Yoo, "Semi-discretization of stochastic partial differential equations on R^1 by a finite-difference method," *Math. Comput.*, vol. 69, no. 230, pp. 653–666, 1999.
90. Y. He and Q.G. Wang, "An improved ILMI method for static output feedback control with application to multivariable PID control," *IEEE Trans. Autom. Control*, vol. 51, pp. 1678–1683, 2006.
91. Y.Y. Cao, J. Lam, and Y.X. Sun, "Static output feedback stabilization: An ILMI approach," *Automatica*, vol. 34, pp. 1641–1645, 1998.
92. D. Huang and S.K. Nguang "Robust H_∞ static output feedback control of fuzzy systems: An ILMI approach," *IEEE Trans. Syst., Man Cybern., B: Cybern.*, vol. 36, pp. 216–222, 2006.
93. F.P. Incropera and D.P. DeWitt, *Introduction to Heat Transfer*, Wiley, New York, 1996.
94. H.J. Gao, Y. Zhao, J. Lam, and K. Chen, "H_∞ fuzzy filtering of nonlinear systems with intermittent measurements," *IEEE Trans. Fuzzy Syst.*, vol. 17, no. 2, pp. 291–300, 2009.
95. C.S. Tseng, "Robust fuzzy filter design for a class of nonlinear stochastic systems," *IEEE Trans. Fuzzy Syst.*, vol. 15, no. 2, pp. 261–274, 2007.
96. J.H. Zhang, Y.Q. Xia, and R. Tao, "New results on H_∞ filtering for fuzzy time-delay systems," *IEEE Trans. Fuzzy Syst.*, vol. 17, no. 1, pp. 128–137, 2009.
97. B. Jiang, Z.H. Mao, and P. Shi, "H_∞-filter design for a class of networked control systems via T-S fuzzy-model approach," *IEEE Trans. Fuzzy Syst.*, vol. 18, no. 1, pp. 201–208, 2010.
98. S.J. Huang, X.Q. He, and N.N. Zhang, "New results on filter design for nonlinear systems with time delay via T-S fuzzy models," *IEEE Trans. Fuzzy Syst.*, vol. 19, no. 1, pp. 193–199, 2011.
99. M. Grimble and A. El Sayed, "Solution to the H_∞ optimal linear filtering problem for discrete-time systems," *IEEE Trans. Acoust., Speech, Signal Process.*, vol. 38, no. 7, pp. 1092–1104, 1990.
100. B.S. Chen, C.L. Tsai, and D.S. Chen, "Robust H_∞ and H_∞ mixed filters for equalization designs of nonlinear communication systems: Fuzzy interpolation approach," *IEEE Trans. Fuzzy Syst.*, vol. 11, no. 3, pp. 384–398, 2003.
101. B.S. Chen, C.L. Tsai, and Y.F. Chen, "Mixed H_2/H_∞ filtering design in multirate transmultiplexer systems: LMI approach," *IEEE Trans. Signal Process.*, vol. 49, no. 11, pp. 2693–2701, 2001.
102. C. Jiang, Q. Zhang, and D. Zou, "Delay-dependent robust filtering for networked control systems with polytopic uncertainties," *Int. J. Innovat. Comput., Inf. Control*, vol. 6, no. 11, pp. 4857–4868, 2010.
103. Z. Deng, P. Shi, H. Yang, and Y. Xiam "Robust H_∞ filtering for nonlinear systems with interval time-varying delays," *Int. J. Innovat. Comput., Inf. Control*, vol. 6, no. 12, pp. 5527–5538, 2010.
104. Y. Liu and W. Wang, "Fuzzy H_∞ filtering for nonlinear stochastic systems with missing measurements," *ICIC Exp. Lett.*, vol. 3, no. 3(B), pp. 739–744, 2009.

105. M. Liu, J. You, and X. Ma, "H_∞ filtering for sampled-data stochastic systems with limited capacity channel," *Signal Process.*, vol. 91, no. 8, pp. 1826–1837, 2011.
106. M. Liu, D.W.C. Ho, and Y. Niu, "Robust filtering design for stochastic system with mode-dependent output quantization," *IEEE Trans. Signal Process.*, vol. 58, no. 12, pp. 6410–6416, 2010.
107. K. Liu, *Stability of Infinite Dimensional Stochastic Differential Equations with Applications*, Chapman & Hall/CRC, London, 2006.
108. P.L. Chow, *Stochastic Partial Differential Equations*, Chapman & Hall/CRC, London, 2007.
109. M.J. Anabtawi and S. Sathananthan, "Stability and convergence results for Itô-type parabolic partial differential equations in Hilbert spaces," *Stoch. Anal. Appl.*, vol. 27, no. 4, pp. 671–693, 2009.
110. Y. Lou, G. Hu, and P.D. Christofides, "Model predictive control of nonlinear stochastic partial differential equations with application to a sputtering process," *Aiche J.*, vol. 54, no. 8, pp. 2065–2081, 2008.
111. H. Yoo, "Semi-discretization of stochastic partial differential equations on R1 by a finite-difference method," *Math. Comput.*, vol. 69, no. 230, pp. 653–666, 1999.
112. C.I. Byrnes, I.G. Lauk´o, D.S. Gilliam, and V.I. Shubov, "Output regulation for linear distributed parameter systems," *IEEE Trans. Autom. Control*, vol. 45, no. 12, pp. 2236–2252, 2000.
113. C.L. Lin and B.S. Chen, "Robust observer-based control of large flexible structures," *J. Dyn. Syst. Meas. Control*, vol. 116, no. 4, pp. 713–722, 1994.
114. J. Keener and J. Sneyd, *Mathematical Physiology*, Springer-Verlag, Berlin, 1998.
115. H.T. Banks, *Modeling and Control in the Biomedical Sciences*, Springer-Verlag, Berlin, 1975.
116. C.V. Pao, *Nonlinear Parabolic and Elliptic Equations*, Plenum, New York, 1992.
117. Y.Y. Chen, Y.T. Chang, and B.S. Chen, "Fuzzy solutions to partial differential equations: Adaptive approach," *IEEE Trans. Fuzzy Syst.*, vol. 17, no. 1, pp. 116–127, 2009.
118. K. Tanaka, T. Ikeda, and H.O. Wang, "Fuzzy regulators and fuzzy observers: Relaxed stability conditions and LMI-based designs," *IEEE Trans. Fuzzy Syst.*, vol. 6, no. 2, pp. 250–265, 1998.
119. K. Tanaka and H.O. Wang, *Fuzzy Control Systems Design and Analysis: A Linear Matrix Inequality Approach*, Wiley, New York, 2001.
120. Q.L. Liang and J.M. Mendel, "Equalization of nonlinear time-varying channels using type-2 fuzzy adaptive filters," *IEEE Trans. Fuzzy Syst.*, vol. 8, no. 5, pp. 551–563, 2000.
121. F.H. Hsiao, J.D. Hwang, C.W. Chen, and Z.R. Tsai, "Robust stabilization of nonlinear multiple time-delay large-scale systems via decentralized fuzzy control," *IEEE Trans. Fuzzy Syst.*, vol. 13, no. 1, pp. 152–163, 2005.
122. J. An, G. Wen, C. Lin, and R. Li, "New results on a delay-derivative-dependent fuzzy H_∞ filter design for T-S fuzzy systems," *IEEE Trans. Fuzzy Syst.*, vol. 19, no. 4, pp. 770–779, 2011.
123. E.S. Tognetti, R.C.L.F. Oliveira, and P.L.D. Peres, "Selective H_2 and H_∞ stabilization of Takagi-Sugeno fuzzy systems," *IEEE Trans. Fuzzy Syst.*, vol. 19, no. 5, pp. 890–900, 2011.
124. C.L. Hwang, "Decentralized fuzzy control of nonlinear interconnected dynamic delay systems via mixed H_2/H_∞ optimization with Smith predictor," *IEEE Trans. Fuzzy Syst.*, vol. 19, no. 2, pp. 276–290, 2011.
125. C. Zhang, G. Feng, H. Gao, and J. Qiu, "H_∞ Filtering for nonlinear discrete-time systems subject to quantization and packet dropouts," *IEEE Trans. Fuzzy Syst.*, vol. 19, no. 2, pp. 353–365, 2011.

126. H. Dong, Z. Wang, D.W.C. Ho, and H. Gao, "Robust H_∞ fuzzy output- feedback control with multiple probabilistic delays and multiple missing measurements," *IEEE Trans. Fuzzy Syst.*, vol. 18, no. 4, pp. 712–725, 2010.
127. J. Qiu, G. Feng, and H. Gao, "Fuzzy-model-based piecewise H_∞ static-output- feedback controller design for networked nonlinear systems," *IEEE Trans. Fuzzy Syst.*, vol. 18, no. 5, pp. 919–934, 2010.
128. S.H. Kim and P. Park, "State-feedback-control design for discrete-time fuzzy systems using relaxation technique for parameterized LMI," *IEEE Trans. Fuzzy Syst.*, vol. 18, no. 5, pp. 985–993, 2010.
129. S.H. Kim, "Improved approach to robust H_∞ stabilization of discrete-time T-S fuzzy systems with time-varying delays," *IEEE Trans. Fuzzy Syst.*, vol. 18, no. 5, pp. 1008–1015, 2010.
130. P.L. Chow, "Stability of nonlinear stochastic-evolution equations," *J. Math. Anal. Appl.*, vol. 89, no. 2, pp. 400–419, 1982.
131. K. Zhou and P.P. Khargonekar, "Robust stabilization of linear systems with norm bounded time varying uncertainty," *Syst. Control Lett.*, vol. 10, no. 1, pp. 17–20, 1998.
132. G. Evans, J. Blackledge, and P. Yardley, *Numerical Methods for Partial Differential Equations*, Springer-Verlag, Berlin, 2000.
133. F.P. Incropera and D.P. DeWitt, *Introduction to Heat Transfer*, Wiley, New York, 1996.
134. A. Pikovsky, M. Rosenblum, and J. Kurths, *Synchronization: A Universal Concept in Nonlinear Sciences*, Cambridge, UK, Cambridge University Press, 2002.
135. F. Varela, J. Lachaux, E. Rodriguez, and J. Martinerie, "The brainweb: Phase synchronization and large-scale integration," *Nat. Rev. Neurosci.*, vol. 2, no. 4, pp. 229–239, 2001.
136. M. Bier, B. Bakker, and H. Westerhoff, "How yeast cells synchronize their glycolytic oscillations: A perturbation analytic treatment," *Biophys. J.*, vol. 78, no.3, pp. 1087–1093, 2000.
137. G. Kozyreff, A. Vladimirov, and P. Mandel, "Global coupling with time delay in an array of semiconductor laser," *Phys. Rev. Lett.*, vol. 85, no. 18, pp. 3809–3812, 2000.
138. Q. Li and D. Rus, "Global clock synchronization in sensor networks," *IEEE Trans. Comput.*, vol. 55, no. 2, pp. 214–226, 2006.
139. K. Lian, C. Chiu, T. Chiang, and P. Liu, "LIM-based fuzzy chaotic synchronization and communications," *IEEE Trans. Fuzzy Syst.*, vol. 9, no. 4, pp. 539–553, 2001.
140. K. Lian, T. Chiang, C. Chiua, and P. Liu, "Synthesis of fuzzy model-based designs to synchronization and secure communications for chaotic systems," *IEEE Trans. Syst., Man Cybern., B: Cybern.*, vol. 31, no. 1, pp. 66–83, 2001.
141. M. Chen, "Chaos synchronization in complex networks," *IEEE Trans. Circuits Syst., I: Reg. Pap.*, vol. 55, no. 5, pp. 1335–346, 2008.
142. L. Pecora and T. Carroll, "Master stability functions for synchronized coupled systems," *Phys. Rev. Lett.*, vol. 80, no. 10, pp. 2109–2112, 1998.
143. C.W. Wu and L. Chua, "Synchronization in an array of linear coupled dynamical systems," *IEEE Trans. Circuits Syst., I: Fund. Theory Appl.*, vol. 42, no.8, pp. 430–447, 1995.
144. T. Nishikawa and A. Motter, "Synchronization is optimal in nondiagonalizable networks," *Phys. Rev. E*, vol. 73, no. 6, p. 065106, 2006.
145. A. Motter, C. Zhou, and J. Kurthus, "Network synchronization, diffusion and the paradox of hetergeneity," *Phys. Rev. E*, vol. 71, no. 1, p. 016116, 2005.
146. D. Watts and S. Strogatz, "Colletive dynamics of small-world networks," *Nature*, vol. 393, no. 6684, pp. 440–442, 1998.

147. A. Barabasi and R. Albert, "Emergency of scaling in random networks," *Science*, vol. 286, no. 5439, pp. 509–512, 1999.
148. M. Barahona and L. Pecora, "Synchronization in small-world systems," *Phys. Rev. Lett.*, vol. 89, no. 5, p. 054101, 2002.
149. X. Wang and G. Chen, "Synchronization in small-world dynamical networks," *Int. J. Bifurcation Chaos*, vol. 12, no. 1, pp. 187–192, 2002.
150. J. Lu, X. Yu, G. Chen, and D. Cheng, "Characterizing the synchronizability of small-world dynamical networks," *IEEE Trans. Circuits Syst., I: Reg. Pap.*, vol. 51, no. 4, pp. 787–796, 2004.
151. C. Yin, B. Wang, W. Wang, and G. Chen, "Geographical effect on small-world network synchronization," *Phys. Rev. E*, vol. 77, no. 2, p. 027102, 2008.
152. X. Wang and G. Chen, "Synchronization in scale-free dynamical networks: Robustness and fragility," *IEEE Trans. Circuits Syst., I: Fund. Theory Appl.*, vol. 49, no. 1, pp. 54–62, 2002.
153. Z. Fei, H. Gao, and W.X. Zheng, "New synchronization stability of complex networks with an internal time-varying coupling delay," *IEEE Trans. Circuits Syst., II: Exp. Briefs*, vol. 56, no. 6, pp. 499–503, 2009.
154. W. He and J. Cao, "Exponential synchronization of hybrid coupled networks with delay coupling," *IEEE Trans. Neural Netw.*, vol. 21, no. 4, pp. 571–583, 2010.
155. J. Liang, Z. Wang, Y. Liua, and X. Liu, "Robust synchronization of an array of coupled stochastic discrete-time delayed neural networks," *IEEE Trans. Neural Netw.*, vol. 19, no. 11, pp. 1910–1921, 2008.
156. A. Kruszewski, W. Jianga, E. Fridman, J.P. Richard, and A. Toguyeni, "A switched system approach to exponential stabilization through communication network," *IEEE Control Syst. Technol.*, 2011, doi: 10.1109/TCST.2011.2159793.
157. B. Liu, D.J. Hill, and J. Yao, "Global uniform synchronization with estimated error under transmission channel noise," *IEEE Trans. Circuits Syst., I: Reg. Pap.*, vol. 56, no. 12, pp. 2689–2702, 2009.
158. A. Fradkov, B. Andrievsky, and B.J. Evans, "Synchronization of passifiable Lurie systems via limited-capacity communication channel," *IEEE Trans. Circuits Syst., I: Reg. Pap.*, vol. 56, no. 2, pp. 430–439, 2009.
159. J.G. Lu and D.J. Hill, "Global asymptotical synchronization of chaotic Lur's systems using sampled data: A linear matrix inequality approach," *IEEE Trans. Circuits Syst., II: Exp. Briefs*, vol. 55, no. 6, pp. 586–590, 2008.
160. M. Zhong and Q.L. Han, "Fault-tolerant master-slave synchronization for Lur's systems using time-delay feedback control," *IEEE Trans. Circuits Syst., I: Reg. Pap.*, vol. 56, no. 7, pp. 1391–1404, 2009.
161. Y. Hou, T. Liao, and J. Yan, "Synchronization of chaotic systems using output feedback control design," *Phys. A*, vol. 379, no. 1, pp. 81–89, 2007.
162. S.M Lee, D.H. Ji, J.H. Park, and S.C. Won, "Synchronization of chaotic systems via dynamic feedback approach," *Phys. Lett. A*, vol. 372, no. 29, pp. 4905–4912, 2008.
163. J.H. Park, D.H. Ji, S.C. Won, and S.M. Lee, "Synchronization of time- delayed chaotic systems," *Appl. Math. Compt.*, vol. 204, no. 1, pp. 170–177, 2008.
164. C.K. Ahn, "T-S fuzzy synchronization for chaotic systems via delayed output feedback control," *Nonlinear Dyn.*, vol. 59, no. 4, pp. 535–543, 2010.
165. B.S. Chen, C.H. Chiang, and S.K. Nguang, "Robust synchronization design of nonlinear coupled network via fuzzy interpolation methods," *IEEE Trans. Circuits Syst., I: Reg. Pap.*, vol. 58, no. 2, pp. 349–362, 2011.
166. H. Huang and G. Feng, "Robust synchronization of chaotic Lurie's systems," *Chaos*, vol. 18, no. 3, p. 033113, 2008.

167. J.H. Park, D.H. Ji, S.C. Won, and S.M. Lee, "Adaptive synchronization of unified chaotic systems," *Mod. Phys. Lett. B*, vol. 23, no. 9, pp. 1157–1169, 2009.
168. H.R. Karimi and H. Gao, "New delay-dependent exponential synchronization for uncertain neural networks with mixed time delays," *IEEE Trans. Syst., Man Cybern., B: Cybern.*, vol. 40, no. 1, pp. 173–185, 2010.
169. C.K. Ahn, "An approach to antisynchronization for chaotic systems," *Phys. Lett. A*, vol. 373, no. 20, pp. 1729–1733, 2009.
170. K. Liu, *Stability of Infinite Dimensional Stochastic Differential Equations with Applications*, New York, Chapman & Hall/CRC, 2006.
171. P.L. Chow, *Stochastic Partial Differential Equations*, New York, Chapman & Hall/CRC, 2007.
172. B.S. Chen and Y.T. Chang, "Fuzzy state space modeling and robust observer- based control design for nonlinear partial differential systems," *IEEE Trans. Fuzzy Syst.*, vol. 17, no. 5, pp. 1025–1043, 2009.
173. Y.T. Chang and B.S. Chen, "A fuzzy approach for robust reference tracking control design of nonlinear distributed parameter time-delay systems and its application," *IEEE Trans. Fuzzy Syst.*, vol. 18, no. 6, pp. 1041–1057, 2010.
174. C.I. Byrnes, I.G. Lauk'o, D.S. Gilliam, and V.I. Shabov, "Output regulation for linear distributed parameter systems," *IEEE Trans. Autom. Control*, vol. 45, no. 12, pp. 2236–2252, 2000.
175. C.L. Lin and B.S. Chen, "Robust observer-based control of large flexible structures," *J. Dyn. Syst. Maes. Control*, vol. 116, pp. 713–722, 1994.
176. H.T. Banks, *Modeling and Control in the Biomedical Sciences*, New York, Springer-Verlag, 1975.
177. C.V. Pao, *Nonlinear Parabolic and Elliptic Equations*, New York, Plenum Press, 1992.
178. J.C. Strikwerda, *Finite Difference Schemes and Partial Differential Equations*, 2nd ed, Philadelphia, PA, SIAM, 2004.
179. G. Evans, J. Blackledge, and P. Yardley, *Numerical Methods for Partial Differential Equations*, Springer-Verlag, 2000.
180. Y.Y. Chen, Y.T. Chang, and B.S. Chen, "Fuzzy solutions to partial differential equations: Adaptive approach," *IEEE Trans. Fuzzy Syst.*, vol. 17, no. 1, pp. 116–127, 2009.
181. E. Fridman and Y. Orlov, "An LMI approach to boundary control of semilinear parabolic and hyperbolic systems," *Automatica*, vol. 45, no. 9, pp. 2060–2066, 2009.
182. A. Isidori and A. Astolfi, "Disturbance attenuation and control via measurement feedback in nonlinear systems," *IEEE Trans. Autom. Control*, vol. 37, no. 9, pp. 1283–1293, 1992.
183. H. Gao, J. Lam, L. Xie, and C. Wang, "New approach to mixed filtering for polytopic discrete-time systems," *IEEE Trans. Signal Process.*, vol. 53, no. 8, pp. 3183–3192, 2005.
184. B.S. Chen, W.H. Chen, and H. Wu, "Robust global linearization filter design for nonlinear stochastic systems," *IEEE Trans. Circuits Syst., I: Reg. Pap.*, vol. 56, no. 7, pp. 1441–1454, 2009.
185. C.L. Chen, G. Feng, D. Sun, and X.P. Guan, "Output feedback control of discrete-time fuzzy systems with application to chaos control," *IEEE Trans. Fuzzy Syst.*, vol. 13, no. 4, pp. 531–543, 2005.
186. D. Huang and S.K. Nguang, "Robust output feedback control of fuzzy systems: An LMI approach," *IEEE Trans. Syst., Man Cybern., B. Cybern.*, vol. 36, no. 1, pp. 216–222, 2006.
187. B.S. Chen, C.T. Tseng, and H.C. Wang, "Mixed fuzzy output feedback control for nonlinear uncertain systems: LMI approach," *IEEE Trans. Fuzzy Syst.*, vol. 8, no. 3, pp. 249–265, 2000.

188. S. Mikhailov, "Analysis of united boundary-domain integro-differential and integral equations for a mixed BVP with variable coefficient," *Math. Meth. Appl. Sci.*, vol. 29, pp. 715–739, 2006.
189. J. Chang and B. Guo, "Identification of variable spacial coefficients for a beam equation from boundary measurements," *Automatica*, vol. 43, pp. 732–737, 2007.
190. S. Boyd, L. El Ghaoui, E. Feron, and V. Balakrishnan, *Linear Matrix Inequalities in Systems and Control Theory*, Philadelphia, PA, SIAM, 1994.
191. J. Oostveen, *Strongly Stabilizable Distributed Parameter Systems*, Philadelphia, PA, SIAM, 2000.
192. J.G. Lu, "Global exponential stability and periodicity of reaction-diffusion delayed recurrent neural networks with Dirichlet boundary conditions," *Chaos, Solitons Fractals*, vol. 35, pp. 116–125, 2008.
193. F. John, *Partial Differential Equations*, New York, Springer-Verlag, 1982.
194. S.N. Alturi and A.K. Amos, *Large Space Structures: Dynamics and Control*, New York, Springer-Verlag, 1988.
195. M.J. Balas, "Feedback control of alexible systems," *IEEE Trans. Autom. Control*, vol. 23, pp. 673–679, 1978; Ibid., vol. 27, p. 522, 1982.
196. B.J. Benhabib, R.P. Iwens, and R.L. Jackson, "Stability of large space structure control systems using positivity concepts," *J. Guid. Control*, vol. 4, pp. 487–494, 1981.
197. J. Bontsema and R.F. Curtain, "A note on spillover and robustness for flexible systems," *IEEE Trans. Autom. Control*, vol. 33, pp. 567–569, 1988.
198. B.S. Chen and S.S. Wang, "The stability of feedback control with nonlinear saturating actuator: time domain approach," *IEEE Trans. Autom. Control*, vol. 33, p. 483–487, 1988.
199. R. Curtain and A. Pritchrad, *Functional Analysis in Modern Applied Mathematics*, New York, Academic Press, 1977.
200. C.A. Desoer and M. Vidyasagar, *Feedback Systems: Input- Output Properties*, New York, Academic Press, 1975.
201. D.J. Inman, *Vibration with Control, Measurement, and Stability*, Englewood Cliffs, NJ, Prentice Hall, 1989.
202. T. Kato, *Perturbation Theory for Linear Operators*, New York, Springer-Verlag, 1966.
203. R.L. Kosut, H. Salzwedel, and A. Emani-Naeini, "Robust control of flexible spacecraft," *J. Guidance Control Dynam.*, vol. 6, pp. 104–111, 1983.
204. R.V. Patel, M. Toda, and B. Sridhar, "Robustness of linear quadratic state feedback designs in the presence of system uncertainty," *IEEE Trans. Autom. Control*, vol. 22, pp. 945–949, 1977.
205. V.M. Popov, *Hyperstability of Control Systems*, New York, Springer-Verlag, 1973.
206. A.J. Pritchard and A. Townley, "Robust compensator design via structured stability radii," *Syst. Control Lett.*, vol. 11, pp. 33–37, 1988.
207. L. Qiu and E.J. Davison, Proc. 25th IEEE Conference on Decision and Control, Athens, pp. 751–755, 1986.
208. D.A. Wagie and R.E. Skelton, Proc. American Control Conference, San Diego, CA, pp. 135–140, 1984.
209. S.J. Wang and J.M. Cameron, "Dynamics and control of a large space antenna," *J. Guid. Control Dynam.*, vol. 7, p. 69–76, 1984.
210. W. Wonham, *Linear Multivariable Control*, New York, Springer-Verlag, 1974.
211. J.S.-C. Yuan and M.E. Stieber, *J. Guid. Control Dynam.*, vol. 9, pp. 228–234, 1986.
212. M.J. Balas, "Feedback control of flexible systems," *IEEE Trans. Autom. Control*, vol. AC-23, no. 4, pp. 673–679, 1978.

213. R.J. Benhabib, R.P. Iwens, and R.L. Jackson, "Stability of large space structure control systems using positivity concepts," *J. Guid. Control*, vol. 4, no. 5, pp. 487–494, 1981.
214. M.J. Balas, "Toward a more practical control theory for distributed parameter systems," *Control Dynam. Syst.*, vol. 18, pp. 361–421, 1982.
215. B.S. Chen, C.L. Lin, and F.B. Hsiao, "Robust observer-based control of a vibrating beam," *J. Mech. Sci. Eng.*, vol. 205, pp. 77–89, 1991.
216. R.L. Kosut, H. Salzwedel, and A. Emami-Naeini, "Robust control of flexible spacecraft," *J. Guid. Control Dynam.*, vol. 6, no. 2, pp. 104–111, 1983.
217. J.S.-C. Yuan and M.E. Stieber, "Robust beam-pointing and attitude control of a flexible spacecraft," *J. Guid. Control Dynam.*, vol. 9, no. 2, pp. 228–234, 1986.
218. J. Bontsema and R.F. Curtain, "A note on spillover and robustness for flexible systems," *IEEE Trans. Autom. Control*, vol. 33, no. 6, pp. 567–569, 1988.
219. L. Meirovitch and H. Baruh, "Effect of damping on observation spillover instability," *J. Optim. Theory Appl.*, vol. 35, no. 1, pp. 31–44, 1981.
220. G. Skidmore, W.L. Hallauer, Jr., and L.C. Mesquita, "Experimental-Theoretical Study of Active Vibration Control," Proc. of 1st International Modal Analysis Conference, 1982.
221. M. Vidyasagar, *Control System Synthesis: A Factorization Approach*, M.I.T. Press, Cambridge, MA, 1985.
222. C.A. Desoer and M. Vidyasagar, *Feedback Systems: Input-Output Properties*, Academic Press, London, 1975.
223. J. Bontsema, R.F. Curtain, and J.M. Schumacher, "Robust control of flexible structures: A case study," *Automatica*, vol. 24, no. 2, pp. 177–186, 1988.
224. K. Zhou and P.P. Khargonekar, "An algebraic Riccati equation approach to H∞ optimization," *Syst. Control Lett.*, vol. 11, pp. 85–89, 1988.
225. I. Bar-Kana, H. Kaufman, and M.J. Balas, "Model reference adaptive control of large structural systems," *J. Guid. Control Dynam.*, vol. 6, no. 2, pp. 112–118, 1983.
226. R. Courant and D. Hilbert, *Methods of Mathematical Physics*, Wiley-Interscience, New York, 1953.
227. J.N. Franklin, *Matrix Theory*, Prentice-Hall, Englewood Cliffs, NJ, 1968.
228. L. Guo and S.A. Billings, "Identification of partial differential equation models for continuous spatio-temporal dynamical systems," *IEEE Trans. Circuits Syst., II: Exp. Briefs*, vol. 53, pp. 657–661, 2006.
229. J.W. Wang, H.N. Wu, and H.X. Li, "Distributed fuzzy control design of nonlinear hyperbolic PDE systems with application to nonisothermal plug-flow reactor," *IEEE Trans. Fuzzy Syst.*, vol. 19, no. 3, pp. 514–526, 2011.
230. P.D. Christofides, *Nonlinear and Robust Control of PDE Systems: Methods and Applications to Transport-Reaction Processes*, Birkhauser, Basel, 2001.
231. D. Fan and Y. Wang, "Global numerical boundary condition based PDE solution techniques for open-region electromagnetic field problems," *IEEE Trans. Antennas Propag.*, vol. 41, pp. 253–260, 1993.
232. M. Lysaker, A. Lundervold, and X.C. Tai, "Noise removal using fourth-order partial differential equation with applications to medical magnetic resonance images in space and time," *IEEE Trans. Image Process.*, vol. 12, pp. 1579–1590, 2003.
233. R.V.D. Berg, E. Lefeber, and K. Rooda, "Modeling and control of a manufacturing flow line using partial differential equations," *IEEE Trans. Control Syst. Technol.*, vol. 16, pp. 130–136, 2008.
234. J. Keener and J. Sneyd, *Mathematical Physiology*, Springer-Verlag, Berlin, 1998.
235. H.T. Banks, *Modeling and Control in the Biomedical Sciences*, Springer-Verlag, Berlin, 1975.

236. C.L. Lin and B.S. Chen, "Robust observer-based control of large flexible structures," *J. Dyn. Syst. Meas. Control*, vol. 116, pp. 713–722, 1994.
237. K. Liu, *Stability of Infinite Dimensional Stochastic Differential Equations with Applications*, Chapman & Hall/CRC, London, 2006.
238. P. L. Chow, *Stochastic Partial Differential Equations*, Chapman & Hall/CRC, London, 2007.
239. M. Garetto and E. Leonardi, "Analysis of random mobility models with partial differential equations," *IEEE Trans. Mobile Comput.*, vol. 6, pp. 1204–1217, 2007.
240. Y. Lou, G. Hu, and P.D. Christofides, "Model predictive control of nonlinear stochastic partial differential equations with application to a sputtering process," *Aiche J.*, vol. 54, pp. 2065–2081, 2008.
241. H. Yoo, "Semi-discretization of stochastic partial differential equations on R1 by a finite-difference method," *Math. Comput.*, vol. 69, no. 230, pp. 653–666, 1999.
242. P.L. Chow, *Probabilistic Analysis and Related Topics*, Academic, vol. I, pp. 1–43, Cambridge, MA, 1978.
243. D. Dawson, "Stochastic evolution equation," *Math. Biosci.*, vol. 15, pp. 287–316, 1972.
244. W. Fleming, *Distributed Parameter Stochastic Systems in Population Biology*, Springer, vol. 107, Berlin, 1975.
245. H.R. Karimi, "Robust delay-dependent H_∞ control of uncertain time-delay systems with mixed neutral, discrete, and distributed time-delays and Markovian switching parameters," *IEEE Trans. Circuits Syst., I: Reg. Pap.*, vol. 58, pp. 1910–1923, 2011.
246. B.S. Chen, C.H. Chiang, and S.K. Nguang, "Robust H_∞ synchronization design of nonlinear coupled network via fuzzy interpolation method," *IEEE Trans. Circuits Syst., I: Reg. Pap.*, vol. 58, pp. 349–362, 2011.
247. W.Y. Chiu and B.S. Chen, "Multi-source prediction under nonlinear dynamics in WSNs Using a robust fuzzy approach," *IEEE Trans. Circuits Syst., I: Reg. Pap.*, vol. 58, pp. 137–149, 2011.
248. W. Paszke, K. Gałkowski, E. Rogers, and J. Lam, "H_2 and mixed H_2/H_∞ stabilization and disturbance attenuation for differential linear repetitive processes," *IEEE Trans. Circuits Syst., I: Reg. Pap.*, vol. 55, pp. 2813–2826, 2008.
249. S. Xu, J. Lam, and X. Mao, "Delay-dependent H_∞ control and filtering for uncertain Markovian jump systems with time-varying delays," *IEEE Trans. Circuits Syst., I: Reg. Pap.*, vol. 54, pp. 2070–2077, 2007.
250. L. Zhang, B. Huang, and J. Lam, "LMI synthesis of H_2 and mixed H_2/H_∞ controllers for singular systems," *IEEE Trans. Circuits Syst., II: Analog Digit. Signal Process.*, vol. 50, pp. 615–626, 2003.
251. R. Wang, B. Wang, G.P. Liu, W. Wang, and D. Rees, "H_∞ controller design for networked predictive control systems based on the average dwell-time approach," *IEEE Trans. Circuits Syst., II: Exp. Briefs*, vol. 57, pp. 310–314, 2010.
252. B.S. Chen and W. Zhang, "Stochastic H_2/H_∞ control with state-dependent noise," *IEEE Trans. Automat. Control*, vol. 49, no. 1, pp. 45–57, 2004.
253. W. Zhang and B.S. Chen, "State feedback H_∞ control for a class of nonlinear stochastic systems," *SIAM J. Control Optim.*, vol. 44, no. 6, pp. 1973–1991, 2006.
254. Z. Wang, F. Yang, D.W.C. Ho, and X. Liu, "Robust H_∞ control for networked systems with random packet losses," *IEEE Trans. Syst., Man Cybern., B: Cybern.*, vol. 37, pp. 916–924, 2007.
255. B.S. Chen, W.H. Chen, and H.L. Wu, "Robust H_2/H_∞ global linearization filter design for nonlinear stochastic systems," *IEEE Trans. Circuits Syst., I: Reg. Pap.*, vol. 56, no. 7, pp. 1441–1454, 2009.
256. W.H. Chen and B.S. Chen, "Robust filter for linear stochastic partial differential systems via a set of sensor measurements," *IEEE Trans. Circuits Syst., I: Reg. Pap.*, vol. 59, pp. 1251–1264, 2012.

257. B. Shen, Z. Wang, and X. Liu, "A stochastic sampled-data approach to distributed H_∞ filtering in sensor networks," *IEEE Trans. Circuits Syst., I: Reg. Pap.*, vol. 58, pp. 2237–2246, 2011.
258. B.S. Chen, W.H. Chen, and W. Zhang, "Robust filter for nonlinear stochastic partial differential systems in sensor signal processing: Fuzzy approach," *IEEE Trans. Fuzzy Syst.*, vol. 20, no.5, pp. 957–970, 2012.
259. P.L. Chow, "Stability of nonlinear stochastic-evolution equations," *J. Math. Anal. Appl.*, vol. 89, pp. 400–419, 1982.
260. K. Zhou and P.P. Khargonekar, "Robust stabilization of linear systems with norm bounded time varying uncertainty," *Syst. Control Lett.*, vol. 10, pp. 17–20, 1998.
261. M. Kamrani and S.M. Hosseini, "The role of coefficients of a general SPDE on the stability and convergence of a finite difference method," *J. Comput. Appl. Math.*, vol. 234, pp. 1426–1434, 2010.
262. A.M. Davie and J.G. Gaines, "Convergence of numerical schemes for the solution of parabolic stochastic partial differential equations," *Math. Comput.*, vol. 70, no. 233, pp. 121–134, 2001.
263. J.C. Strikwerda, *Finite Difference Schemes and Partial Differential Equations*, SIAM, Philadelphia, PA, 2004.
264. G. Evans, J. Blackledge, and P. Yardley, *Numerical Methods for Partial Differential Equations*, Springer-Verlag, Berlin, 2000.
265. Y. He and Q.G. Wang, "An improved ILMI method for static output feedback control with application to multivariable PID control," *IEEE Trans. Automat. Control*, vol. 51, pp. 1678–1683, 2006.
266. Y.Y. Cao, J. Lam, and Y.X. Sun, "Static output feedback stabilization: An ILMI approach," *Automatica*, vol. 34, pp. 1641–1645, 1998.
267. F.P. Incropera and D.P. DeWitt, *Introduction to Heat Transfer*, Wiley, New York, 1996.
268. M.J. Balas, "Feedback control of flexible systems," *IEEE Trans. Autom. Control*, vol. AC-23, no. 4, pp. 673–679, 1978.
269. C.L. Lin and B.S. Chen, "Robust observer-based control of large flexible structures," *ASME J. Dyn. Syst. Meas. Control*, vol. 116, pp. 713–722, 1994.
270. P.D. Christofides, *Nonlinear and Robust Control of PDE Systems: Methods and Applications to Transport-Reaction Processes*, Birkhauser, Basel, 2001.
271. P.D. Christofides, "Robust control of parabolic PDE systems," *Chem. Eng. Sci.*, vol. 53, pp. 2949–2965, 1998.
272. J. P. Kernevez and D. Thomas, "Numerical analysis and control of some biochemical systems, " *Appl. Math. Optim.*, vol. 1, pp.222–285, 1975.
273. H.T. Banks, *Modeling and Control in the Biomedical Sciences*, Springer-Verlag, Berlin, 1975.
274. C.V. Pao, *Nonlinear Parabolic and Elliptic Equations*, Plenum, New York, 1992.
275. M.J. Balas, "The Galerkin method and feedback control of linear distributed parameter systems," *J. Math. Anal. Appl.*, vol. 91, pp. 527–546, 1983.
276. M.J. Balas, "Finite-dimensional control of distributed parameter systems by Galerkin approximation of infinite dimensional controllers," *J. Math. Anal. Appl.*, vol. 114, pp. 17–36, 1986.
277. R.F. Curtain and H. Zwart, *An Introduction to Infinite-Dimensional Linear Systems Theory*, Springer-Verlag, Berlin, 1995.
278. J. Robinson, *Infinite-Dimensional Dynamical Systems*, Cambridge University Press, London, 2001.
279. M.J. Balas, "Nonlinear finite-dimensional control of a class of nonlinear distributed parameter systems using residual mode filters: A proof of local exponential stability," *J. Math. Anal. Appl.*, vol. 162, pp. 63–70, 1991.

280. J. Baker and P.D. Christofides, "Finite-dimensional approximation and control of non-linear parabolic PDE systems," *Int. J. Control*, vol. 73, no. 5, pp. 439–456, 2000.
281. A. Armaou and P.D. Christofides, "Dynamic optimization of dissipative PDE systems using nonlinear order reduction," *Chem. Eng. Sci.*, vol. 57, no. 24, pp. 5083–5114, 2002.
282. B.S. Chen, C.S. Tseng, and H.J. Uang, "Mixed H_2/H_∞ fuzzy output feedback control design for nonlinear dynamic systems: An LMI approach," *IEEE Trans. Fuzzy Syst.*, vol. 8, no. 3, pp. 249–265, 2000.
283. B.S. Chen, C.S. Tseng, and H.J. Uang, "Robustness design of nonlinear uncertain system via fuzzy linear control," *IEEE Trans. Fuzzy Syst.*, vol. 7, no. 5, pp. 571–585, 1999.
284. C.S. Tseng, B.S. Chen, and H.J. Uang, "Fuzzy tracking control design for nonlinear dynamic systems via T-S fuzzy model," *IEEE Trans. Fuzzy Syst.*, vol. 9, no. 3, pp. 381–392, 2001.
285. T.H.S. Li and S.H. Tsai, "T-S fuzzy bilinear model and fuzzy controller design for a class of nonlinear systems," *IEEE Trans. Fuzzy Syst.*, vol. 15, no. 3, pp. 494–506, 2007.
286. T.H.S. Li and J.J. Lin, "Composite fuzzy control of nonlinear singularly perturbed systems," *IEEE Trans. Fuzzy Syst.*, vol. 15, no. 2, pp. 176–187, 2007.
287. K.Y. Lian, C.S. Chiang, and P. Liu, "Secure communication for chaotic systems with robust performance via fuzzy observer-based design," *IEEE Trans. Fuzzy Syst.*, vol. 9, no. 1, pp. 212–220, 2001.
288. C.S. Tseng and C.K. Hwang, "Fuzzy observer-based fuzzy control design for nonlinear systems with persistent bounded disturbances," *Fuzzy Sets Syst.*, vol. 158, no. 2, pp. 164–179, 2007.
289. Y.Y. Chen, Y.T. Chang, and B.S. Chen, "Fuzzy solutions to partial differential equations: Adaptive approach," *IEEE Trans. Fuzzy Syst.*, vol. 17, no. 1, pp. 116–127, 2009.
290. K. Yuan, H.X. Li, and J. Cao, "Robust stabilization of the distributed para- meter system with time delay via fuzzy control," *IEEE Trans. Fuzzy Syst.*, vol. 16, no. 3, pp. 567–584, 2008.
291. H.N. Wu and H.X. Li, "H_∞ fuzzy observer-based control for a class of nonlinear distributed parameter systems with control constraints," *IEEE Trans. Fuzzy Syst.*, vol. 16, no. 2, pp. 502–516, 2008.
292. V. Thome Garlerkin, *Finite Element Methods for Parabolic Problems*, Springer-Verlag, Berlin, 1997.
293. A. van der Schaft, *L_2-Gain and Passivity Techniques in Nonlinear Control*, Springer-Verlag, Berlin, 2000.
294. A. van der Schaft, "Robust stabilization of nonlinear systems via stable kernel representation with L_2-gain bounden uncertainty," *Syst. Control Lett.*, vol. 24, pp. 75–81, 1995.
295. G. Dong, *Nonlinear Partial Differential Equations of Second Order*, American Mathematical Society, Providence, RI, 1991.
296. S.G. Gao, N.W. Rees, and G. Feng, "Analysis and design for a class of complex control systems part I: Fuzzy modelling and identification," *Automatica*, vol. 33, pp. 1017–1028, 1997.
297. S.G. Gao, N.W. Rees, and G. Feng, "Universal fuzzy controllers for a class of nonlinear systems," *Fuzzy Sets Syst.*, vol. 122, pp. 117–123, 2001.
298. K. Tanaka and H.O. Wang, *Fuzzy Control Systems Design and Analysis: A Linear Matrix Inequality Approach*, Wiley, New York, 2001.
299. K.M. Passino and S. Yurkovich, *Fuzzy Control*, Addison-Wesley, Boston, MA, 1998.
300. R. Johansson, *System Modeling and Identification*, Prentice-Hall, Englewood Cliffs, NJ, 1993.
301. L. Ljung, *System Identification: Theory for The User*, 2nd ed., Prentice-Hall, Englewood Cliffs, NJ, 1999.

302. L.X. Wang, *A Course in Fuzzy Systems and Control*, Prentice-Hall, Englewood Cliffs, NJ, 1997.
303. X.J. Zeng and M.G. Singh, "Approximation theory for fuzzy systems SISO case," *IEEE Trans. Fuzzy Syst.*, vol. 2, no. 2, pp. 162–176, 1994.
304. L.W. Johnson and R.D. Riess, *Numerical Analysis*, Addison-Wesley, Boston, MA, 1982.
305. L.C. Evans, *Partial Differential Equations*, American Mathematical Society, Providence, RI, 1998.
306. J.C. Willems, "Dissipative dynamical systems Part I: General theory," *Arch. Ration. Mech. Anal.*, vol. 45, pp. 321–351, 1972.
307. S. Boyd, L.E. Ghaoui, E. Feron, and V. Balakrishnan, "Linear Matrix Inequalities in System and Control Theory," *Soc. Ind. Appl. Math.*, 1994.
308. S.G. Cao, N.W. Rees, and G. Feng, "Analysis and design of a class of continuous time fuzzy control systems," *Int. J. Control*, vol. 64, pp. 1069–1087, 1996.
309. K. Tanaka, T. Hori, and H.O. Wang, "A multiple Lyapunov function approach to stabilization of fuzzy control systems," *IEEE Trans. Fuzzy Syst.*, vol. 11, no. 4, pp. 582–589, 2003.
310. G. Feng, C.L. Chen, D. Sun, and Y. Zhu, "H_∞ controller synthesis of fuzzy dynamic systems based on piecewise Lyapunov functions and bilinear matrix inequalities," *IEEE Trans. Fuzzy Syst.*, vol. 13, no. 1, pp. 94–103, 2005.
311. G. Feng, "H_∞ controller design of fuzzy dynamic systems based on piecewise Lyapunov functions," *IEEE Trans. Syst. Man Cybern., B: Cybern.*, vol. 34, no. 1, pp. 283–292, 2004.
312. K. Tanaka, T. Ikeda, and H.O. Wang, "Fuzzy regulators and fuzzy observers: Relaxed stability conditions and LMI-based designs," *IEEE Trans. Fuzzy Syst.*, vol. 6, no. 2, pp. 250–265, 1998.
313. K. Tanaka, T. Ikeda, and H.O. Wang, "Robust stabilization of a class of uncertain nonlinear systems via fuzzy control: Quadratic stabilizability, H_∞ control theory, and linear matrix inequalities," *IEEE Trans. Fuzzy Syst.*, vol. 4, no. 1, pp. 1–13, 1996.
314. W.J. Wang and C.H. Sun, "Relaxed stability and stabilization conditions for a TS fuzzy discrete system," *Fuzzy Sets Syst.*, vol. 156, no. 2, pp. 208–225, 2005.
315. W.J. Wang and C.H. Sun, "A relaxed stability criterion for TS fuzzy discrete systems," *IEEE Trans. Syst. Man Cybern., B: Cybern.*, vol. 34, no. 5, pp. 2155–2158, 2004.
316. M. Kovara and M. Stingl, "PENNON: A code for convex nonlinear and semi- definite programming," *Optim. Methods Softw.*, vol. 18, pp. 317–333, 2003.
317. E. Kim and S. Kim, "Stability analysis and synthesis for an affine fuzzy control system via LMI and ILMI: A continuous case," *IEEE Trans. Fuzzy Syst.*, vol. 10, no. 3, pp. 391–400, 2002.
318. B.S. Chen, C.S. Tseng, and H.C. Wang, "Mixed H_2/H_∞ fuzzy output feedback control for nonlinear uncertain systems: LMI approach," *IEEE Trans. Fuzzy Syst.*, vol. 8, no. 3, pp. 249–265, 2000.
319. J.C. Lo and M.L. Lin, "Observer-based robust H_∞ control for fuzzy systems using two-step procedure," *IEEE Trans. Fuzzy Syst.*, vol. 12, no. 3, pp. 350–359, 2004.
320. D. Huang and S.K. Nguang, "Robust H_∞ static output feedback control of fuzzy systems: An ILMI approach," *IEEE Trans. Syst. Man Cybern., B: Cybern.*, vol. 36, no. 1, pp. 216–222, 2006.
321. D.W.C. Ho and G. Lu, "Robust stabilization for a class of discrete-time non- linear systems via output feedback: The unified LMI approach," *Int. J. Control*, vol. 76, pp. 105–115, 2003.
322. Z. Wang, F. Yang, D.W.C. Ho, and X. Liu, "Robust H_∞ control for networked systems with random packet losses," *IEEE Trans. Syst. Man Cybern., B: Cybern.*, vol. 37, no. 4, pp. 916–924, 2007.

323. P. Gahinet, A. Nemirovski, A.J. Laub, and M. Chilali, *LMI Control Toolbox for Use with MATLAB*, Natick, MA, The MathWorks, Inc, 2004.
324. P.D. Christofides, *Nonlinear and Robust Control of PDE Systems: Methods and Applications to Transport-Reaction Processes*, Birkhauser, Basel, 2001.
325. J. Keener and J. Sneyd, *Mathematical Physiology*, Springer-Verlag, Berlin, 1998.
326. C.V. Pao, *Nonlinear Parabolic and Elliptic Equations*, Plenum, New York, 1992.
327. M.J. Balas, "Feedback control of flexible systems," *IEEE Trans. Autom. Control*, vol. AC-23, no. 4, pp. 673–679, 1978.
328. C.L. Lin and B.S. Chen, "Robust observer-based control of large flexible structures," *J. Dyn. Syst. Meas. Control*, vol. 116, pp. 713–722, 1994.
329. C. Koch, *Biophysics of Computation: Information Processing in Single Neurons*, Oxford University Press, Oxford, 1999.
330. R.F. Curtain and H. Zwart, *An Introduction to Infinite-Dimensional Linear Systems Theory*, Springer-Verlag, Berlin, 1995.
331. J. Robinson, *Infinite-Dimensional Dynamical Systems*, Cambridge University Press, London, 2001.
332. B. van Keulen, *H∞-Control for Distributed Parameter Systems: A State-Space Approach*, Birkhauser, Basel, 1993.
333. M.J. Balas, "Nonlinear finite-dimensional control of a class of nonlinear distributed parameter systems using residual mode filters: A proof of local exponential stability," *J. Math. Anal. Appl.*, vol. 162, pp. 63–70, 1991.
334. J. Baker and P.D. Christofides, "Finite-dimensional approximation and control of non-linear parabolic PDE systems," *Int. J. Control*, vol. 73, no. 5, pp. 439–456, 2000.
335. C.S. Tseng, B.S. Chen, and H.J. Uang, "Fuzzy tracking control design for nonlinear dynamic systems via T-S fuzzy model," *IEEE Trans. Fuzzy Syst.*, vol. 9, no. 3, pp. 381–392, 2001.
336. L.X. Wang, *A Course in Fuzzy Systems and Control*, Prentice-Hall, Englewood Cliffs, NJ, 1997.
337. C.L. Hwang, Y.M. Chen, and C. Jan, "Trajectory tracking of large-displacement piezoelectric actuators using a nonlinear observer-based variable structure control," *IEEE Trans. Control Syst. Technol.*, vol. 13, no. 1, pp. 56–66, 2005.
338. K.Y. Lian and J.J. Liou, "Output tracking control for fuzzy systems via output feedback design," *IEEE Trans. Fuzzy Syst.*, vol. 14, no. 5, pp. 628–639, 2006.
339. K.Y. Lian, C.S. Chiu, T.S. Chiang, and P. Liu, "LMI-based fuzzy chaotic synchronization and communications," *IEEE Trans. Fuzzy Syst.*, vol. 9, no. 4, pp. 539–553, 2001.
340. K. Tanaka, T. Ikeda, and H.O. Wang, "Fuzzy regulators and fuzzy observers: Relaxed stability conditions and LMI-based designs," *IEEE Trans. Fuzzy Syst.*, vol. 6, no. 2, pp. 250–265, 1998.
341. K. Tanaka and H.O. Wang, *Fuzzy Control Systems Design and Analysis: A Linear Matrix Inequality Approach*, Wiley, New York, 2001.
342. H.O. Wang, K. Tanaka, and M.F. Griffin, "An approach to fuzzy control of nonlinear systems: Stability and design issues," *IEEE Trans. Fuzzy Syst.*, vol. 4, no. 1, pp. 14–23, 1996.
343. S.S. Chen, Y.C. Chang, S.F. Su, S.L. Chung, and T.T. Lee, "Robust static output-feedback stabilization for nonlinear discrete-time systems with time delay via fuzzy control approach," *IEEE Trans. Fuzzy Syst.*, vol. 13, no. 2, pp. 263–272, 2005.
344. J. Qiu, G. Feng, and J. Yang, "A new design of delay-dependent robust H_∞ filtering for discrete-time T-S fuzzy systems with time-varying delay," *IEEE Trans. Fuzzy Syst.*, vol. 17, no. 5, pp. 1044–1058, 2009.

345. M. Chen, G. Feng, H. Ma, and G. Chen, "Delay-dependent H_∞ filter design for discrete-time fuzzy systems with time-varying delays," *IEEE Trans. Fuzzy Syst.*, vol. 17, no. 3, pp. 604–616, 2009.
346. B. Chen and X. Liu, "Fuzzy guaranteed cost control for nonlinear systems with time-varying delay," *IEEE Trans. Fuzzy Syst.*, vol. 13, no. 2, pp. 238–249, 2005.
347. H. Zhang, Y. Wang, and D. Liu, "Delay-dependent guaranteed cost control for uncertain stochastic fuzzy systems with multiple time delays," *IEEE Trans. Syst. Man Cybern., B: Cybern.*, vol. 38, no. 1, pp. 126–140, 2008.
348. Y.C. Chang, "Adaptive fuzzy-based tracking control for nonlinear SISO systems via VSS and H_∞ approaches," *IEEE Trans. Fuzzy Syst.*, vol. 9, no. 2, pp. 278–292, 2001.
349. Y.C. Chang, "Intelligent robust tracking control for a class of uncertain strict feedback nonlinear systems," *IEEE Trans. Syst. Man Cybern., B: Cybern.*, vol. 39, no. 1, pp. 142–155, 2009.
350. F.H. Hsiao, C.W. Chen, Y.W. Liang, S.D. Xu, and W.L. Chiang, "T-S fuzzy controllers for nonlinear interconnected systems with multiple time delays," *IEEE Trans. Circuits Syst., I: Reg. Pap.*, vol. 52, no. 9, pp. 1883–1893, 2005.
351. F.H. Hsiao, J.D. Hwang, C.W. Chen, and Z.R. Tsai, "Robust stabilization of nonlinear multiple time-delay large-scale systems via decentralized fuzzy control," *IEEE Trans. Fuzzy Syst.*, vol. 13, no. 1, pp. 152–163, 2005.
352. C.S. Tseng, "A novel approach to H_∞ decentralized fuzzy-observer-based fuzzy control design for nonlinear interconnected systems," *IEEE Trans. Fuzzy Syst.*, vol. 16, no. 5, pp. 1337–1350, 2008.
353. Y.Y. Chen, Y.T. Chang, and B.S. Chen, "Fuzzy solutions to partial differential equations: Adaptive approach," *IEEE Trans. Fuzzy Syst.*, vol. 17, no. 1, pp. 116–127, 2009.
354. H.N. Wu and H.X. Li, "H_∞ fuzzy observer-based control for a class of nonlinear distributed parameter systems with control constraints," *IEEE Trans. Fuzzy Syst.*, vol. 16, no. 2, pp. 502–516, 2008.
355. K. Yuan, H.X. Li, and J. Cao, "Robust stabilization of the distributed parameter system with time delay via fuzzy control," *IEEE Trans. Fuzzy Syst.*, vol. 16, no. 3, pp. 567–584, 2008.
356. B.S. Chen and Y.T. Chang, "Fuzzy state space modeling and robust stabilization design for nonlinear partial differential systems," *IEEE Trans. Fuzzy Syst.*, vol. 17, no. 5, pp. 1025–1043, 2009.
357. C.I. Byrnes, I.G. Laukó, D.S. Gilliam, and V.I. Shubov, "Output regulation for linear distributed parameter systems," *IEEE Trans. Autom. Control*, vol. 45, no. 12, pp. 2236–2252, 2000.
358. F. Fröhlich and S. Jezernik, "Feedback control of Hodgkin—Huxley nerve cell dynamics," *Control Eng. Pract.*, vol. 13, pp. 1195–1206, 2005.
359. A.L. Hodgkin and A.F. Huxley, "A quantitative description of membrane current and its application to conduction and excitation in nerve," *J. Physiol.*, vol. 177, pp. 500–544, 1952.
360. N. Chakravarthy, S. Sabesan, K. Tsakalis, and L. Iasemidis, "Controlling epileptic seizures in a neural mass model," *J. Comb. Optim.*, vol. 17, no. 1, pp. 98–116, 2009.
361. A.W.L. Chiu and B.L. Bardakjian, "Control of state transitions in an in silico model of epilepsy using small perturbations," *IEEE Trans. Biomed. Eng.*, vol. 51, no. 10, pp. 1856–1859, 2004.
362. N. Chakravarthy, K. Tsakalis, S. Sabesan, and L. Iasemidis, "Homeostasis of brain dynamics in epilepsy: A feedback control systems perspective of seizures," *Ann. Biomed. Eng.*, vol. 37, no. 3, pp. 565–585, 2009.
363. E. Kim and S. Kim, "Stability analysis and synthesis for an affine fuzzy control system via LMI and ILMI: A continuous case," *IEEE Trans. Fuzzy Syst.*, vol. 10, no. 3, pp. 391–400, 2002.

364. B.S. Chen, C.S. Tseng, and H.C. Wang, "Mixed H_2/H_∞ fuzzy output feedback control for nonlinear uncertain systems: LMI approach," *IEEE Trans. Fuzzy Syst.*, vol. 8, no. 3, pp. 249–265, 2000.
365. D. Huang and S.K. Nguang, "Robust H_∞ static output feedback control of fuzzy systems: An ILMI approach," *IEEE Trans. Syst. Man Cybern., B: Cybern.*, vol. 36, no. 1, pp. 216–222, 2006.
366. M. Kocvara and M. Stingl, "PENNON: A code for convex nonlinear and semidefinite programming," *Optim. Method Softw.*, vol. 18, pp. 317–333, 2003.
367. C. Lin, Q.G. Wang, T.H. Lee, and Y. He, "Design of observer-based H_∞ control for fuzzy time-delay systems," *IEEE Trans. Fuzzy Syst.*, vol. 16, no. 2, pp. 534–543, 2008.
368. M. de Oliveira, J. Bernussou, and J. Geromel, "A new discrete-time robust stability condition," *Syst. Control Lett.*, vol. 37, pp. 261–265, 1999.
369. D.V. Vavoulis, V.A. Straub, I. Kemenes, J. Feng, and P.R. Benjamin, "Dynamic control of a central pattern generator circuit: A computational model of the snail feeding network," *Eur. J. Neurosci.*, vol. 25, pp. 2805–2818, 2007.
370. T. Takahata, S. Tanabe, and K. Pakdaman, "White-noise simulation of the Hodgkin–Huxley model," *Biol. Cybern.*, vol. 86, pp. 403–417, 2002.
371. T. Nowotny, V.P. Zhigulin, A.I. Selverston, H.D.I. Abarbanel, and M.I. Rabinovich, "Enhancement of synchronization in a hybrid neural circuit by spike- timing dependent plasticity," *J. Neurosci.*, vol. 23, no. 30, pp. 9776–9785, 2003.
372. T.H.S. Li and S.H. Tsai, "T-S fuzzy bilinear model and fuzzy controller design for a class of nonlinear systems," *IEEE Trans. Fuzzy Syst.*, vol. 15, no. 3, pp. 494–506, 2007.
373. T.H.S. Li and K.J. Lin, "Composite fuzzy control of nonlinear singularly perturbed systems," *IEEE Trans. Fuzzy Syst.*, vol. 15, no. 2, pp. 176–187, 2007.
374. J.C. Strikwerda, *Finite Difference Schemes and Partial Differential Equations*, SIAM, Philadelphia, PA, 2004.
375. G. Evans, J. Blackledge, and P. Yardley, *Numerical Methods for Partial Differential Equations*, Springer-Verlag, Berlin, 2000.
376. P. Lancaster and M. Tismenetsky, *The Theory of Matrices: With Application*, Academic, Cambridge, MA, 1985.
377. J.E. Marsden and M.J. Hoffman, *Elementary Classical Analysis*, Freeman, New York, 1993.
378. S. Boyd, L.E. Ghaoui, E. Feron, and V. Balakrishnan, *Linear Matrix Inequalities in System and Control Theory*, SIAM, Philadelphia, PA, 1994.
379. K. Tanaka, T. Hori, and H.O. Wang, "A multiple Lyapunov function approach to stabilization of fuzzy control systems," *IEEE Trans. Fuzzy Syst.*, vol. 11, no. 4, pp. 582–589, 2003.
380. S.G. Cao, N.W. Rees, and G. Feng, "Analysis and design of a class of continuous time fuzzy control systems," *Int. J. Control*, vol. 64, pp. 1069–1087, 1996.
381. W.J. Wang and C.H. Sun, "Relaxed stability and stabilization conditions for a T-S fuzzy discrete system," *Fuzzy Sets Syst.*, vol. 156, no. 2, pp. 208–225, 2005.
382. W.J. Wang and C.H. Sun, "A relaxed stability criterion for T-S fuzzy discrete systems," *IEEE Trans. Syst. Man Cybern., B: Cybern.*, vol. 34, no. 5, pp. 2155–2158, 2004.
383. J.G. Verwer and J.M. Sanz-Serna, "Convergence of method of lines approximations to partial differential equations," *Computing*, vol. 33, pp. 297–313, 1984.
384. K. Liu, *Stability of Infinite Dimensional Stochastic Differential Equations with Applications*, Chapman & Hall/CRC, London, 2006.
385. P.L. Chow, *Stochastic Partial Differential Equations*, Chapman & Hall/CRC, London, 2007.

386. M. Kamrani and S.M. Hosseini, "The role of coefficients of a general SPDE on the stability and convergence of a finite difference method," *J. Comput. Appl. Math.*, vol. 234, no. 5, pp. 1426–1434, 2010.
387. M.J. Anabtawi and S. Sathanathan, "Stability and convergence results for Itô-type parabolic partial differential equations in hilbert spaces," *Stoch. Anal. Appl.*, vol. 27, pp. 671–693, 2009.
388. P.L. Chow, "Stability of nonlinear stochastic-evolution equations," *J. Math. Anal. Appl.*, vol. 89, pp. 400–419, 1982.
389. Y. Lou, G. Hu, and P.D. Christofides, "Model predictive control of nonlinear stochastic partial differential equations with application to a sputtering process," *Aiche J.*, vol. 54, pp. 2065–2081, 2008.
390. A.M. Davie and J.G. Gaines, "Convergence of numerical schemes for the solution of parabolic stochastic partial differential equations," *Math. Comput.*, vol. 70, pp. 121–134, 2001.
391. I. Gyögy, "Lattice approximations for stochastic quasi-linear parabolic partial differential equations driven by space-time white noise I," *Potential Anal.*, vol. 9, pp. 1–25, 1998.
392. H. Yoo, "Semi-discretization of stochastic partial differential equations on R1 by a finite-difference method," *Math. Comput.*, vol. 69, pp. 653–666, 1999.
393. P.L. Lions and P.E. Souganidis, "Uniqueness of weak solutions of fully non- linear stochastic partial differential equations," *Compt. Rendus Acad. Sci. Ser. I: Math.*, vol. 331, pp. 783–790, 2000.
394. P. Balasubramaniam and C. Vidhya, "Global asymptotic stability of stochas- tic BAM neural networks with distributed delays and reaction–diffusion terms," *J. Comput. Appl. Math.*, vol. 234, no. 12, pp. 3458–3466, 2010.
395. P. Balasubramaniam and C. Vidhya, "Exponential stability of stochastic reaction–diffusion uncertain fuzzy neural networks with mixed delays and Markovian jumping parameters," *Exp. Syst. Appl.*, vol. 39, no. 3, pp. 3109–3115, 2012.
396. P.D. Christofides, *Nonlinear and Robust Control of PDE Systems: Methods and Applications to Transport-Reaction Processes*, Birkhauser, Basel, 2001.
397. J. Keener and J. Sneyd, *Mathematical Physiology*, Springer-Verlag, Berlin, 1998.
398. C.V. Pao, *Nonlinear Parabolic and Elliptic Equations*, Plenum Press, New York, 1992.
399. P.L. Chow, "Stochastic partial differential equations in turbulence-related problems," in *Probabilistic Analysis and Related Topic*, Academic Press, New York, Vol. I, pp. 1–43, 1978.
400. R.F. Curtain and H. Zwart, *An Introduction to Infinite-Dimensional Linear Systems Theory*, Springer-Verlag, Berlin, 1995.
401. C.I. Byrnes, I.G. Laukó, D.S. Gilliam, and V.I. Shubov, "Output regulation for linear distributed parameter systems," *IEEE Trans. Automat. Contr.*, vol. 45, pp. 2236–2252, 2000.
402. M.J. Balas, "Nonlinear finite-dimensional control of a class of nonlinear distributed parameter systems using residual mode filters: A proof of local exponential stability," *J. Math. Anal. Appl.*, vol. 162, pp. 63–70, 1991.
403. J. Baker and P.D. Christofides, "Finite-dimensional approximation and control of non-linear parabolic PDE systems," *Int. J. Control*, vol. 73, pp. 439–456, 2000.
404. H.N. Wu and H.X. Li, "H_∞ fuzzy observer-based control for a class of non- linear distributed parameter systems with control constraints," *IEEE Trans. Fuzzy Syst.*, vol. 16, pp. 502–516, 2008.
405. K. Yuan, H.X. Li, and J. Cao, "Robust stabilization of the distributed parameter system with time delay via fuzzy control," *IEEE Trans. Fuzzy Syst.*, vol. 16, pp. 567–584, 2008.

406. B.S. Chen and Y.T. Chang, "Fuzzy state space modeling and robust stabilization design for nonlinear partial differential systems," *IEEE Trans. Fuzzy Syst.*, vol. 17, pp. 1025–1043, 2009.
407. H.N. Wu, J.W. Wang, and H.-X. Li, "Design of distributed H_∞ fuzzy controllers with constraint for nonlinear hyperbolic PDE systems," *Automatica*, vol. 48, no. 10, pp. 2535–2543, 2012.
408. H.N. Wu, J.W. Wang, and H.-X. Li, "Exponential stabilization for a class of nonlinear parabolic PDE systems via fuzzy control approach," *IEEE Trans. Fuzzy Syst.*, vol. 20, no. 2, pp. 318–329, 2012.
409. B.S. Chen, C.S. Tseng, and H.J. Uang, "Robustness design of nonlinear uncertain system via fuzzy linear control," *IEEE Trans. Fuzzy Syst.*, vol. 7, pp. 571–585, 1999.
410. K. Tanaka and H.O. Wang, *Fuzzy Control Systems Design and Analysis: A Linear Matrix Inequality Approach*, New York, Wiley, 2001.
411. H.O. Wang, K. Tanaka, and M.F. Griffin, "An approach to fuzzy control of nonlinear systems: Stability and design issues," *IEEE Trans. Fuzzy Syst.*, vol. 4, pp. 14–23, 1996.
412. T.H.S. Li and K.J. Lin, "Composite fuzzy control of nonlinear singularly perturbed systems," *IEEE Trans. Fuzzy Syst.*, vol. 15, pp. 176–187, 2007.
413. B.S. Chen, C.H. Chiang, and S.K. Nguang, "Robust H_∞ synchronization design of nonlinear coupled network via fuzzy interpolation method," *IEEE Trans. Circuits Syst., I: Reg. Pap.*, vol. 58, pp. 349–362, 2011.
414. S.K. Nguang, P. Shi, and S. Ding, "Fault detection for uncertain fuzzy systems: An LMI approach," *IEEE Trans. Fuzzy Syst.*, vol. 15, pp. 1251–1262, 2007.
415. S.K. Nguang and P. Shi, "Robust H_∞ output feedback control design for fuzzy dynamic systems with quadratic D stability constraints: An LMI approach," *Inform. Sci.*, vol. 176, pp. 2161–2191, 2006.
416. Y.Y. Chen, Y.T. Chang, and B.S. Chen, "Fuzzy solutions to partial differential equations: Adaptive approach," *IEEE Trans. Fuzzy Syst.*, vol. 17, pp. 116–127, 2009.
417. Y.T. Chang and B.S. Chen, "A fuzzy approach for robust reference- tracking-control design of nonlinear distributed parameter time-delayed systems and its application," *IEEE Trans. Fuzzy Syst.*, vol. 18, pp. 1041–1057, 2010.
418. G. Da Prato and J. Zabczyk, *Stochastic equations in infinite dimensions*, Cambridge, NY, Cambridge University Press, 1992.
419. R.F. Curtain, "Stability of stochastic partial-differential equation," *J. Math. Anal. Appl.*, vol. 79, pp. 352–369, 1981.
420. A. Jentzen and P.E. Kloeden, "The numerical approximation of stochastic partial differential equations," *Milan J. Math.*, vol. 77, pp. 205–244, 2009.
421. J.G. Verwer and J.M. Sanz-Serna, "Convergence of method of linear approximations to partial differential equations," *Computing*, vol. 33, pp. 297–313, 1984.
422. L. Zambotti and P. di Milano, "Itô-Tanaka's formula for stochastic partial differential equations driven by addative space-time white noise," *Stoch. Partial Differ. Equ. Appl.*, vol. VII, pp. 337–347, 2006.
423. J.C. Strikwerda, *Finite Difference Schemes and Partial Differential Equations*, 2nd ed, SIAM, Philadelphia, PA, 2004.
424. G. Evans, J. Blackledge, and P. Yardley, *Numerical Methods for Partial Differential Equations*, Springer-Verlag, Berlin, 2000.
425. Z. Kamont and K. Kropielnicka, "Numerical method of lines for parabolic functional differential equations," *Appl. Anal.*, vol. 88, pp. 1631–1650, 2009.
426. S. Boyd, L.E. Ghaoui, E. Feron, and V. Balakrishnan, *Linear Matrix Inequalities in System and Control Theory*, SIAM, Philadelphia, PA, 1994.
427. B.S. Chen and W.H. Zhang, "Stochastic H_2/H_∞ control with state- dependent noise," *IEEE Trans. Autom. Control*, vol. 49, pp. 45–57, 2004.

428. W.H. Zhang and B.S. Chen, "State feedback H_∞ control for a class of nonlinear stochastic systems," *SIAM J. Control Optim.*, vol. 44, pp. 1973–1991, 2006.
429. A.L. Hodgkin and A.F. Huxley, "A quantitative description of membrane current and its application to conduction and excitation in nerve," *J. Phys.*, vol. 177, pp. 500–544, 1952.
430. J. Wang, L.Q. Chen, and X.Y. Fei, "Analysis and control of the bifurcation of Hodgkin-Huxley model," *Chaos Solitons Fractals*, vol. 31, pp. 247–256, 2007.
431. J. Wang, L.Q. Chen, and X.Y. Fei, "Bifurcation control of the Hodgkin-Huxley equations," *Chaos Solitons Fractals*, vol. 33, pp. 217–224, 2007.
432. D.M. Durand and M. Bikson, "Suppression and control of epileptiform activity by electrical stimulation: A review," *Proc. IEEE*, vol. 89, pp. 1065–1082, 2001.
433. F. Fröhlich and S. Jezernik, "Feedback control of Hodgkin-Huxley nerve cell dynamics," *Control Eng. Pract.*, vol. 13, pp. 1195–1206, 2005.
434. L. Ding and C. Hou, "Stabilizing control of Hopf bifurcation in the Hodgkin-Huxley model via washout filter with linear control term," *Nonlinear Dyn.*, vol. 60, pp. 131–139, 2010.
435. P. Hanggi, G. Schmid, and I. Goychuk, "Excitable membranes: Channel noise, synchronization, and stochastic resonance," *Adv. Solid State Phys.*, vol. 42, pp. 359–370, 2002.
436. G. Schmid, I. Goychuk, and P. Hanggi, "Channel noise and synchronization in excitable membranes," *Phys. A*, vol. 325, pp. 165–175, 2003.
437. R.F. Fox and Y.N. Lu, "Emergent collective behavior in large numbers of globally coupled independently stochastic ion channels," *Phys. Rev. E*, vol. 49, pp. 3421–3431, 1994.
438. J. Moehlis, "Canards for a reduction of the Hodgkin-Huxley equations," *J. Math. Biol.*, vol. 52, pp. 141–153, 2006.
439. D.V. Vavoulis, V.A. Straub, I. Kemenes, J. Feng, and P.R. Benjamin, "Dynamic control of a central pattern generator circuit: A computational model of the snail feedingnetwork," *Eur. J. Neurosci.*, vol. 25, pp. 2805–2818, 2007.
440. T. Takahata, S. Tanabe, and K. Pakdaman, "White-noise simulation of the Hodgkin-Huxley model," *Biol. Cybern.*, vol. 86, pp. 403–417, 2002.
441. Y. Xie, L. Chen, Y.M. Kang, and K. Aihara, "Controlling the onset of Hopf bifurcation in the Hodgkin-Huxley model," *Phys. Rev. E*, vol. 77, p. 061921, 2008.
442. W.C. Troy, "The bifurcation of periodic solutions in the Hodgkin-Huxley equations," *Q. Appl. Math.*, vol. 36, pp. 73–83, 1978.
443. J. Rinzel and R.N. Miller, "Numerical calculation of stable and unstable pe- riodic solutions to the Hodgkin-Huxley equations," *Math. Biosci.*, vol. 49, pp. 27–59, 1980.
444. L.B. Jackson, *Signals, Systems, and Transforms*, Addison-Wesley, Boston, MA, 1991.
445. B.S. Chen, Y.C. Chang, and T.C. Lee, "Adaptive control in robotic systems with H_∞ tracking performance," *Automatica*, vol. 33, pp. 227–234, 1997.
446. B.S. Chen, C.S. Tseng, and H.J. Uang, "Mixed H_2/H_∞ fuzzy output feedback control design for nonlinear dynamic systems: An LMI approach," *IEEE Trans. Fuzzy Syst.*, vol. 8, pp. 249–265, 2000.
447. C.S. Tseng and B.S. Chen, "Robust fuzzy observer-based fuzzy control design for nonlinear discrete-time systems with persistent bounded disturbances," *IEEE Trans. Fuzzy Syst.*, vol. 17, pp. 711–723, 2009.
448. C.S. Tseng, B.S. Chen, and Y.F. Li, "Robust fuzzy observer-based fuzzy control design for nonlinear systems with persistent bounded disturbances: A novel decoupled approach," *Fuzzy Sets Syst.*, vol. 160, pp. 2824–2843, 2009.

449. H. Li, H. Liu, S. Hand, and C. Hilton, "Design of robust H_∞ controller for a half-vehicle active suspension system with input delay," *Int. J. Syst. Sci.*, vol. 44, pp. 625–640, 2011.
450. M.H. Asemani and V.J. Majd, "A robust observer-based controller design for uncertain T-S fuzzy systems with unknown premise variables via LMI," *Fuzzy Sets Syst.*, vol. 212, pp. 21–40, 2013.
451. H.G. Harno and I.R. Petersen, "Robust H_∞ control via a stable decentralized nonlinear output feedback controller," *Int. J. Robust Nonlinear Control*, vol. 24, pp. 191–213, 2014.
452. B.S. Chen, C.C. Wu, and Y.W. Chen, "Human walking gait with 11-DOF humanoid robot through robust neural fuzzy networks tracking control," *Int. J. Fuzzy Syst.*, vol. 15, pp. 22–35, 2013.
453. J.B. Qiu, G. Feng, and H.J. Gau, "Static-output-feedback H_∞ control of continuous-time T-S fuzzy affine systems via piecewise Lyapunov functions," *IEEE Trans. Fuzzy Syst.*, vol. 21, pp. 245–261, 2013.
454. W.H. Zhang and B.S. Chen, "H-representation and applications to generalized Lyapunov equations and linear stochastic systems," *IEEE Trans. Autom. Control*, vol. 57, pp. 3009–3022, 2012.
455. W.H. Zhang, B.S. Chen, L. Sheng, and M. Gao, "Robust H2/H∞ filter design for a class of nonlinear stochastic systems with state-dependent noise," *Math. Probl. Eng.*, 2012.
456. M.E. Shaikin, "Stochastic H_2/H_∞-control for a dynamical system with internal noises multiplicative with respect to state, control, and external disturbance," *Autom. Remote Control*, vol. 74, pp. 426–441, 2013.
457. J. Lian, Z. Feng, and P. Shi, "Robust H_∞ filtering for a class of uncertain stochastic hybrid neutral systems with time-varying delay," *Int. J. Adapt. Control Signal Process.*, vol. 27, pp. 462–477, Jun 2013.
458. H. De Jong, "Modeling and simulation of genetic regulatory systems: A literature review," *J. Comput. Biol.*, vol. 9, pp. 67–103, 2002.
459. H. Kitano, "Biological robustness," *Nat. Rev. Genet.*, vol. 5, pp. 826–837, 2004.
460. B.S. Chen and C.H. Wu, "A systematic design method for robust synthetic biology to satisfy design specifications," *BMC Syst. Biol.*, vol. 3, 2009.
461. C.H. Wu, W.H. Zhang, and B.S. Chen, "Multiobjective H_2/H_∞ synthetic gene network design based on promoter libraries," *Math. Biosci.*, vol. 233, pp. 111–125, 2011.
462. B.S. Chen and S.J. Ho, "The stochastic evolutionary game for a population of biological networks under natural selection," *Evol. Bioinf.*, vol. 10, pp. 17–38, 2014.
463. B.S. Chen and C.W. Li, "Robust observer-based tracking control of Hodgkin-Huxley neuron systems under environmental disturbances," *Neural Comput.*, vol. 22, pp. 3143–3178, 2010.
464. L.X. Wang, *A Course in Fuzzy Systems and Control*, Prentice Hall PTR, Englewood Cliffs, NJ, 1997.
465. K. Tanaka and H.O. Wang, *Fuzzy Control Systems Design and Analysis: A Linear Matrix Inequality Approach*, John Wiley & Sons, New York, 2004.
466. C.T. Sun and E. Mizutani, *Neuro-Fuzzy and Soft Computing: A Computational Approach to Learning and Machine Intelligence*, Prentice Hall, Englewood Cliffs, NJ, 1997.
467. T. Takagi and M. Sugeno, "Fuzzy identification of systems and its applications to modeling and control," *IEEE Trans. Syst., Man Cybern., B: Cybern.*, vol. SMC-15, pp. 116–132, 1985.
468. P.D. Christofides, *Nonlinear and Robust Control of PDE Systems: Methods and Applications to Transport-Reaction Processes*, Springer, Berlin, 2001.

469. U. Tyn Myint and L. Debnath, *Partial Differential Equations for Scientists and Engineers*, 3rd ed, Dover Publications Inc., New York, 1987.
470. I. Lasiecka and R. Triggiani, *Control Theory for Partial Differential Equations: Volume 1, Abstract Parabolic Systems: Continuous and Approximation Theories*, Cambridge University Press, London, 2000.
471. I. Lasiecka and R. Triggiani, *Control Theory for Partial Differential Equations: Volume 2, Abstract Hyperbolic-Like Systems Over a Finite Time Horizon: Continuous and Approximation Theories*, Cambridge University Press, London, 2000.
472. R. Johansson, *System Modeling and Identification*, Prentice-Hall, Englewood Cliffs, NJ, 1993.
473. T. Bechtold, E.B. Rudnyi, and J.G. Korvink, *Fast Simulation of Electro-Thermal MEMS: Efficient Dynamic Compact Models*, Springer, Berlin, 2006.
474. P. Esposito and Y. Guo, *Mathematical Analysis of Partial Differential Equations Modeling Electrostatic MEMS*, American Mathematical Society, Providence, RI, 2010.
475. M.I. Younis, *MEMS Linear and Nonlinear Statics and Dynamics*, Springer, Berlin, 2011.
476. E.B. Magrab, *Vibrations of Elastic Systems: With Applications to MEMS and NEMS*, Springer, Berlin, 2012.
477. B.S. Chen and Y.T. Chang, "Fuzzy state-space modeling and robust observer- based control design for nonlinear partial differential systems," *IEEE Trans. Fuzzy Syst.*, vol. 17, pp. 1025–1043, 2009.
478. Y.T. Chang and B.S. Chen, "A fuzzy approach for robust reference-tracking- control design of nonlinear distributed parameter time-delayed systems and its application," *IEEE Trans. Fuzzy Syst.*, vol. 18, pp. 1041–1057, 2010.
479. H.N. Wu, J.W. Wang, and H.X. Li, "Exponential stabilization for a class of nonlinear parabolic PDE systems via fuzzy control approach," *IEEE Trans. Fuzzy Syst.*, vol. 20, pp. 318–329, 2012.
480. J.W. Wang, H.N. Wu, and H.X. Li, "Distributed proportional-spatial derivative control of nonlinear parabolic systems via fuzzy PDE modeling approach," *IEEE Trans. Syst., Man Cybern., B: Cybern.*, vol. 42, pp. 927–938, 2012.
481. B. Luo and H.N. Wu, "Approximate optimal control design for nonlinear one-dimensional parabolic PDE systems using empirical eigen functions and neural network," *IEEE Trans. Syst., Man Cybern., B: Cybern.*, vol. 42, pp. 1538–1549, 2012.
482. H.N. Wu, J.W. Wang, and H.X. Li, "Fuzzy boundary control design for a class of nonlinear parabolic distributed parameter systems," *IEEE Trans. Fuzzy Syst.*, vol. 22, pp. 642–652, 2014.
483. B.S. Chen, W.H. Chen, and W.H. Zhang, "Robust filter for nonlinear stochastic partial differential systems in sensor signal processing: Fuzzy approach," *IEEE Trans. Fuzzy Syst.*, vol. 20, pp. 957–970, 2012.
484. Y.T. Chang, S.J. Ho, and B.S. Chen, "Robust stabilization design of nonlinear stochastic partial differential systems: Fuzzy approach," *Fuzzy Sets Syst.*, vol. 248, pp. 61–85, 2014.
485. Y.Y. Chen, Y.T. Chang, and B.S. Chen, "Fuzzy solutions to partial differential equations: Adaptive approach," *IEEE Trans. Fuzzy Syst.*, vol. 17, pp. 116–127, 2009.
486. J.W. Wang, H.N. Wu, and H.X. Li, "Distributed fuzzy control design of nonlinear hyperbolic PDE systems with application to nonisothermal plug-flow reactor," *IEEE Trans. Fuzzy Syst.*, vol. 19, pp. 514–526, 2011.
487. S. Boyd, L.E. Ghaoui, E. Feron, and V. Balakrishnan, *Linear Matrix Inequalities in System and Control Theory*, SIAM, Philadelphia, PA, 1994.
488. L.E. Payne and H.F. Weinberger, "An optimal Poincaré inequality for convex domains," *Arch. Rational Mech. Anal.*, vol. 5, pp. 286–292, 1960.

489. F. Gesztesy, B. Simon, W. Schlag, P. Perry, P. Deift, and C. Galvez, *Spectral Theory and Mathematical Physics: A Festschrift in Honor of Barry Simon's 60th Birthday: Quantum Field Theory, Statistical Mechanics, and Nonrelativistic Quantum Systems*, American Mathematical Society, Providence, RI, 2007.
490. L. Brasco and A. Pratelli, "Sharp stability of some spectral inequalities," *Geom. Funct. Anal.*, vol. 22, pp. 107–135, 2012.
491. B.G. Pachpatte, "On Poincaré-type integral inequalities," *J. Math. Anal. Appl.*, vol. 114, pp. 111–115, 1986.
492. D. Saoudi, M. Chadli, C. Mechmeche, and N.B. Braiek, "Unknown input observer design for fuzzy bilinear system: An LMI approach," *Math. Problems Eng.*, pp. 1–21, 2012.
493. D. Saoudi, M. Chadli, and N.B. Braiek, "Robust estimation design for unknown inputs fuzzy bilinear models: application to faults diagnosis," *Complex System Modelling and Control through Intelligent Soft Computations*, vol. 319, pp. 655–685, 2014.
494. D. Saoudi, M. Chadli, and N.B. Braiek, "Robust H_∞ fault detection for fuzzy bilinear systems via unknown input observer," Proc. 22nd Mediterranean Conference on Control and Automation MED, Palermo, Italy, 2014.
495. J.G. VanAntwerp and R.D. Braatz, "A tutorial on linear and bilinear matrix inequalities," *J. Process Control*, vol. 10, pp. 363–385, 2000.
496. J.C. Lo and M.L. Lin, "Existence of similarity transformation converting BMIs to LMIs," *IEEE Trans. Fuzzy Syst.*, vol. 15, pp. 840–851, 2007.
497. C.S. Tseng, B.S. Chen, and H.J. Uang, "Fuzzy tracking control design for nonlinear dynamic systems via T-S fuzzy model," *IEEE Trans. Fuzzy Syst.*, vol. 9, pp. 381–392, 2001.
498. J.C. Lo and M.L. Lin, "Observer-based robust H_∞ control for fuzzy systems using two-step procedure," *IEEE Trans. Fuzzy Syst.*, vol. 12, pp. 350–359, 2004.
499. H.N. Wu and H.X. Li, "H_∞ fuzzy observer-based control for a class of nonlinear distributed parameter systems with control constraints," *IEEE Trans. Fuzzy Syst.*, vol. 16, pp. 502–516, 2008.
500. J.D. Murray, *Mathematical Biology I: An Introduction*, 3rd ed, Springer, Berlin, 2002.
501. J.D. Murray, *Mathematical Biology II: Spatial Models and Biomedical Applications*, 3rd ed, Springer, Berlin, 2003.
502. Y. Demirel, *Nonequilibrium Thermodynamics: Transport and Rate Processes in Physical, Chemical and Biological Systems*, 2nd ed, Elsevier, London, 2007.
503. N. Shanks, "Modeling biological systems: The Belousov-Zhabotinsky reaction," *Found. Chem.*, vol. 3, pp. 33–53, 2001.
504. P.D. Christofides, *Nonlinear and Robust Control of PDE Systems: Methods and Applications to Transport-Reaction Processes*, Springer, Berlin, 2001.
505. U. Tyn Myint and L. Debnath, *Partial Differential Equations for Scientists and Engineers*, 3rd ed, Dover Publications Inc., New York, 1987.
506. I. Lasiecka and R. Triggiani, *Control Theory for Partial Differential Equations: Volume 1, Abstract Parabolic Systems: Continuous and Approximation Theories*, Cambridge University Press, London, 2000.
507. I. Lasiecka and R. Triggiani, *Control Theory for Partial Differential Equations: Volume 2, Abstract Hyperbolic-Like Systems Over a Finite Time Horizon: Continuous and Approximation Theories*, Cambridge University Press, London, 2000.
508. U. Tyn Myint and L. Debnath, *Linear Partial Differential Equations for Scientists and Engineers*, 4th ed, Springer Science & Business Media, Berlin, 2007.
509. T. Bechtold, E.B. Rudnyi, and J.G. Korvink, *Fast Simulation of Electro-Thermal MEMS: Efficient Dynamic Compact Models*, Springer, Berlin, 2006.

510. P. Esposito and Y. Guo, *Mathematical Analysis of Partial Differential Equations Modeling Electrostatic MEMS*, American Mathematical Society, Providence, RI, 2010.
511. M.I. Younis, *MEMS Linear and Nonlinear Statics and Dynamics*, Springer, Berlin, 2011.
512. E.B. Magrab, *Vibrations of Elastic Systems: With Applications to MEMS and NEMS*, Springer, Berlin, 2012.
513. Y. Demirel, *Nonequilibrium Thermodynamics: Transport and Rate Processes in Physical, Chemical and Biological Systems*, 2nd ed, Elsevier, London, 2007.
514. J.D. Murray, *Mathematical Biology I: An Introduction*, 3rd ed, Springer, Berlin, 2002.
515. J.D. Murray, *Mathematical Biology II: Spatial Models and Biomedical Applications*, 3rd ed, Springer, Berlin, 2003.
516. Y.Y. Chen, Y.T. Chang, and B.S. Chen, "Fuzzy solutions to partial differential equations: Adaptive approach," *IEEE Trans. Fuzzy Syst.*, vol. 17, pp. 116–127, 2009.
517. B.S. Chen and Y.T. Chang, "Fuzzy state-space modeling and robust observer- based control design for nonlinear partial differential systems," *IEEE Trans. Fuzzy Syst.*, vol. 17, pp. 1025–1043, 2009.
518. Y.T. Chang and B.S. Chen, "A fuzzy approach for robust reference-tracking- control design of nonlinear distributed parameter time-delayed systems and its application," *IEEE Trans. Fuzzy Syst.*, vol. 18, pp. 1041–1057, 2010.
519. B.S. Chen, W.H. Chen, and W.H. Zhang, "Robust filter for nonlinear stochastic partial differential systems in sensor signal processing: Fuzzy approach," *IEEE Trans. Fuzzy Syst.*, vol. 20, pp. 957–970, 2012.
520. Y.T. Chang, S.J. Ho, and B.S. Chen, "Robust stabilization design of nonlinear stochastic partial differential systems: Fuzzy approach," *Fuzzy Sets Syst.*, vol. 248, pp. 61–85, 2014.
521. J.W. Wang, H.N. Wu, and H.X. Li, "Distributed fuzzy control design of nonlinear hyperbolic PDE systems with application to nonisothermal plug-flow reactor," *IEEE Trans. Fuzzy Syst.*, vol. 19, pp. 514–526, 2011.
522. H.N. Wu, J.W. Wang, and H.X. Li, "Exponential stabilization for a class of nonlinear parabolic PDE systems via fuzzy control approach," *IEEE Trans. Fuzzy Syst.*, vol. 20, pp. 318–329, 2012.
523. J.W. Wang, H.N. Wu, and H.X. Li, "Distributed proportional-spatial derivative control of nonlinear parabolic systems via fuzzy PDE modeling approach," *IEEE Trans. Syst., Man Cybern., B: Cybern.*, vol. 42, pp. 927–938, 2012.
524. H.N. Wu, J.W. Wang, and H.X. Li, "Fuzzy boundary control design for a class of nonlinear parabolic distributed parameter systems," *IEEE Trans. Fuzzy Syst.*, vol. 22, pp. 642–652, 2014.
525. J.W. Wang and H.N. Wu, "Fuzzy output tracking control of semi-linear first-order hyperbolic PDE systems with matched perturbations," *Fuzzy Sets Syst.*, vol. 254, pp. 47–66, 2014.
526. B. Luo and H.N. Wu, "Approximate optimal control design for nonlinear one-dimensional parabolic PDE systems using empirical eigenfunctions and neural network," *IEEE Trans. Syst., Man Cybern., B: Cybern.*, vol. 42, pp. 1538–1549, 2012.
527. M. Arcak, "Certifying spatially uniform behavior in reaction-diffusion PDE and compartmental ODE systems," *Automatica*, vol. 47, pp. 1219–1229, 2011.
528. K.N. Wu and B.S. Chen, "Synchronization of partial differential systems via diffusion coupling," *IEEE Trans. Circuits Syst., I: Reg. Pap.*, vol. 59, pp. 2655–2668, 2012.
529. H. De Jong, "Modeling and simulation of genetic regulatory systems: A literature review," *J. Comput. Biol.*, vol. 9, pp. 67–103, 2002.
530. H. Kitano, "Biological robustness," *Nat. Rev. Genet.*, vol. 5, pp. 826–837, 2004.

531. C.H. Wu, W.H. Zhang, and B.S. Chen, "Multiobjective H_2/H_∞ synthetic gene network design based on promoter libraries," *Math. Biosci.*, vol. 233, pp. 111–125, 2011.
532. B.S. Chen and S.J. Ho, "The stochastic evolutionary game for a population of biological networks under natural selection," *Evol. Bioinf.*, vol. 10, pp. 17–38, 2014.
533. B.S. Chen and C.W. Li, "Robust observer-based tracking control of Hodgkin-Huxley neuron systems under environmental disturbances," *Neural Comput.*, vol. 22, pp. 3143–3178, 2010.
534. L.X. Wang, *A Course in Fuzzy Systems and Control*, Prentice Hall PTR, Englewood Cliffs, NJ, 1997.
535. K. Tanaka and H.O. Wang, *Fuzzy Control Systems Design and Analysis: A Linear Matrix Inequality Approach*, John Wiley & Sons, New York, 2004.
536. C.T. Sun and E. Mizutani, *Neuro-Fuzzy and Soft Computing: A Computational Approach to Learning and Machine Intelligence*, Prentice Hall, Englewood Cliffs, NJ, 1997.
537. T. Takagi and M. Sugeno, "Fuzzy identification of systems and its applications to modeling and control," *IEEE Trans. Syst., Man Cybern., B: Cybern.*, vol. SMC-15, pp. 116–132, 1985.
538. B.S. Chen, C.S. Tseng, and H.J. Uang, "Robustness design of nonlinear dynamic systems via fuzzy linear control," *IEEE Trans. Fuzzy Syst.*, vol. 7, pp. 571–585, 1999.
539. J.B. Qiu, G. Feng, and H.J. Gau, "Static-output-feedback H_∞ control of continuous-time T-S fuzzy affine systems via piecewise Lyapunov functions," *IEEE Trans. Fuzzy Syst.*, vol. 21, pp. 245–261, 2013.
540. H.G. Li, H.G. Liu, H.J. Gao, and P. Shi, "Reliable fuzzy control for active suspension systems with actuator delay and fault," *IEEE Trans. Fuzzy Syst.*, vol. 20, pp. 342–357, 2012.
541. W.H. Zhang and B.S. Chen, "State feedback H_∞ control for a class of nonlinear stochastic systems," *SIAM J. Control Optim.*, vol. 44, pp. 1973–1991, 2006.
542. W.H. Zhang, B.S. Chen, and C.S. Tseng, "Robust H_∞ filtering for nonlinear stochastic systems," *IEEE Trans. Signal Process.*, vol. 53, pp. 589–598, 2005.
543. J.J.E. Slotine and W.P. Li, *Applied Nonlinear Control*, Prentice Hall, Englewood Cliffs, NJ, 1991.
544. L.E. Payne and H.F. Weinberger, "An optimal Poincaré inequality for convex domains," *Arch. Rational Mech. Anal.*, vol. 5, pp. 286–292, 1960.
545. F. Gesztesy, B. Simon, W. Schlag, P. Perry, P. Deift, and C. Galvez, *Spectral Theory and Mathematical Physics: A Festschrift in Honor of Barry Simon's 60th Birthday: Quantum Field Theory, Statistical Mechanics, and Nonrelativistic Quantum Systems*, American Mathematical Society, Providence, RI, 2007.
546. L. Brasco and A. Pratelli, "Sharp stability of some spectral inequalities," *Geom. Funct. Anal.*, vol. 22, pp. 107–135, 2012.
547. H.A. Levine and H.F. Weinberger, "Inequalities between Dirichlet and Neumann eigenvalues," *Arch. Rational Mech. Anal.*, vol. 94, pp. 193–208, 1986.
548. M.S. Ashbaugh and R.D. Benguria, "Universal bounds for the low eigenvalues of Neumann Laplacians in *N*-dimensions," *Siam J. Math. Anal.*, vol. 24, pp. 557–570, 1993.
549. D.S. Grebenkov and B. Nguyen, "Geometrical structure of Laplacian eigenfunctions," *SIAM Rev.*, 2013, pp. 601–667.
550. D.A. Kopriva, *Implementing Spectral Methods for Partial Differential Equations: Algorithms for Scientists and Engineers*, Springer, Berlin, 2009.

551. A. Henrot, *Extremum Problems for Eigenvalues of Elliptic Operators*, Springer, Berlin, 2006.
552. S. Boyd, L.E. Ghaoui, E. Feron, and V. Balakrishnan, *Linear Matrix Inequalities in System and Control Theory*, SIAM, Philadelphia, PA, 1994.
553. B.S. Chen, H.C. Lee, and C.F. Wu, "Pareto optimal filter design for non- linear stochastic fuzzy systems via multiobjective H2/H∞ optimization," *IEEE Trans. Fuzzy Syst.*, vol. 23, no. 2, pp. 387–399, 2014.
554. B.S. Chen and S.J. Ho, "Multiobjective tracking control design of T-S fuzzy systems: Fuzzy Pareto optimal approach," *Fuzzy Sets Syst.*, vol. 290, pp. 39–55, 2016.

Index

For Product Safety Concerns and Information please contact our EU
representative GPSR@taylorandfrancis.com
Taylor & Francis Verlag GmbH, Kaufingerstraße 24, 80331 München, Germany

www.ingramcontent.com/pod-product-compliance
Lightning Source LLC
Chambersburg PA
CBHW060744220326
41598CB00022B/2315

* 9 7 8 1 0 3 2 1 3 4 4 5 1 *